D0903063

Biology of Plant Litter Decomposition

Biology of Plant Litter Decomposition

Edited by

C. H. DICKINSON

Department of Plant Biology
University of Newcastle upon Tyne

and

G. J. F. PUGH

Department of Botany
University of Nottingham

Volume 2

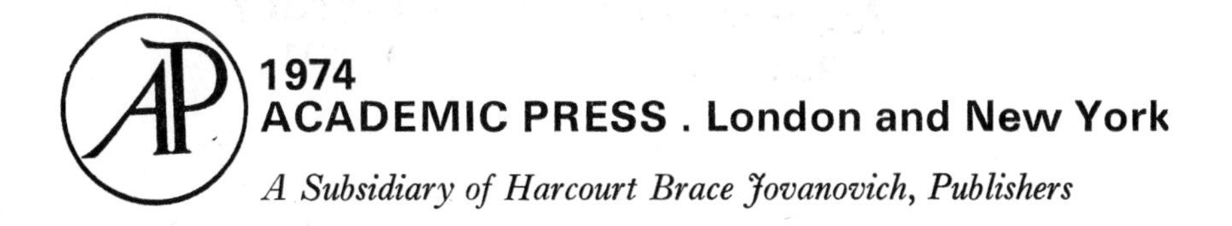

1974
ACADEMIC PRESS . London and New York

A Subsidiary of Harcourt Brace Jovanovich, Publishers

ACADEMIC PRESS INC. (LONDON) LTD.
24/28 Oval Road,
London NW1

United States Edition published by
ACADEMIC PRESS INC.
111 Fifth Avenue
New York, New York 10003

Library of Congress Catalog Card Number: 73-9457
ISBN: 0-12-215002-3

Text set in 11/12 pt. Monotype Imprint, printed by letterpress, and bound in Great Britain at The Pitman Press, Bath

Contributors

*MARY K. BELL, *Science Faculty, Open University, Walton Hall, Bletchley, Bucks., England.*

A. J. BIDDLESTONE, *Department of Chemical Engineering, The University of Birmingham, England.*

T. CROSS, *Postgraduate School of Studies in Biological Sciences, University of Bradford, Bradford, England.*

C. H. DICKINSON, *Department of Plant Biology, The University, Newcastle upon Tyne, England.*

C. A. EDWARDS, *Rothamsted Experimental Station, Harpenden, Herts., England.*

EVA EKLUND, *Department of Microbiology, University of Helsinki, SF-00710 Helsinki 71, Finland.*

R. S. FORBES, *Department of Extra-mural Studies, The Queen's University of Belfast, Northern Ireland.*

*JULIET C. FRANKLAND, *Institute of Terrestrial Ecology, Merlewood Research Station, Grange-over-Sands, Lancs., England.*

M. GOODFELLOW, *Department of Microbiology, The University, Newcastle upon Tyne, England.*

K. R. GRAY, *Department of Chemical Engineering, The University of Birmingham, England.*

T. R. G. GRAY, *Department of Botany, University of Liverpool, England.*

H. G. GYLLENBERG, *Department of Microbiology, University of Helsinki, SF-00710, Helsinki 71, Finland.*

D. J. L. HARDING, *Department of Biological Sciences, Wolverhampton Polytechnic, England.*

*V. JENSEN, *Department of Microbiology, Royal Veterinary and Agricultural University, Copenhagen, Denmark.*

E. B. G. JONES, *Department of Biological Sciences, Portsmouth Polytechnic, England.*

*AINO A. KÄÄRIK, *The Royal College of Forestry, Box S-104 05, Stockholm, Sweden.*

* Contributors to Volume 1.

M. LADLE, *Freshwater Biological Association, River Laboratory, East Stoke, Wareham, Dorset, England.*

*B. C. LODHA, *Department of Botany, University of Rajasthan, Jaipur, India.*

J. R. LOFTY, *Rothamsted Experimental Station, Harpenden, Herts., England.*

C. F. MASON, *School of Biological Sciences, University of East Anglia, Norwich, England.*

*C. S. MILLAR, *Forestry Department, University of Aberdeen, Scotland.*

E. J. PERKINS, *University of Strathclyde, Marine Laboratory, Dalandhui House, Garelochhead, Dunbartonshire, Scotland.*

G. J. F. PUGH, *Department of Botany, University of Nottingham, England.*

*J. E. SATCHELL, *The Nature Conservancy, Merlewood Research Station, Grange-over-Sands, Lancs., England.*

J. D. STOUT, *Soil Bureau, Department of Scientific and Industrial Research, Lower Hutt, New Zealand.*

R. A. STUTTARD, *Department of Biological Sciences, Wolverhampton Polytechnic, England.*

D. C. TWINN, *May and Baker Ltd., Ongar Research Station, Ongar, Essex, England.*

*J. S. WAID, *Department of Botany, University of Canterbury, Christchurch, New Zealand.*

S. T. WILLIAMS, *Department of Botany, University of Liverpool, England.*

L. G. WILLOUGHBY, *Freshwater Biological Association, Ferry House, Ambleside, Westmorland, England.*

* Contributors to Volume 1.

Preface

In the past much emphasis has been placed on finding out what organisms occur in particular processes such as litter decomposition and in habitats such as soil. Within some disciplines the emphasis is moving towards autecological studies. Recent volumes have tended to be concerned with particular groups of organisms in specific habitats, such as fungi in soil, animals in soil and marine bacteria. We now see the need to synthesize the knowledge that has been obtained during studies of specific groups of organisms and to draw attention to their interrelationships in plant litter decomposition.

A central process in the life cycle of all green plants is the decomposition of their remains. During this decomposition many complementary and/or competing organisms are active. Frequently the processes of decomposition begin before the plant part senesces and the sequence of organisms involved is related to the type of plant material and the environment. Decomposition is then conditioned by the nature of the plant tissues, the range of organisms able to decompose these tissues, and the environment.

The arrangement of the chapters in this book follows the above pattern, and the quantity of material involved has made it necessary to divide the book into two volumes. In Part I, which constitutes Volume 1, the primary emphasis is placed on the type of litter. In this context litter is taken to include all plant remains, which range from still standing dead trees to the decomposing hyphae of fungi and cells of bacteria. Herbivore dung is considered in this section as digested litter.

The organisms involved in decomposition processes are discussed in Part II, which forms the first part of Volume 2. The treatment of each group is not meant to be exhaustive and may be vulnerable to criticism by specialists in the appropriate disciplines. However, our intention has been that the chapters indicate the range of structure and function of the organisms concerned. Biotrophic and necrotrophic parasites of higher plants, by their activities, are involved in the early stages of decomposition. It has therefore been found appropriate, when considering certain groups of organisms, at least to mention such parasites in the appropriate chapters.

The second part of Volume 2, Part III, is concerned with the environmental conditions under which breakdown occurs over the whole global surface. Terrestrial, freshwater and marine environments are considered separately. A further two aspects are anthropocentric: agriculture, with an emphasis on the importance of the saprophytic activity of plant pathogenic fungi; and the increasingly important composting of urban waste. We have allowed a small amount of overlap between certain chapters where we felt it desirable to give a balanced account within the particular treatment.

The editors wish to express their thanks to the chapter authors for their co-operation. Mrs S. J. Dickinson prepared the systematic index. Mrs M. Beck, Mr G. Maggs, Mrs V. Ross and Mrs B. Wallace have assisted in the preparation of this volume, and efficient secretarial assistance has been provided by Mrs J. Hall. We have also been greatly encouraged and assisted by Academic Press.

C. H. Dickinson
G. J. F. Pugh

September, 1973

Contents of Volume 2

Contents of Volume 1

8

Bacteria

H. G. Gyllenberg and Eva Eklund

Department of Microbiology
University of Helsinki
SF-00710 Helsinki 71
Finland

I. Introduction

The gross function of a system can usually be expressed in simple general terms, and biological systems are no exception. The decomposition of plant litter can be described with a few equations which indicate that organic matter is ultimately broken down to carbon dioxide and water. However, this formalization does not tell us how the system operates and which control mechanisms keep the system in its dynamic state. Understanding a biological system requires knowledge about the organisms involved in the system, their characteristics and abilities, as well as their relationships and interactions. Taxonomy is the tool to produce much of this knowledge, and particularly to arrange the knowledge in a way which makes it available and which maximizes its content of relevant information. Any attempt to survey the bacteria involved in plant litter decomposition thus necessarily requires a taxonomic approach.

Taxonomy is based on three elements: classification, nomenclature, and identification. In this context the problems of identification are less significant as compared with classification and nomenclature, and the discussion below is thus restricted to these two elements. Classification and nomenclature of bacteria operate with the *species* as the basic unit. True, this concept is by no means clear, as is obvious from the definition proposed by Buchanan (1957) which states that "a species of a bacterium is the type culture or specimen together with all other cultures or specimens regarded by an investigator as sufficiently like the type (or sufficiently closely related to it) to be grouped with it". This definition, like other attempts to define the bacterial species, gives rise to several controversial questions: Who is a competent investigator? What is meant by "like" or "closely related"? And how should "sufficient" be interpreted?

However, Buchanan's definition cannot be regarded as an inconsistent one. It expresses what every microbiologist knows or intuitively feels, namely that classifications of bacteria are never static and final, but dynamic and continuously changing. Moreover, Buchanan's definition can be applied to the different "technical" species that have been introduced into microbiology. Ravin (1963) defined three technically different species: (a) the *taxospecies*, which is a tight phenetic cluster produced when individual strains that have a high proportion of characters in common are grouped together; (b) the *genospecies*, which is a group of organisms which show mutual exchange of genes; and (c) the *nomenspecies*, which can be defined as a group of organisms that bear a certain binomial name. All these types of species have been used in the literature on soil bacteria, and hence also on litter-decomposing bacteria. The nomenspecies is distinctly connected with traditional taxonomy which applies a "downward" classification (dividing groups into subgroups and these into subgroups, etc.) into usually strictly monothetic groups. As characteristics which define groups one may find quite ordinary determinative features of the type "production of hydrogen sulphide" or "utilization of pentoses", etc. However, some extraordinary biochemical ability may also be used for group definition (e.g. *Pseudomonas methanica*) as well as specific ecological features as the habitat (e.g. *Enterobacter cloacae*). The taxospecies is due to application of numerical taxonomy which implies an "upward" classification (collection of individual strains into primary groups which are combined into larger groups etc.) into polythetic groups. Group definition in simple terms is difficult, and usually the circumscription of taxospecies is performed with a conventional classification into nomenspecies as reference. The same also concerns genospecies. Since exchange of genes is the basic criterion for the definition of a genospecies, reference strains are likely to be necessary. Again the reference system is usually provided by a conventional nomenspecies classification. In addition to exchange of genes, a narrow range of GC- (guanine and cytosine) content in the DNA is indicative of a genospecies. A better alternative criterion for a genospecies is a high DNA homology. However, the necessary technique is more laborious than that for determination of the GC content. Information is thus usually limited to homology with a given reference organism. It follows then that application of DNA homology data involves merely a kind of verification of the nomenspecies by independent methods rather than the building up of a "pure" genospecies classification.

It is quite obvious that the three types of species described above often give rise to considerably different taxonomic practices when applied to certain groups of organisms. To take the Enterobacteriaceae as an example, the number of recognized nomenspecies is quite high. Utilization of the taxospecies approach already reduces the number of formal taxa, but since exchange of genes is known to take place between members of many taxospecies, a classification on the basis of genospecies would reduce the number of recognizable "species" still more. Alternatively,in groups of organisms which have been studied more superficially (i.e. the number of recorded characteristics is low), only a few still heterogeneous nomenspecies and/or taxospecies have been described. Most probably such "under-classified" taxa may contain several genospecies. There occur, of course, other cases where the three species concept corresponds quite well. For further discussion of this topic reference may be made to the paper of Jones and Sneath (1970).

As far as the taxonomy of litter-decomposing bacteria is concerned, the general approach seems still to follow traditional lines with the application of the nomenspecies as the basic unit. However, numerical methods have been employed quite frequently in recent years in soil microbiology (and the microbiology of decomposing organic matter) to verify earlier classifications and to increase distinction and usefulness of classifications. Similarly, genetic and chemical methods have been applied to characterize bacterial DNA for the same purposes.

Although a traditional classification into monothetic groups still provides the basic reference system for soil bacteria (and bacteria from corresponding environments), there is much complementary information available from application of modern techniques. This permits discussion of many groups of bacteria in quite distinct terms. For other groups, unfortunately, the picture is still unclear and sometimes even confusing, and hence, for these groups, the basis for a detailed discussion is non-existent.

II. Taxonomy

A. General

The bacteria which are involved to a significant extent in the decomposition of plant litter can be classified in the orders Pseudomonadales,

Eubacteriales, Actinomycetales, and Myxobacterales. Since the Actinomycetales will be treated in a separate chapter of this book (Goodfellow and Cross, Chapter 9), the following survey is limited to the other orders. This does not imply that these orders are discussed in full detail, the order structure has been chosen only for the sake of obtaining a logical disposition of the material. The species, as already indicated in the introduction, constitutes the basic unit in the taxonomy of bacteria, and main emphasis will therefore be given to the species level, and whenever necessary, to the generic level. Taxa of higher hierarchical ranks will be largely ignored.

B. Pseudomonadales

From the point of view of decomposition processes the most significant group in the Pseudomonadales is the genus *Pseudomonas*, which can be defined as consisting of Gram-negative, polarly flagellated rods with an oxidative sugar metabolism. Some *Pseudomonas* types are able to use denitrification as an alternative anaerobic mechanism for respiration. Organisms with a fermentative carbohydrate metabolism are definitively excluded from the genus. In addition, most pseudomonads are oxidase positive (oxidase test by Kovacs, 1956). Exceptions are *P. maltophila* (Stanier *et al.*, 1966) and certain plant pathogenic pseudomonads (Lelliott *et al.*, 1966).

A positive arginine dihydrolase reaction (Thornley, 1960), production of protocatechuic acid from quinic acid (Stewart, 1965), and production of fluorescent diffusible pigments are all characters which when they occur in combination with the primary criteria for *Pseudomonas* are strongly indicative for the classification of an organism as a member of this genus. However, these characters are not ultimate criteria. The genus *Pseudomonas* comprises non-pigmented as well as fluorescent organisms, and some pseudomonads lack a constitutive arginine dihydrolase system (cf. Stanier *et al.*, 1966). The degradation of quinic acid to protocatechuic acid is also a common, but not a universal feature, among the organisms classified as *Pseudomonas*.

The definition of *Pseudomonas* is thus relatively broad, and includes organisms which are mutually quite different. This is also indicated by the average range of the DNA base composition (De Ley, 1964). The range for *Pseudomonas* extends from 60% to almost 70% GC, which is a wide range as compared for instance with *Rhizobium* (60–62% GC). The fact that *Pseudomonas* seems to constitute a heterogeneous group has introduced the question as to whether it is justifiable to exclude some organisms from the genus. Davis and Park (1962) suggested the exclusion of *Comamonas*, i.e. organisms which otherwise fit the basic definition of *Pseudomonas*, but are completely inactive against sugars. *Xanthomonas* has been separated on similar grounds (plant pathogenity, carotenoid pigments; Dowson, 1939). However, the separation of different groups from *Pseudomonas* only on the basis of one or two functional characteristics does not contribute to the homogeneity of *Pseudomonas*. The situation is similar for *Methanomonas* (ability to utilize methane as sole source of carbon), *Alginomonas* (capable of alginic acid degradation), *Cellvibrio* (capable of cellulose degradation), and *Hydrogenomonas* (ability to grow chemolithotrophically with hydrogen). Stanier *et al.* (1966; cf. also Hendrie *et al.*, 1968) have pointed out that all the organisms which have been classified into the genera mentioned above and further genera of the same kind could well be included in *Pseudomonas*, If a subclassification of *Pseudomonas* is felt advisable it should be based on other criteria, and it should, of course, create groups that are more homogeneous than is *Pseudomonas*.

In the seventh edition of Bergey's Manual (1957), the genus *Pseudomonas* comprised not less than 149 separate "species". Since then investigations using various named *Pseudomonas* cultures and/or fresh isolates of *Pseudomonas* from different environments, and statistical treatment of the results, have in general yielded only a few major clusters in which some of earlier named "species" have been combined (Rhodes, 1959; Lysenko, 1961; Colwell and Liston, 1961; Gyllenberg, 1965; Stanier *et al.*, 1966; Gyllenberg and Eklund, 1967).

Stanier *et al.* (1966), who have performed the most comprehensive study, accept four major groups of pseudomonads, namely the fluorescent group, the pseudomallei group, the acidovorans group, and the alcaligenes group. In addition to these groups only four further species are defined (*P. multivorans*, *P. stutzeri*, *P. maltophilia* and *P. lemoignei*). The fluorescent group constitutes one of the most extensively studied groups of bacteria. It includes the pathogens, *P. aeruginosa*, *P. fluorescens* and *P. putida*; for the last two species a number of biotypes have been defined. The fluorescent group is characterized by the ability to produce water-soluble green fluorescent pigments (pyoverdin), and to use different organic compounds as carbon and energy sources. Amongst other groups a higher potential capacity for breakdown of organic compounds is shown only by *P. multivorans* which seems, however, to be either geographically restricted or unable to compete successfully with fluorescent pseudomonads under ordinary soil conditions.

As already mentioned, a number of minor groups of bacteria have been separated from *Pseudomonas* on different grounds. Two of the groups, *Cellvibrio* and *Alginomonas*, are certainly involved in the degradation of complex organic molecules. Others, like *Methanomonas*, utilize and hence also decompose lower hydrocarbons. As a whole, these organisms

(*Pseudomonas* included) are characterized by relatively rapid growth, and versatility both as regards available nutrients and environmental factors.

The genera *Aeromonas* and *Vibrio* consist of polarly flagellated, oxidase-positive but fermentative organisms. An identification key for these organisms is presented by Bain and Shewan (1968).

C. Eubacteriales

1. Sporeformers

The members of the genus *Bacillus* can be defined as rod-shaped-organisms which produce endospores. They are usually Gram-positive, produce catalase, and are capable of sporulation under aerobic conditions (Wolf and Barker, 1968).

Non-sporulating variants have been described in several of the species. Biochemical reactions may in such cases need a supplementary study of the cell antigens (cf. Baillie, 1967; Norris and Wolf, 1961). The non-sporulating bacilli seem to constitute a link to *Lactobacillus* with *L. plantarum* as an obvious counterpart (Davis, 1964).

The present classification of the genus *Bacillus* is mainly based on the extensive report of Smith, Gordon and Clark (1952). Since then repeated approaches to the classification of bacilli have in general verified the conclusions arrived at by Smith *et al.*, and proposed changes are not essential. However, a new edition of their monograph (Gordon *et al.*, 1973) is due out shortly and it will widely be appreciated. In fact only minor modifications have been suggested to the characters applied for the definition of species, and the introduction of new taxonomic methods supplies additional detailed information.

Three subgroups can be recognized in the genus *Bacillus*; they are distinguished on the basis of the form of spores, the size of spores in relation to the vegetative cells, the thickness of the spore wall and the nutritional requirements of the vegetative culture. In the following description of the subgroups the modifications proposed by Wolf and Barker (1968) are noted. The first group (oval or cylindrical spores; sporangia not swollen; thin spore wall) comprises *B. subtilis* and variants, *B. cereus* and variants, *B. megaterium*, *B. licheniformis*, *B. coagulans*, *B. firmus* and *B. lentus*. The second group (oval or cylindrical spores; sporangia swollen; thick spore wall) includes *B. stearothermophilus*; some strains of *B. coagulans*, *B. laterosporus*, *B. polymyxa*, *B. macerans*, *B. larvae*, *B. popillae*, *B. lentimorbus*, *B. brevis*, *B. pulvifaciens*, *B. pantothenticus*, *B. alvei* and the *B. circulans* complex. *B. pantothenticus* cultures form approximately equal numbers of ellipsoidal and round spores. They are therefore intermediate between the groups 2 and 3. The third group (spherical spores; swollen sporangia;

complex nutritional requirements) is composed, besides the intermediate *B. pantothenticus* of only two species, namely *B. sphaericus* and *B. pasteurii.*

The major taxonomic problem relating to the genus *Bacillus* concerns the unclear relations of *B. circulans* to several other organisms mentioned above as members of the second subgroup of *Bacillus*. Intermediate strains, especially between *B. circulans*, on one hand, and *B. alvei*, *B. brevis* and *B. laterosporus*, on the other, are common. It is worth emphasizing that the relationship between *B. circulans* and *B. macerans* is also very close. All this may point to high variability and even genetic instability in the second subgroup of *Bacillus*, which can be considered as an indication of a state of active development (cf. *Lactobacillus plantarum*; Davis, 1964). This is of particular interest since active decomposers are found in this group as well as organisms which grow under extreme conditions of temperature.

Records of the thermophilic bacilli in the literature go back to the last century, and several types and "species" have been described (cf. Allen, 1953). In the monograph of Smith *et al.* (1952), the number of separate thermophilic species was, however, reduced to only two, viz. *B. stearothermophilus* and *B. coagulans*. Darland and Brock have recently (1971) described a new thermophilic species—*B. acidocaldarius*—which differs from all known bacilli in its very high GC value (62%). Therefore, it seems questionable whether this organism can be included in the genus *Bacillus* at all. The problem of the temperature requirements of thermophilic bacteria under natural conditions have been much discussed. Most probably these organisms may show alternative nutritional requirements, etc., as a response to changing environmental conditions, which would explain their occurrence also in habitats where extremely high temperatures are improbable. In any case, the resistance of spores to external environmental effects is correlated to the temperature relations of the organisms. Hence, thermophilic bacilli produce highly thermoresistant spores, which may be an ecologically significant feature, particularly under natural conditions.

In 1966 Larkin and Stokes published a report on psychrophilic bacilli. In another paper (Larkin and Stokes, 1967) four psychrophilic species were described, namely *B. psychrosaccharolyticus*, *B. insolitus*, *B. globisporus* and *B. psychrophilus*. Except for *B. insolitus* which was described as related to *Sarcina ureae*, the other psychrophilic species seemed to be related either to the *B. circulans* complex as such or at least to the second subgroup of bacilli. This is also obvious from the later work of Laine (1970) in which the observations of Larkin and Stokes were largely confirmed. In addition, Laine was able to isolate a few gas-forming psychrophilic bacilli which thus bear a resemblance to *B. polymyxa*. A further psychrophilic type, quite frequent in nature, described by Laine, also seems to belong to the second subgroup of bacilli. Laine's data (1970) indicate that psychrophilic bacilli occur frequently in different natural habitats, among them plant materials. The ability of these organisms to maintain their metabolic activity at low temperatures points to their contribution to natural decomposition processes. The resistance of the spores of psychrophiles is quite low, and dependent on the temperature where sporulation takes place. Laine (1970) reported about three times higher D-values for psychrophilic spores produced at 20°C as compared with spores produced at 2°C.

As an illustration of further problems in the taxonomy of bacilli, the status of certain varieties of *B. subtilis* and *B. cereus* can be mentioned. For traditional reasons, particularly due to their specific pathogenity, *B. anthracis* and *B. thuringiensis* could well be considered as separate species. From strictly taxonomic points of view these organisms fit the definition of a variety, and should be regarded as varieties. *Sarcina ureae*, mentioned above, is a sporeformer, and available information indicates that, in the opinion of some workers, it should be classified among the bacilli, but its relation to the different subgroups of bacilli is not clear. In this connection the results of Rogers *et al.* (1970), who describe coccoid mutants of *B. subtilis* and *B. laterosporus*, are of special interest. On the other hand, the present definition of the genus *Bacillus* and the meaning of the generic name denote a rod-shaped micro-organism.

As compared with the genus *Bacillus*, the taxonomic information available for *Clostridium* is much less detailed. This group of strictly anaerobic sporeformers is not very homogeneous, and the genus includes organisms which cover a wide ecological and physiological range. A common general feature is, however, the versatile ability to attack different organic compounds as well as the high rate of the breakdown processes when they take place under optimal conditions.

From the point of view of plant litter decomposition, the pectinolytic and the cellulolytic species deserve attention. In addition to *Clostridium felsinum* which is usually mentioned in connection with flax retting, several other pectinolytic species have been described (cf. Lanigan, 1959). The separation of these species from each other is mainly based on differences in spore morphology, pigmentation, and the volatile compounds they produce from glucose. The pectinolytic clostridia reveal fermentation patterns varying from the butyric type to the acetobutylicum type, as well as intermediates. *C. felsinum* represents the acetobutylicum type, whereas *C. flavum* brings about a butyric acid fermentation. *C. laniganii* constitutes an intermediate to the two main metabolic types.

The cellulolytic clostridia can be divided into mesophiles and thermophiles. The organisms are however, obviously closely related, and this is confirmed by the type of fermentation. Both mesophiles and thermophiles decompose cellulose with vigorous production of hydrogen, carbon dioxide and organic acids. The borderline between the mesophiles and the thermophiles remains somewhat indistinct due to the fact that some mesophilic species may grow at temperatures around 50°C, whereas the temperature optima of the thermophilic species remain around 60°C or even lower. From the ecological point of view an obvious difference exists. The thermophilic species seem to occur frequently in soil and on decomposing plant material, whereas the main habitat of the mesophilic species is the intestinal tract of herbivorous animals. However, the intestinal bacteria may be disseminated in soil by manure, etc. The most frequently encountered mesophilic species are *C. cellulosolvens*, *C. cellobioparum*, *C. omelianskii*, and *C. dissolvens*, whereas *C. thermocellum* and *C. thermocellulaseum* are the main thermophilic species. Differentiation of species is based on the ability to ferment glucose, motility, pigmentation and spore morphology.

2. *Enterobacteriaceae*

The Enterobacteriaceae are of interest in connection with plant litter decay (a) because organisms related to "*Aerobacter*" *aerogenes* occur frequently in nature and some seem to be active decomposers of complex organic compounds under natural conditions, and (b) because there are, in the genus *Erwinia*, organisms which colonize and decompose plant materials. Moreover the taxonomic problems relating to the correct classification and nomenclature of these organisms are quite intricate.

A comprehensive review of the genus *Erwinia* has been recently published by Starr and Chatterjee (1972). The obvious conclusions which can be drawn from this paper, and the background literature, are that the organisms involved can be divided into three main groups, namely (a) the amylovora group, (b) the carotovora group and (c) the herbicola group. Of these organisms the herbicola group is interesting because these organisms frequently colonize seeds, leaves and other living plant materials, and they thus constitute an "indigenous" population available to break down litter immediately it is formed. The organisms in the carotovora group are pectinolytic, and it has been suggested that these bacteria could be collected in the genus *Pectobacterium* (Waldee, 1945). In addition to the three main groups, some erwinias remain unclassified. These may be classified in further "species" or collected into a common group. The taxonomic problem follows from the fact that the different groups of erwinias mentioned above show affinities to other genera and tribes within the Enterobacteriaceae. Obviously the separation of the carotovora group into a new genus (*Pectobacterium*) solely on the basis of the pectolytic activity does not seem well motivated. Another alternative is to transfer the herbicola group to the genus *Enterobacter* (under the name *Enterobacter agglomerans*, as suggested by Ewing and Fife (1972). However, the carotovora group could then also be included in *Enterobacter* on similar grounds, which would limit *Erwinia* to concern only the amylovora group. Starr and Chatterjee (1972) conclude that the erwinias are enterobacteria without any doubt, but that the question of retaining the genus as it now stands or of distributing its members among other genera of Enterobacteriaceae must await acquisition of additional information. Gardner and Kado (1972) have recently compared base sequence homologies of the DNA of *Erwinia* species and other members of Enterobacteriaceae. Their results do not support some of the recently proposed taxonomic divisions within the genus *Erwinia*, particularly certain combinations of species. Their DNA homology data together with GC values show that erwinias really form a loosely combined group of bacteria, often with no greater affinities to each other than to other enterobacteria. Gardner and Kado do not however support the including of the herbicola group in *Klebsiella* as proposed by Ewing and Fife (1972) since they found only 10% relative homology between *E. herbicola* and *Klebsiella aerogenes*. On the other hand, the homologies between *E. herbicola* and other erwinias were not much greater.

Aerobacter aerogenes was allocated to the genus *Klebsiella* in the classification scheme which Kauffmann published in 1959. This, however, led to misinterpretations since the earlier name was frequently applied to organisms which should be classified as *Klebsiella pneumoniae*. Kauffmann (1954) also introduced the two new genera *Cloaca* and *Hafnia*. A rearrangement was suggested by Hormaeche and Edwards (1960) who classified all these organisms in the genus *Enterobacter*, with *E. cloacae* as the type species, and *E. aerogenes* and *E. liquefaciens* as further members of the genus. As mentioned above, it has since been proposed that *Erwinia herbicola* should be grouped here and, consequently, renamed as *Enterobacter agglomerans* (Ewing and Fife, 1972).

The validity of Hormaeche's and Edwards' (1960) classification is, however, questionable in the light of a recent numerical analysis performed by Bascomb *et al.* (1971). These authors found that *E. aerogenes* showed strong affinity to *Klebsiella*, whereas *E. liquefaciens* was most closely related to *Serratia*. Again, more information is obviously needed before the taxonomic position of the organisms in question can be finally settled.

3. Other Eubacteria

The taxonomy of the genus *Agrobacterium* has been discussed by De Ley *et al.* (1966). These authors suggest that only three species can be recognized in the genus, viz. *A. radiobacter* (the crown-gall bacterium is considered as a variety *A. radiobacter* var. *tumefaciens*), *A. rhizogenes* and *A. pseudotsugae*. The last organism was represented by only one strain in the study of De Ley *et al.* (1966), and its relation to the genus *Agrobacterium* could not be positively defined. *A. radiobacter* and the crown gall variety show, on the basis of the GC percentage, a very close relationship to *Rhizobium leguminosarum*, but *A. radiobacter* can be separated from the latter by its ability to produce 3-ketoglucosides. Rhizobia are reported to show negative 3-ketoglucoside reactions (*Bernaerts* and *De Ley*, 1963), but the fact that the same authors noted later that *R. leguminosarum* produces some 3-ketolactose from lactose actually leaves open the extent to which *A. radiobacter* and *R. leguminosarum* can be reliably distinguished.

The taxonomic position of *Achromobacter* is still confusing. The problem has been discussed in detail by Ingram and Shewan (1960). The difficulty is that in earlier literature achromogenic pseudomonads were frequently referred to as "achromobacteria". At the same time the definition of the genus was unclear since no type cultures of *A. liquefaciens*, which was assigned as the type species of *Achromobacter*, were available. However, later Tulecke *et al.* (1965) were able to isolate an organism which corresponded almost exactly to the original description of *A. liquefaciens*. Tulecke *et al.* suggested that the genus *Achromobacter* would be retained with the re-isolated, motile, peritrichously flagellated strain as type species. On the basis of a broad computer survey on *Acinetobacter* and related genera, Thornley (1967, 1968) is of the opinion that the name *Alcaligenes* should be retained for the non-motile, Gram-variable, coccoid bacteria usually referred to as the *Achromobacter-Alcaligenes*-group. The human and animal pathogens also described as "achromobacteria", are quite different from this group. As shown by De Ley *et al.* (1967) these organisms form a distinct cluster which includes *Acinetobacter anitratum*. Another main group has been described by these authors. It includes the *Alcaligenes* group referred to above, and as a further subgroup certain strains of marine origin. The taxonomic position of "*Achromobacter*" seems still to be confusing.

There has been much disagreement concerning the correct taxonomic position of the genera *Corynebacterium* and *Mycobacterium*. Waksman and Henrici (1943) made a compromise and excluded *Corynebacterium* from the Actinomycetales, but retained *Mycobacterium* among the Actinomycetes. Jensen (1952) defined two "levels" of corynebacteria, viz. *sensu stricto* and *sensu lato*, the former including mainly the diphtheroids of animal origin whereas the latter comprises the diphtheroids, the plant pathogenic corynebacteria, and the saprophytic group. For Jensen's saprophytic group of corynebacteria Conn and Dimmick (1947) proposed the generic name *Arthrobacter* with *A. globiformis* as the type organism. However, the organisms related to *A. globiformis* are characterized by weak (if any) ability to attack carbohydrates. Clark (1952) thus suggested that saprophytes which utilize carbohydrates vigorously, and some of which are cellulolytic, should be included in the genus *Cellulomonas*. Cummins and Harris (1958) noted that the animal strains of corynebacteria were characterized by a distinctive pattern of sugar and amino acid components in their cell walls, very similar to the pattern of cell wall components found in strains of *Mycobacterium* and *Nocardia*. In a later paper, Cummins and Harris (1959) analysed the cell walls of seven *Arthrobacter* species. Their conclusion was that there cannot be any close relationship between these organisms and the corynebacteria proper. The results also indicated heterogeneity in the genus *Arthrobacter*. Five of the species studied fell in the same group as *A. globiformis*, whereas *A. simplex* and *A. tumescens* differed from other species and grouped together.

Jensen (1952) put forward the hypothesis that the arthrobacteria form a primitive and versatile group from which other groups of corynebacteria have developed by loss of characters. They may also be ancestors of the Actinomycetes, with some of which they share certain features. Recent investigations have shown, however, that the arthrobacteria are not very closely related to diphtheroids of animal origin, and have confirmed that the borderline between the arthrobacteria and certain groups of Actinomycetales is far from being distinct and easily definable (Gordon, 1966; Bousfield, 1972).

Application of numerical taxonomy (Nigel da Silva and Holt, 1965; Goodfellow, 1967) has shown that phenetic similarities between members of *Arthrobacter* may be fairly low (e.g. between *A. tumescens* and *A. globiformis*). Zagallo and Wang (1962) found that *A. globiformis* and *A. ureafaciens* rely mainly on the operation of the Embden-Meyerhof-Parnas pathway, whereas *A. simplex* requires intermediate formation of gluconate.

Skyring and Quadling (1970) drew attention to the fact that the arthrobacteria can be divided into two groups on the basis of nutritional requirements. *A. terregens*, *A. flavescens* and *A. duedecades* are nutritionally exacting, and thus separable from the nutritionally non-exacting species. Further work (Skyring and Quadling, 1969) has indicated certain difficulties in the practical differentiation between *Agrobacterium*, *Rhizobium* and *Arthrobacter*, which also show, however, similarities in the GC content of their DNA. Principal component analysis of representative material of cultures from these three genera has shown (Skyring *et al.* 1971) that separation of the genera can be based on the abilities to use various carbohydrates and amino acids or to grow under various conditions.

As regard other eubacteria, it may be mentioned that taxonomic surveys of micrococci have recently been made by Baird-Parker (1963) and Rosypal *et al.* (1966). The taxonomic position of *Flavobacterium* is very problematic. The genus includes two main groups of organisms: (a) organisms with GC values of 30–40% and (b) organisms with GC values >50%. The first group seems to represent atypical cytophagas, and the other group may also partly be allocated in some other genus. McMeekin *et al.* (1972) have suggested a simple scheme for the tentative classification of yellow pigmented rods in "low GC types" versus "high GC types". Bousfield (1972) has shown that some flavobacteria of the "high GC type" (% GC 65–67) obviously are coryneforms. These organisms could either be placed in *Arthrobacter* or in a new genus which would replace the rearranged genus *Brevibacterium*. The type species of the last mentioned genus seems to be better placed in *Arthrobacter*, and then the name *Brevibacterium* would be illegitimate.

The taxonomic position of the "low GC" flavobacteria is discussed under the heading Myxobacterales. Some of these bacteria seem actually to be atypical cytophagas (cf. e.g. Mitchell *et al.*, 1969).

D. Myxobacterales

The myxobacteria are usually described very superficially in textbooks and monographs with some general references to their strange morphology and their supposed importance in decomposition processes. During the last decade there has been an obvious increase in the interest devoted to these organisms which is reflected in many suggestions for taxonomic rearrangement.

In Bergey's Manual, 7th edition (1957) the order is divided into five families on the basis of occurrence and characteristics of resting cells, cysts, and fruiting bodies. A successive change to higher complexity of these bodies is obvious. *Cytophaga* occurs only in the vegetative state, *Sporocytophaga* forms resting cells, but they are not condensed into fruiting bodies. In Myxococcaceae and Archangiaceae the resting cells are aggregated into special fruiting bodies, and, finally, in Sorangiaceae and in Polyangiaceae the primary resting cells are contained in secondary cysts which are attached to fruiting bodies. This logical spectrum of increasing morphological complexity was not, however, followed in the classification key of Myxobacterales in the last edition of Bergey's Manual. For some less obvious reason *Sporocytophaga* was included in the family Myxococcaceae which was listed as the last family after the cyst-producing types:

Family I.	Cytophagaceae:	No resting cells or fruiting bodies.
Family II.	Archangiaceae:	Resting cells on fruiting bodies.
Family III.	Sorangiaceae:	Resting cells in cysts on fruiting bodies
Family IV.	Polyangiaceae:	Resting cells in cysts on fruiting bodies
Family V.	Myxococcaceae	(including *Sporoctyophaga:* free resting cells): resting cells on fruiting bodies.

The invalidity of this classification became immediately obvious when GC analyses were applied to the group. The GC contents divide the order into two quite distinct groups with *Cytophaga* and *Sporocytophaga* in one group (GC contents in the range of 35–42%) and the fruiting body forming organisms in the other group (GC contents in the range of 68–72%). The difference is so clear-cut that the relation between both the groups already looks somewhat dubious. However, for the sake of the historical background a discussion of the cytophagas is included under the present heading.

It has been confirmed that some of the bacteria, earlier classified as "flavobacteria", a vaguely defined group of organisms, actually belong to the *Cytophaga-Sporocytophaga* group (Hendrie *et al.*, 1968; Mitchell *et al.*, 1969; Weeks, 1969), and this group is obviously very closely related to several other types of "gliding bacteria" (Lewin and Lounsbery, 1969; Mandel and Lewin, 1969; Fager, 1969; and Colwell, 1969). Lewin (1969) suggested that all these organisms should be collected into one family, Cytophagaceae (or Flexibacteriaceae) including the genera: *Cytophaga, Flexibacter, Saprospira, Flexitrix, Herpetosiphon, Sporocytophaga*, "*Sphaerocytophaga*" (or *Fusobacterium*) and *Microscilla*. In view of the distribution of GC content within these genera, the proposed classification seems questionable. The data of Mandel and Lewin (1969) indicated following figures for the GC contents: *Cytophaga* 34–41%, *Flexibacter* 30–47%, *Microscilla* 32–45%, *Saprospira* 35–48%, and *Herpetosiphon* 45–53%. The numerical analysis of Colwell (1969) indicates possibilities for some rearrangement. Since high similarity values (>80%) were found for strains assigned to *Cytophaga*, *Flexibacter*, and *Microscilla*, particularly with strains in the range of 35–40% GC, several strains labelled *Flexibacter* or *Microscilla* may in fact belong to *Cytophaga*.

For the myxobacteria which produce fruiting bodies a taxonomic revision has been proposed by McCurdy (1970). The major point in McCurdy's scheme is a redefinition of the family Polyangiaceae to include organisms with cylindrical cells. As a consequence the organisms earlier classified as Sorangiaceae are transferred to Polyangiaceae, and a new family, Cystobacteraceae, is proposed for sporangial myxobacteria ("sporangial" here refers to the formation of cysts containing resting cells or microcysts). McCurdy's revised scheme follows the logic of the gradual morphological change from one group to next, and seems to constitute a more reliable approach to the taxonomy of the fruiting myxobacteria than the classification followed in the seventh edition of Bergey's Manual. McCurdy's (1970) scheme is presented here for comparison:

I. Vegetation cells tapered, microcysts (slime encapsulated myxospores) produced.
   A. Microcysts spherical or oval. Myxococcaceae.

B. Microcysts rod-shaped.
1. Microcysts not in sporangia. Archangiaceae.
2. Microcysts in sporangia. Cystobacteriaceae.

II. Vegetative cells of uniform diameter with blunt, rounded ends. Myxospores resemble vegetative cells. Polyangiaceae.

III. Ecological Aspects of Bacterial Litter Decomposition

As a result of varying conditions plant litter decomposition may show different characteristics in different ecosystems, but in most cases it is characterized by periodicity, the accumulation of a huge biomass of decomposer micro-organisms and the resynthesis of organic compounds which may be less degradable than the original matter. Since the microbial biomass is also broken down successively the decomposition process is perhaps most correctly described as a chain process which ultimately combines with the nutrient cycles and even the primary production processes.

The successions of micro-organisms in litter decomposition are primarily determined and controlled by the abilities of the micro-organisms present to utilize available organic compounds, but partly also by the response of the organisms to changes in the environment caused by the decomposition process. The subject is extensive (cf. Alexander, 1964, 1968) and the following brief survey is presented in order to indicate the role of the bacteria involved in these processes.

The first compounds which are utilized in dead organic litter are water soluble nitrogenous compounds, sugars, and various organic acids. However, these compounds are readily transported by rainwater into deeper litter layers. It seems likely, therefore, that these compounds are: (a) utilized *in situ* by the phylloplane population, the development of which on the surface of living leaves and needles is controlled by exuded sugars and easily available nitrogenous compounds; and (b) extracted from the surface litter and transported to deeper litter layers where they are consumed either by the organisms originating from the litter itself or by the indigenous bacterial population of soil.

The bacteria which inhabit the surface of leaves and other aerial plant organs are mainly chromogenics, and among them *Erwinia herbicola* seems to be particularly significant. As with the enterobacteria in general, *E. herbicola* is stimulated by sugars, and consequently by the exudates in the phylloplane of living plants, and comparable extracted compounds. Stout (1958, 1960, 1961), in a series of extensive investigations of the bacterial populations of natural grasslands, cultivated pastures, and various forest habitats, reported an abundant occurrence of "flavobacteria" with an oxidative sugar metabolism in the phylloplane of grasses and in the upper litter layers in forests. This points to the particular importance of the phylloplane as a source of litter-colonizing organisms of the *Flavobacterium-Flexibacter-Cytophaga* complex. Actually the chromogenic bacteria can be followed into deeper litter layers, where they are mixed with the indigenous population, but may occur abundantly as decomposers of cellulose and other polymeric compounds.

The soil bacteria are usually divided into two functional groups: (a) the indigenous organisms, whose numbers in soil are supposed to remain unaffected by the litter, and (b) the zymogenous organisms, which are actively involved in litter decomposition and which, therefore, increase to very high numbers during periods of litter decomposition. The zymogenous organisms do not of course disappear completely from the soil when decomposition slows down, and probably the leaching of water-soluble compounds from recently fallen litter may cause an activation of the retained zymogenes, which take over the next step of the degradation. This concerns the more easily decomposable polymeric compounds, including the pectins, hemicelluloses, cellulose and chitin.

The abilities to degrade these polymeric compounds are widespread among fungi and Actinomycetes but among bacteria these activities are restricted to a few genera or only a few species in certain genera. In Pseudomonadales cellulolytic organisms are found in the genus *Cellvibrio*. However, as pointed out in the taxonomic survey, the validity of this genus is unclear, and the only conclusion thus is that cellulolytic pseudomonads occur. Among the coryneforms, *Cellulomonas* represents a homogeneous taxonomic group (Nigel da Silva and Holt, 1965). In Eubacteriales other cellulose-decomposers are found among the sporeformers. It seems, however, that the activity of these organisms is mainly restricted to extreme environmental conditions. The cellulolytic clostridia in soil are thermophiles. Mesophilic species may be introduced by the faeces of herbivorous animals, but this route of colonization is hardly very important except for extensively manured field soils.

Among the aerobic bacilli there are active decomposers of starch, pectins and xylans. Mishustin and Mirsoeva (1968) are of the opinion that in soils where biological processes are at equilibrium, most of the bacilli are present as spores but that sporeformers multiply during later stages of mineralization of plant litter. It is interesting to note also that bacilli may be connected with the breakdown and recirculation of biomass. *Bacillus circulans* for example, was found to produce 1-3-glucanase lytic against fungal and yeast cell walls (Horikoshi *et al.*, 1963; Tanaka and Phaff, 1965). In Myxobacterales, i.e. the complex of fruiting "slime bacteria", flexibacteria, and flavobacteria, the ability to decompose polymeric organic compounds seems to be more evenly distributed among the various organisms. The ubiquitous distribution in living and dead plant material of the "*Flavobacterium-Flexibacter-Cytophaga*" group was noted above, and fruiting myxobacteria are widespread in soil, on bark and decaying material as well as on dung (Peterson, 1969). Among the *Cytophaga*-species for instance *C. hutchinsonii*, *C. aurantiaca*, *C. rubra*, and *C. tenuissima* are reported to be active cellulose decomposers. *Cytophaga johnsonii* is known to be a chitinoclastic organism. Other types of cytophagas show activity against various organic compounds and they may be highly proteolytic, pectinolytic and amylolytic. In addition to *Cytophaga* and *Sporocytophaga* several other fruiting myxobacteria are cellulose decomposers. Moreover, these organisms are known to be highly bacteriolytic, mycolytic and nematodolytic. However, there seems to be a certain specificity between the attacked organism and the lytic myxobacterium (Stanier, 1942, 1947; Dworkin, 1966; Webley and Jones, 1971).

Although bacteria which are capable of degrading various polymeric organic compounds are widespread in nature, the existing environmental conditions may limit bacterial activity. In most environments fungi obviously constitute the primary decomposer population of plant material, whereas bacteria appear as a secondary population. This particularly concerns forest soils, where the predominance of fungi is well documented. The predominance of fungi as decomposing agents in forest soils is partly due to high acidity, but also to the fact that tannic compounds inhibit bacterial enzymes (Basaraba and Starkey, 1966; Henis *et al.*, 1964).

Bacteria exert considerable activity in the breakdown of fungal mycelium and similar materials. Different techniques to elucidate microbial activities in soil (such as the examination of buried cellulose films), have indicated that bacteria are particularly abundant around fungal hyphae (Tribe, 1957). In addition to the myxobacteria other bacteria, especially certain Streptomycetes, are strongly mycolytic.

Most reports on bacteria in connection with litter decomposition are descriptive and information is given only of relative numbers of various bacteria in decaying organic matter. For a few groups of bacteria more exact information is available. First, it is obvious (e.g. Rovira and Sands, 1971) that fluorescent pseudomonads are distributed unevenly in soil, and that these bacteria are associated particularly with organic matter. In the presence of readily utilizable nutrients fluorescent pseudomonads usually show a very rapid increase in numbers under soil conditions, but as soon as maximum cell density is reached they disappear rapidly (Eklund, 1970; Eklund and Sinda, 1971). This may point to an efficient utilization of useful substrates, which obviously causes a change in the environment so that the organisms are suppressed. The bacilli are another group of bacteria associated with the decay of plant residues. In this connection *Bacillus circulans* is worth special attention; it seems to occur frequently in acid forest soils (Holding *et al.*, 1965; Goodfellow *et al.*, 1968). As emphasized in the taxonomic survey, *B. circulans* constitutes a highly variable and unstable group. The high variability explains how *B. circulans* can adapt itself to extreme environmental pressures such as high or low temperatures, and acid conditions. In soil it has been observed (Hill and Gray, 1967) that *B. circulans* frequently encapsulates on organic material. This mechanism may help to attach the organism to soil particles which may protect it against desiccation, ingestion by protozoa or lysis by myxobacteria (cf. Gray and Williams, 1971).

Pseudomonads can be considered as particularly important (cf. Dagley, 1967) in the decomposition of the aromatic compounds released in other stages of litter degradation. Their versatility and adaptibility to varying environmental conditions may encourage these activities. Other bacteria which have been reported as decomposers of aromatic compounds are *Agrobacterium*, *Arthrobacter*, *Achromobacter* and *Cellulomonas* (Sundman, 1965; Kunc, 1971; Stevenson, 1967).

As emphasized by Clark and Paul (1970) many micro-organisms utilize humic materials as sources of nitrogen. Accordingly, proteolytic bacteria are capable of utilization of polypeptide chains bound to the aromatic skeleton (cf. Haider *et al.*, 1965).

As a general rule both determinants of litter decomposition (the occurrence of various organic compounds and the prevailing environmental conditions) are interdependent. As in any ecosystem (and also in non-biological systems) a sudden change in prevailing conditions is likely to be controlled and reversed. A description of plant litter decomposition in these very general and simplified terms would indicate that in the first stages there occur non-exacting organisms which are versatile in breaking down organic compounds. Since low temperatures and availability of adequate oxygen can be expected for most environments, aerobic psychrophiles can be predicted as frequent members of the population. In fact, strong fungal activity is indicative of the initial stages of decomposition, and among bacteria, pseudomonads are found as well as yellow pigmented aerobic bacteria.

Strong aerobic breakdown of organic constituents immediately causes several changes in the environment. First the availability of oxygen is likely to decrease, the liberated energy is partly changed to heat effecting an increase in temperature, the production of microbial biomass introduces a new source of potentially decomposable material and may also decrease the availability of organic nitrogen compounds. Moreover, decomposition may give rise to specific by-products which are left untouched by the initial population, and other intermediates of the decomposition process may exert direct inhibitory effects on certain initial decomposers. The population responds gradually to these changes. Fermentative processes replace the initial oxidative breakdown, typical mesophiles and even thermophiles increase in numbers. Proteolysis and ammonification are needed for nitrogen mobilization; mycolytic organisms appear as well as other specialized organisms.

The general changes described are indicated by the appearance and disappearance of several individual organisms. For many pseudomonads the rapid increase to high numbers at initial stages of decomposition can be easily verified. They grow rapidly and, because of their wide variety of inducible enzymes, are versatile utilizers of different kinds of carbon sources. The competing capacity of fluorescent pseudomonads is further increased by the production of substances suppressing the growth of some other soil bacteria (Sieburth, 1967; Chan and Katznelson, 1961; Katznelson and Rouatt, 1957). As soon as the peak has been reached the curve usually declines very rapidly. Organisms which are able to utilize fermentative metabolic pathways, such as enterobacteria, bacilli and corynebacteria, replace pseudomonads. Streptomycetes, myxobacteria and other mycolytic organisms also increase in numbers during the secondary stage.

Since the decomposing plant litter ecosystem does not operate as a homogeneous entity, the changes described above cannot be considered as an overall process. A better model is to look at the ecosystem as composed by innumerable "microecosystems", which may well be at different stages of development. Thus the aerobic stage may continue in some loci, whereas others have moved to the secondary stage, and some loci finally may comprise strictly anaerobic conditions, indicated by the occurrence of clostridia and facultatively anaerobic bacteria, such as enterobacteria. Considering plant litter decomposition as a "total process", however, the initial stages are characterized by the occurrence of relatively few types of organisms, which are uniform in their response to the environment and in other behaviour. As decomposition proceeds, the microbial population becomes more heterogeneous, which also increases the probability of direct synergistic and antagonistic interrelations between the different organisms involved. This particularly concerns the bacteria, which are represented by only a few species at the beginning, but by a wide variety of different organisms at the stages of most vigorous decomposition.

IV. Concluding Remarks

The environment where litter decomposition takes place constitutes a system of extreme complexity. Physical, chemical and biological factors and agents interact on many different levels and in many different ways. The functions of the system can be formalized in a few simple equations, but it remains very difficult to imagine how the system actually operates to bring about its general overall functions. Separate details are well known, others can be assumed with satisfactory certainty, but the combination of the high number of various processes and events to a general and still detailed picture is almost impossible at our present level of knowledge. It follows then that any attempt to assess the role and significance of bacteria in the decomposition of litter must be uncertain. It is possible to present lists of organisms which are potentially involved in decomposition of cellulose, pectins and aromatic compounds, but these lists do not tell us very much about the actual contribution of the organisms to specific degradation processes under various conditions.

However general ecological principles also hold true for the bacteria. Ecosystems are often schematically described in terms of production, consumption, and decomposition. The alternative is to distinguish between the production which relies on photosynthesis (or to some extent on chemosynthesis) and the heterotrophic food chains in which the organogenic material produced is decomposed stepwise to carbon dioxide, water, and inorganic nitrogen compounds. This approach defines the simple fact that in a particular stage of the food or decomposition chain, decomposition is always balanced by the production of organic biomass of a new kind, which implies that every consumer (or decomposer) has its decomposer. Hence the biomass produced in plant litter decomposition is also decomposed. The role of bacteria in plant litter decomposition is thus partly a direct breakdown of litter constituents, and partly an indirect degradation of the organic material which accumulates as a result of litter decomposition. This underlines the importance of bacteria as mycolytic agents. The reasoning presented above may pose the question: How is the ultimate bacterial biomass decomposed? The answer is: partly by other bacteria (as mentioned above myxobacteria are very active bacteriolytic agents) and partly by autolysis. The significance of autolytic processes increases with decreasing size of the organisms, and bacteria which in this respect represent the ultimate end of the food chains are obviously decomposed mainly by autolysis in natural environments.

The taxonomy of the bacteria which are involved in plant litter decomposition may seem confusing, since the organisms in question are distributed throughout the whole bacterial classification system. The conclusion to be drawn from this is, however, that litter decomposing abilities have developed independently in different groups of bacteria, e.g. in groups which are as far as from each other (from the taxonomic point of view) as pseudomonads on the one hand and streptomycetes on the other. This illustrates the strength of the selective environmental forces and the key role of litter decomposition in various ecosystems.

The taxonomic survey of the bacteria involved in litter decomposition may have shown that further work is needed in this field of research. Although, for the sake of convenient transfer of scientific information, taxonomic reference systems based on the nomenspecies concept are still given preference, more detailed elucidation of taxonomic relationship as well as re-evaluation of the positions of certain groups of bacteria call for application of modern tools, like numerical classification methods and studies of similarities or even homology of the DNA. Recent progress in the clarification of the close relationship of bacteria earlier referred to as *Cytophaga*, *Flexibacter*, and *Flavobacterium*, respectively, illustrates the possibilities opened up by taxonomic innovations.

In the earlier literature particular ecological characteristics have frequently been attached to given taxonomic groups of bacteria. Recent observations have shown that such a view is too schematic. Accordingly, there is documentation to show occurrence of acidophilic streptomycetes, and psychrophilic bacilli and clostridia. It seems worth emphasizing in this connection that some groups of bacteria are well established with easily separable subgroups, whereas other groups still seem to be in a highly unstable state which is characterized by the frequent occurrence of intermediates and varieties.

Adaptations to various environmental conditions and versatile abilities to participate in different decomposition processes are found especially in unstable groups. The *Bacillus circulans* group provides a good example of a group "in active development" which thus constitutes a particularly attractive research target from the point of view of plant litter decomposition. The genus *Clostridium* calls for taxonomic re-evaluation, and the complex *Erwinia-Enterobacter-Serratia* still seems to constitute an intricate taxonomic problem. The same concerns the organisms related to *Pseudomonas*. The actual taxonomic position of several groups, which have been separated from *Pseudomonas* for different, not always very clear, reasons (e.g. *Xanthomonas*, *Cellvibrio*, *Comamonas*, *Methanomonas*, *Hydrogenomonas*), calls for renewed studies.

References

ALEXANDER, M. (1964). *A. Rev. Microbiol.* **18,** 217–252.

ALEXANDER, M. (1968). *In* "Festskrift til Hans Laurits Jensen," pp. 41–47. Godgaard Nielsens Bogtrykkeri. Lemwig, Denmark.

ALLEN, M. B. (1953). *Bact. Rev.* **17,** 125–173.

BAILLIE, A. (1967). *J. appl. Bact.* **30,** 230–238.

Bain, N. and Shewan, J. M. (1968). *In* "Identification Methods for Microbiologists," Part B (B. M. Gibbs and D. A. Shapton, eds), pp. 79–84. Academic Press, London and New York.

Baird-Parker, A. C. (1963). *J. gen. Microbiol.* **30**, 409–427.

Basaraba, J. and Starkey, R. L. (1966). *Soil Sci.* **101**, 17–23.

Bascomb, S., Lapage, S. P., Willcox, W. R. and Curtis, M. A. (1971). *J. gen. Microbiol.* **66**, 279–295.

Bergey's Manual of Determinative Bacteriology (1957). 7th edition. (R. S. Breed, E. G. D. Murray and N. R. Smith, eds). Baillière, Tindall and Cox, London.

Bernaerts, M. J. and De Ley, J. (1963). *Nature, Lond.* **197**, 406–407.

Bousfield, I. J. (1972). *J. gen. Microbiol.* **71**, 441–455.

Buchanan, R. E. (1957). *In* "Bergey's Manual of Determinative Bacteriology." (R. S. Breed, E. G. D. Murray and N. R. Smith, eds), 7th edition, pp. 15–28. Baillière, Tindall and Cox, London.

Chan, E. C. S. and Katznelson, H. (1961). *Can. J. Microbiol.* **7**, 759–767.

Clark, F. E. (1952). *Int. Bull. bact. Nomencl. Taxon.* **2**, 45–56.

Clark, F. E. and Paul, E. A. (1970). *Adv. Agron.* **22**, 375–435.

Colwell, R. R. (1969). *J. gen. Microbiol.* **58**, 207–215.

Colwell, R. R. and Liston, J. (1961). *J. Bact.* **82**, 1–14.

Conn, H. J. and Dimmick, I. (1947). *J. Bact.* **54**, 291–303.

Cummins, C. S. and Harris, H. (1958). *J. gen. Microbiol.* **18**, 173–189.

Cummins, C. S. and Harris, H. (1959). *Nature, Lond.* **184**, 831–832.

Dagley, S. (1967). *In* "Soil Biochemistry." (A. D. McLaren and G. H. Peterson, eds), Vol. I, pp. 287–317. Marcel Dekker, New York.

Darland, G. and Brock, T. D. (1971). *J. gen. Microbiol.* **67**, 9–15.

Davis, G. H. G. (1964). *J. gen. Microbiol.* **34**, 177–184.

Davis, G. H. G. and Park, R. W. A. (1962). *J. gen. Microbiol.* **27**, 101–119.

De Ley, J. (1964). *A. Rev. Microbiol.* **18**, 17–46.

De Ley, J., Bernaerts, M., Rassel, A. and Guilmot, J. (1966). *J. gen. Microbiol.* **43**, 7–17.

De Ley, J., Bain, N. and Shewan, J. M. (1967). *Nature, Lond.* **214**, 1037–1038.

Dowson, W. J. (1939). *Zentbl. Bakt. ParasitKde, Abt. II* **100**, 177–193.

Dworkin, M. (1966). *A. Rev. Microbiol.* **20**, 75–106.

Eklund, E. (1970). *Acta Agric. scand. Suppl.* **17**, 57 pp.

Eklund, E. and Sinda, E. (1971). *Pl. Soil* **35**, 495–504.

Ewing, W. H. and Fife, M. A. (1972). *Int. J. Syst. Bact.* **22**, 4–11.

Fager, E. W. (1969). *J. gen. Microbiol.* **58**, 179–187.

Gardner, J. M. and Kado, C. I. (1972). *Int. J. syst. Bact.* **22**, 201–209.

Goodfellow, M. (1967). *Can. J. Microbiol.* **13**, 1365–1374.

Goodfellow, M., Hill, I. R. and Gray, T. R. G. (1968). *In* "The Ecology of Soil Bacteria." (T. R. G. Gray and D. Parkinson, eds), pp. 500–514. Liverpool University Press.

Gordon, R. E. (1966). *J. gen. Microbiol.* **43**, 329–343.

Gordon, R. E., Haynes, W. C. and Hor-Nay Pang, C. (1973). *In* "The Genus *Bacillus*," Agriculture Handbook No. 427, in press.

Gray, T. R. G. and Williams, S. T. (1971). "Soil Micro-organisms." Oliver and Boyd, Edinburgh.
Gyllenberg, H. G. (1965). *Ann. Med. exp. Fenn.* **43,** 82–90.
Gyllenberg, H. G. and Eklund, E. (1967). *Annales Academiae Scientiarum fennicae. Ser. A, IV Biologica no.* **113,** 19 pp.
Haider, K., Frederick, L. R. and Flaig, W. (1965). *Pl. Soil* **22,** 49–64.
Hendrie, M. S., Mitchell, T. G. and Shewan, J. M. (1968). *In* "Identification Methods for Microbiologists," Part B. (B. M. Gibbs and D. A. Shapton, eds), pp. 67–78. Academic Press, London and New York.
Henis, Y., Tagari, H. and Volcani, R. (1964). *Appl. Microbiol.* **12,** 204–209.
Hill, I. R. and Gray, T. R. G. (1967). *J. Bact.* **93,** 1888–1896.
Holding, A. J., Franklin, D. A. and Watling, R. (1965). *J. Soil Sci.* **16,** 44–59.
Horikoshi, K., Koffler, H. and Arima, K. (1963). *Biochim. biophys. Acta* **73,** 267–275.
Hormaeche, E. and Edwards, P. R. (1960). *Int. Bull. bact. Nomencl. Taxon.* **10,** 71–74.
Ingram, M. and Shewan, J. M. (1960). *J. appl. Bact.* **23,** 323–378.
Jensen, H. L. (1952). *A. Rev. Microbiol.* **6,** 77–90.
Jones, D. and Sneath, P.H.A. (1970). *Bact. Rev.* **34,** 40–81.
Katznelson, H. and Rouatt, J. W. (1957). *Can. J. Microbiol.* **3,** 265–275.
Kauffmann, F. (1954). "Enterobacteriaceae," 2nd edition. E. Munksgaard, Copenhagen.
Kauffmann, F. (1959). *Int. Bull. bact. Nomencl. Taxon.* **9,** 1–6.
Kovacs, N. (1956). *Nature, Lond.* **178,** 703.
Kunc, F. (1971). *Folia Microbiol., Praha* **16,** 41–50.
Laine, J. J. (1970). *Annales Academiae Scientiarum fennicae, Ser. A, IV Biologica no.* **169,** 36 pp.
Lanigan, G. W. (1959). *J. Bact.* **77,** 1–9.
Larkin, J. M. and Stokes, J. L. (1966). *J. Bact.* **91,** 1667–1671.
Larkin, J. M. and Stokes, J. L. (1967). *J. Bact.* **94,** 889–895.
Lelliott, R. A., Billing, E. and Hayward, A. C. (1966). *J. appl. Bact.* **29,** 470–489.
Lewin, R. A. (1969). *J. gen. Microbiol.* **58,** 189–206.
Lewin, R. A. and Lounsbery, D. M. (1969). *J. gen. Microbiol.* **58,** 145–170.
Lysenko, O. (1961). *J. gen. Microbiol.* **25,** 379–408.
Mandel, M. and Lewin, R. A. (1969). *J. gen. Microbiol.* **58,** 171–178.
McCurdy, H. D. Jr. (1970). *Int. J. Syst. Bact.* **20,** 283–296.
McMeekin, T. A., Stewart, D. B., Murray, J. G. (1972). *J. appl. Bact.* **35,** 129–137.
Mishustin, E. N. and Mirsoeva, V. A. (1968). *In* "The Ecology of Soil Bacteria." (T. R. G. Gray and D. Parkinson, eds), pp. 458–473. Liverpool University Press.
Mitchell, T. A., Hendrie, M. S. and Shewan, J. M. (1969). *J. appl. Bact.* **32,** 40–50.
Nigel da Silva, G. A. and Holt, J. G. (1965). *J. Bact.* **90,** 921–927.
Norris, J. R. and Wolf, J. (1961). *J. appl. Bact.* **24,** 42–56.

PETERSON, J. E. (1969). *J. appl. Bact.* **32,** 5–12.
RAVIN, A. W. (1963). *Adv. Genet.* **10,** 61.
RHODES, M. E. (1959). *J. gen. Microbiol.* **21,** 221–263.
ROGERS, H. J., MCCONNELL, M. and BURDETT, I. D. J. (1970). *J. gen. Microbiol.* **66,** 297–308.
ROSYPAL, S., ROSYPALOVÁ, A. and HOŘEJŠ, J. (1966). *J. gen. Microbiol.* **44,** 281–292.
ROVIRA, A. D. and SANDS, D. C. (1971). *J. appl. Bact.* **34,** 253–259.
SIEBURTH, J. MCN. (1967). *J. Bact.* **93,** 1911–1916.
SKYRING, G. W. and QUADLING, C. (1969). *Can. J. Microbiol.* **15,** 141–158.
SKYRING, G. W. and QUADLING, C. (1970). *Can. J. Microbiol.* **16,** 95–106.
SKYRING, G. W., QUADLING, C. and ROUATT, J. W. (1971). *Can. J. Microbiol.* **17,** 1299–1311.
SMITH, N. R., GORDON, R. E. and CLARK, F. E. (1952). *U.S. Dept. Agric. Agriculture Monograph no.* **16.** Washington D.C.
STANIER, R. Y. (1942). *Bact. Rev.* **6,** 143–196.
STANIER, R. Y. (1947). *J. Bact.* **53,** 297–315.
STANIER, R. Y., PALLERONI, N. J., DOUDOROFF, M. (1966). *J. gen. Microbiol.* **43,** 159–271.
STARR, P. M. and CHATTERJEE, A. K. (1972). *A. Rev. Microbiol.* **26,** 389–426.
STEVENSON, I. L. (1967). *Can. J. Microbiol.* **13,** 205–211.
STEWART, D. J. (1965). *Nature, Lond.* **208,** 200.
STOUT, J. D. (1958). *N.Z. Jl agric. Res.* **1,** 943–957.
STOUT, J. D. (1960). *N.Z. Jl agric. Res.* **3,** 214–224.
STOUT, J. D. (1961). *N.Z. Jl agric. Res.* **4,** 1–30.
SUNDMAN, V. (1965). *Acta polytech. scand. Ser.* (*a*) *Chem. Met. no.* **40,** 116 pp.
TANAKA, H. and PHAFF, H. J. (1965). *J. Bact.* **89,** 1570–1580.
THORNLEY, M. J. (1960). *J. appl. Bact.* **23,** 37–52.
THORNLEY, M. J. (1967). *J. gen. Microbiol.* **49,** 211–257.
THORNLEY, M. J. (1968). *In* "Identification Methods for Microbiologists," Part B. (B. M. Gibbs and D. A. Shapton, eds), pp. 29–50. Academic Press, London and New York.
TRIBE, H. T. (1957). *In* "Microbial Ecology." *Seventh Symp. Soc. gen. Microbiol.* (R. E. O. Williams and C. C. Spicer, eds), pp. 287–298. Cambridge University Press.
TULECKE, W. ORENSKI, S. W., TAGGART, R. and COLAVITO, L. (1965). *J. Bact.* **89,** 905–906.
WAKSMAN, S. A. and HENRICI, A. T. (1943). *J. Bact.* **46,** 337–341.
WALDEE, E. L. (1945). *Iowa St. Coll. J. Sci.* **19,** 435–484.
WEBLEY, D. M. and JONES, D. (1971). *In* "Soil Biochemistry." (A. D. McLaren and J. Skujins, eds), Vol. II, pp. 446–485. Marcel Dekker, New York.
WEEKS, O. B. (1969). *J. appl. Bact.* **32,** 13–18.
WOLF, J. and BARKER, N. (1968). *In* "Identification Methods for Microbiologists," Part B. (B. M. Gibbs and D. A. Shapton, eds), pp. 93–109. Academic Press, London and New York.
ZAGALLO, A. C. and WANG, C. H. (1962). *J. gen. Microbiol.* **29,** 389–401.

9

Actinomycetes

M. Goodfellow

Department of Microbiology
University of Newcastle upon Tyne
England

and

T. Cross

Postgraduate School of Studies in Biological Sciences
University of Bradford
England

I. Introduction

The Actinomycetes are prokaryotic bacteria with elongaged cells or filaments (0·5–2·0 μm dia.) usually showing some degree of true branching. They produce a variety of spore types which includes the endospore, long regarded as the typical spore structure of the Eubacteriales. Some genera, such as *Streptomyces* and *Micromonospora*, form an extensive branched mycelium composed of individual hyphae subdivided by infrequent cross walls and containing many distinct and separate nuclear elements, most probably represented by a circular genophore. Taxa with well developed mycelia have many characteristics in common with the nocardioform bacteria, organisms forming a mycelium which sooner or later fragments into coccoid or rod-like elements. In turn, there is no sharp distinction between these nocardioform bacteria and actinomycetes such as *Mycobacterium* that occur as single or occasionally as branched cells but which only rarely form a primary mycelium. The actinomycetes are predominantly aerobic, heterotrophic and saprophytic; they have a cell wall structure characteristic of Gram-positive bacteria and frequently show the presence of lytic viruses (actinophage).

The actinomycetes are grouped together mainly on morphological grounds but they cannot be satisfactorily distinguished from coryneform genera such as *Arthrobacter* and *Corynebacterium* which also have a tendency to produce branched elements. There is, therefore, some doubt as to whether the actinomycetes form a natural group or merely a convenient artificial taxon. The little evidence that there is suggests that the actinomycetes form a reasonably tight group which can be usefully distinguished from the other prokaryotic orders. For example, actinomycete DNA is rich in guanine and cytosine and all but a few thermophilic strains have a base composition within the range 62–78·5% (De Ley, 1970). The evidence from DNA homology experiments (Tewfik and Bradley, 1967; Yamaguchi, 1967; Farina and Bradley, 1970), the theoretical predictions of homology (De Ley, 1970) and numerical taxonomy (Sneath, 1970) all present the actinomycetes as a related group of bacteria. However, within the order, there is considerable morphological and physiological diversity. The ideal classification in which microbial taxa are discrete and readily distinguishable probably does not exist. Thus we find that certain actinomycetes and coryneform taxa, such as *Arthrobacter*, share common features and bridge the divisions that would simplify their taxonomy.

The actinomycetes are a very successful group of bacteria and they are usually considered to play a significant role in the breakdown of complex organic compounds along with other saprophytic organisms, and in the cycling of nitrogen, sulphur and potassium. The actinomycetes appear to have a number of properties which favour them in competition with other microbes and ensure their survival under unfavourable environmental conditions.

These properties can be summarized as follows:

(1) They are capable of producing a variety of spores (Cross, 1970) which facilitate their rapid dispersal in aquatic habitats (actinoplane zoospores), their dispersal in air and survival in soil (streptomycete arthrospores) and may even ensure viability over many decades (thermoactinomycete endospores). These spores can be produced after a very short period of vegetative growth (Mayfield *et al.*, 1972) or by the subdivision of the mycelium into fragmentation spores.

(2) The actinomycetes are nutritionally versatile, being able to grow both on rich substrates and on those containing a minimum or even an apparent lack of nutrients. For example, a common isolation medium consisting of tap water agar (Lechevalier, 1964) will allow the development of taxa capable of growing on richly supplemented media such as blood agar.

(3) The order contains organisms able to attack and metabolize a wide variety of substrates including naturally occurring and synthetic compounds that are usually resistant to microbial decomposition.

(4) Following active periods of growth they produce a wide variety of secondary metabolites which include antibiotically active compounds which might give them an advantage in certain microsites.

(5) The ability to form a mycelium allowing radial growth facilitates colonization of organic debris away from the initial growth centre.

These properties are important in the role of actinomycetes as soil organisms and will be considered in relation to their activities in plant litter.

II. Classification

Actinomycetes have a widespread distribution and are not uncommon in litter. Since taxonomists have been unable to produce a stable classification or workable identification schemes for the actinomycetes the microbial ecologist does well to characterize his isolates even to the generic level. Sometimes there are difficulties in allocating strains to well established genera such as *Streptomyces*, *Nocardia* and *Mycobacterium*, and the sub-generic classification of these taxa is difficult even for the specialist. An even greater problem is that of assigning a specific name to isolates of genera such as the *Streptomyces* where over 400 species descriptions exist resulting from the absence of generally accepted and reliable diagnostic characters. Although there are fewer species in the more recently described genera, some of these are based on the examination of only a few strains and must be treated with caution until additional isolates have been examined. The inability of workers to identify many of their isolates has prevented detailed studies on the importance of actinomycetes in litter decomposition. It is hoped that the recent progress made in the taxonomy of these organisms will now make such investigations possible.

The past decade has been an active one for taxonomy and a lot of effort has been directed towards establishing a better classification of the actinomycetes. The early genera have been circumscribed more carefully, new genera have been proposed to clarify the taxonomy and many nomen-species have been reduced to synonyms of well described taxospecies. One of the fruits of improved classification is that good characters become available for the recognition and differentiation of taxa. It is still too early to claim that the actinomycetes are well classified but it would appear that an acceptable outline of the major groups is emerging (Cross and Goodfellow, 1973).

Early classifications were based primarily on morphological and staining properties but at present weight is also given to chemical characters and the nature of the spores. Numerical taxonomic studies, with their emphasis on using many different kinds of characters, have made an impact, particularly at the subgeneric level, and notably in the genera *Nocardia* and *Mycobacterium* (Goodfellow, 1971; Goodfellow *et al.*, 1972; Wayne *et al.*, 1971; Kubica *et al.*, 1972; Kubica and Silcox, 1973; Kubica *et al.*, 1973). Phenetic data on the genus *Streptomyces* (Silvestri *et al.*, 1962) suggest that a natural classification of this taxon is possible for there are few major clusters compared with the number of specific names applied to streptomycete strains. Despite these studies many taxa are still recognized by morphological criteria, but such characters are no longer used uncritically, and other features may be examined to confirm an identification.

The simplest morphology, found in the families Actinomycetaceae and Mycobacteriaceae, involves cells which divide without forming branches or filaments. These cells are readily seen on rich media or in smears prepared from colonies but many strains will display a rudimentary branched mycelium when grown undisturbed on the surface of agar media containing mineral nutrients. "*Mycobacterium*" *rhodochrous* and *Nocardia* strains usually produce a recognizable mycelium which later fragments, colonies of the former may carry a few feeble aerial hyphae but in *Nocardia* the aerial hyphae range from being sparse and invisible to the naked eye, to completely covering the substrate mycelium with a white down. In *Nocardia* short chains of arthrospores may be found on the aerial hyphae.

Most genera form an extensive permanent mycelium which securely anchors colonies to agar and usually carries aerial hyphae which branch and project above the colony. The genus *Micromonospora* produces smooth, leathery colonies which characteristically lack aerial hyphae. Single spores are formed laterally on the substrate hyphae; they become embedded within the colony or form a black shiny layer on its surface. Spores are often borne on aerial hyphae in other genera, giving the colonies a powdery or velvety surface texture. The individual hyphae are surrounded by a thin fibrous sheath which has hydrophobic properties. The sheath surrounds the chains of arthrospores formed by the regular septation of preformed hyphae in taxa such as *Streptomyces* and *Streptoverticillium* and usually remains around the individual spores when they become detached. This sheath also appears to form the spore vesicle or sporangium found in the genera classified in the family Actinoplanaceae. One, two, several or many motile or non-motile spores mature within the sac-like sheath which at one time was equated with the fungal sporangium. The spores are heat sensitive, show a variable resistance to desiccation and are thick-walled or thin-walled and motile. They are usually formed by the subdivision of preformed hyphae or singly by the walling off of a hyphal bud. A specialized spore with all the properties of a typical bacterial endospore (Cross *et al.*, 1968) is characteristic of the genus *Thermoactinomyces*. The diversity and properties of the spores are of value in classification for they provide good characters. Methods for determining the aerial mycelium morphology have been described by Williams and Cross (1971); the morphology of several common genera can be seen in Figs 1–8 (see end of chapter).

The use of chemical characters in classification has helped to reduce the heavy reliance placed on morphological features. Actinomycetes can be subdivided into nine broad groups (Table I) based on the distribution of certain amino acids and sugars found in major amounts in the cell wall peptidoglycan (Cummins, 1962; Lechevalier and Lechevalier, 1970). Although cell wall analyses provide information of great taxonomic value they are time-consuming and require expensive equipment, and cannot be recommended when large numbers of isolates have to be examined. In many instances, however, the analysis of whole-cell hydrolysates is sufficient for identification. Using such hydrolysates, simple chromatographic techniques are available for the detection of sugars (Lechevalier, 1968), and for the presence and kind of diaminopimelic acid (DAP) (Becker *et al.*, 1964; Murray and Proctor, 1965). The rapid detection of amino acids is most useful for it allows streptomycetes with their high content of LL-DAP to be distinguished from taxa which contain strains with the *meso*-isomer or alternative diamino acids such as lysine, ornithine or diaminobutyric acid.

The nocardias, mycobacteria, corynebacteria *sensu stricto* and strains of the *rhodochrous* complex cannot be separated on the basis of wall sugar and amino acid components. They all have a Type IV cell wall and form a spectrum of morphological types ranging from bacilli, through forms with a rudimentary primary mycelium to those with an extensive substrate mycelium carrying aerial hyphae and spore chains. These organisms also contain mycolic acids which are high molecular weight β-hydroxy acids with a long chain in the α position (Asselineau, 1966). At present three types of mycolic acids are recognized; mycolic acids *sensu stricto* with carbon skeletons of 80–90 carbon atoms are associated with *Mycobacterium*, nocardomycolic acids with 50–60 carbon atoms with *Nocardia* and corynemycolic acids with approximately 30 carbon atoms are characteristic of *Corynebacterium*. Relatively simple techniques are now available for

TABLE I. Cell wall, sugar, and fatty acid composition in actinomycetes and related genera

Cell wall type	Cell wall constituents	Whole cell sugars	Genera
I	gly. LL-DAP	NC	*Chainia*, *Elytrosporangium*, *Microellobosporia*, *Sporichthya*, *Streptomyces*, *Streptoverticillium*
II	gly. *meso*-DAP	xylose arabinose	*Actinoplanes*, *Ampullariella*, *Dactylosporangium*, *Micromonospora*
III	*meso*-DAP	madurose	*Actinomadura* (*madurae* type), *Dermatophilus*, *Microbispora*, *Planomonospora*, *Spirillosopra*, *Streptosporangium*
	meso-DAP	no sugar	*Actinomadura* (*dassonvillei* type), *Geodermatophilus*, *Thermoactinomyces*, *Thermomonospora*
IV	*meso*-DAP	arabinose galactose	Characteristic lipids. 1. Mycolic *Mycobacterium* 2. Nocardomycolic "*Mycobacterium*" *rhodochrous*, *Nocardia* 3. Corynemycolic acids *Corynebacterium sensu stricto* Uncharacterized lipids. *Micropolyspora*, *Pseudonocardia*, *Saccharomonospora*
V	lys. orn. glyc.[a]	NC	*Actinomyces* (*israeli* type), some cellulomonads, some corynebacteria
VI	lys. asp. glyc.[a]	NC	*Actinomyces* (*bovis* type), *Arthrobacter* (*globiformis* type), *Oerskovia*, *Rothia*
VII	DAB lys. asp. glyc.[a]	NC	*Erysipelothrix*, some corynebacteria
VIII	asp. glyc.[a]	NC	*Cellulomonas*
IX	*meso*-DAP	NC	*Mycoplana* (Gram-negative)

[a] glycine usually present in these groups. NC No characteristic sugar pattern.

DAB 2:4 diamino-butyric acid; DAP diaminopimelic acid.

Based on data published in Lechevalier and Lechevalier (1970); Lechevalier *et al.* (1971).

separating the three kinds of mycolic acids associated with Type IV cell walls (Lechevalier *et al.*, 1971; Mordarska *et al.*, 1972; Kanetsuna and Bartoli, 1972).

A classification based primarily on morphological, chemical and spore characters is presented in Table II, though it is stressed that it includes several genera which so far have not been recovered from litter. This classification should not be seen as a final or conclusive one but is merely a product of its time based on what are currently accepted as good characters.

TABLE II. Proposed classification of the actinomycetes

Family	Characteristics	Genera
Actinomycetaceae	Branched filaments which fragment. No aerial mycelium. Cell walls variable but include Types V and VI. Facultative or strict anaerobes. Catalase negative and positive.	*Actinomyces* *Agromyces*[a] *Arachnia* *Bacterionema* *Bifidobacterium* *Rothia*
	Group A	
Actinoplanaceae	No aerial mycelium. Cell wall Type II. Zoospores in vesicle (sporangium).	*Actinoplanes*[b] *Ampullariella*[b] *Dactylosporangium*
	Group B	
	Aerial mycelium. Cell wall Type III. Spores either motile or non-motile formed within vesicle (sporangium).	*Planobispora* *Planomonospora* *Spirillospora*[b] *Streptosporangium*[b]
Dermatophilaceae	Mycelium dividing in all planes. No aerial mycelium. Cell wall Type III. Form motile elements (zoospores).	*Dermatophilus* *Geodermatophilus*
Frankiaceae	Obligate symbionts forming root nodules on non-leguminous plants which can fix nitrogen.	*Frankia*
Micromonosporaceae	Substrate mycelium only. Cell wall Type II. Form single heat sensitive spores.	*Micromonospora*[b]

Family	Characteristics	Genera
Mycobacteriaceae	Occur as rods, branched filaments, occasionally as a rudimentary mycelium Strongly acid-fast. Cell wall Type IV. High lipid content which includes mycolic acids.	*Mycobacterium*[b] *Mycococcus*[a]
Nocardiaceae	Substrate mycelium fragments. Form nocardomycolic acids (lipid LCN-A) Cell wall Type IV Chains of spores may be formed on aerial and substrate mycelium.	*Gordona*[a] "*Mycobacterium*" *rhodochrous* *Micropolyspora* *Nocardia*[b]
Strepto-mycetaceae	Non-fragmenting substrate mycelium and aerial mycelium. Aerobic. Cell wall Type I. Spores carried on aerial mycelium but may also be seen on substrate mycelium Chains of arthrospores formed within a sheath. Spores mostly non-motile.	*Chainia* *Elytrosporangium* *Intrasporangium*[a] *Kitasatoa*[a] *Microellobosporia* *Streptomyces*[b] *Streptoverticillium*
Thermo-actinomycetaceae	Substrate and aerial mycelium which do not fragment. Cell wall Type III. Form heat resistant endospores.	*Thermoactinomyces*[b]
Thermomono-sporaceae	Substrate and aerial mycelium. Substrate mycelium may fragment. Cell wall Type III (some IV). Spores single, in pairs or in short chains, enclosed within a sheath.	*Actinomadura*[b] *Microbispora*[b] *Microtetraspora*[b] *Saccharomonospora* *Thermomonospora*[b]
Not assigned	1. Mycelium rudimentary, breaking into motile elements. Gram-negative. Cell wall Type IX.	*Mycoplana*[a]

Family	Characteristics	Genera
Not assigned (*contd.*)	2. No aerial mycelium formed. Substrate mycelium breaks into motile elements (fragmentation spores). Cell wall Type VI.	*Oerskovia*[b]
	3. No aerial mycelium. Substrate mycelium fragments, bears single spores. Cell wall contains lysine but no aspartic acid.	*Promicromonospora*[b]
	4. Non-fragmenting substrate and aerial mycelium. Cell wall Type IV. Spores formed on substrate and aerial mycelium.	*Pseudonocardia*[a]
	5. No substrate mycelium. Cell wall Type I. Holdfast supports short chain of zoospores.	*Sporichthya*

[a] Taxonomy uncertain. [b] Detected in litter.

III. Identification

Precise ecological work is dependent upon sound taxonomy. When the microbial ecologist eventually turns his attention to the role of actinomycetes in litter he may find a tabular key (Newell, 1970) invaluable for identification to the genus level. Such a key (Table III) is offered in the hope that it will stimulate specialists and non-specialists alike to identify the actinomycetes they encounter during their studies. Identification to the genus level is likely to be sufficient in initial studies but more detailed work requires diagnostic tables (Cross and Goodfellow, 1973) which are essential for the recognition of species.

IV. Isolation

The various isolation techniques available were designed to allow the growth of a wide range of actinomycetes and to discourage other bacteria and fungi. Isolation may involve the prior treatment of samples before transfer to a selective medium formulated to encourage the growth of

Table III

Statement of characters and character variants

1. Organism in the form of a stable branching mycelium (myc), a mycelium exhibiting obvious fragmentation (frag) or mainly in the form of bacilli (bac) which may occasionally exhibit branching or even form a rudimentary mycelium.
2. Aerial mycelium formed on colonies growing on suitable media (AM), otherwise colonies of primary or substrate mycelium (SM) or vegetative cells (VC).
3. Spores may be absent (0), single (1), in pairs (2), in chains of four (4), several in a short chain (2–10) or many in long chains or aggregates (∞). They may be single, borne in chains (ch) or enclosed within a spore vesicle, a sac or sporangium (sv).
4. Spores may be chains of non-motile arthrospores (arth), motile zoospores (zoo), non-motile aplanospores (apl) or resistant endospores (end).
5. Organisms aerobic (O), anaerobic (NO) or facultative anaerobes (FNO).
6. Cells acid fast (AF), partially acid fast (PAF) or non-acid fast (NAF).
7. Cell wall type according to Lechevalier's classification (I–IX).
8. Free fatty acids of high molecular weight, when present, may be mycolic acids (myco) or nocardo-mycolic acids (LCN-A).
9. Optimum growth temperatures normally below 45°C (mes) or above 45°C (ther).

TABLE III. Tabular key for genera of Actinomycetales in soil and litter

1	2	3	4	5	6	7	8	9	Genus
myc/frag	AM	2–10–∞ ch	arth	O	NAF	III	—	mes	*Actinomadura*
myc	SM	∞ sv	zoo	O	NAF	II	—	mes	*Actinoplanes*
frag	SM/VC	0	0	O/FNO	NAF	VII	ND	mes	*Agromyces*
myc	SM	∞ sv	zoo	O	NAF	II	ND	mes	*Ampullariella*
myc	AM	∞ ch	arth	O	NAF	I	ND	mes	*Chainia*
myc	SM	2–10 sv	zoo	O	NAF	II	ND	mes	*Dactylosporangium*
myc	AM	∞ ch	arth/zoo	O	NAF	I	ND	mes	*Elytrosporangium*
frag	SM	∞ sv	zoo	O	NAF	III	ND	mes	*Geodermatophilus*
bac	VC	0	0	O	PAF	IV	LCN-A	mes	*Gordona*
myc	AM	∞ ch	arth/zoo	O	NAF	I	ND	mes	*Kitasatoa*
myc	AM	2 ch	arth	O	NAF	III	—	mes	*Microbispora*
myc	AM	2–10 sv	apl	O	NAF	I	—	mes	*Microellobosporia*
myc	SM	1	apl	O/NO	NAF	II	—	mes	*Micromonospora*
myc	AM	2–10 ch	arth	O	NAF	IV	ND	mes/ther	*Micropolyspora*
myc	AM	4 ch	arth	O	NAF	III	ND	mes	*Microtetraspora*
bac	VC	0	0	O	AF	IV	myco	mes	*Mycobacterium*
bac/frag	SM/VC	0	0	O	PAF	IV	LCN-A	mes	"*Mycobacterium*" *rhodochrous*
bac	VC	0	0	O	NAF	ND	ND	mes	*Mycococcus*
frag	SM	0	zoo	O	NAF	IX	ND	mes	*Mycoplana*
frag	SM/AM	0–2–10 ch	arth	O	AF/NAF	IV	LCN-A	mes	*Nocardia*
frag	SM/VC	0	zoo	O	NAF	VI	—	mes	*Oerskovia*
myc	AM	2 sv	zoo	O	NAF	III	ND	mes	*Planobispora*
myc	AM	1 sv	zoo	O	NAF	III	ND	mes	*Planomonospora*
frag	SM	1	apl	O	NAF	ND	ND	mes	*Promicromonospora*
myc	AM	2–10–∞ ch	arth	O/FNO	NAF	IV	—	ther	*Pseudonocardia*
myc	AM	1	apl	O	NAF	IV	ND	ther/mes	*Saccharomonospora*
myc	AM	∞ sv	zoo	O	NAF	III	—	mes	*Spirillospora*
myc	AM	2–10 ch	zoo	O	NAF	I	ND	mes	*Sporichthya*
myc	AM	∞ ch	arth	O	NAF	I	—	mes/ther	*Streptomyces*
myc	AM	∞ sv	apl	O	NAF	III	—	mes/ther	*Streptosporangium*
myc	AM	2–10–∞ ch	arth	O	NAF	I	—	mes	*Streptoverticillium*
myc	AM	1	end	O	NAF	III	—	ther	*Thermoactinomyces*
myc	AM	1	apl	O	NAF	III	—	ther	*Thermomonospora*

ND = no data available

actinomycetes. Such media may also contain inhibitors which suppress the growth of associated bacteria and fungi.

Lists of suitable media for the growth of actinomycetes can be found in the papers of Nuesch (1965) and Williams and Cross (1971). Most of these media were devised for the selective isolation of streptomycetes and are probably quite unsuitable for the detection of some of the other taxa. For example, a commonly used agar containing glycerol and asparagine favours the growth of typical streptomycete colonies but nocardioform strains and micromonosporae are rarely seen. At the moment, the most extensively used media are casein-starch-nitrate agar (Küster and Williams, 1964) and agars which contain colloidal chitin as the sole carbon and nitrogen source (Lingappa and Lockwood, 1962). Colloidal chitin has many advantages when screening samples for actinomycetes; even those genera unable to degrade chitin produce colonies which can be counted and studied with the aid of a microscope. On some plates, however, bacteria may be found to inhibit the growth of actinomycetes, and streptomycetes sometimes interfere with slower growing nocardioform bacteria.

Nocardioform bacteria are readily isolated from soil by enrichment culture (Fredricks, 1967; Kvasnikov *et al.*, 1971) and nocardias have been recovered from a variety of habitats using paraffin baiting techniques (Gordon and Hagan, 1936; Mishra and Randhawa, 1969). The Actinoplanaceae can be isolated from aquatic and soil habitats by baiting with pollen and keratin (Couch, 1949; Karling, 1954; Willoughby, 1968, 1969*a*), these organisms have been recovered in large numbers from leaf washings plated onto chitin agar (Willoughby, 1969*b*). Though very useful, the enrichment and baiting techniques give little information on the numbers of organisms occurring in a habitat.

Prevention of fungal growth can readily be achieved by the addition of the antifungal antibiotics cycloheximide (Actidione) or the polyenes pimaricin and nystatin. The use of antibiotics for the prevention of bacterial growth has been less successful for optimal combinations and concentrations also reduce the total number of actinomycetes (Williams and Davies, 1965; Davies and Williams, 1970). Anti-bacterial antibiotics do, however, appear to be useful in the isolation of specific components of the actinomycete population as in the case of *Thermoactinomyces* (Cross, 1968) or *Streptomyces* sp. F1 (Williams and Mayfield, 1971). The Actinoplanaceae have been isolated by plating leaf washings onto media containing tellurite as a selective inhibitor (Willoughby, 1971).

When attempting to gain a complete picture of the actinomycete flora of a particular habitat it is advisable to use methods which favour the recovery of specific taxa. Spores of *Micromonospora* tend to be more resistant to environmental extremes and it has been found that the treatment of soil suspensions with chlorine or heating at 45°C for two hours (Burman *et al.*, 1969) can have a marked effect on recovery. Mycobacteria and related organisms are resistant to strong alkali and pretreatment of samples with 4% sodium hydroxide liquor and 5% oxalic acid, and with 8% sodium hydroxide, have been used to isolate a wide range of slow and fast growing mycobacteria and "*Myco.*" *rhodochrous* respectively (Beerwerth and Schürmann, 1969; Tsukamura, 1971). Nonomura and O'Hara (1969, 1971) found that heating dried soil in a hot air oven at 100–120°C reduced the number of bacteria subsequently growing on isolation plates but allowed the recovery of taxa such as *Microbispora*, *Microtetraspora* and *Streptosporangium* which form spores having a greater resistance to dry heat. Colonies of these genera are rarely encountered on dilution plates seeded with suspensions of untreated soil.

The thermophilic actinomycetes are common in litter and additional isolation plates should always be incubated at high temperatures (40–60°C) to detect this component (Gregory and Lacey, 1962). At these temperatures *Thermoactinomyces* spores can be isolated very readily, and almost completely free from competing bacteria, using a selective medium containing novobiocin and cycloheximide (Cross, 1968).

Improvements in isolation techniques will probably reveal further taxa and provide information on their frequency and distribution in litter. It is also worth considering the use of an Andersen sampler (Andersen, 1958) for dislodging and isolating actinomycetes from litter samples using the wind tunnel technique (Gregory and Lacey, 1962, 1963). This method has been used very successfully for determining the actinomycete flora of hay and grain samples and seems ideal for sampling litter.

V. Actinomycetes in Litter

Research on the activities of micro-organisms in any habitat occurs in two stages, often well separated in time. Initially there is a lengthy pioneer period during which the micro-organisms are counted and identified, and this may be followed by the more difficult task of determining the roles, if any, of the indigenous microflora and -fauna.

Litter has been extensively examined using mycological and bacteriological techniques but relatively little information is available on the actinomycete component. We suspect that the low numbers and the few species reported from litter partly reflects the use of inappropriate isolation techniques and the lack of any sustained studies by microbiologists acquainted with the order. However, the limited evidence available does suggest that the actinomycetes form a characteristic component of the flora playing a minor yet important role in the cycling of organic material and minerals.

A. Mesophilic Actinomycetes

Most of the actinomycetes isolated from litter have been identified as streptomycetes but it is likely that this apparent pre-eminence reflects the selectivity of isolation media for these organisms. Jensen (1971) estimated that the number of streptomycetes accounted for less than 2% of the total bacterial flora of *Fagus* litter, a result in good agreement with the earlier findings of Wolniewicz-Czerwinska (1956). Low counts of actinomycetes have also been observed in *Juncus* litter (Latter and Cragg, 1967), highly acid peat soils (Jensen, 1928, 1930; Waksman and Purvis, 1932; Ishizawa and Araragi, 1970), and in acid, as opposed to alkaline and neutral, soil (Gillespie, 1918; Corke and Chase, 1964; Davies and Williams, 1970). These findings have suggested that acid conditions are one of the prime factors limiting the presence of actinomycetes (Brock, 1969).

Some streptomycetes can only grow within an acidic pH range and Jensen (1928) classified a number of these, isolated from peat, in the taxon *Streptomyces acidophilus*. Recently many acidophilic streptomycetes have been recovered from the acidic horizons of *Pinus nigra* forest soils using an acidified starch casein medium (Williams *et al.*, 1971). These strains form a sizeable proportion of the microbial population of *Pinus* litter (Table IV) and will not grow on isolation media adjusted to a pH of 6·8–7·0. Using morphological and pigmentation characters the acidophilic strains were classified into nine taxa, eight of which were found at the Delamere site and only one, *Streptomyces* sp. W1, in the acidic horizons of the Freshfield soil. It seems likely that the acidophilic streptomycetes at Freshfield, unlike those in the natural podsol soil at Delamere, have not been there long enough for adaptive radiation and speciation to have occurred. In direct observation studies Williams *et al.* (1971) detected strains of *Streptomyces* sp. W1 growing on *Pinus* needles and humus particles in the A_1 horizon at Freshfield. It now seems probable that the streptomycete Gray and Baxby (1968) observed growing on chitin strips buried in *Pinus* litter belonged to the same acidophilic species for this organism also failed to grow on neutral media. We can conclude that acidophilic actinomycetes are probably more common in litter than hitherto suspected and were not detected in the past because they failed to grow on the neutral isolation media used.

There are some data that suggest that actinomycetes, other than streptomycetes, are common in litter. Numerous strains of *Streptosporangium* have been reported in acidic *Quercus* and *Fagus* forest litter (van Brummelen and Went, 1957) and micromonosporae appear to be able to tolerate the high moisture contents of certain peat soils (Ishizawa and Araragi, 1970). Poorly aerated peats contained very high populations of

Mim. chalcea in contrast to better aerated sites where *Stm. griseus* and *Streptomyces* sp. G predominated. Ruddick and Williams (1972) found that strains of *Stm. griseus* were common on the bodies of the indigenous litter fauna, and formed up to 20% of the streptomycete population in *Pinus* litter. Recently strains of oerskoviae and promicromonosporae have

TABLE IV. Dilution plate counts of streptomycetes, bacteria and fungi in the horizons of two *Pinus* forest soils

Horizon	Mean pH	Streptomycetes[a]: Acid medium	Streptomycetes[a]: Neutral medium	Bacteria	Fungi[a]
Freshfield soil[c] — Number per g dry wt ($\times 10^4$)					
Litter	3·6	80	0·6	1083	225
A_1	4·2	0·6	8	1810[b]	29·8
C	7·8	0	37·5	3080	1·3
Delamere soil[c]					
Litter	3·9	1·3	2·1	—	—
A_1	3·7	1·9	0·1	—	8·9
A_2	4·0	2·1	0·5	—	3·0
B_1	3·6	2·5	0·1	—	4·5
B_2	4·4	2·5	0·05	—	0·7

[a] Taken from Williams *et al.* (1971) and Goodfellow *et al.* (1967).
[b] Unpublished data (M.G.).
[c] Detailed descriptions of these soils have been given by Goodfellow *et al.* (1967), Kendrick and Burges (1962) and Burges (1963).

been isolated from decaying organic matter (Lechevalier, 1972), and although mycobacteria have not been specifically isolated from litter they do seem to be part of the normal bacterial flora of soil (Beerwerth, 1971; Beerwerth and Schürmann, 1969).

There is very little information about the occurrence of actinomycetes on litter in the mineral horizons of soil or in aquatic ecosystems. It is well known that the addition of organic material to soil causes an increase in the number of actinomycetes (Fousek, 1912; Singh, 1937), and there is some evidence that streptomycetes are common in soil rich in plant roots (Conn, 1916; Küster, 1967). These findings suggest that actinomycetes are able to colonize litter in soil. Further evidence comes from the direct observation studies of Williams *et al.* (1970) for they detected streptomycetes bearing long chains of spores growing on *Pinus* rootlets, dead fungal hyphae, and on heavily colonized chitin amendments. Lake muds rich in organic matter have been shown to contain large populations of *Micromonospora* (Colmer and McCoy, 1943, 1950; Umbreit and McCoy, 1941).

Actinomycetes may participate in the decomposition of litter in aquatic ecosystems. During a study on agarolytic bacteria present in sea water Humm and Shepard (1946) isolated three agar-digesting actinomycetes. Two of these were classified in the genus *Nocardia* (*Proactinomyces*) and the other in the genus *Streptomyces* (*Actinomyces*). Chesters *et al.* (1956) isolated 17 different alginate and laminarin decomposing strains from the decaying thalli of brown algae putrifying on the sea shore, from lagoon sediments, and from salt marsh soil and beach sand. Many of the streptomycetes were able to degrade insoluble laminarin and use it as a sole source of carbon for energy and growth. In parallel studies Siebert and Schwarz (1956) isolated one nocardia and six streptomycete strains from decaying seaweed. These workers have also reported that marine muds amended with cellulose show an increase in the number of actinomycetes. Their results indicate a possible role for actinomycetes in the marine environment, that of aiding the decomposition of seaweeds and cellulose rich material washed into the sea. The actinomycetes isolated in such surveys were from samples collected from inshore sites and could arise from soil contamination rather than from the marine environment. It is also possible that terrigenous streptomycete spores could contaminate the seaweed in the sea but only grow when cast up on the drift line.

The recent work of Weyland (1969) suggests that the actinomycetes may form part of the marine microflora. Using media supplemented with sodium chloride he was able to isolate many strains of *Microbispora*, *Micromonospora*, *Nocardia* and *Streptomyces* from water and mud samples collected up to 60 km from the nearest land. Many of his isolates would not grow on standard actinomycete culture media lacking sodium chloride. Further evidence supporting a degradative role for actinomycetes in the marine ecosystem is provided by the *in vitro* studies of Willingham *et al.* (1966). They obtained an index of substrate degradation for actinomycetes, bacteria and fungi isolated from the sea, by measuring oxygen consumption during pre-determined incubation periods with various substrates suspended in sterile sea water. The actinomycetes were most active in degrading lignin and starch, and the decomposition of cellulose by individual actinomycetes was only exceeded by a mixed population of fungi.

Chandramohan *et al.* (1972) have detected cellulolytic activity in marine streptomycetes.

Actinomycetes have also been detected on decaying plant material in freshwater habitats. It is possible that some of the observed species represent a wash-in component, derived from the surrounding soil, able to survive for long periods as spores. There is, however, strong evidence to show that actinomycetes, particularly members of the Actinoplanaceae, can play an important role in the decomposition of leaves and twigs in streams and freshwater lakes (Willoughby, 1968, 1969*a*, *b*, 1971) (see Chapter 21).

B. Thermophilic Actinomycetes

Thermophilic micro-organisms are common in wet hay stacks, compost heaps and other decomposing vegetable matter where temperatures of up to 60°C are common. The thermophilic actinomycetes form a significant component of this population (Miehe, 1907; Lieske, 1921; Waksman *et al.*, 1939; Corbaz *et al.*, 1963) and may be visible as chalk white deposits on grass composts (Forsyth and Webley, 1948) or as clouds of fine dust released from mouldy hays (Gregory and Lacey, 1963). Their ecological distribution has been reviewed by Cross (1968) and a picture emerges of a cosmopolitan group present in a wide range of soils at most altitudes and latitudes.

The ability to grow at temperatures above 40° is found in several actinomycete genera, especially in species of *Thermoactinomyces*, *Thermomonospora*, *Streptomyces*, *Pseudonocardia* and *Micropolyspora*. Many strains in these taxa possess amylolytic and cellulolytic properties (Henssen, 1957*a*, *b*; Fergus, 1969; Kuo and Hartman, 1966; Stutzenberger *et al.*, 1970; Stutzenberger, 1971) and show excellent growth on laboratory media containing plant extracts and structural polymers. A useful working hypothesis is to consider that growth and sporulation of these strains occur at high temperatures in habitats such as composts and wet hay stacks, and that spores, whose dormancy can only be broken by meeting high temperature conditions, are widely dispersed. Suitable high temperatures may occur naturally in heaps of leaves and fallen grasses in low-lying tropical and sub-tropical regions, and at the surface of litter layers in tropical areas.

Growth studies using a polythermostat have shown that many thermophilic species will grow at temperatures between 30° and 35°C and the use of nutrient deficient media (e.g. inorganic salts + starch, or casein agars) often tends to lower the minimal growth temperature (Cross, 1968). *Thermoactinomyces vulgaris* strains have been grown at 27°C after five weeks of incubation (Locci *et al.*, 1968). With reference to these studies it is interesting to note that the use of a perfusion isolation technique has shown that insolated soils in Austria attain temperatures above 30°C for several days during the summer, so allowing the growth of thermophilic fungi (Eggins *et al.*, 1972). The same fungal species recovered from adjacent shaded sites using Warcup plate techniques appeared to be inactive and could not be isolated using screened substrate immersion tubes. It now seems likely that thermophilic actinomycetes can grow in litter and surface layers of soil in temperate regions as long as the sites are exposed to sunlight; this view

TABLE V. Dilution plate counts of *Thermoactinomyces vulgaris* in soil litter and mud samples collected in Wharfedale, Yorkshire and the English Lake District

Sample	*T. vulgaris* per g dry wt
Surface peat beneath *Calluna vulgaris*	15×10^2
Soil from upland sheep pasture on limestone	27×10^2
Soil from upland sheep pasture on millstone grit	124×10^2
Pteridium aquilinum litter	32×10^2
Soil beneath *Pteridium aquilinum* litter	0
Pinus sylvestris litter	75×10^2
Soil beneath *Pinus sylvestris* litter	31×10^2
Organic mud taken from margin of small, alkaline, eutrophic lake	997×10^2
Equisetum fluviatile litter from margin of lake	519×10^2
Devoke water lake mud[a]	83×10^2
Thirlmere lake mud[a]	1000×10^2
Fixed sand dunes with *Festuca, Plantago, Galium, Viola*	26×10^2
Sand dune pasture with dominant *Festuca*	28×10^2

[a] Data from Cross and Johnston (1971).

lends support to the earlier suggestions of Apinis (1965). Temperatures over 30°C are likely to be the norm for litter in tropical and sub-tropical latitudes.

There is strong circumstantial evidence to show that thermophilic actinomycetes can grow in litter in temperate regions. Counts of *Tha. vulgaris* obtained from a variety of samples, using the selective medium of Cross (1968), are given in Table V. It can be seen that this organism was present in soil, leaf litters and also in mud samples taken from the bottom of Thirlmere and Devoke Water in the English Lake District. These lakes are largely surrounded either by coniferous plantations or by *Pteridium aquilium* covered hills. In order to explain the distribution pattern of thermoactinomycetes it was originally suggested that growth and spore production were restricted to high temperature habitats such as overheated hay. Spores were dispersed directly by wind, or indirectly by man and cattle, to the soil of pastures from which they could be washed into streams and later accumulate in lake muds. However, it now seems probable that a proportion of the *Tha. vulgaris* spores surviving in these lake muds are derived from the active growth of organisms in the soil and litter of the surrounding countryside.

VI. Activity in Litter

The extent of our knowledge concerning the numbers and types of actinomycetes in litter can be seen to be very limited. We have even less information on the role of those organisms that have been encountered, and for some we can only postulate activities based upon the properties similar organisms exhibit under laboratory conditions. It is generally accepted that actinomycetes can utilize a wide range of residues from plants and animals usually considered to contribute an important part of the organic matter fraction.

Streptomycetes and actinoplanae observed growing on roots, twigs and leaf debris must have derived some, if not most of their nutrients from the plant remains. Good evidence for the utilization of plant residues by *Streptomyces flavovirens* has been provided by Grossbard (1971) who buried labelled plant leaves and stems in soil and measured the evolution of radioactive CO_2, or the presence of ^{14}C atoms in hyphae and spores by an autoradiographic technique. It was also found that some streptomycetes derived their nutrients from living or dead hyphae of the primary fungal colonizers. The lysis of fungal hyphae by streptomycetes has frequently been reported (Rehm, 1959; Skujins *et al.*, 1965; Jones *et al.*, 1968; Howard and Gupta, 1971) and these organisms may be important in the decomposition of the basidiomycete mycelium commonly found in A_0 horizons. It does, however, seem likely that the vegetative growth phase of streptomycetes will be short and soon superseded by the spore phase. Indeed these organisms appear to exist for long periods in soil, and possibly in litter, as dormant spores (Lloyd, 1969; Williams *et al.*, 1971), only growing when a suitable food source presents itself in the near vicinity. The competitive ability of streptomycetes in normal, non-sterile soil appears to be limited and their continued presence and abundance presumably depends to a large extent on their ability to form spores. These spores can remain viable in dry soil for more than 14 years (Kalakoutskii and Pouzharitskaya, 1973). The longevity of the spores of other non-endospore producing genera is not known but indirect evidence from the examination of deep mud cores show that sediment samples deposited between 50 and

100 years ago contain viable *Streptomyces*, *Streptosporangium* and *Micromonospora* spores. Streptomycete spores are able to germinate and grow at relative humidities between 91·5% and 99% (Jagnow, 1957), so giving them an intermediate moisture requirement between the bacteria and fungi. Streptomycete species will also grow slowly at temperatures between 5° and 10°C and certain strains are active at 0°C (Haines, 1932).

The common actinomycetes, such as streptomycetes, nocardias and micromonosporae, can generally utilize cellulose (Waksman, 1919; Reese and Levinson, 1952; Chien, 1960; Griffiths and Jones, 1963; Hardisson and Villanueva, 1964; Fergus, 1969; Ishizawa and Araragi, 1970); chitin (Veldkamp, 1955; Jeuniaux, 1955; Okafor, 1966; Gray and Baxby, 1968), cutin (Gray and Lowe, 1967), keratin (Noval and Nickerson, 1959; Kuchaeva *et al.*, 1963; Goodfellow, 1971), pectin (Bilimoria and Bhat, 1961; Knösel, 1970), xylan (Sørensen, 1957; Kusakabe *et al.*, 1969; Iizuka and Kawaminami, 1969) and oxalic acid (Müller, 1950; Jagnow, 1957). In addition, nocardias and strains of the *rhodochrous* complex can degrade a large range of carbon compounds including long chain fatty acids and hydrocarbons (Raymond and Jamison, 1971; Raymond *et al.*, 1967; Nolof, 1962; Goodfellow, 1971).

Little is known about the ways in which lignin and humus are degraded but there is some evidence to suggest that the actinomycetes are involved. Mixed bacterial populations, giving high counts of "*Myco.*" *rhodochrous* and *Myco. fortuitum*, are able to grow on substrates containing lignin (Sundman *et al.*, 1964), and nocardias have been implicated in the decomposition of humic acids (Küster, 1950, 1952). In addition humic and fulvic acids are known to encourage the growth, and induce sporangia formation, in strains of the Actinoplanaceae (Willoughby *et al.*, 1968; Willoughby and Baker, 1969). The streptomycetes may also contribute to the formation of humus (Küster, 1967; Kutzner, 1968) though little is known of this complex process. It is usually conceded that the formation of humic acid precursors is a result of microbial syntheses, though the subsequent processes, resulting in the production of the highly polymerized humus, are chemical. After autolysis many streptomycetes form dark brown pigments which have properties similar to those of humic acids extracted from soil (Flaig *et al.*, 1952; Bremner *et al.*, 1955).

The actinomycetes are probably an ephemeral vegetative component of the soil and litter microflora. They can thus be compared with many other groups of soil bacteria showing rapid growth when suitable food materials such as plant litter are presented. That they can germinate and grow on fresh plant material is shown by the increase in actinomycete numbers following the green manuring of soil, but they probably have a lower competitive ability than the "sugar" fungi and rapidly growing bacteria.

However, many of the associated bacterial species have a limited survival phase when the nutrient supply becomes exhausted and when adverse environmental conditions occur. The actinomycetes can survive as spores and have a marginal advantage over some of the associated fungi, for the spores can later germinate and exhibit limited growth when adjacent to autolysing fungal hyphae or more recalcitrant plant debris. We can conclude that the actinomycetes play a minor role in the decomposition of the total litter added to soil (Gray and Williams, 1971) but form an integral part of a balanced biological community which is invariably characterized by an extremely diverse flora.

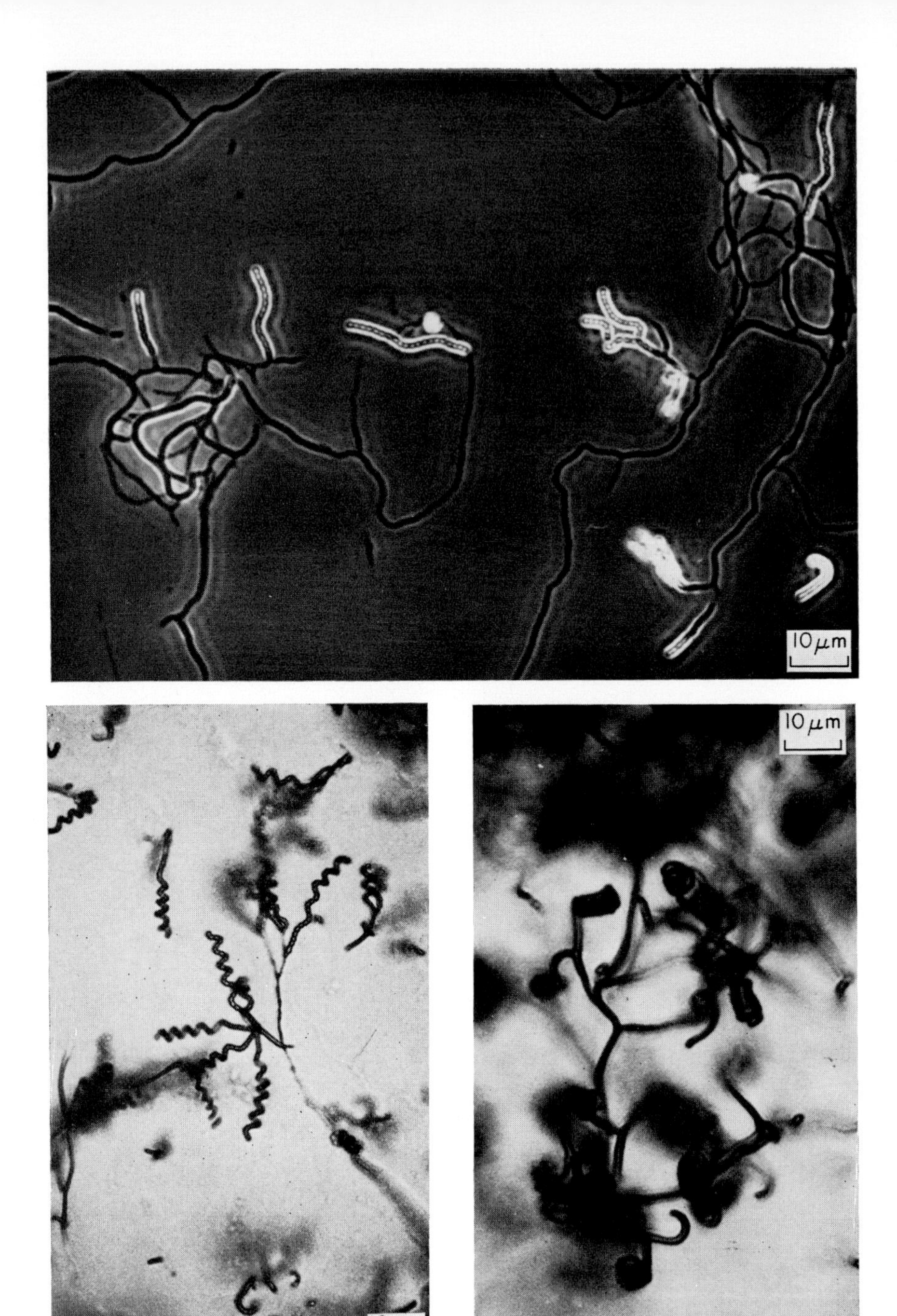

Fig. 1. *Streptomyces*

(a) Inclined cover slip culture of *Stm. griseus* showing limited substrate mycelium and the straight to wavy chains of arthrospores.

(b) *Stm. viridochromogenes* growing on the surface of oatmeal agar. Branched aerial mycelium with regular open-spiral spore chains.

(c) *Stm. lusitanus* var. *tetracyclini* exhibiting tight spring-like spirals on the surface of malt extract–yeast extract agar.

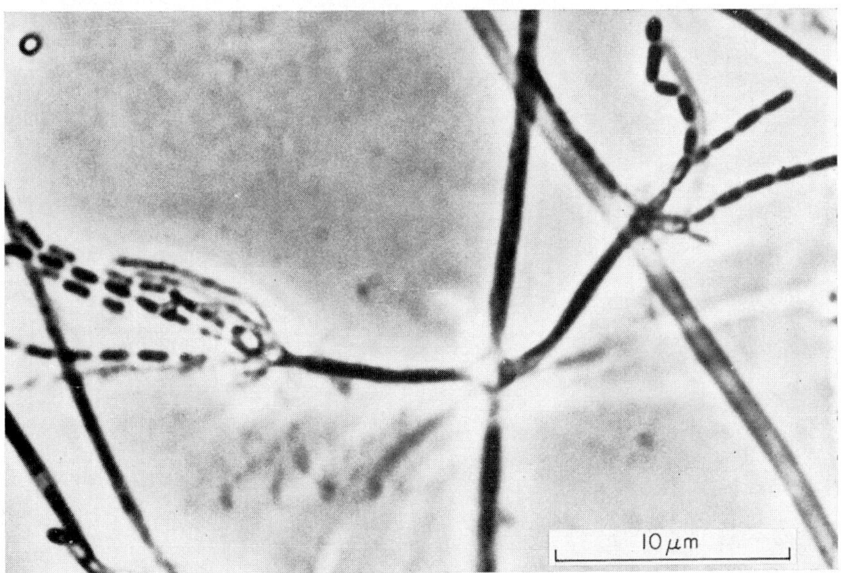

Fig. 2. *Streptoverticillium griseoverticillatus*

(a) Aerial hyphae bearing whorls of lateral branches at regular intervals, each branch terminates in an umbel of short spore chains. Agar plate culture.

(b) Inclined cover-slip culture at higher magnification showing a node with two lateral branches terminating in an umbel of spore chains. The mature individual spores are held within a thin fibrous sheath.

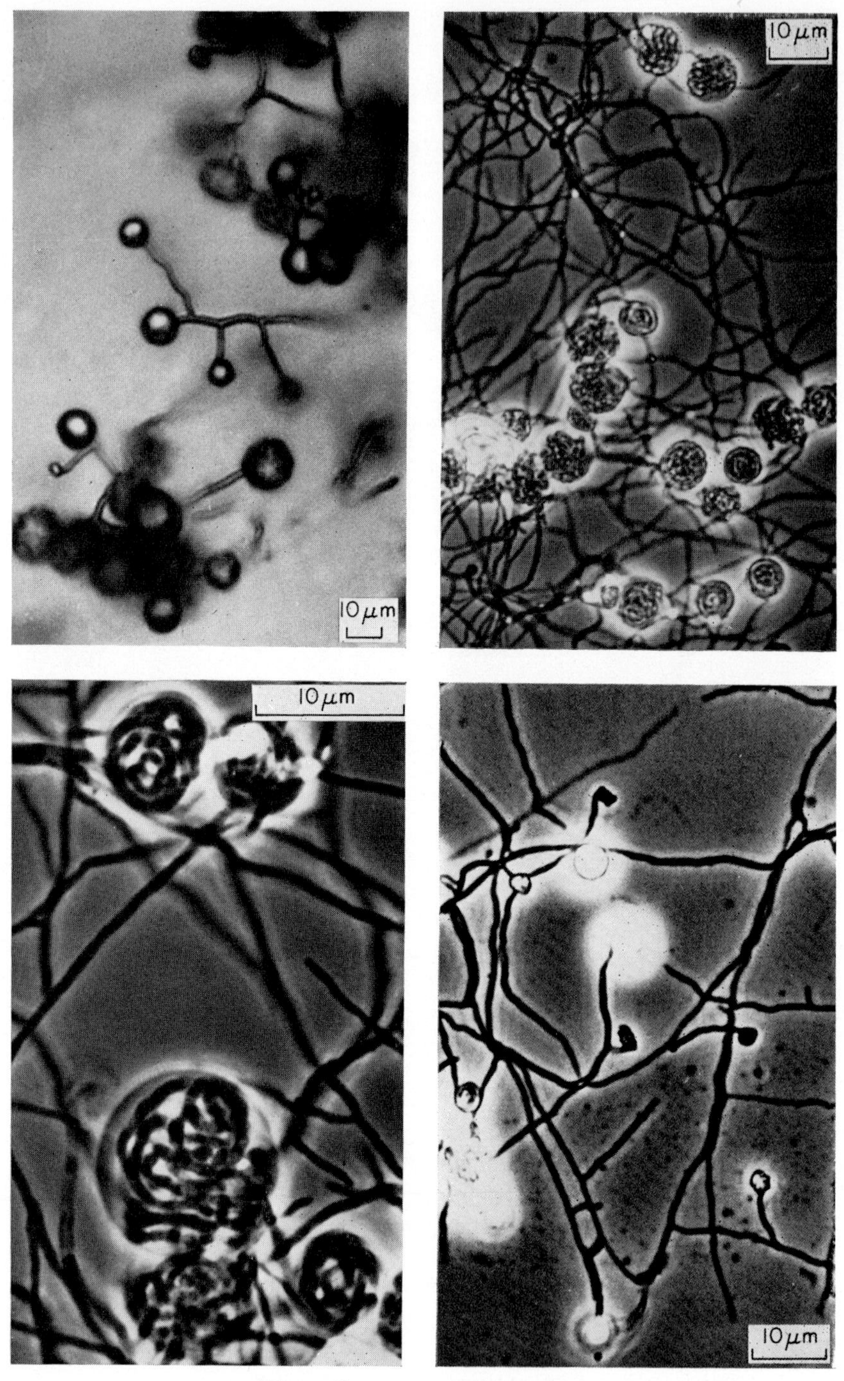

Fig. 3. *Streptosporangium roseum*

(a) Aerial hyphae bearing spherical spore vesicles. Culture on oatmeal +0·5% yeast extract agar.
(b) Inclined cover slip culture showing coiled spore chain within spore vesicle.
(c) Spore vesicle enclosing the coiled spore chain at a higher magnification.
(d) Aerial hyphae terminating in a spring-like spiral not enclosed within a vesicle.

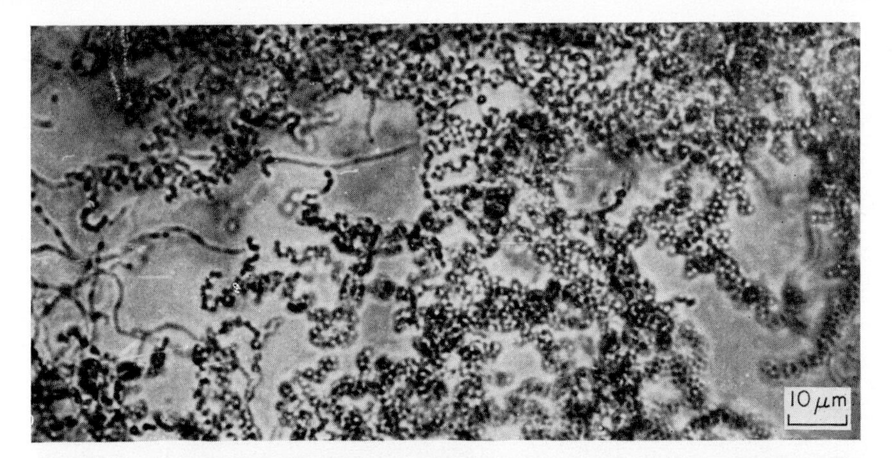

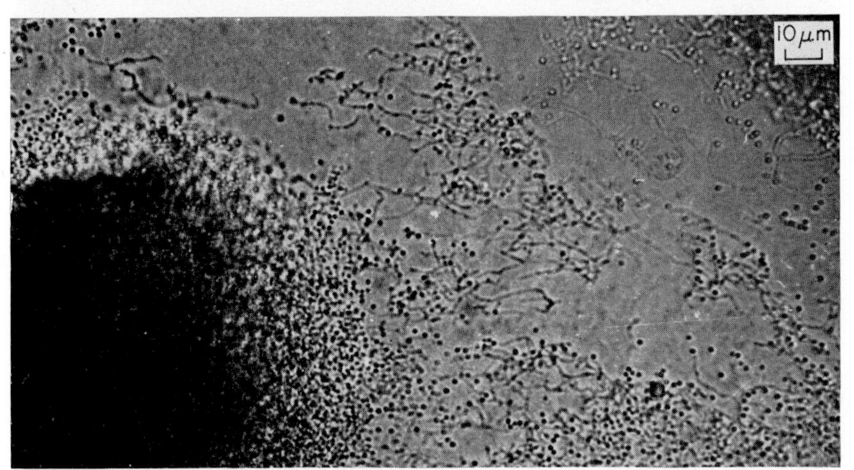

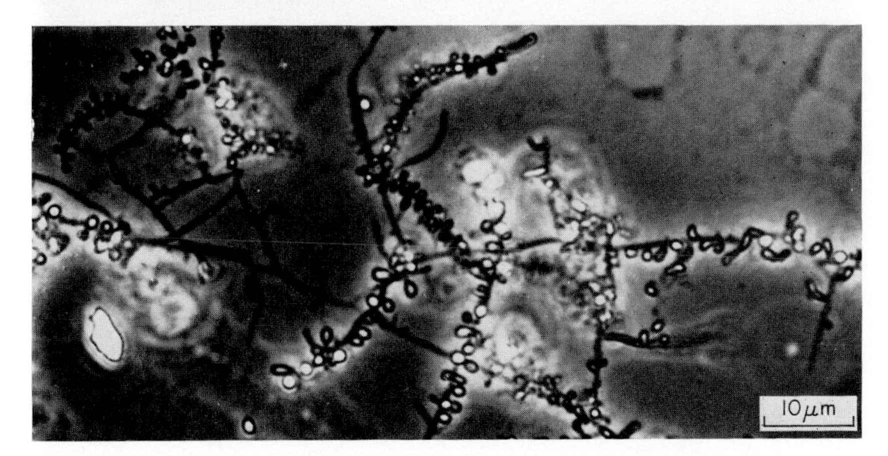

Fig. 4

(a) Slide culture of a nocardioform actinomycete growing on the surface of a thin layer of potato dextrose agar showing the transformation of the mycelium into coccoidal elements.

(b) Slide culture of a *Micromonospora* sp. growing near the surface of a thin layer of colloidal chitin agar. The fine substrate hyphae bear single spores which may appear as dense grape-like clusters in some species.

(c) *Saccharomonospora* (*Thermomonospora*) *viridis*. Aerial mycelium which bears many lateral single spores on short simple sporophores.

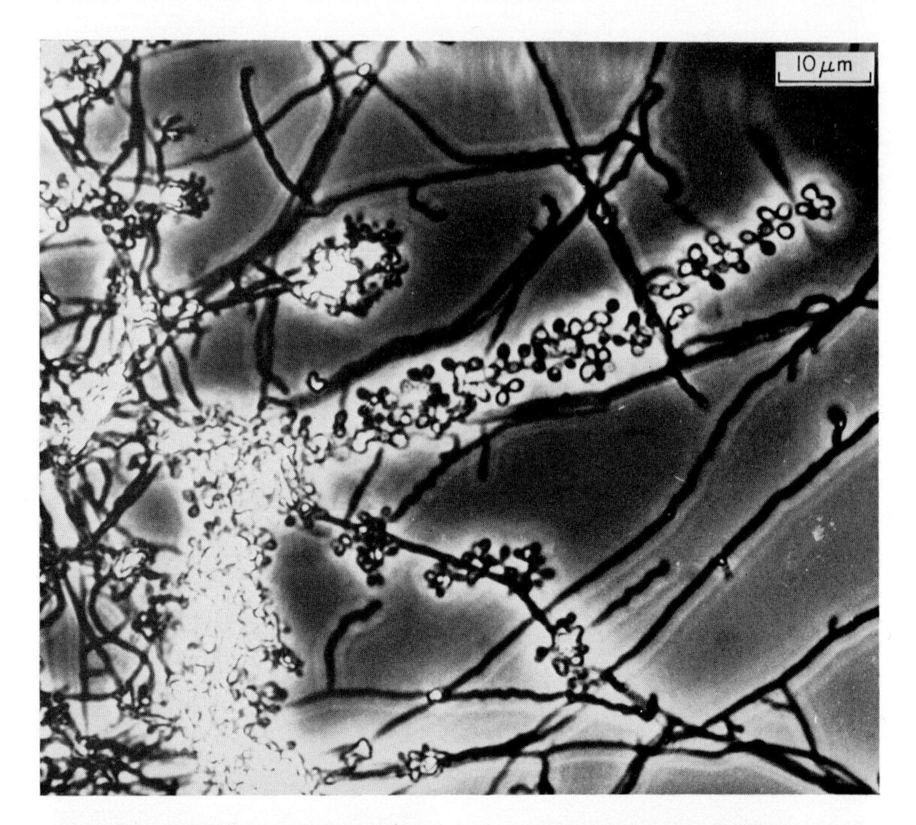

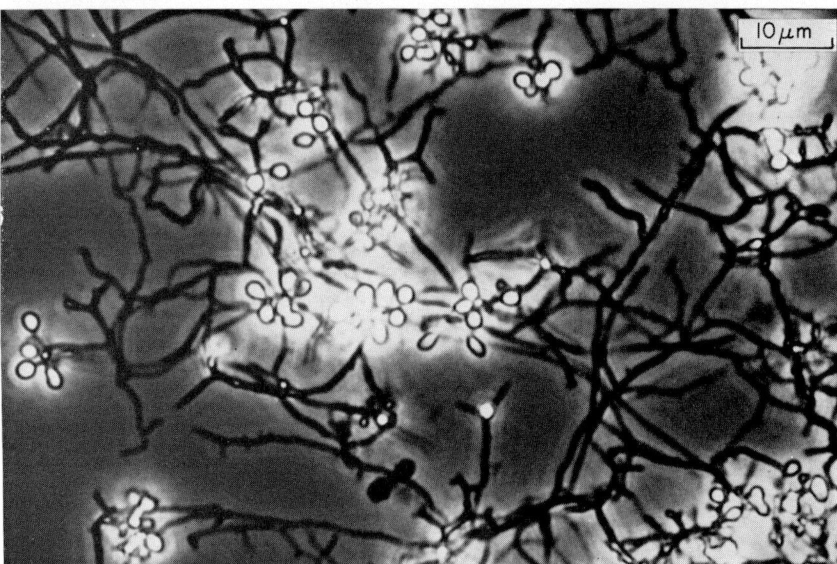

Fig. 5. *Thermomonospora*, inclined cover slip cultures

(a) Aerial mycelium with many lateral spores in the form of a spike. The lateral sporophores on the main hyphae may branch to give pairs or clusters of spores. *Thm. curvata* type of morphology.

(b) Aerial mycelium bearing clusters of spores; *Thm. fusca* type of morphology. There seems to be no clear cut morphological distinction between these two species, for discussion see Cross and Lacey (1970).

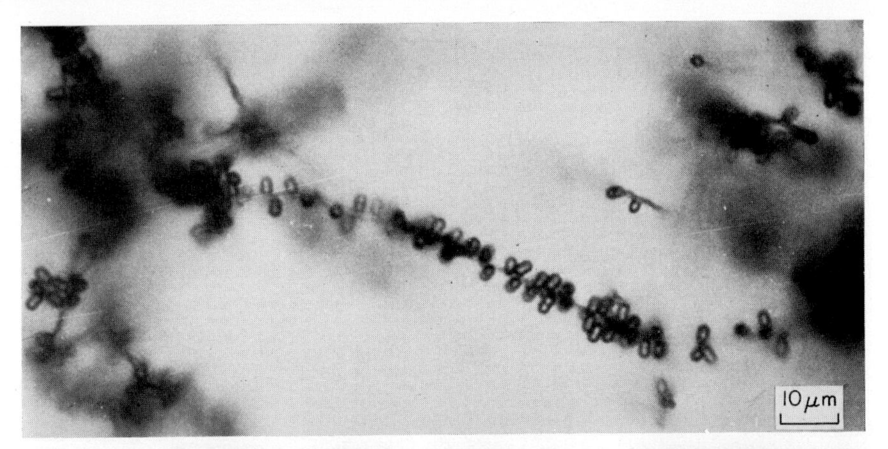

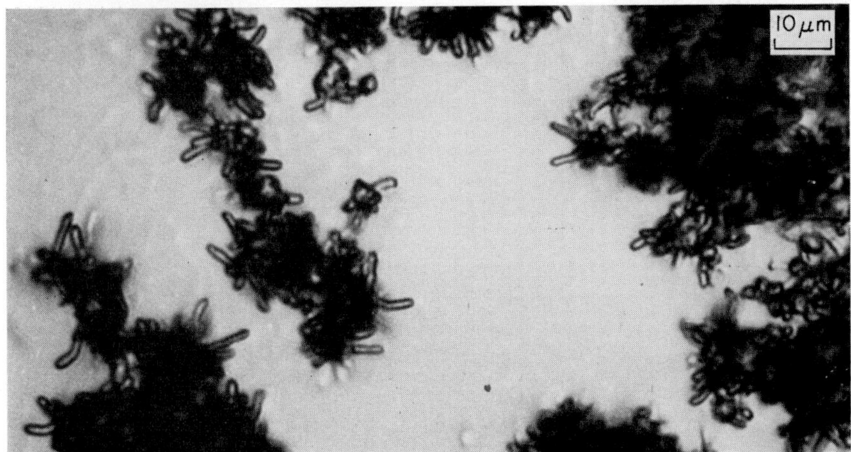

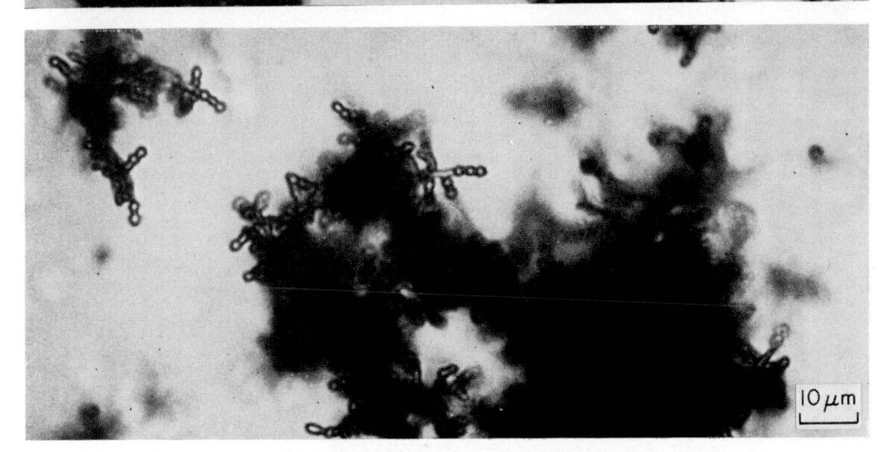

Fig. 6

(a) *Microbispora rosea* aerial mycelium with spores in characteristic longitudinal pairs. Culture on oatmeal +0·5% yeast extract agar.

(b) *Dactylosporangium* sp. Clusters of claviform or finger-like spore vesicles at the surface of oatmeal agar.

(c) *Micropolyspora faeni* isolated from mouldy hay growing on nutrient agar. The limited off-white aerial mycelium bears many short chains of spores, the individual spores are separated by spore pads so giving the chain a definite beaded appearance. Such cultures will also form chains of spores on the substrate hyphae deep within the agar.

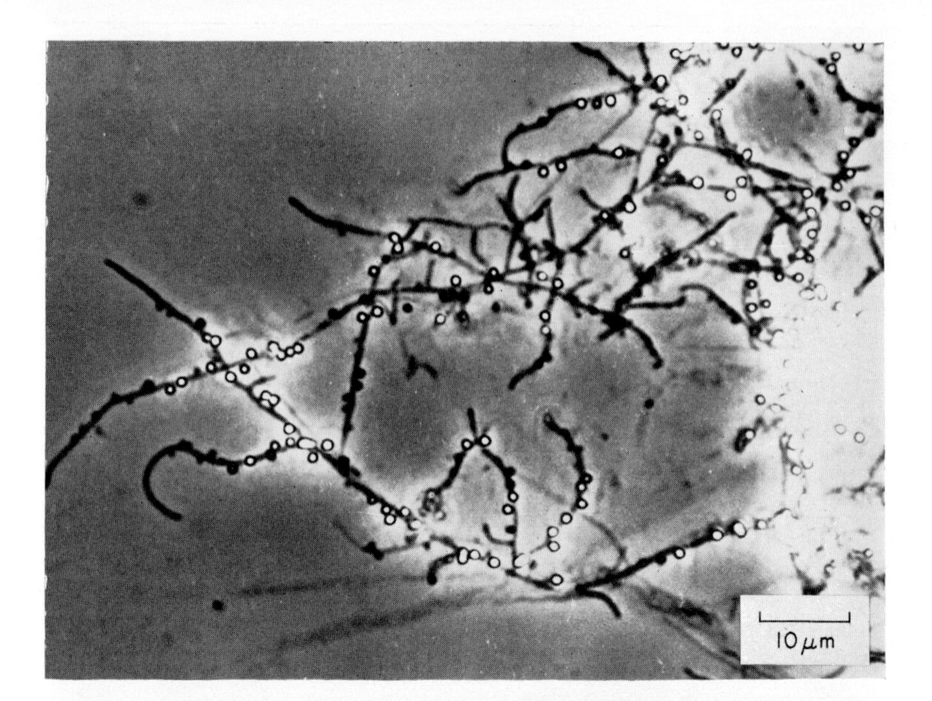

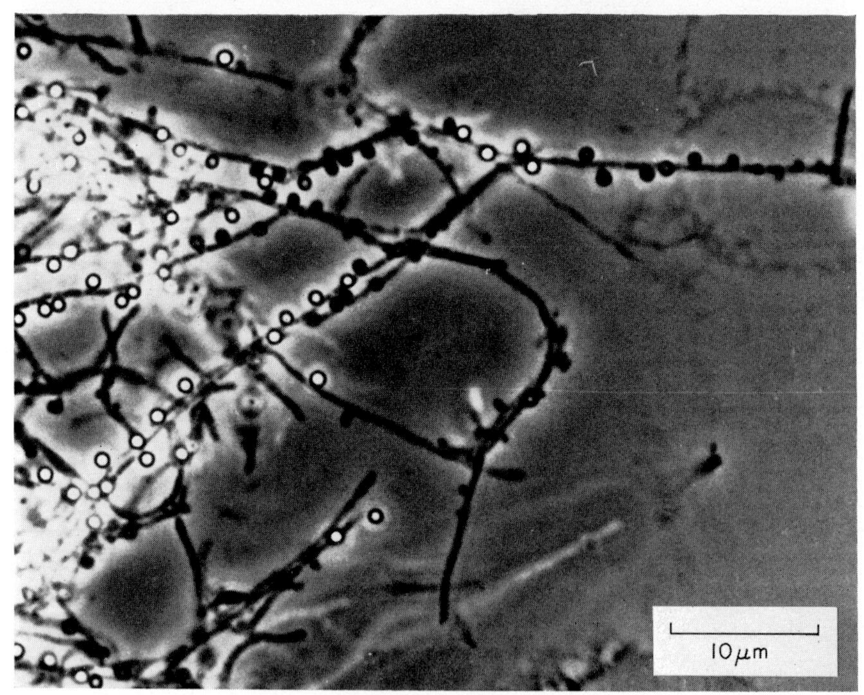

Fig. 7. *Thermoactinomyces vulgaris*

(a) Single phase-bright spores along aerial hyphae. Similar spores are formed on the substrate hyphae by colonies growing on agar.

(b) Inclined cover-slip cultures showing young (phase dark) and mature (phase bright) spores lacking definite sporophores. These spores are endospores, formed within a mother cell, which bears ridges giving the endospore a polygonal outline.

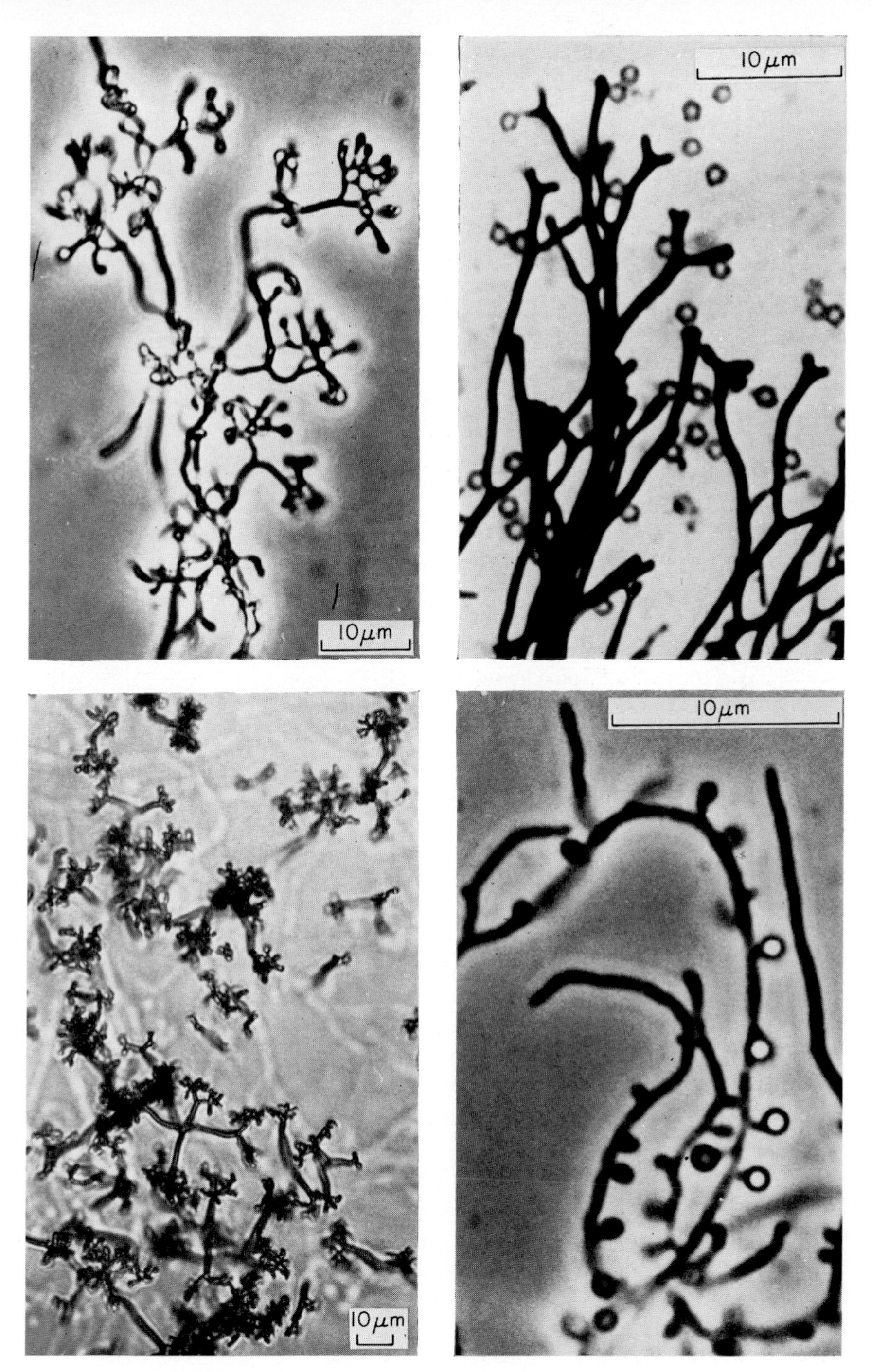

Fig. 8

(a) *Thermoactinomyces* (*Actinobifida*) *dichotomica*, inclined cover slip culture showing dichotomously branched aerial mycelium forming terminal spores.
(b) *Tha. dichotomica* substrate mycelium showing pronounced dichotomous branching and a background of lightly stained mature pentagonal endospores. Slide culture fixed and stained with crystal violet.
(c) *Tha. dichotomica* aerial mycelium growing on corn steep-starch agar.
(d) *Tha. peptonophilus* inclined cover slip culture showing maturation of phase bright endospores which are carried on definite sporophores.

References

ANDERSEN, A. A. (1958). *J. Bact.* **76,** 471–484.

APINIS, A. E. (1965). *In* "Biosoziologie." (R. Tuxen, ed.), pp. 290–303. W. Junk, Der Haag.

ASSELINEAU, J. (1966). "The Bacterial Lipids." Hermann, Paris.

BECKER, B., LECHEVALIER, M. P., GORDON, R. E. and LECHEVALIER, H. A. (1964). *Appl. Microbiol.* **12,** 421–423.

BEERWERTH, W. (1971). *Prax. Pneumol.* (*Tuberkuloseartz*) **25,** 661.

BEERWERTH, W. and SCHÜRMANN, J. (1969). *Zentbl. Bakt. Parasitkde, Abt. I,* **211,** 58–69.

BILIMORIA, M. H. and BHAT, J. V. (1961). *J. Sci. Technol., Cawnpore* **43,** 16–25.

BREMNER, J. M., FLAIG, W. and KÜSTER, E. (1955). *Z. Pflernähr. Düng. Bodenk.* **71,** 58–63.

BROCK, T. D. (1969). *In* "Microbial Growth." (P. M. Meadow and S. J. Pirt, eds), pp. 15–41. Cambridge University Press.

VAN BRUMMELEN, J. and WENT, J. C. (1957). *Antonie van Leeuwenhoek* **23,** 385–392.

BURGES, A. (1963). *In* "Soil Organisms." (J. Doeksen and J. van Drift, eds), pp. 151–157. North-Holland, Amsterdam.

BURMAN, N. P., OLIVER, C. W. and STEVENS, J. K. (1969). *In* "Isolation Methods for Microbiologists." (D. A. Shapton and G. W. Gould, eds), pp. 127–134. Academic Press, London and New York.

CHANDRAMOHAN, D., RAMU, S. and NATARAJAN, R. (1972). *Curr. Sci.* **41,** 245–246.

CHESTERS, C. G. C., APINIS, A. and TURNER, M. (1956). *Proc. Linn. Soc. Lond.* **166,** 87–97.

CHIEN, T. (1960). *Mikrobiologiya* **29,** 104–106. (English translation.)

COLMER, A. R. and MCCOY, E. (1943). *Trans. Wis. Acad. Sci. Arts Lett.* **35,** 187–220.

COLMER, A. R. and MCCOY, E. (1950). *Trans. Wis. Acad. Sci. Arts Lett.* **40,** 49–70.

CONN, H. J. (1916). *J. Bact.* **1,** 197–207.

CORBAZ, R., GREGORY, P. H. and LACEY, M. E. (1963). *J. gen. Microbiol.* **32,** 449–455.

CORKE, C. T. and CHASE, F. E. (1964). *Proc. Soil Sci. Soc. Am.* **28,** 68–70.

COUCH, J. N. (1949). *J. Elisha Mitchell Sci. Soc.* **65,** 315–318

CROSS, T. (1968). *J. appl. Bact.* **31,** 36–53.

CROSS, T. (1970). *J. appl. Bact.* **33,** 95–102.

CROSS, T. and GOODFELLOW, M. (1973). *In* "Actinomycetes: Characteristics and Practical Importance." (F. A. Skinner and G. Sykes, eds), pp. 11–112. Academic Press, London and New York.

CROSS, T. and JOHNSTON, D. W. (1971). *In* "Spore Research 1971." (A. N. Barker, G. W. Gould and J. Wolf, eds), pp. 315–330. Academic Press, London and New York.

Cross, T. and Lacey, J. (1970). *In* "The Actinomycetales." (H. Prauser, ed.), pp. 211–219. Gustav Fischer, Jena.
Cross, T., Walker, P. D. and Gould, G. W. (1968). *Nature, Lond.* **220,** 352–354.
Cummins, C. S. (1962). *J. gen. Microbiol.* **28,** 35–50.
Davies, F. L. and Williams, S. T. (1970). *Soil Biol. Biochem.* **2,** 227–238.
De Ley, J. (1970). *In* "The Actinomycetales." (H. Prauser, ed.), pp. 317–327. Gustav Fischer, Jena.
Eggins, H. O. W., von Szilvinyi, A. and Allsopp, D. (1972). *Int. Biodetn Bull.* **8,** 53–58.
Farina, G. and Bradley, S. G. (1970). *J. Bact.* **102,** 30–35.
Fergus, C. L. (1969). *Mycologia* **61,** 120–129.
Flaig, W., Küster, E., Segler-Holzweissig, G. and Beutelspacher, H. (1952). *Z. Pflemähr. Düng. Bodenk.* **57,** 42–51.
Forsyth, W. G. D. and Webley, D. M. (1948). *Proc. Soc. appl. Bact.* pp. 34–39.
Fousek, A. (1912). *Mitt. Land. Lehrkanz .K. Hochschule Bodenk. Wien* **1,** 217–244.
Fredricks, K. M. (1967). *Antonie van Leewenhoek* **33,** 41–48.
Gillespie, L. J. (1918). *Phytopathology* **8,** 257–269.
Goodfellow, M. (1971). *J. gen. Microbiol.* **69,** 33–80.
Goodfellow, M., Fleming, A. and Sackin, M. J. (1972). *Int. J. Syst. Bacteriol.* **22,** 81–98.
Goodfellow, M., Hill, I. R. and Gray, T. R. G. (1967). *In* "The Ecology of Soil Bacteria." (D. Parkinson and T. R. G. Gray, eds), pp. 500–515. Liverpool University Press.
Gordon, R. E. and Hagan, W. A. (1936). *J. infect. Dis.* **59,** 200–206.
Gray, T. R. G. and Baxby, P. (1968). *Trans. Br. mycol. Soc.* **51,** 293–309.
Gray, T. R. G. and Lowe, W. E. (1967). *Bact. Proc.* 3–4.
Gray, T. R. G. and Williams, S. T. (1971). *In* "Microbes and Biological Productivity." (D. E. Hughes and A. H. Rose, eds), pp. 255–286. Cambridge University Press.
Gregory, P. H. and Lacey, M. E. (1962). *Nature, Lond.* **195,** 95.
Gregory, P. H. and Lacey, M. E. (1963). *J. gen. Microbiol.* **30,** 75–88.
Griffiths, E. and Jones, D. (1963). *Trans. Br. mycol. Soc.* **46,** 285–294.
Grossbard, E. (1971). *J. gen. Microbiol.* **66,** 339–348.
Haines, R. B. (1932). *J. exp. Biol.* **9,** 45–60.
Hardisson, C. and Villanueva, J. R. (1964). *Annls Inst. Pasteur, Paris* **99,** 148–151.
Henssen, A. (1957*a*). *Arch. Mikrobiol.* **26,** 373–414.
Henssen, A. (1957*b*). *Arch. Mikrobiol.* **27,** 63–81.
Howard, D. H. and Gupta, R. K. (1971). *Can. J. Microbiol.* **17,** 521–523.
Humm, H. J. and Shepard, K. S. (1946). *Duke Univ. Marine Sta. Bull.* **3,** 76–80.
Iizuka, H. and Kawaminami, T. (1969). *Agric. biol. Chem.* **33,** 1257–1263.
Ishizawa, S. and Araragi, M. (1970). *Soil Sci. Pl. Nutr.* **16,** 110–120.
Jagnow, G. (1957). *Arch. Mikrobiol.* **26,** 175–191.
Jensen, H. L. (1928). *Soil Sci.* **25,** 225–236.
Jensen, H. L. (1930). *Soil Sci.* **30,** 59–77.

JENSEN, V. (1971). *In* "Ecology of Leaf Surface Micro-organisms." (T. F. Preece and C. H. Dickinson, eds), pp. 463–469. Academic Press, London and New York.

JEUNIAUX, C. (1955). *C. r. Séanc. Soc. Biol.* **149,** 1307–1308.

JONES, D., BACON, J. S. D., FARMER, V. C. and WEBLEY, D. M. (1968). *Antonie van Leeuwenhoek* **34,** 173–182.

KALAKOUTSKII, L. V. and POUZHARITSKAJA, L. M. (1973). *In* "Actinomycetes: Characteristics and Practical Importance." (F. A. Skinner and G. Sykes, eds), pp. 155–178. Academic Press, New York. and London

KANETSUNA, F. and BARTOLI, A. (1972). *J. gen. Microbiol.* **70,** 209–212.

KARLING, J. S. (1954). *Proc. Indiana Acad. Sci.* **63,** 83–86.

KENDRICK, W. B. and BURGES, A. (1962). *Nova Hedwigia* **4,** 313–342.

KNÖSEL, D. (1970). *Phytopath. Z.* **67,** 205–213.

KUBICA, G. P. and SILCOX, V. A. (1973). *J. gen. Microbiol.* **74,** 149–158.

KUBICA, G. P., SILCOX, V. A. and HALL, E. (1973). *J. gen. Microbiol.* **74,** 159–167.

KUBICA, G. P., BAESS, I., GORDON, R. E., JENKINS, P. A., KWAPINSKI, J. B. G., MCDURMONT, C., PATTYN, S. R., SAITO, H., SILCOX, V., STANFORD, J. L., TAKEYA, K. and TSUKAMURA, M. (1972). *J. gen. Microbiol.* **73,** 55–70.

KUO, M. J. and HARTMAN, P. A. (1966). *J. Bact.* **92,** 723–726.

KUCHAEVA, A. G., TAPTYKOVA, S. D., GESHEVA, R. L. and KRASSILNIKOV, N. A. (1963). *Dokl. Akad. Nauk SSSR* **148,** 1400–1402.

KUSAKABE, I., YASUI, I. and KOBAYASHI, T. (1969). *J. agric. chem. Soc. Japan* **43,** 145–000.

KÜSTER, E. (1950). *Arch. Mikrobiol.* **15,** 1–12.

KÜSTER, E. (1952). *Zentbl. Bakt. Parasitkde* (*Abt. I*) **158,** 350–356.

KÜSTER, E. (1967). *In* "Soil Biology." (A. Burges and F. Raw, eds). pp. 111–127. Academic Press, London and New York.

KÜSTER, E. and WILLIAMS, S. T. (1964). *Nature, Lond.* **202,** 928–929.

KUTZNER, H. J. (1968). *Landwirtsch-Forsch.* **21,** 48–61.

KVASNIKOV, E. I., NESTERENKO, O. A., ROMANOVSKAYA, V. A. and KASUMOVA, S. A. (1971). *Microbiology* **40,** 240–246.

LATTER, P. M. and CRAGG, J. B. (1967). *J. Ecol.* **55,** 465–482.

LECHEVALIER, H. A. (1964). *In* "Principles and Applications in Aquatic Microbiology." (H. Heukelekian and N. C. Dondero, eds), pp. 230–253. John Wiley New York.

LECHEVALIER, M. P. (1968). *J. Lab. clin. Med.* **71,** 934–944.

LECHEVALIER, M. P. (1972). *Int. J. Syst. Bact.* **22,** 260–264.

LECHEVALIER, M. P. and LECHEVALIER, H. A. (1970). *Int. J. Syst. Bact.* **20,** 435–443.

LECHEVALIER, M. P., HORAN, A. C. and LECHEVALIER, H. A. (1971). *J. Bact.* **105,** 313–318.

LIESKE, R. (1921). "Morphologie und Biologie der Strahlenpilze." G. Borntraeger, Leipzig.

LINGAPPA, Y. and LOCKWOOD, L. J. (1962). *Phytopathology* **52,** 317–323.

LLOYD, A. B. (1969). *J. gen. Microbiol.* **56,** 165–170.

LOCCI, R., BALDACCI, E. and PETROLINI, B. (1968). *Giorn. Microbiol.* **16,** 103–111.

MAYFIELD, C. I., WILLIAMS, S. T., RUDDICK, S. M. and HATFIELD, H. L. (1972). *Soil. Biol. Biochem.* **4**, 79–91.
MIEHE, H. (1907). "Die Selbsterhizung des Heues." G. Fischer, Jena.
MISHRA, S. K. and RANDHAWA, H. S. (1969). *Appl. Microbiol.* **18**, 686–687.
MORDARSKA, H., MORDARSKI, M. and GOODFELLOW, M. (1972). *J. gen. Microbiol.* **71**, 77–86.
MÜLLER, H. (1950). *Arch. Mikrobiol.* **15**, 137–148.
MURRAY, I. G. and PROCTOR, A. G. (1965). *J. gen. Microbiol.* **41**, 163–167.
NEWELL, I. M. (1970). *Pacific Insects* **12**, 25–37.
NOLOF, G. (1962). *Arch. Mikrobiol.* **44**, 278–297.
NONOMURA, H. and OHARA, Y. (1969). *J. Ferm. Techn.* **47**, 463–469.
NONOMURA, H. and OHARA, Y. (1971). *J. Ferm. Techn.* **49**, 1–7.
NOVAL, J. J. and NICKERSON, W. J. (1959). *J. Bact.* **77**, 251–263.
NUESCH, J. (1965). *Zentbl. Bakt. ParasitKde, Abt.* 1, *Suppl.* **1**, 234–252.
OKAFOR, N. (1966). *J. gen. Microbiol.* **44**, 311–327.
RAYMOND, R. L. and JAMISON, V. W. (1971). *Adv. appl. Microbiol.* **14**, 93–120.
RAYMOND, R. L., JAMISON, V. W. and HUDSON, J. O. (1967). *Appl. Microbiol.* **15**, 857–865.
REESE, E. T. and LEVINSON, H. S. (1952). *Physiologia Pl.* **5**, 345–366.
REHM, H. (1959). *Zentbl. Bakt. ParasitKde, Abt.* 11, **112**, 235–263.
RUDDICK, S. M. and WILLIAMS, S. T. (1972). *Soil Biol. Biochem.* **4**, 93–103.
SIEBERT, G. and SCHWARTZ, W. (1956). *Arch. Hydrobiol.* **52**, 321–366.
SINGH, J. (1937). *Ann. appl. Biol.* **24**, 154–158.
SKUJINS, J. J., POTGIETER, H. J. and ALEXANDER, M. (1965). *Archs Biochem. Biophys.* **111**, 358–364.
SILVESTRI, L., TURRI, M., HILL, L. R. and GILARDI, E. (1962). *In* "Microbial Classification." (G. C. Ainsworth and P. H. A. Sneath, eds), pp. 333–360. Cambridge University Press.
SNEATH, P. H. A. (1970). *In* "The Actinomycetales." (H. Prauser, ed.), pp. 371–377. Gustav Fischer, Jena.
SØRENSEN, H. (1957). *Acta. agric. scand. Suppl.* **1**.
STUTZENBERGER, F. J. (1971). *Appl. Microbiol.* **22**, 147–152.
STUTZENBERGER, F. J., KAUFMAN, A. J. and LOSSIN, R. D. (1970). *Can. J. Microbiol.* **16**, 553–560.
SUNDMAN, V., KUUSI, T., KUHANEN, S. and CARLBERG, G. (1964). *Acta Agric. scand.* **14**, 229–248.
TEWFIK, E. M. and BRADLEY, S. G. (1967). *J. Bact.* **94**, 1994–2000.
TSUKAMURA, M. (1971). *J. gen. Microbiol.* **68**, 15–26.
UMBREIT, W. W. and MCCOY, E. (1941). *In* "Symp. Hydrobiology Univ. Wis." pp. 106–114.
VELDKAMP, H. (1955). *Med. Landbouw., Wageningen* **55**, 127–174.
WAKSMAN, S. A. (1919). *J. Bact.* **4**, 189–216.
WAKSMAN, S. A. and PURVIS, E. R. (1932). *Soil Sci.* **34**, 95–113.
WAKSMAN, S. A., UMBREIT, W. W. and CORDON, T. C. (1939). *Soil Sci.* **47**, 37–61.

WAYNE, L. G., DIETZ, T. M., GERNEZ-RIEUX, C., JENKINS, P. A., KÄPPLER, W., KUBICA, G. P., KWAPINSKI, J. B. G., MEISSNER, G., PATTYN, S. R., RUNYON, E. H., SCHRÖDER, K. H., SILCOX, V. A., TACQUET, A., TSUKAMURA, M. and WOLINSKY, E. (1971). *J. gen. Microbiol.* **66,** 255–271.

WEYLAND, H. (1969). *Nature, Lond.* **223,** 858.

WILLIAMS, S. T. and CROSS, T. (1971). *In* "Methods in Microbiology," Vol. 4. (C. Booth, ed.), pp. 295–334. Academic Press, London and New York.

WILLIAMS, S. T. and DAVIES, F. L. (1965). *J. gen. Microbiol.* **38,** 251–261.

WILLIAMS, S. T., DAVIES, F. L., MAYFIELD, C. I. and KHAN, M. R. (1971). *Soil Biol. Biochem.* **3,** 187–195.

WILLIAMS, S. T. and MAYFIELD, C. I. (1971). *Soil Biol. Biochem.* **3,** 197–208.

WILLIAMS, S. T., HATFIELD, H. L. and MAYFIELD, C. I. (1970). *In* "The Actinomycetales." (H. Prauser, ed.), pp. 379–391. Gustav Fischer, Jena.

WILLINGHAM, C. A., ROACH, A. W. and SILVEY, J. K. G. (1966). *Am. Midl. Nat.* **75,** 232–241.

WILLOUGHBY, L. G. (1968). *Veröff. Inst. Meeresforsch. Bremerh. Sonderband* **3,** 19–26.

WILLOUGHBY, L. G. (1969*a*). *Hydrobiologia* **34,** 465–483.

WILLOUGHBY, L. G. (1969*b*). *Nova Hedwigia* **18,** 45–113.

WILLOUGHBY, L. G. (1971). *Freshwat. Biol.* **1,** 23–27.

WILLOUGHBY, L. G. and BAKER, C. D. (1969). *Verh. Internat. Verein. Limnol.* **17,** 795–801.

WILLOUGHBY, L. G., BAKER, C. D. and FOSTER, S. E. (1968). *Experientia* **24,** 730–731.

WOLNIEWICZ-CZERWINSKA, K. (1956). *Acta Soc. Bot. Pol.* **25,** 111–158.

YAMAGUCHI, T. (1967). *J. gen. appl. Microbiol.* **13,** 63–71.

10

Terrestrial Fungi

G. J. F. Pugh

Department of Botany
University of Nottingham
England

I. Introduction

A very large number of fungi have been described and estimates of the total are open to discussion (see Ainsworth, 1971). The number of species may be at least 100,000 or perhaps as many as 250,000. Of these, a minority in normal circumstances appear to require the presence of a living host. These include various mycorrhizal fungi and the obligate biotrophs, which parasitize their hosts without killing them. The remaining fungi are either unspecialized necrotrophs which are facultative parasites, able to live on the dead remains of the host, or saprophytes. These are the fungi which play a large part in the decomposition of plant litter. In this context, litter is taken to include not only leaves, branches, roots and woody remains, but also man-made litter such as paper and its products, textiles and plastics made from cellulosic materials and worked wood. There is a convergence of naturally occurring and anthropocentric litter, as ultimately the materials are degraded in a similar manner regardless of their immediate origin. The exception to this generalization is found in plastics which have been derived from coal or oil and are thus partly or wholly derived from plant materials, where the synthetic polymers may be highly resistant to microbial attack. Such resistance may be caused by the hydrophobic nature of the polymers together with a high level of intermolecular forces. Ways of avoiding the breakdown of plastics include the incorporation of specific antioxidants. However, if a short-lived polymer is required, then such antioxidants should be avoided. Hydroperoxides in the plastics act as powerful photoactivators for autoxidation in the presence of ultraviolet light, and have the practical effect of reducing chain length so that micro-organisms can attack and degrade these small molecules (Eggins *et al.*, 1971).

On litter derived directly or indirectly from plants fungi are able to grow whenever they possess the necessary enzyme systems and when the environmental conditions allow. Within these conditions must be included the biological component, as the fungi must be good competitive saprophytes (Garrett, 1951) to be able to exploit the substrate fully. The ability of fungi to grow on a substrate will determine their role in the decomposition of that material. The methods used to study the presence or activity of fungi in a particular habitat, however, will influence the results: all isolation techniques are selective to a greater or a lesser degree, while observational techniques which enable a fungus to be seen on the substrate do not always allow the identification of that fungus.

In a historical sense, although fungi have been described on aerial parts of plants for 100 years or more, the main emphasis for many years was on fungi in the soil, in the rhizosphere and in mycorrhizal associations. Only in relatively recent years has attention been directed at leaf-litter fungi and still more recently at phylloplane organisms. These studies have shown that leaves become colonized as soon as they unfold, or even before the bud bursts. Decomposition is thus initiated before the leaf is even mature. Harley (1971) has recently discussed the role of fungi in ecosystems showing the part which they play in decomposition processes, and how they in turn can become sources of nutrients for other organisms. Some aspects of fungal activities are considered in this chapter.

II. Ecology

In studying the ecology of fungi on plant litter, the early emphasis was on soil fungi, where an awareness of substrate relationships gradually evolved. The surface covering of plant litter, however, was relatively neglected: Chesters (1960) complained that litter lying on the soil had scarcely been looked at, indeed that it was "thrust aside in an effort to get at the 'soil'."

A. Fungi on Plants

In studying the ecology of fungi associated with plant material, several processes must be considered. The plant arises from a seed which had a mycoflora. The seed coat fungi may include some potential pathogens: there will certainly be a number of fungal saprophytes (see Malone and Muskett, 1964). The role of these saprophytes in the colonization of the germinating seedling is probably not great under normal agricultural conditions. Their influence on the colonization of roots was studied by Peterson (1959), who found that fungi associated with seeds of barley, flax and wheat played little part in the colonization of roots in the soil, Similarly, Catska *et al.* (1960) reported that soil fungi were more apparent on the roots of wheat seedlings than were seed-borne fungi. Seeds which have been surface sterilized have also been used for studying the fungal colonization of seedling roots: Parkinson *et al.* (1963) found that a mixed fungal population was at first present, but that this rapidly became stabilized into a typical root-surface mycoflora. A similar picture was presented by Stenton (1958), who isolated over 40 different fungi from pea roots; of these the genera *Cylindrocarpon*, *Pythium*, *Fusarium*, *Gliocladium* and *Mortierella* were abundant. He reported that fungi spread rapidly along roots until they reached an already occupied area. However, Taylor and Parkinson (1961) suggested that lateral colonization of roots by fungi in the soil is more important than growth of fungi along the roots.

In soils with an impoverished mycoflora, and where conditions adverse to fungal growth are prevalent, the seed coat fungi may play a more important part in root colonization. Thus Pugh and Dickinson (1965) showed that roots of seedlings of *Halimione portulacoides* (L.) Aell. growing in salt marsh muds were at first dominated by fungi present on the seeds and in the mud, and they suggested that there may be a joint source of inoculum, while Pugh (1960) found that mature roots of *Salicornia stricta* agg. growing in the same salt marsh, were colonized by *Dendryphiella salina* (Sutherland) Pugh & Nicot and a sterile white mycelium. These two fungi were also isolated from the seeds, but not from the muds in which the plants were growing.

The influence of the seed mycoflora on the aerial parts of plants has been relatively little studied. However, Pugh and Buckley (1971*a*) showed that *Aureobasidium pullulans* (de Bary) Arnaud could be isolated from the surface-sterilized seeds of sycamore (*Acer pseudoplatanus* L.), from the testa after removal of the pericarp, and from the embryo. It was also found on the hypocotyl and on the young roots of the seedlings. Buckley and Pugh (1971) showed that *A. pullulans*, *Cladosporium herbarum* Link and *Epicoccum purpurascens* Ehrenb. ex Schlecht. can produce auxins in the laboratory and Pugh (1973) indicated that the presence of the seed coat mycoflora may aid the germination of sycamore seeds.

Aureobasidium pullulans can also be isolated from the first opening of the buds, and is thus present on the leaves as the first of the primary colonizers. It has the ability to decompose pectin (Smit and Weiringa, 1953). It is joined by both yeasts and filamentous fungi which are able to grow in the hostile environment of the leaf surface: the majority are coloured or pigmented, and this is known to afford protection against harmful radiations. Thus *Rhodotorula* and *Sporobolomyces* are common yeasts. *S. roseus* Kluyver & van Niel has been shown by McBride (1972) to affect the leafy surface waxes. Amongst the filamentous fungi, *Cladosporium* spp. (mainl- *C. cladosporioides* (Fresen.) de Vries and *C. herbarum*) become abundant on the mature leaf, at a time more or less corresponding to their peak occurrence in the atmosphere. Many workers have found that *Cladosporium* spp. do not grow on leaves until they become senescent (e.g. Bainbridge and Dickinson, 1972). However Pugh and Buckley (1971*b*) reported the growth and sporulation of *Cladosporium* on green leaves of sycamore from July onwards. Other fungal colonists which arrive on the living leaves later in the year include *Alternaria alternata* (Fr.) Keissl., *Botrytis cinerea* Pers. ex Fr., *Epicoccum purpurascens*, as well as pycnidial and perithecial forms which remain as sterile hyphae on the living leaves and develop their fructifications after leaf-fall. Thus Dickinson (1965), working with phylloplane fungi on *Halimione portulacoides*, included *Ascochytula obionis* (Jaap) Died. in this third category.

The colonization of sycamore leaves from the bursting of the buds, throughout maturity and senescence, until they are decomposing on the ground, is summarized in Table I. After the arrival of *Cladosporium* and *Epicoccum*, *A. pullulans* appears to be present only as chlamydospores and microsclerotia on the leaf surface; after the leaf has been lying on the ground over winter, *Cladosporium* also is present in the form of microsclerotia. Pugh and Buckley (1971*b*) suggested that isolations of these fungi from old leaf litter and from the soil may indicate the survival of such resting stages rather than any activity on the part of the fungi. Last and Deighton (1965) reviewed in detail the occurrence of fungi on leaf surfaces,

TABLE I. Stages in the succession of fungi on leaves of *Acer pseudoplatanus* (from Buckley, 1971)

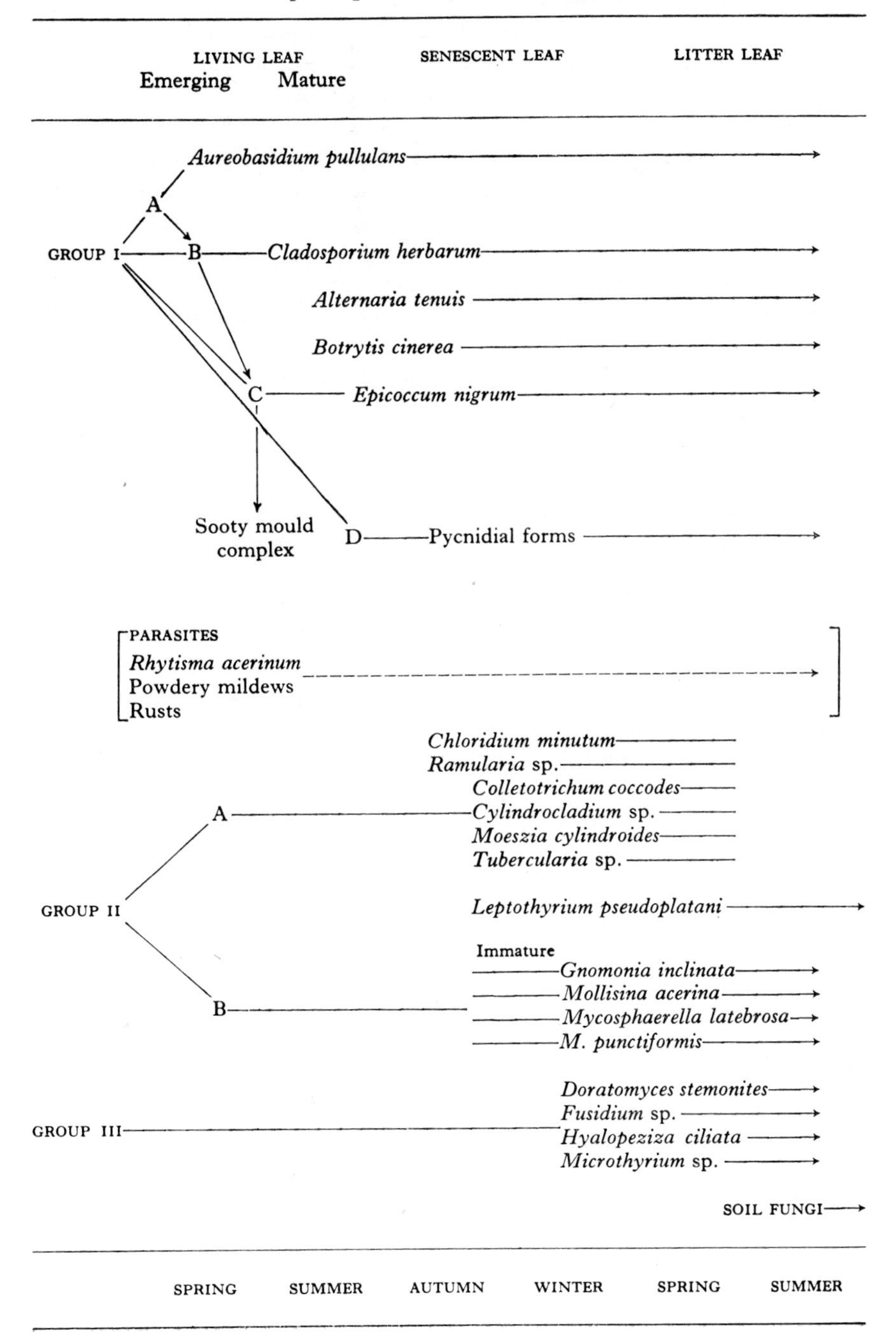

regarded the aspergilli as being more prevalent in tropical soils and the penicillia as more abundant in temperate regions. The cosmopolitan nature of these genera in the soil was borne out 56 years later by Domsch and Gams (1972), who carried out a massive series of isolations in two areas of northern Germany. They obtained 23,500 isolates, covering 209 species. Their survey of the literature on these species showed "more and more soil fungi to be ubiquitous, with a virtually worldwide distribution."

During the course of the floristic studies of the soil mycoflora the question was raised as to what the fungi were doing in the soil: were they merely there as propagules, or were they actively growing? In attempts to answer this question a number of techniques were used to distinguish between these states. Each in turn presented a situation equivalent to the "Principle of Uncertainty", i.e. would the result have been the same if the technique had not been applied, or did the application of the technique so alter conditions that fungal growth became possible? Thus those techniques which involve the insertion of a glass slide or a nutrient substrate into the soil provide new uncolonized space, as well as altering the nutritional status, and the physical characteristics of the soil such as the water regime. The nutrients are also attractive to members of the soil fauna, which carry spores on and in their bodies and thus are a source of inoculum. Techniques which involve taking soil samples to the laboratory normally involve incubating soil crumbs on nutrient media, where spores can germinate. The use of embedding methods (e.g. Nicholas *et al.*, 1965) showed the presence of hyphae in the soil, but could not distinguish between living and dead mycelium and normally cannot be used to identify the fungi. The hyphal isolation method (Warcup, 1957) can be used to indicate species which at some point in time have grown in that soil, but as Warcup pointed out "the presence of viable hyphae need not be synonymous with active growth in soil" immediately prior to isolation.He also emphasized that the method of investigation will greatly influence the picture obtained of the types of activity of the fungi in the soil. Further evidence for such influence was given by Pugh (1969) who showed that the use of cellulose agar (Eggins and Pugh, 1962) eliminated members of the Phycomycetes from the fungi isolated from the soils studied, and also more than doubled the isolation rate of the Dematiaceae. Other selective techniques have been devised specificially to isolate particular ecological or nutritional groups of fungi. There are comprehensive lists of methods used in soil mycology in Johnson and Curl (1972), and critical evaluations are given by Warcup (1960, 1967), and Parkinson *et al.* (1971).

The present situation is that the generally used techniques cannot distinguish between dormant and actively growing fungi in an unaltered soil. However, in certain situations, such as the growth of fungi along an elongating root, or in the use of labelled isotopes incorporated into a medium to trace the spread of hyphae able to translocate those isotopes, fungal growth can be measured without altering the micro-environment. Other methods, such as respirometry or analyses of enzyme production, can indicate total activity within a complete soil sample or the activity of individual organisms in a previously sterilized soil.

Harley and Waid (1955) summed up the difficulty of use of techniques by emphasizing that the methods must be able to "distinguish between organisms which are vegetatively active and playing a part in the soil processes and those which exist as dormant or inactive forms as spores and other propagules." Another point of view was expressed by Pugh (1969) who said that while the phase of active growth is important, spores and other propagules are formed in the soil as a result of previous activity and are normally a concomitant of, and evidence for that activity. This view had arisen from the transient nature of the substrates described as "finite and exhaustible" by Garrett (1951), which are colonized by a succession of fungi able to grow on the available part of the substrate and which then sporulate and die down; the inability of techniques to distinguish active growth, and from the earlier use of the term "fungal potential", which was defined as "active mycelium plus propagules which are the result of previous activity, held dormant by such factors as mycostasis" (Pugh, 1963). Within the soil the amount of organic matter normally diminishes with increasing depth, as does the frequency of isolation of fungi. However this is not merely a reflection of the relationship between organic content and fungi, as a similar decrease in fungal isolations also occurs in peat soils. The growth of fungi on the organic particles led Chesters (1948) to devise his soil-washing technique, which enabled the organic fraction to be separated from the mineral particles, while the propagules were generally removed with the washing water. This technique was elaborated by Williams *et al.* (1965) and has since been regularly used to study the fungi on organic particles in a variety of soils. This emphasis on the substrate has also resulted in the development of various media and methods for the selective isolation of nutritional groups of fungi. Of these the cellulose-decomposers have received much attention if only because cellulose forms such a large proportion of plant material. Amongst the methods are those of Tribe (1957) who buried strips of "Cellophane" attached to cover slips, Webley and Duff (1962), who used kaolinite pellets in which individual substances, including cellulose, had been incorporated, and Eggins and Pugh (1962) who incorporated ball-milled cellulose powder into an agar medium so that cellulolytic activity could be detected by a clearing of the previously opaque medium.

The picture of the soil that has emerged is of a system of organic microhabitats dispersed through an essentially mineral matrix. Each microhabitat has its mycoflora which may be active, quiescent or dormant. However, there is other organic matter in the form of humus which may be deposited on the mineral particles, and stem structures (rhizomes, bulbs, corms) present at various depths, and as a major influence there are roots of plants growing down through the soil profile, which may have mycorrhizas growing into the litter layers. Each of these living structures can profoundly affect the microflora in the soil. Their exudates include amino acids and sugars, while sloughed-off materials include root cap cells, root hairs and external tissues after secondary thickening (see Dickinson, Chapter 20). Micro-organisms grow and flourish on these additional substrates, creating the rhizosphere effect. Associations between soil micro-organisms and roots have been considered by Parkinson (1967), and the decomposition of roots is reviewed by Waid (this volume Chapter 6).

III. Biology

A. The Thallus

The basic vegetative form of the fungi is a thallus which may be a structurally simple cell, or a system of branched hyphae which may fuse or anastomase to form complex sporophores. The form of the thallus can conveniently be divided into three groups: (i) the unicellular, uninucleate thallus as in the chytrids and the yeasts; (ii) the unicellular, multinucleate thallus (coenocyte) found in filamentous, phycomycetous fungi such as *Saprolegnia* and *Mucor*; (iii) the multicellular thallus. This may have a single haploid nucleus in each cell; or a single diploid nucleus in specialized reproductive cells such as the ascus initial and the young basidium; it may have two compatible but heterocaryotic nuclei, as in the dicaryotic mycelia of Basidiomycetes; or there may be several nuclei in each cell. These may be genetically different (heterocaryotic) or similar (homocaryotic). Somatic variations can arise in heterocaryotic mycelia when different nuclear combinations occur in branches or as a result of anastomoses between hyphae. They can also arise during sporogenesis. Such changes can result in great differences in the physiology and metabolism of morphologically similar isolates of a given fungus. Morphological change can also occur and can be seen as a "sector" in a culture growing on agar media.

1. Hyphal Morphology

In filamentous fungi the basic structural unit is the hypha. Hyphae arise as tubular outgrowths from the germinating spore, and grow by extension of the hyphal apex. The wall of the hyphae is composed mainly of polysaccharides, which account for 80–90% of the wall. In addition there are proteins, lipids and polyphosphates. Peberdy (1972) has reviewed fungal wall structure, an understanding of which is necessary in his work on fungal protoplasts. He has indicated that there are at least 11 monosaccharides present in the walls. Of these, D-glucose is present as 1-3β and 1-6β linkages in glucans, 1-6β polymers of *N*-acetylglucosamine in chitin and D-mannose with α and β links mainly 1-3 and 1-6 in mannans. Most fungi have chitin, which accounts for up to 60% of the wall dry weight. Cellulose is found only in the walls of members of the Oomycetes.

As the hyphae grow they produce septate or non-septate filaments which normally branch in a racemose manner, and which collectively form the mycelium. In some fungi the hyphae become aggregated into cord-like mycelial strands, which are formed around a central hypha by branches which interweave and anastomose, or rhizomorphs, in which closely associated parallel unbranched hyphae grow as a single unit from an apical meristematic region. Garrett (1970) has calculated that there will be several thousand hyphae in a rhizomorph 1 mm in diameter. He and his co-workers have studied the morphogenesis and behaviour of strands and rhizomorphs and they found that the leading hyphae, around which strands of *Serpula* (*Merulius*) *lacrymans* (Pers.) ex Gray form exude nutrients which attract the investing hyphae, and this close association leads on to the development of anastomoses between compatible hyphae.

Townsend (1954) studied the development of rhizomorphs of 16 species of Basidiomycetes and arranged them into a series showing increasing differentiation. Pugh (1969) interpolated this scheme into an earlier one of Garrett (1944), to form a continuum from individual hyphae of *Rhizoctonia solani* Kuhn to complex, zoned rhizomorphs of *Armillaria mellea* (Fr.) Quel. (Table II), via hyphal strands and rhizomorphs which showed increasing degrees of differentiation. Garrett (1970) has pointed out that the production of rhizomorph initials increases with added glucose concentration at least to the level of 2% whereas strand formation is delayed by the presence of nutrients above a low concentration.

The biological advantages of strand and rhizomorph production are probably associated with translocation of nutrients from a food base, with increased competitiveness in saprophytes and virulence in parasites, as a greater enzyme concentration or penetration ability can be brought to bear on the substrate, and certainly with an increase in the rate of growth: rhizomorphs of *A. mellea* were reported by Garrett (1970) to grow 5–13 times faster than unorganized hyphae. A further advantage is conferred on these rhizomorphs, the rind of which becomes pigmented with melanin. This does not occur when the rhizomorph is bathed in a film of water (i.e.

when the oxygen tension is low). When rhizomorphs grow into dry soil, the melanin is able to form a mycostatic layer over the apex, and growth ceases.

2. Asexual Dispersal Spores

In the great majority of fungi the living mycelium grows over or through its substrate, exuding enzymes and absorbing soluble nutrients. Under these conditions asexual spores quickly form and are normally able to be dispersed. The method of formation of the spores is used as a criterion in

TABLE II. Runner hyphae, mycelial strands and rhizomorphs (from Garrett, 1944, and Townsend, 1954)

I. Hyphae and mycelial strands	
1. *Rhizoctonia solani*	individual hyphae
2. *Ophiobolus graminis*	individual hyphae and occasional strands
3. *Helicobasidium purpureum*	undifferentiated strands
4. *Phymatotrichum omnivorum*	strands with distinct differentiation
II. Rhizomorphs	
5. *Fomes lignosus*	undifferentiated rhizomorphs
6. *Collybia platyphylla*	different arrangement of similar hyphae
7. *Psalliota campestris*	hyphae of different sizes forming zones
8. *Armillaria mellea*	hyphae differ in size and structure forming distinct zones

the classification of fungi. The spores in fungi may be unicellular or multicellular, with each cell able to produce a germ tube: such multicellular spores may be considered as groups of one-celled spores (Sussman and Halvorson, 1966). The spores may be formed within a sporangium (sporangiospores): these may be motile, with one flagellum or two, or they may be non-motile. Conidia are asexual dispersal spores which arise singly or successively on hyphae, and are thus distinct from sporangiospores which arise simultaneously and endogenously. Uredospores and aecidiospores may be regarded as specialized conidia in the Rust Fungi (Uredinales). Teleutospores (Uredinales), which are not dispersed, and brand spores (Ustilaginales) are biologically similar in that nuclear fusion occurs in them prior to basidium formation and both act as resting spores but their ontogeny is very different (see Moore, 1972).

3. Dispersal

Spores become disseminated by many agencies. Amongst them are air currents; water, including raindrops; animals, including man, insects and earthworms; and soil. Dispersal has been fully discussed by Ingold (1971).

Air movements are by far the most important, both for localized dispersal, for example in convection currents in a room, and also for long distance transport. Within the atmosphere fungal spores form the largest proportion of the air spora, and they have to withstand many adverse conditions which can affect their viability. Gregory (1973) has discussed the survival of spores in the atmosphere and said that exposure to radiations, particularly near the ultraviolet wavelengths, was a serious hazard: in the upper atmosphere when spores are not sheltered by clouds they receive relatively high dose rates. Many fungi which sporulate on the aerial parts of plants, including pathogens, have spores which are protected from the harmful effects of radiations by pigmented walls or by the presence of carotenoids. These would fit into Gregory's group of "hardy microbes"; others which do not appear to have any such protection were regarded as "tender pathogens", a term which could be expanded to include saprophytes such as *Aureobasidium pullulans* and the hyaline yeasts.

Other adverse factors affecting spores include the risk of desiccation, which is greatest in daytime and in the lower air layers. At higher altitudes and at the lower night temperatures evaporation is reduced. Temperature fluctuations were also considered, but were thought to be not great, as the minute size of the spores would allow them to equilibriate quickly with the surrounding air. The combination of low temperatures and desiccation may help combat radiation damage to spores. At the same time "lowering the water content of organisms mitigates the lethal effects of temperature extremes" (Sussman and Halvorson, 1966), although they point out that dehydration can also be injurious or even lethal.

Gregory (1973) has discussed at length the ways in which spores are removed from the atmosphere. Three common phenomena are sedimentation under the influence of gravity in still air; impaction of spores being moved in air currents which go around an object while the spore is carried on to the object by its momentum; and boundary layer exchange when eddies of spore-laden air get into the relatively still air near the ground, or when they are injected into the spore-free laminar air layer immediately around an object. The spores then settle out of the air by sedimentation. In addition to these methods of deposition, the effect of rain on the air spora is very great and has been described as "rain scrubbing". This was regarded as "an ideal method of deposition for air-dispersed soil fungi" (Hirst, 1959), although the biological significance of rain deposition of spores of leaf pathogens was not clear, as they could be washed off the leaf surface. They could, however, be retransported to the leaves in splash droplets.

B. Resting Propagules

As a fungus grows on its substrate it produces spores which are usually asexual sporangiospores or conidia and which disperse the fungus spatially. The fungus may also grow out from its substrate by means of hyphae, or mycelial strands or rhizomorphs. However, it may also, or alternatively, produce a resting state, which extends the duration of the fungus in time (Garrett, 1956). The resting form of the fungus may arise as a result of sexual reproduction or it may represent a changed state of the vegetative mycelium.

Spores associated with sexual reproduction vary greatly in form, size and method of production. In the Saprolegniales (water moulds) the oospores are formed within oogonia; the zygospore of the Mucorales is produced from the fusion of two multinucleate gametangia; ascospores are formed endogenously after nuclear fusion and meiosis in the ascus initial; basidiospores arise exogenously on a basidium in which the dicaryotic nuclei have fused and divided meiotically.

Other types of resting spore associated with sexual reproduction are the teleutospore and the brand spore. Teleutospores are formed on the host towards the end of the growing season by members of the Uredinales, and act as overwintering spores; brand spores (Ustilaginales) form on or in the host and may be dispersed. Both are dicaryotic and germinate to produce a basidium from which basidiospores arise.

Within an individual hypha the apex contains dense, nonvacuolate cytoplasm. This soon becomes vacuolate, and as the hypha ages the cytoplasm dies leaving a cylindrical shell of wall material. Parts of the living mycelium may form chlamydospores, which represent cells in which the cytoplasm has become condensed and the wall thickened. These are generally intercalary, but can be terminal and are found in such common genera as *Fusarium*, *Mucor* and *Trichoderma*. They can be seen as remnants of once-living hyphae, although they may be undetected inside organic particles in the soil.

Sclerotia are more complex but biologically similar structures. They may act simply as vegetative resting organs and give rise to hyphae in due course. Alternatively they may germinate to produce asexual spores, as in *Botrytis* spp., or they may develop to produce stromata in which organs of sexual reproduction are formed, as in *Claviceps purpurea* (Fr.) Tul., or sporophores as in *Sclerotinia sclerotiorum* (Lib.) de Bary.

Sclerotia consist of a mass of living hyphae, and broadly, can be grouped into those with no definite pattern of organization of the hyphae, so that a loose structure is formed in which there are no well defined zones, and those with a surface covering, or rind. The internal tissues may be further differentiated into a cortex and medulla. Townsend and Willetts

(1954) gave an account of earlier studies on sclerotial structure and described the development of sclerotia in a range of fungi, distinguishing three methods of formation. Their "loose" type arose from an aggregation of adjacent hyphae and their side branches of short, barrel-shaped cells which have dense contents and no vacuoles. These are found in *Rhizoctonia solani*. Their other two types both have a rind at maturity. In their "strand" type, found in *Sclerotinia gladioli* Drayton, numerous lateral branches develop from one or more axis hyphae, and may lie parallel with them or at right angles to them. They may branch again before fusing and coalescing with each other. At maturity there is a medulla of large thin-walled cells enclosed within the rind. In *Phymatotrichum omnivorum* (Shear) Duggar both cortex and medulla are present. Their "terminal" type was described from *Botrytis allii* Munn., and they found a somewhat similar development in *B. cinerea* and at least initially in *Sclerotium cepivorum* Berk. The hyphal apex branches repeatedly and dichotomously, with the formation of cross-walls, until the hyphal mass coalesces to form the sclerotium, in which there are rind, a narrow cortex of pseudoparenchyma and a large central medulla of loosely arranged filaments. More recently, Coley-Smith and Cooke (1971) have reviewed the structure, survival and germination of sclerotia, which, as they pointed out, have been most studied in plant pathogenic fungi.

C. Dormancy

While many spores are able to germinate as soon as they become detached from their parent hyphae, others remain in a dormant state until they are able to germinate. Two types of dormancy, constitutive and exogenous, have been recognized and these apply to both sexually produced and vegetative resting stages. These have been defined by Sussman and Halvorson (1966) as: Constitutive dormancy: a condition in which development is delayed due to an innate property of the dormant stage such as a barrier to the penetration of nutrients, a metabolic block, or the production of a self-inhibitor. Exogenous dormancy: a condition in which development is delayed because of unfavourable chemical or physical conditions of the environment".

Once a resting structure has overcome any constitutive dormancy it may then be held in a state of exogenous dormancy until there are favourable environmental conditions. Most work in this field has been carried out with plant parasites, in which survival times of several years may be encountered. In considering the role of fungi in the breakdown of plant remains, one aspect of exogenous dormancy has received considerable attention. This is the phenomenon of mycostasis.

D. Mycostasis

The soil can be regarded as a habitat in which spores form as the result of the activity of the parent mycelia, and into which spores fall or are washed after being dispersed. Sussman and Halvorson (1966) mentioned a number of observations of the presence of substances which inhibit spore germination in the soil. Dobbs and Hinson (1953) described the phenomenon of spores remaining ungerminated in the soil. They showed that this was a widespread occurrence which affected the spores of several different species. Later workers have shown that mycostasis (also referred to as fungistasis) affects fungal spores on a worldwide basis: for example Jackson (1958) in Nigeria, Lingappa and Lockwood (1961) in Michigan and Knaphus (1964) in Iowa, U.S.A., Cooke (1967) in Iceland. The possible causes of inhibition have also received attention (see Forbes, Chapter 23). It has been known that the addition of glucose will stimulate germination of the spores as will exudates from roots; but mycostasis is more than an absence of nutrients: there is an inhibitory factor present which can be overcome by the addition of a suitable substrate. Inhibition can be overcome by heating the soil, and there is some evidence that an inhibitor can be extracted from soil, although results have been erratic. Among the possible inhibitors must be considered antibiotic metabolites, which could be staling products. Balis and Kouyeas (1968) have suggested that a volatile factor is involved, and that it could be an unsaturated hydrocarbon or its derivative. Certainly volatile inhibitors can be produced which affect radial extension growth of fungi. Dick and Hutchinson (1966) showed that volatile metabolites of two yeasts and of *Fomes annosus* (Fr.) Cooke produced significant inhibition of growth of various test fungi.

The leaching of nutrients from the spore could encourage the growth of competing organisms. Lingappa and Lockwood (1964) reported a 7-fold increase in both bacterial numbers and soil respiration when living fungal spores, or water used to wash them, were added to the soil. This implies that the exudates from the spores could act as a trigger for the germination of those spores if the exudates are not leached away or utilized by competing micro-organisms.

The type of mycostasis which may be caused by or associated with the presence of soil micro-organisms was referred to as "microbial mycostasis" by Dobbs and Gash (1965). This can be removed by autoclaving and by the use of sterilizing agents, and it does not return as long as the soil remains sterile. Similarly, it is generally absent from sterile or near-sterile subsoils. There is, however, a "residual mycostasis", which they found in several calcareous soils in North Wales. This was thermostable, and was not overcome by the addition of sugars. In these basic soils the inhibition of germination may be caused by unchelated iron, or by the presence of calcium carbonate.

E. The Fate of Hyphae and Spores

Although hyphae may be locally or seasonally abundant in particular situations, they do not accumulate in the soil or in leaf litter. There is evidence that hyphae with pigmented walls are more frequently seen in soil sections than normal isolation techniques would indicate (Burges, 1958), but Pugh (1969) suggested that this discrepancy may have been caused by the type of isolation medium used.

Many members of the soil microfauna feed on hyphae and spores, and appear to have feeding preferences. Heal (1963) showed that amoebae fed on yeasts and on 2 of 16 species of fungi which he tested. The role of fungi as food for nematodes is discussed by Twinn (Chapter 13) and for micro-arthropods by Harding and Stuttard (Chapter 15). Hyphae which are not eaten by these or other animals may autolyse, or may be utilized by soil micro-organisms. Bumbieris and Lloyd (1967) showed that in a wet soil (25% moisture content) the cell wall and its contained protoplasm disappeared concurrently, with bacteria following protozoa. In a dry soil (13% moisture content) the protoplasm was lost, leaving the hyphal walls; this was associated with Actinomycetes. In intermediate soils (19% moisture content) Actinomycetes followed bacteria on the hyphae.

Lloyd and Lockwood (1966) found that hyphae were completely lysed within 4–8 days of being added to soil, and Old (1967) showed that after 10 days only 20% of hyphal fragments of *Cochliobolus sativus* (Ito & Kurib.) Dreshs. ex Dastur survived in soil, compared with 73% of conidia which were still viable. However, few conidia survived 50 days in natural soils. A detailed account of the fate of residues of Actinomycetes, algae, bacteria and fungi in the soil has been given by Webley and Jones (1971).

F. Predacious Fungi

At various stages in the decomposition of plants, the litter becomes invaded by nematodes. In leaves of *Typha latifolia* L. for example, Pugh and Mulder (1971) found that nematodes were plentiful in the final stages of decomposition. Certain fungi prey on nematodes, trapping them in specialized, sticky branches, in complex loops, or in constricting nooses. Once the eelworm is caught, its body is invaded by hyphae, which can be seen ramifying inside. It has been suggested that a toxin is produced by the fungus which paralyses or kills the nematodes (see Olthof and Estey, 1963). Typical predacious fungi are species of *Arthrobotrys*, *Dactylaria* and

Dactylella, which are members of the Hyphomycetes, and members of the Zoopagales (Zygomycotina), which shed sticky spores. In *Arthrobotrys conoides* Drechsler, the formation of the complex loops requires the presence of CO_2 as well as "nemin", a substance produced by nematodes. Studies on the succession of predacious fungi have been carried out by Cooke (1963), while Duddington (1968) gave an account of their activities.

IV. Classification

The classification of fungi which occur in terrestrial situations virtually involves the classification of all fungi except the few which are restricted to aquatic environments (see Jones, Chapter 11). No system of classification can hope to be ideal, but each can be adapted to serve a particular purpose. Thus for purposes of identification a taxonomic approach will normally be appropriate; this may make use of arbitrarily selected characters as criteria, and result in an artificial classification; or it may aim at a classification which makes use of natural relationships and phylogeny. On the other hand there may be an ecological approach, in which organisms are grouped on their patterns of occurrence; or a biological classification based on their pattern of behaviour or a physiological approach making use for example, of nutritional attributes.

A. Taxonomy

The criteria used in fungal taxonomy are essentially all morphological, and may be selected from different stages in the life cycle. Ideally all observations of the selected criteria should be carried out on the same substrate or host under the same conditions and on the fungus at a specified age. In practice this ideal is not essential except for particular genera (e.g. *Penicillium*, *Aspergillus*). Chesters (1968) has said that criteria are not all equally valuable at a given taxonomic level. He stated that morphological criteria should show considerable constancy, as well as definite degrees of difference which would allow effective separation of taxa; the variability of each criterion should be known and accurate observations should be possible for each criterion.

Traditionally the fungi have been classified into Phycomycetes, with hyphae which characteristically lack septa, except those that cut off reproductive structures or are found in older mycelia. In some of the simpler forms there may be no vegetative thallus, or only a rudimentary one. The other three sub-divisions, Ascomycetes, Basidiomycetes and Deuteromycetes (Fungi Imperfecti) all possess septate hyphae and are distinguished by their sexually produced spores, which are endogenous, exogenous or absent respectively. The Phycomycetes have long been recognized as a heterogeneous collection of fungi (see von Arx, 1967). Recently Webster (1970) has treated the Mastigomycotina, with a motile stage in their life cycles, as of equal rank with the Zygomycotina (with non-motile spores). Ascomycotina (Ascomycetes) and Basidiomycotina (Basidiomycetes).

Within the Ascomycetes, the system of classification has seen many changes since Miller (1949) based his revision on the type of ascus and on the ontogeny of the ascocarp, drawing attention to earlier work which recognized that loculate stromata were of different origin from perithecia.

The validity of the Basidiomycetes as an entity has been questioned by Moore (1972) who has proposed the erection of a new taxon, the Ustomycota, to be of equal rank with the Ascomycota and Basidiomycota, to accommodate the ustilaginaceous fungi.

In the Deuteromycetes, separation of form-genera has been based on the work of Saccardo (1886). The relative ease with which fungi can be placed in their pigeon-holes has made this artificial system widely used. Searches for a more natural system have centred around the method of spore production, and have resulted in the adoption of many criteria which were not previously used.

All fungi are dependent, directly or indirectly, on green plants or their products. Plant-pathogenic fungi include the biotrophs, which are generally obligate parasites and which do not normally kill the host, and necrotrophs, or facultative parasites. These can cause severe damage to the plant, and can live on the dead remains of the plant. The saprophytes grow on the dead remains of plants and animals. Some of them, however, may act as weak parasites or primary colonizers of still living plant tissues in which natural resistance has been lowered, for example, when they become moribund.

1. Mastigomycotina

The lower members of this division live in water and are generally regarded as aquatic. However, there is normally sufficient water present in the soil for their growth and reproduction. Among the plant parasitic Chytridiomycetes in soil are *Olpidium brassicae* (Woron.) Dang., which is common on roots of cabbages, and which has been implicated in the spread viruses causing big-vein disease of lettuce; and *Synchytrium endobioticum* (Sch.) Percival, which causes potato wart disease. *Rhizophlyctis* and *Nowakowskiella* are widespread saprophytes in the soil. The former occurs on cellulosic substrates and has been shown able to decompose cellulose in culture. In the Blastocladiaceae, *Allomyces*, and *Blastocladiella* can be isolated from the soil by the use of suitable baits, such as rose hips and hawthorn fruits.

Members of the Oomycetes, which have a cellulosic cell wall include the Saproleginales (the water moulds), which are discussed by Jones (Chapter 11) and the Peronosporales. Here there are many plant parasites, including the members of the Peronosporaceae (downy mildews), as well as *Pythium* and *Phytophthora*. Members of these two genera can also grow saprophytically in the soil.

2. Zygomycotina

These fungi, with no motile stages in their life-histories, are typical terrestrial fungi. In the Mucorales species of *Absidia*, *Mucor*, *Mortierella* and *Zygorrhynchus* form a significant part of the soil mycoflora, and have also been regularly isolated from plant litter. Although they have been isolated from "Cellophane" and on carboxy-methyl-cellulose, they are not normally regarded as decomposers of natural cellulose. Species of *Mortierella* are often associated with decomposition of chitin, and several members of the Mucorales (e.g. *Pilobolus*, *Pilaira*) occur regularly on dung (see Lodha, Chapter 7). They are generally considered as members of the "sugar fungi" (Burges, 1939), growing as primary colonists on newly available, simple, soluble carbohydrates. Some species can also grow as "secondary sugar fungi" (see Hudson, 1968) when they utilize the glucose produced by cellulose-decomposing species in excess of their own ability to absorb the products of decomposition. Some genera such as *Chaetocladium*, *Piptocephalis* and *Syncephalis* are parasitic on other members of the Mucorales. *Syncephalastrum* is typically saprophytic, but may also act as a parasite on other fungi. *Syzygites megalocarpus* Ehrenb. (= *Sporodinia grandis*) can frequently be found on decaying toadstools.

While members of the Mucorales are widely distributed on a worldwide basis, *Choanephora* and *Cunninghamella* are found in the soil in tropical and sub-tropical areas.

3. Ascomycotina

(a) *Hemiascomycetes*. The yeasts constitute the majority of the Hemiascomycetes which are concerned with the early stages of colonization of higher plants, forming a large proportion of the fungal population on leaves and fruits. Reviews have recently been published on their taxonomy by Kreger-van Rij (1969), their distribution in terrestrial and aquatic environments by Carmo-Sousa (1969), and their presence on living plants by Last and Price (1969).

(b) *Plectomycetes*. The parasitic Erysiphales grow almost exclusively on the surfaces of their hosts, obtaining their nutrients from haustoria which are formed in the epidermal cells. It would be interesting to know if the numerous minute punctures, produced by the infection pegs, have any stimulating effect on the eventual rate of decomposition of the host leaves.

Equally the mat of mycelium formed on the plant surface may keep potential saprophytes off the leaves and so retard decomposition.

Aspergillus and *Penicillium* are normally included in the Deuteromycetes. Those species which produce perfect states, however, are good examples of the Eurotiales, and they therefore will be considered here. Apart from the industrial and medical usage of their fermentation products, species of these genera are important economically for their deterioration of food, textiles and other substrates important to man. They are also widespread and abundant in the soil, with *Aspergillus* more common in warmer soils and *Penicillium* more frequent in temperate and wetter soils. Both genera owe their ubiquity to several features which they possess in common: an ability to grow on substrates with a relatively low moisture content (e.g. 82% r.g. for *P. expansum* Link), over a wide range of temperature (−3°C–35°C for *P. expansum*) and pH (e.g. *in vitro* from 2·0–10·0 in *P. cyclopium* Westling) and on a range of substrates; a high level of sporulation; rapid germination of conidia; antibiotic metabolites. Amongst these metabolites, penicillin is important medicinally, but aflatoxin, produced by *Aspergillus flavus*, for example on groundnuts, may be carcinogenic.

One of the common species of *Penicillium*, *P. cyclopium* Westling, may be taken as an example of a widespread species with a remarkable range of attributes. Domsch and Gams (1972) have recorded its worldwide occurrence in acid leaf litter and calcareous habitats, in saline estuaries and polluted streams, as well as in sand-dunes and arable soils. On higher plants, it has been found on roots, on wood pulp, hay and straw, and is frequent on stored wheat grain. It has been described as osmophilic and can grow at 81–84% r.h., and it has been shown to utilize pectin, starch, xylan, carboxymethyl cellulose, and filter paper, although its role as a cellulose decomposer is uncertain. An important attribute in decomposition of plant litter is its ability to grow on 5% tannin as the sole C source. It produces many metabolities which have antibiotic characteristics, and it is antagonistic to numerous fungi and bacteria *in vitro*.

There are many other common species of *Aspergillus* and *Penicillium* with a comparable range of attributes which would account for their ubiquity.

Other members of the Eurotiales include *Ceratocystis ulmi* (Buism.) C. Moreau, the causal agent of Dutch Elm Disease, and in the family Gymnoascaceae, *Byssochlamys fulva* Olliver & G. Smith, a soil-borne fungus which can cause spoilage in canned fruit: it is able to survive high temperatures and can grow at low oxygen tensions.

(c) *Pyrenomycetes*. Some members of the Pyrenomycetes are parasitic on plants (e.g. *Claviceps purpurea*, *Epichloe typhina* (Fr.) Fr.) or on animals (e.g. *Cordyceps militaris* (L.) Link on caterpillars). Others are widespread saprophytes, which are often or normally recorded on dung (e.g. *Sordaria fimicola* (Rob.) Ces. & de Not.), *Podospora minuta* (Fr.) Niessl., *Poronia punctata* (L.) Fr.) while members of the genus *Chaetomium* are regularly isolated from soil, seeds and from almost any substrate which contains cellulose. Consequently it is a fungus of great economic importance, for example on textiles and military equipment. *Hypocrea pulvinata* Fuckel grows on the dead sporophores of *Piptoporus betulinus* (Bull.) Karst., a member of the *Polyporaceae* which grows on birch trees (*Betula* spp.).

The great majority of the other members of the Pyrenomycetes occur on wood (see Kaarik, Chapter 5). Many of them are primary colonists and amongst these are some parasites, such as *Nectria galligena* Bres. on apple and pear, *Daldinia concentrica* (Bolt.) Ces. & de Not. mainly on ash, and *Ustulina deusta* (Fr.) Petrak on beech.

Amongst the Pyrenomycetes which occur on wood, some are confined to one type of tree. Thus *Xylosphaera carpophila* (Pers.) Dum. is confined to fallen beech mast, *Cryptosphaeria eunomia* (Fr.) Fuckel is found on *Fraxinus*, *Diatrypella quercina* (Pers. ex Fr.) Cooke on *Quercus*, *Eutypa acharii* Tulasne on *Acer*, *Hypoxylon fragiforme* (Pers. ex Fr.) Kickx on *Fagus* and *Quaternaria quaternata* (Pers. ex Fr.) Schr. on *Fagus*.

Other species occur predominantly on one host, but are recorded on others, including *Diatrype disciformis* (Hoffm. ex Fr.) Fr. which is typically found on *Fagus*, but has been collected on *Quercus*, *Acer* and locally *Betula*, and *Hypoxylon fuscum* (Pers. ex Fr.) Fr. is normally collected on *Corylus*, but also occurs on *Alnus*. Yet other species are plurivorous, occurring on many types of wood. These include *Diatrype stigma* (Hoffm. ex Fr.) Fr., *Hypoxylon rubiginosum* (Pers. ex Fr.) Fr., *H. semi-immersum* Nitschke and *Xylosphaera hypoxylon* (Linn.) Dum.

These fungi are predominantly colonists of newly dead tissues and are "fungi of first incidence". Their stromata containing the perithecia are immersed in or erumpent from the bark. Some, however occur on both bark and decorticated wood (e.g. *Diatrype stigma*, *Hypoxylon multiforme* (Fr.) Fr.) while others only occur on the bare wood (e.g. *H. semi-immersum*).

A comprehensive account of the taxonomy and habitats of these fungi and of the Discomycetes has been given by Dennis (1960).

4. Basidiomycotina

The most commonly seen members of the Basidiomycotina on and associated with plant remains are members of the Agaricales (toadstools) and Aphyllophorales or Polyporales (bracket fungi). Both have the hymenium of basidia, upon which the spores are produced, freely exposed at maturity and they are consequently grouped together in the Hymenomycetes. In the Gasteromycetes, the hymenium is enclosed within the sporophore. Here are included the Lycoperdales (puff-balls), Sclerodermatales (earth balls), Phallales (stink horns) and Nidulariales (bird's nest fungi).

Some of the toadstools form symbiotic, mycorrhizal associations with the roots of higher plants, especially with trees (see Harley, 1969). Many of the bracket fungi are pathogens on trees (e.g. *Fomes annosus*, *Piptoporus betulinus* and *Polyporus squamosus* Fr.), as are some toadstools (e.g. *Armillaria mellea*) and cause considerable economic damage in the timber industry. *Serpula lacrimans* has been described as one of the most serious causes of timber deterioration in buildings, as it is the causal agent of dry rot (see Cartwright and Findlay, 1958).

Although toadstools and other Basidiomycetes are plentiful in fields and woods, and their mycelia can be seen in great quantities in leaf litter, these fungi are rarely recorded during normal isolation procedures. However, Warcup (1959) has isolated mycelia from soils and roots, using Warcup's (1957) hyphal isolation method and also a sieving and a flotation method to obtain sclerotia, and from direct plating of rhizomorphs. Warcup and Talbot (1962) identified 20 Basidiomycetes representing 18 genera which were isolated by these methods.

The fungi were mostly members of the Hymenomycetes, from all three orders recognized by Webster (1970), and two—*Cyathus stercoreus* (Schur.) de Toni and *Sphaerobolus stellatus* Tode ex Pers.—were members of the Nidulariales in the Gasteromycetes.

The occurrence of toadstools in woods has been recorded by Hora (1959), and in an apple orchard by Bond (1972). They undoubtedly play a large part in the breakdown of litter in these localities. *Coprinus cinereus* (Sch.) Gray and other species of *Coprinus* and of *Panaeolus* regularly occur on dung (see Lodha, Chapter 7).

Members of the Aphyllophorales are generally to be found on wood where they are the causal agents of brown rots (e.g. *Trametes serialis* Fr.) where the cellulose and associated polysaccharides have been destroyed, leaving the lignin, and white rots where the lignin has been degraded. The lignin-decomposers also attack the cellulose slowly and Cartwright and Findlay (1958) said that *Polystictus hirsutus* Fr. was able to attack cellulose more quickly than *Trametes pini* (Thore) Fr.

5. *Deuteromycotina*

The generally accepted grouping of the Deuteromycotina (Fungi Imperfecti) separates those species in which there is a protective covering around the conidiophores (Coelomycetes) from the Hyphomycetes, in which the conidiophores may or may not be grouped together, but in which there is no surrounding wall.

Members of the Coelomycetes which possess pycnidia are placed in the Sphaeropsidales, and these are frequently found on newly dead plant tissues. They can also be isolated from the soil. Grove (1935, 1937) gave a comprehensive account of the occurrence of these fungi on plants and Morgan-Jones *et al.* (1972) gave recent descriptions of selected genera.

Within the Hyphomycetes, the artificial system of Saccardo (1886) has long been used for purposes of identification and classification. There have been various attempts made to find a more natural system of classification (see Mason, 1937; Hughes, 1953; Tubaki, 1958). Arising from these, Barron (1968) published a comprehensive account of soil-borne Hyphomycetes arranged in groups which are distinguished from each other by the ontogeny of the conidia, regardless of their pigmentation. Saccardo's system, with emphasis on conidium pigmentation, had some ecological significance, as fungi with dark hyphae and/or conidia tend to predominate in soils subjected to high levels of insulation (Nicot, 1960) and on plant parts above the ground (Pugh and Williams, 1968; see also Ellis, 1971).

Members of the Deuteromycotina are very common and widespread. In his "Manual of Fungi", Gilman (1957) indicated that in the soil they comprised almost 57% of the species recorded, mainly drawing his results from the use of non-selective media and the soil dilution technique. With the introduction of a selective medium their occurrence rose to over 80% of the total number of isolations (see Pugh, 1969).

The observed life cycle of the Fungi Imperfecti is one in which conidia are produced on a mycelium, are dispersed and germinate to produce a new mycelium. The stage of sexual reproduction either does not occur, as in the self-incompatible strains which arise from single ascospores of *Neurospora crassa* Shear & Dodge, for example, or the sexual and asexual reproductive phases of the life cycle have not yet been associated. There is scope for further studies in this field, but it will not be straightforward, as several imperfect spore-forms may be associated with one generic perfect state, and conversely, one imperfect state may occur in several perfect genera (see Pugh, 1969). Webster (1970) omitted consideration of the Deuteromycotina from his account of the fungi, preferring to deal with imperfect states as part of the life-cycle of the perfect states which he discussed. There is a long way to go before this aim can be fulfilled even reasonably completely, and then, as Mason (1937) put it, "If we could consistently predict the Perfect genus by an inspection of its conidia, there would be no need for an Imperfect classification at all."

B. Biological Classification

The behaviour of fungi can be used to place them into categories which are essentially nutritional, such as parasites and saprophytes, or into primary or secondary colonizers of substrates. There are also broad ecological groups which result from the interaction of the behaviour of the fungi and the environments in which they can grow.

In studies of fungi in the soil the concept arose of substrate relationships, with fungi colonizing the most readily available foods first, and being succeeded in waves by other species able to decompose in turn the more recalcitrant substrates. Thus the simple soluble carbohydrates would be used by the "sugar fungi", which would be followed by those able to attack cellulose, lignin and humus. The "sugar fungi", with their ability to "flare-up" rapidly on newly added substrates fall into the category of zymogenous organisms as described in Winogradsky's ecological grouping of soil bacteria. Those fungi able to utilize the more stable substrates such as cellulose and lignin, would grow more slowly and steadily over a longer period of time, and would form the bulk of his autochthonous flora. While such a general succession can be seen during the colonization of dung, Harper and Webster (1964) showed that the fungal sequence was a reflexion of the minimum time from the commencement of growth of the fungi until the appearance of the sporing structures. The succession on leaves and litter frequently omits the "sugar-fungi". Indeed the primary colonizers can produce a wide range of enzymes, including cellulase, cutinase, pectinase and protease, as well as enzymes for sugars and other carbohydrates (see Pugh *et al.*, 1972). Griffin (1972) has also criticized the division of soil fungi into zymogenous and autochthonous categories; as he said, "all soil fungi for which we have much information are zymogenous and one could be forgiven for doubting if truly autochthonous forms exist".

Another form of biological classification was used by Reinking and Manns (1933) to separate "soil-inhabiting" and "soil-invading" species of *Fusarium*. Garrett (1956) recognized that the "soil invaders" were really root-inhabiting species, and put forward a scheme in which he showed the ecological relationships between root-inhabiting, root-disease and soil-inhabiting fungi.

Physiological criteria can also be used to distinguish broad ecological groups of fungi. Thus while most fungi are mesophilic in their temperature requirements, the thermophiles grow at high temperatures and the psychrophiles grow in cold conditions.

V. Physiology

A. Introduction

The decomposition of a substrate by a fungus requires the presence of both, in an environment which is suitable for the fungus to grow, produce the necessary enzymes and absorb soluble nutrients. In considering the role of fungi on plant litter the nutrients in that litter will normally be sufficient for growth, even if the balance of elements may not be optimal at any given moment, while in the soil there may well be a starvation diet away from the influence of living roots. The limiting factors will vary with the geographical and topical locality of the substrate. On a leaf surface, desiccation and harmful radiations may be most important; in the soil, the water regime, the prevailing temperature, oxygen supply, hydrogen ion concentration and the presence of fungistatic or fungitoxic substances will all help determine whether the fungus can grow. Its saprophytic survival will depend on its competitive ability as well as on endogenous characters such as the ability to produce enzymes in what is almost certainly a far from ideal environment. Indeed, it has been said that fungal ecology is the study of fungal physiology under the worst possible conditions.

B. Carbon : Nitrogen Ratio

Carbon is plentiful in plant remains, particularly in cellulose, which may form 15–60% of the dry weight, hemicelluloses (10–30%) and lignin (5–30%), as well as being present in soluble carbohydrates. Nitrogen represents a much smaller part of the plant remains, being present mainly in proteins and amino acids. Succulent, herbaceous plants will have relatively more nitrogen; woody plants and particularly trees will have relatively less. Older plants generally will have relatively more carbon and less nitrogen than younger plants. In the decomposition of litter, the availability of nitrogen will determine the speed of the mineralization processes. Litter with a good supply of nitrogen (i.e. C:N ratio of 30:1 as in leguminous plants) will decompose quickly, but when the supply is poor (i.e. a C:N ratio of >200:1 in wood) the rate will be slow and nitrogen may be taken up from the surrounding soil. In the composting of plant material additional nitrogen is necessary for rapid decomposition. As mineralization proceeds, the C:N ratio eventually falls to a level of about 10:1 because carbon is given off as CO_2 while the nitrogen which has been utilized by the fungi is bound up in the mycelium and will be used by succeeding organisms after the death of the hyphae. This subject has been discussed amongst others by Burges (1958) and Alexander (1961).

The type of growth of the fungi is determined by the nature of the substrate, so that they tend to grow rapidly, vegetatively, on a new substrate, and to sporulate as the nutrient status of the substrate is reduced. Often a high level of nitrogen suppresses the production of spores. It has also been shown to curtail the survival time of six root rot fungi in straws buried in the soil (Garrett, 1972). A study of fungal preferences for different C:N ratios was carried out by Kaufman and Williams (1963), who found that more fungi were isolated from soils with a narrow C:N ratio, where species of *Botryotrichum*, *Chaetomium*, *Masoniella*, *Stysanus* and members of the *Penicillium rugulosum* series were more prevalent. In soils with a wide C:N ratio they more frequently found species of *Actinomucor*, *Coniothyrium*, *Phoma*, *Trichoderma* and the *Penicillium funiculosum* Thom series. *Circinella* sp., *Humicola* sp. and *P. lilacinum* Wehmer series were favoured by specific C:N ratios.

The source of nitrogen can also be important. Most fungi can utilize ammonium salts or nitrates (although some of the water moulds do not use nitrate nitrogen). Ammonium salts are generally more readily used, and the addition of ammonia to a growth medium in which *Scopulariopsis brevicaulis* (Sacc.) Bain. was being grown with nitrate suppressed the nitrate assimilation. The whole question of the utilization of nitrogen sources by fungi has been fully discussed by Nicholas (1965).

C. Fungal Metabolites

During their growth fungi produce many metabolities which may be useful to man (e.g. penicillin) or harmful (e.g. aflatoxin). They have been shown to produce auxins (I.A.A. and I.A.N.) when grown *in vitro* in the presence of tryptophan (see Buckley and Pugh, 1971). Auxin production *in vivo* could have widespread effects in the ecosystem.

The range of substrates upon which fungi can grow is dependent on their enzyme systems. While virtually all fungi can utilize glucose, increasing complexity of the substrate requires the production of particular enzymes, the demonstration of which, in the laboratory at least, indicates the potential to use that substrate in the field. As cellulose forms such a large part of plant remains and particularly of man-made litter such as textiles and paper products, cellulases are very important in nature. The occurrence of cellulose-decomposing fungi has been studied by many workers. Earlier studies were collated by Siu (1951), while some methods for detecting cellulases are given in Kelman (1967). Other enzyme systems that are important in plant litter decomposition include cutinase and pectinase, which are found, for example, in plant pathogens and phylloplane fungi, amylase, lipase, proteinases and saccharase which degrade plant cell contents. The decomposition of lignin (a highly complex phenolic polymer) requires a phenol oxidase, and phenolases, produced by white-rot fungi are involved in the breakdown of lignin and other polyphenols (Hurst and Burges, 1967). Polyphenols and particularly tannins, are responsible for the unpalatability of many leaves to the soil fauna. They are broken down to catechols and gallic acid, for example by *Penicillium expansum* Link.

D. Radiation

1. Light

The general conditions for growth of fungi in the laboratory, on a bench, or in an incubator, normally do not make allowance for the effects of light. In certain circumstances, however, light can play an important part in the life cycle of fungi. Most coprophilous fungi, for example, have a mechanism which ensures that their spores are discharged towards a light source. In *Ascophanus* the asci are orientated towards light; in *Pilobolus* the sporangiophore grows so that the sporangium is discharged towards a light source; in *Sordaria* the neck of the perithecium grows so that the ascospores are similarly discharged. The germination of spores is increased by light in most cases, although the germ tube of *Botrytis cinerea* is negatively phototropic. Light at high intensities generally retards mycelial growth, although there are isolated reports of stimulation. Some hyphae are positively phototropic, and may show zonation if they are grown in an alternating light and dark regime (e.g. *Alternaria brassicae* (Berk.) Sacc.). In other fungi, such as *Fusarium* and *Trichoderma*, sporulation is dependent upon an exposure to light. Page (1965) has briefly reviewed the main effects of light on fungal growth.

2. Ultraviolet Radiations

Ultraviolet radiations present in daylight are generally harmful and can be lethal to fungi at a wavelength of around 2600 Å. The mechanism of this action and its genetic effects have been discussed (Pomper, 1965). In fungal ecology the lethal action can be avoided if the fungus is pigmented, or if it produces its spores in a protective covering. Nicot (1960) drew attention to the prevalence of strongly pigmented members of the phragmo- and dictyospored Dematiaceae, members of the Sphaeropsidales, and dark sterile hyphae in desert sands in North Africa. In a study of fungi on *Salsola kali* L. growing in coastal sands in Britain, Pugh and Williams (1968) showed that dematiaceous fungi (*Alternaria*, *Botrytis*, *Cladosporium*, *Epicoccum*, *Stemphylium*) predominated on the aerial parts of the plant, while *Cephalosporium* and *Fusarium* were more common on the roots. They postulated that the pigmented fungi on the aerial parts were protected against, or derived benefit from, the ultraviolet radiations to which they were exposed. Similarly on leaf surfaces, Pugh and Buckley (1971*b*) have indicated that the phylloplane fungi *Alternaria*, *Botrytis*, *Cladosporium* and *Epicoccum* are protected by their pigmentation, *Sporobolomyces* may derive benefit from the presence of carotenoids and from its growth pattern, with outer cells protecting inner cells in a clump, while *Aureobasidium* on an exposed surface forms pigmented microsclerotia which are resistant to u.v., and can also be isolated from within the veins, where it will be protected by the cells of the leaf. Hyaline spores are more susceptible to u.v. damage than are pigmented spores, where the melanin in the wall acts as a screen. In multicellular spores, such as those of *Alternaria*, *Epicoccum* and *Stemphylium*, some cells are always shielded by others.

E. Temperature

The temperature relationships of fungi have been discussed by Togashi (1949). The great majority of fungi are mesophilic in their temperature requirements, with a minimum temperature for growth generally between 5°C and 10°C, a maximum between 30°C and 40°C, and an optimum between 25°C and 35°C. Outside these temperature ranges, some fungi, the psychrophiles, can grow at low temperatures. These have been found on frozen foods at temperatures as low as −6°C (*Cladosporium herbarum*), while *Alternaria tenuis* auct. and *Botrytis cinerea* have been recorded at −2°C. Pugh *et al.* (1972) have drawn attention to the ability of the common phylloplane fungi to grow at low temperatures, and Deverall (1968) has discussed the growth of fungi in cold environments.

At the other extreme, the thermophiles have been defined by Cooney and Emerson (1964) as having a maximum growth temperature at or above 50°C and minima above 20°C. They regarded as thermotolerant those species which had a maximum temperature near 50°C but a minimum below 20°C. This group more or less coincides with the psychrotolerant thermophiles of Apinis (1963) and includes such common fungi as *Aspergillus fumigatus* Fresen. The thermophiles occur in a number of agricultural and industrial processes, being found for example on stored grain, fermenting cocoa, and in accumulations of organic material such as composts (see Chang and Hudson, 1967).

F. Water Relationship

The influence of water on fungi in the soil has been thoroughly discussed by Griffin (1972). In dry litter on the soil surface, or still attached to the plant in an aerial environment, however, the moisture levels may fluctuate greatly and at their lower levels may be limiting to fungal germination and growth. A similar limiting condition is aimed at in substrates useful to man (paper products, textiles, timber and stored food grains, for example). In these situations a small variation in moisture level can be sufficient to allow germination to occur. In the case of dry rot in building timbers the moisture level around floor joists can be increased by insufficient ventilation, or by a circumvention of the damp course (by a flower border being banked up above its level, for example). Once the water content of the wood reaches fibre saturation point (about 27% of the dry weight of the wood) germination of spores of *Serpula lacrimans* can occur. The fungus can then grow onto and into dry wood, obtaining its water by breaking down the cellulose, and producing excess water which forms drops on the mycelium. This makes eradication of this fungus much more difficult than the wet rot fungi such as *Coniophora cerebella* Pers. which rely on an external supply of water, and can therefore be controlled by rectifying conditions so that the wood can dry out.

Species of *Penicillium* and particularly of *Aspergillus* are extreme xerophiles, able to grow at lower moisture levels than other fungi. There is a direct relationship between the osmotic potential of the hyphae and their ability to grow at a low relative humidity. Thus the suction pressure of hyphae of species of these two genera has been calculated at 217 atm (see Hawker, 1950) and *Aspergillus* has been shown to grow on media with an osmotic pressure of 100 atm. Osmophilic fungi, particularly yeasts, are other examples of organisms adapted to low moisture regimes, where there is a state of physiological drought. In terms of the deterioration of fabrics, paper and cereal grains in storage, a rise in the level of the relative humidity to about 80% can be sufficient to allow fungal germination. In goods packed in non-porous material a fall in temperature can raise the humidity above the danger level.

There have been various studies on the effects of moisture levels on spore germination, hyphal growth and sporulation, particularly in plant-pathogenic fungi.

Similar detailed studies are needed for many of the fungi which are involved in the saprophytic breakdown of plant litter. These could, with advantage, be extended into autecological studies of individual species. In this respect the compilation of available knowledge about more than 200 species of soil fungi by Domsch and Gams (1972) provides an easily accessible source of information and points the way for future development.

References

Ainsworth, G. C. (1971). "Dictionary of the Fungi," 6th edition. Commonwealth Mycological Institute, Kew.

Alexander, M. (1961). "Introduction to Soil Microbiology." John Wiley, New York.

Apinis, A. E. (1963). *In* "Soil Organisms." (J. Doeksen and J. van der Drift, eds), pp. 427–438. North-Holland, Amsterdam.

Arx, J. A. von (1967). "Pilzekunde. Eine Kurzer Abriss der Mykologie unter besonder Berücksichtigung der Pilze in Reinkultur." Cramer, Lehre.

Bainbridge, A. and Dickinson, C. H. (1972). *Trans. Br. mycol. Soc.* **59**, 31–41.

BALIS, C. and KOUYEAS, V. (1968). *Ann. Inst. Phytopath. Benaki N.S.* **8,** 145–149.
BARRON, G. L. (1968). "The Genera of Hyphomycetes from Soil." Williams and Wilkins, Baltimore.
BOND, T. E. T. (1972). *Trans. Br. mycol. Soc.* **58,** 403–416.
BUCKLEY, N. G. (1971). Ph.D. Thesis, University of Nottingham.
BUCKLEY, N. G. and PUGH, G. J. F. (1971). *Nature, Lond.* **231,** 332.
BUMBIERIS, M. and LLOYD, A. B. (1967). *Aust. J. biol. Sci.* **20,** 103.
BURGES, N. A. (1939). *Broteria* **8,** 64–81.
BURGES, N. A. (1958). "Micro-organisms in the Soil." Hutchinson, London.
CARMO-SOUSA, L. DO (1969). *In* "The Yeasts," Vol. 1. (A. H. Rose and J. S. Harrison, eds), pp. 79–105. Academic Press, London and New York.
CARTWRIGHT, K. ST. G. and FINDLAY, W. P. K. (1958). "Decay of Timber and its Prevention." H.M.S.O., London.
CATSKA, V., MACURA, J. and VAGNEROVA, K. (1960). *Folia microbiol., Praha* **5,** 320–330.
CHANG, Y. and HUDSON, H. J. (1967). *Trans. Br. mycol. Soc.* **50,** 649–666.
CHESTERS, C. G. C. (1948). *Trans. Br. mycol. Soc.* **30,** 100–117.
CHESTERS, C. G. C. (1960). *In* "The Ecology of Soil Fungi," pp. 223–238. Liverpool University Press.
CHESTERS, C. G. C. (1968). *In* "The Fungi," Vol. 3. (G. C. Ainsworth and A. S. Sussman, eds), pp. 517–542. Academic Press, New York.
COLEY-SMITH, J. R. and COOKE, R. C. (1971). *A. Rev. Phytopath.* **9,** 65–92.
COOKE, R. C. (1963). *Nature, Lond.* **197,** 205.
COOKE, R. C. (1967). *Nature, Lond.* **213,** 295–296.
COONEY, D. G. and EMERSON, R. (1964). "Thermophilic Fungi." Freeman, San Francisco.
DENNIS, R. W. G. (1960). "British Cup Fungi." Ray Society, London.
DEVERALL, B. J. (1968). *In* "The Fungi," Vol. 3. (G. C. Ainsworth and A. S. Sussman, eds), pp. 129–135. Academic Press, London and New York.
DICK, C. M. and HUTCHINSON, S. A. (1966). *Nature, Lond.* **211,** 868.
DICKINSON, C. H. (1965). *Trans. Br. mycol. Soc.* **48,** 603–610.
DICKINSON, C. H. (1971). *In* "Ecology of Leaf Surface Micro-organisms." (T. F. Preece and C. H. Dickinson, eds), pp. 129–137. Academic Press, London and New York.
DOBBS, C. G. and GASH, M. J. (1965). *Nature, Lond.* **207,** 1354–1356.
DOBBS, C. G. and HINSON, W. H. (1953). *Nature, Lond.* **172,** 197–199.
DOMSCH, K. H. and GAMS, W. (1972). "Fungi in Agricultural Soils." Longmans, London.
DUDDINGTON, C. L. (1968). *In* "The Fungi," Vol. 3. (G. C. Ainsworth and A. S. Sussman, eds), pp. 239–252. Academic Press, London and New York.
EGGINS, H. O. W. and PUGH, G. J. F. (1962). *Nature, Lond.* **193,** 94–95.
EGGINS, H. O. W., MILLS, J., HOLT, A. and SCOTT, G. (1971). *In* "Microbial Aspects of Pollution (G. Sykes and F. A. Skinner, eds), pp. 267–269. Academic Press, London and New York.
ELLIS. M. B. (1971). "Dematiaceous Hyphomycetes." Commonwealth Mycological Institute, Kew.

Garrett, S. D. (1944). "Root Disease Fungi." Chronica Botanica Co., Waltham, Mass.
Garrett, S. D. (1951). *New Phytol.* **50,** 149–166.
Garrett, S. D. (1955). *Trans. Br. mycol. Soc.* **38,** 1–9.
Garrett, S. D. (1956). "Biology of Root-Infecting Fungi." Cambridge University Press.
Garrett, S. D. (1970). "Pathogenic Root-Infecting Fungi." Cambridge University Press.
Garrett, S. D. (1972). *Trans. Br. mycol. Soc.* **59,** 445–452.
Gilman, J. C. (1957). "A Manual of Soil Fungi." Iowa State College Press.
Gregory, P. H. (1973). "Microbiology of the Atmosphere." Leonard Hill, London.
Griffin, D. M. (1972). "Ecology of Soil Fungi." Chapman and Hall, London.
Grove, W. B. (1935). "British Stem and Leaf Fungi," Vol. 1. Cambridge University Press.
Grove, W. B. (1937). "British Stem and Leaf Fungi," Vol. II. Cambridge University Press.
Harley, J. L. (1969). "The Biology of Mycorrhiza." Leonard Hill, London.
Harley, J. L. (1971). *J. Ecol.* **59,** 653–668.
Harley, J. L. and Waid, J. S. (1955). *Trans. Br. mycol. Soc.* **38,** 104–118.
Harper, J. E. and Webster, J. (1964). *Trans. Br. mycol. Soc.* **47,** 511–530.
Hawker, L. E. (1950). "Physiology of Fungi." London University Press.
Heal, O. W. (1963). *In* "Soil Organisms." (J. Doeksen and J. van der Drift, eds), pp. 289–296. North-Holland, Amsterdam.
Hirst, J. M. (1959). *In* "Plant Pathology, Problems and Progress," pp. 1908–1958. University of Wisconsin Press, Madison.
Hora, F. B. (1959). *Trans. Br. mycol. Soc.* **42,** 1–14.
Hudson, H. J. (1968). *New Phytol.* **67,** 837–874.
Hughes, S. J. (1953). *Can. J. Bot.* **31,** 577–659.
Hurst, H. M. and Burges, N. A. (1967). *In* "Soil Biochemistry," Vol. 1. (A. D. McLaren and G. H. Peterson, eds), pp. 260–286. Marcel Dekker, New York.
Ingold, C. T. (1971). "Fungal Spores." Oxford University Press.
Jackson, R. M. (1958). *J. gen. Microbiol.* **18,** 248–258.
Johnson, L. F. and Curl, E. A. (1972). "Methods for Research on the Ecology of Soil-borne Plant Pathogens." Burgess, Minneapolis.
Kaufman, D. D. and Williams, L. E. (1963). *Phytopathology* **53,** 956–960.
Kelman, A. (1967). "Sourcebook of Laboratory Exercises in Plant Pathology." Freeman, San Francisco.
Knaphus, G. (1964). *Diss. Abstr.* **25,** 2699.
Kreger-van Rij, N. J. W. (1969). *In* "The Yeasts," Vol. 1. (A. H. Rose and J. S. Harrison, eds), pp. 5–78. Academic Press, London and New York.
Last, F. T. and Deighton, F. R. (1965). *Trans. Br. mycol. Soc.* **48,** 83–99.
Last, F. T. and Price, D. (1969). *In* "The Yeasts," Vol. 1. (A. H. Rose and J. S. Harrison, eds), pp. 183–218. Academic Press, London and New York.
Lindsey, B. I. (1973). Ph.D. Thesis, University of Nottingham.

LINGAPPA, B. T. and LOCKWOOD, J. L. (1961). *J. gen. Microbiol.* **26,** 473–485.
LINGAPPA, B. T. and LOCKWOOD, J. L. (1964). *J. gen. Microbiol.* **35,** 215–227.
LLOYD, A. B. and LOCKWOOD, J. L. (1966). *Phytopathology* **56,** 595–602.
MALONE, J. P. and MUSKETT, A. E. (1964). Seed-borne Fungi. *Proc. Int. Seed Testing Ass.* **29,** 179–384.
MASON, E. W. (1937). *Mycol. Paper* 3. Commonwealth Mycological Institute, Kew.
MCBRIDE, R. P. (1972). *Trans. Br. mycol. Soc.* **58,** 329–331.
MILLER, J. H. (1949). *Mycologia* **41,** 99–127.
MOORE, R. T. (1972). *Antonie van Leeuwenhoek* **38,** 567–584.
MORGAN-JONES, G., NAG RAJ, T. R. and KENDRICK, B. (1972). *Icones genera coelomycetarum.* University of Waterloo Biology Series.
NICHOLAS, D. J. D. (1965). *In* "The Fungi," Vol. 1. (G. C. Ainsworth and A. S. Sussman, eds), pp. 349–376. Academic Press, London and New York.
NICHOLAS, D. P., PARKINSON, D. and BURGES, N. A. (1965). *J. Soil Sci* .**16,** 258–269.
NICOT, J. (1960). *In* "Ecology of Soil Fungi." (D. Parkinson and J. S. Waid, eds), pp. 94–97. Liverpool University Press.
OLD, K. M. (1967). *Trans. Br. mycol. Soc.* **50,** 615–624.
OLTHOF, T. H. A. and ESTEY, R. H. (1963). *Nature, Lond.* **197,** 514–515.
PAGE, R. M. (1965). *In* "The Fungi," Vol. 1. (G. C. Ainsworth and A. S. Sussman, eds), pp. 559–574. Academic Press, London and New York.
PARKINSON, D. (1967). *In* "Soil Biology." (N. A. Burges and F. Raw, eds), pp. 449–478. Academic Press, London and New York.
PARKINSON, D., GRAY, T. R. G. and WILLIAMS, S. T. (1971). "Methods of Studying the Ecology of Soil Micro-organisms." Blackwell, Oxford.
PARKINSON, D., TAYLOR, G. S. and PEARSON, R. (1963). *Pl. Soil* **19,** 332–349.
PEBERDY, J. F. (1972). *Sci. Prog., Oxf.* **60,** 73–86.
PETERSON, E. A. (1959). *Can. J. Microbiol.* **5,** 579–582.
POMPER, S. (1965). *In* "The Fungi," Vol. 1. (G. S. Ainsworth and A. S. Sussman, eds), pp. 575–598. Academic Press, London and New York.
PUGH, G. J. F. (1960). *In* "The Ecology of Soil Fungi." (D. Parkinson and J. S. Waid, eds), pp. 202–208. Liverpool University Press.
PUGH, G. J. F. (1963). *In* "Soil Organisms." (J. Doeksen and J. van der Drift, eds), pp. 439–445. North-Holland, Amsterdam.
PUGH, G. J. F. (1969). *In* "The Soil Ecosystem." (J. G. Sheals, ed.), pp. 119–130. Systematics Association Publ. No. **8.**
PUGH, G. J. F. (1973). *In* "Ecology of Seeds." (W. Heydecker, ed.), pp. 337–345. Butterworths, London.
PUGH, G. J. F. and BUCKLEY, N. G. (1971*a*). *Trans. Br. mycol. Soc.* **57,** 227–231.
PUGH, G. J. F. and BUCKLEY, N. G. (1971*b*). *In* "Ecology of Leaf Surface Micro-organisms." (T. F. Preece and C. H. Dickinson, eds), pp. 431–445. Academic Press, London and New York.
PUGH, G. J. F., BUCKLEY, N. G. and MULDER, J. L. (1972). *Symp. biol. Hung.* **11,** 329–333.
PUGH, G. J. F. and DICKINSON, C. H. (1965). *Trans. Br. mycol. Soc.* **48,** 381–390.

PUGH, G. J. F. and MULDER, J. L. (1971). *Trans. Br. mycol. Soc.* **57**, 273–282.
PUGH, G. J. F. and WILLIAMS, G. M. (1968). *Trans. Br. mycol. Soc.* **51**, 389–396.
REINKING, O. A. and MANNS, M. M. (1933). *Zentbl. Bakt. ParasitKde Abt. II*, **6**, 23–75.
SACCARDO, P. A. (1886). "Sylloge Fungorum" **4**, Pavia.
SIU, R. G. H. (1951). "Microbial Decomposition of Cellulose." Reinhold, New York.
SMIT, J. and WEIRINGA, K. T. (1953). *Nature, Lond.* **171**, 794–795.
STENTON, H. (1958). *Trans. Br. mycol. Soc.* **41**, 74–80.
SUSSMAN, A. S. and HALVORSON, H. O. (1966). "Spores, Their Dormancy and Germination." Harper and Row, New York.
TAYLOR, G. S. and PARKINSON, D. (1961). *Pl. Soil* **15**, 261–267.
TOGASHI, K. (1949). "Biological Characters of Plant Pathogens: Temperature Relations." Meikundo, Tokyo.
TOWNSEND, B. B. (1954). *Trans. Br. mycol. Soc.* **37**, 222–233.
TOWNSEND, B. B. and WILLETTS, H. J. (1954). *Trans. Br. mycol. Soc.* **37**, 213–221.
TRIBE, H. T. (1957). *In* "Microbial Ecology." *7th Symp. Soc. gen. Microbiol.* 287–298.
TUBAKI, K. (1958). *J. Hattori Bot. Lab.* **20**, 142–244.
WAKSMAN, S. A. (1916). *Soil Sci.* **2**, 102–155.
WARCUP, J. H. (1957). *Trans. Br. mycol. Soc.* **40**, 237–260.
WARCUP, J. H. (1959). *Trans. Br. mycol. Soc.* **42**, 45–52.
WARCUP, J. H. (1960). *In* "The Ecology of Soil Fungi." (D. Parkinson and J. S. Waid, eds), pp. 3–21. Liverpool University Press.
WARCUP, J. H. (1967). *In* "Soil Biology." (N. A. Burges and F. Raw, eds), pp. 51–110. Academic Press, London and New York.
WARCUP, J. H. and TALBOT, P. H. B. (1962). *Trans. Br. mycol. Soc.* **45**, 495–518.
WEBLEY, D. M. and DUFF, R. B. (1962). *Nature, Lond.* **194**, 364–365.
WEBLEY, D. M. and JONES, D. (1971). *In* "Soil Biochemistry," Vol. 2. (A. D. McLaren and J. J. Skujins, eds), pp. 446–485. Marcel Dekker, New York.
WEBSTER, J. (1970). "Introduction to Fungi." Cambridge University Press.
WILLIAMS, S. T., PARKINSON, D. and BURGES, N. A. (1965). *Pl. Soil* **22**, 167–186.

11

Aquatic Fungi: Freshwater and Marine

E. B. G. Jones

Department of Biological Sciences
Portsmouth Polytechnic
England

I. Introduction

The sea covers nearly three-quarters of the earth's surface and the amount of water in lakes, rivers and streams is not insignificant. These are environments often rich in organic matter, such as decaying leaves, twigs and particulate matter, and run-off from land, frequently high in nitrogen and phosphates. Addition of domestic and agricultural sewage and other uncolonized materials to rivers and sea water are common. Knowledge of the turn-over and productivity of aquatic habitats, especially lotic waters, is in its infancy and we know little of the role of fungi in this situation. Sparrow (1968) recently stated that "the ecology of fresh-water fungi has not attained the degree of prominence and sophistication reached for example by the ecology of soil fungi". These comments could also apply to the marine fungi.

II. Taxonomic Groups Found in Aquatic Environments

A. Marine

The term "marine" will be interpreted in a broad sense and a marine fungus defined as one capable of producing successive generations by sexual and asexual means in natural oceanic waters and oceans diluted by freshwater or on intertidal substrates. The test of whether an organism is marine depends on the frequency of recovery from that habitat. If it is isolated predominantly or frequently from that habitat then it can be argued it has a role in the ecology of that habitat.

Over 300 fungi have been described as marine while Roth *et al.* (1964) and Steele (1967) have isolated a number of saprobic fungi in coastal and pelagic waters of the Atlantic and Pacific. Most major groups of the fungi are represented in the sea and new genera and species are constantly added to the list.

1. *Basidiomycetes*

Two higher Basidiomycetes have been repeatedly collected (Doguet, 1962, 1967; Byrne and Jones, 1972), and an increasing number of Heterobasidiomycetous yeasts have been isolated from sea water by Fell and his co-workers (Fell, 1967, 1970; Fell *et al.*, 1969).

Digitatitspora marina (Corticiaceae) and *Nia vibrissa* (Melanogastraceae) have appendaged basidiospores, lack sterigmata and are passively released into the surrounding water. In *N. vibrissa* the basidiospore is attached to the basidium by a peg-like structure or pedicel.

Heterobasidiomycetous yeasts known to occur in sea water have been assigned to the Ustilaginaceae as they form large terminal or intercalary teliospores. These germinate to produce a 1–4-celled promycelium with lateral and terminal sporidia. Some species conjugate to produce a dicaryotic mycelium with clamp connections, e.g. *Leucosporidium scottii* (Fell *et al.*, 1969), while others produce teliospores without apparently mating and lack clamp connections, e.g. *Leucosporidium antarcticum* (Fell *et al.*, 1969).

2. Ascomycetes

Kohlmeyer and Kohlmeyer (1971*b*) recorded 135 Ascomycetes while new species have recently been described by Schmidt (1969), Jones and Le Campion-Alsumard (1970), Kohlmeyer and Kohlmeyer (1971*a*) and Schaumann (1972). Most of these are lignicolous, although a number have been found growing on algae, mangroves, shells of molluscs, the chitinous tubes of hydrozoa and keratin tubes of annelids. All the major divisions of the Ascomycetes, with the exception of the Discomycetes, have marine representatives.

(a) *Euascomycetidae*. The Plectomycetes have two marine representatives, *Amylocarpus encephaloides* and *Eiona tunicata* (Kohlmeyer and Kohlmeyer, 1971*b*). Both have cleistothecia, deliquescing asci and ascospores that are appendaged.

Many Pyrenomycetes are found in the sea and species have been assigned to the following families: Hypocreaceae, Erysiphaceae, Sphaeriaceae, Polystigmataceae and the Halosphaeriaceae. The majority (47) belong in the latter family, originally established by Müller and von Arx (1962) and validated and emended by Kohlmeyer (1972*c*); the spores of some recently described species are illustrated in Fig. 1. All members of this family are marine, have deliquescing asci and most of the species have appendaged ascospores.

Four genera of the Hypocreaceae have marine representatives, *Halonectria milfordensis*, *Hydronectria tethys*, *Trailia ascophylli* and *Orcadia ascophylli*. The latter has operculate asci (Kohlmeyer, 1972*b*) and its position within the Hypocreaceae is questionable.

Schmidt (1969) has recently described *Crinigera maritima* (Erysiphales) from decaying *Fucus vesiculosus* and wood. The cleistothecium is thick-walled and has a number of bifurcating appendages. The asci are clavate, appear thin-walled and deliquesce at maturity to release hyaline bicelled ascospores with subterminal appendages (see Fig. 1).

The Loculoascomycetes are well represented in intertidal habitats (Kohlmeyer, 1969*b*) and a number have been collected on substrates continuously submerged in the sea (Jones, 1968; Byrne and Jones, 1972). The majority are members of the Pleosporaceae and belong to well known terrestrial genera: *Pleospora*, *Leptosphaeria*, *Sphaerulina* and *Didymella*. Other genera including *Paraliomyces*, *Kymadiscus*, *Manglicola* and *Pontoreia*, appear to have only marine species.

(b) *Hemiascomycetidae*. The last decade has seen the publication of an ever-increasing number of papers on the biology and ecology of yeasts in aquatic environments (Meyers *et al.*, 1967; Ahearn *et al.*, 1968*b*; van Uden and Fell, 1968). Some 50 marine-occurring yeasts have been reported from oceanic waters. Many of these have been shown to have a

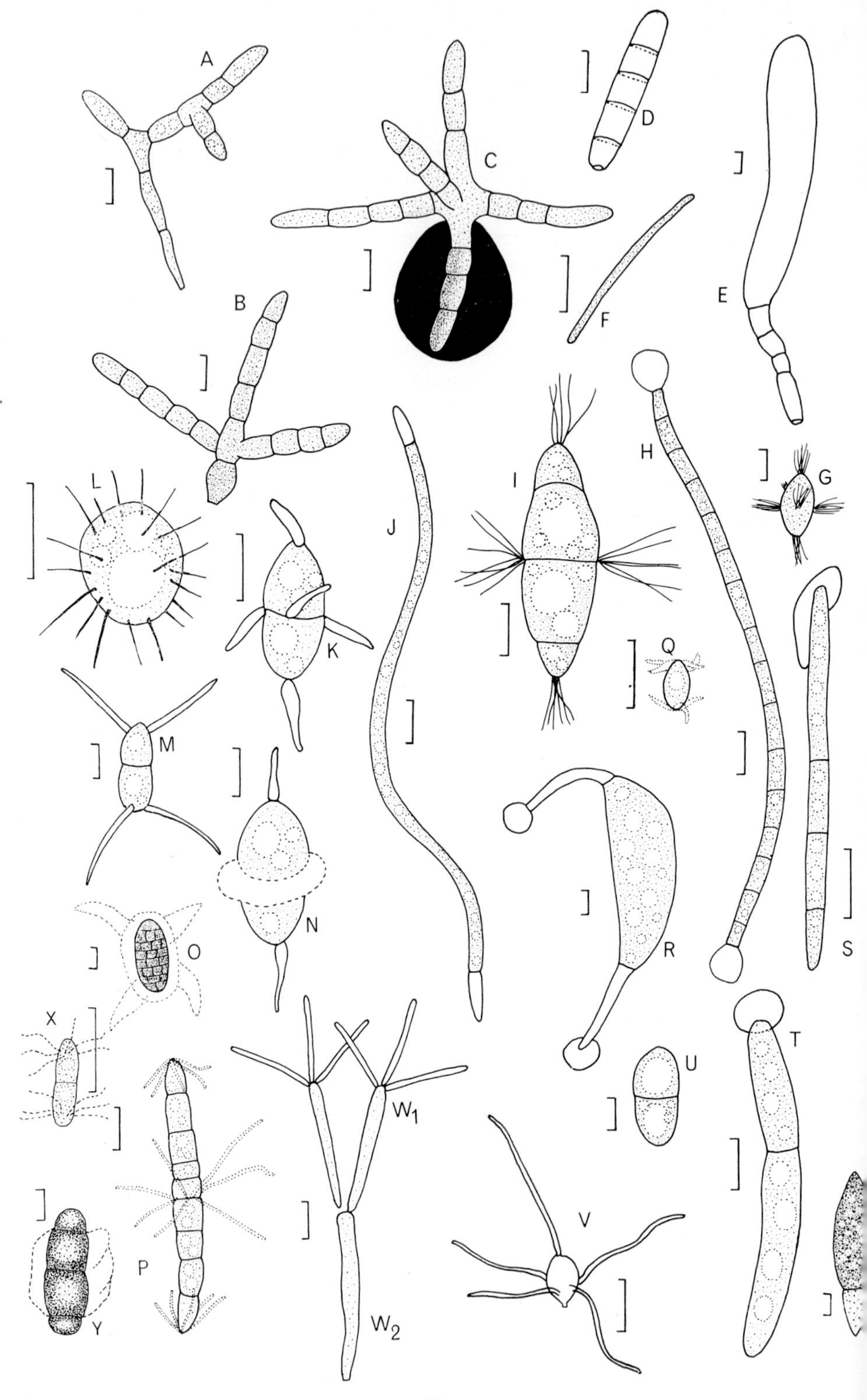

Fig. 1.

heterobasidiomycetous stage in their life history and these have already been discussed above.

The existence of "marine yeasts" has been questioned. Most are known from terrestrial and freshwater environments. However, these species have been found not only in coastal and estuarine waters but also in deep oceanic waters (Fell, 1967). Various workers (Ahearn *et al.*, 1968*b*; Fell, 1967) have shown that concentrations of yeasts decrease with the lowered organic content of water, increasing salinity, and remoteness from land.

Fell (1966) and Fell and Statzell (1971) have described two monotypic yeast-like genera from marine waters: *Sterigmatomyces* and *Sympodiomyces*. Sexual stages and blastospores have not been observed in either. *Sympodiomyces* differs from *Sterigmatomyces* in having a true branched mycelium, but this is somewhat restricted in development. Both produce conidia on short conidiophores or sterigmata and are therefore placed in the Fungi Imperfecti.

3. Fungi Imperfecti

Kohlmeyer and Kohlmeyer (1971*b*) list 51 Deuteromycetes and many of these are illustrated in the "Icones Fungorum Maris" (Kohlmeyer and Kohlmeyer, 1964–1969). The majority belong in well known terrestrial genera (*Alternaria*, *Camarosporium*, *Dendryphiella*, *Diplodia*, *Humicola*, *Monodictys*, *Stemphylium*, *Sporidesmium*), while others appear to be exclusively marine (*Asteromyces*, *Cirrenalia*, *Cremasteria*, *Orbimyces*, *Varicosporina*, *Zalerion*).

4. Phycomycetes

Sparrow (1960) listed 61 marine Phycomycetes while Johnson and Sparrow (1961) recognized 59 species with 32 doubtful species. Few new species have been described since 1961 with the exception of members of the Thraustochytriaceae. This family is of particular interest since it contains only marine forms; they are extremely common saprophytes,

---

Fig. 1. Spore morphology of some marine fungi. Fungi Imperfecti: A, *Varicosporina ramulosa*; B, *Clavariopsis bulbosa*; C, *Orbimyces spectabilis*; D, *Dendryphiella salina*; E, *Sporidesmium salinum*. Ascomycetes: F, *Halonectria milfordensis*; G, *Nautosphaeria cristaminuta*; H, *Lindra inflata*; I, *Corollospora cristata*; J, *Lulworthia purpurea*; K, *Ceriosporopsis calyptrata*; L, *Amylocarpus encephaloides*. M, *Halosphaeria quadri-cornuta*; N, *Halosphaeria torquata*; O, *Pleospora gaudefroyi*; P, *Corollospora pulchella*; Q, *Eiona tunicata*; R, *Corollospora tubulata*; S, *Haligena unicaudata*; T, *Halosphaeria cucullata*; U, *Nais inornata*. X, *Crinigera maritima*; Y, *Carbosphaerella leptosphaerioides*; Z, *Manglicola guatemalensis*. Basidiomycetes: V, *Nia vibrissa*; W, *Digitatispora marina* (1, basidiospore; 2, basidium). Scale by each spore equals 10 μm.

which combine a primitive "chytrid like" thallus with a saprolegniaceous mode of asexual reproduction. The genus *Thraustochytrium* was established by Sparrow in 1936 when he described *T. proliferum* from fronds of the alga *Bryopsis*. Nearly 20 years later Kobayashi and Ookybo (1953) described a non-proliferating form, *Thraustochytrium globosum* and a new genus *Japonochytrium* and more recently 3 new genera and 14 new species have been found (Table I, Figs 2 and 3).

TABLE I. The genera and species of the order Thraustochytriales

Fungus	Authority	Type
Thraustochytrium proliferum	Sparrow (1936)	Proliferous
T. globosum	Kobayashi & Ookubo (1953)	Non-proliferous
T. pachyderma	Scholz (1958)	Non-proliferous
T. roseum	Goldstein (1963*a*)	Non-proliferous
T. motivum	Goldstein (1963*b*)	Proliferous
T. aureum	Goldstein (1963*c*)	Proliferous
T. aggregatum	Ulken (1965)	Non-proliferous
T. visurgense	Ulken (1965)	Non-proliferous
T. striatum	Schneider (1967)	Non-proliferous
T. kineii	Gaertner (1967*b*)	Proliferous
T. coniccum	(in press)	
Japonochytrium marinum	Kobayashi & Ookubo (1953)	Non-proliferous
Schizochytrium aggregatum	Goldstein & Belsky (1964)	Non-proliferous
Aplanochytrium kerguelensis	Bahnweg & Sparrow (1972)	Non-proliferous
Althornia crouchii	Jones & Alderman (1971)	Non-proliferous

Many of the morphological characters used in the identification of species of *Thraustochytrium* have been described from cultures grown on pollen grains. The work of Booth and Miller (1968) has shown that many of these characters, such as sporangial size, zoospore behaviour prior to release, pigmentation and shape and size of zoospores are often substrate dependent and as such are not useful characters upon which to base an identification. They suggest there are two complexes within the family. These are the proliferous *Thraustochytrium motivum–aureum* complex and the non-proliferous *T. roseum–visurgense* complex.

Members of the Labyrinthulales are often isolated from decaying marine organisms and these are currently receiving much attention (Perkins, 1972; Porter, 1969). Gaertner (1968*b*), Goldstein and Moriber (1966) and Poyton (1970) have isolated *Dermocystidium*-like organisms from sediments and seawater samples. Poyton (1970) has placed these in a new genus *Hyalochlorella* and regards the organism as a colourless counterpart of *Chlorella* (see Fig. 3d).

B. Brackish Water

A universally acceptable definition of a brackish water or estuarine fungus is lacking but Höhnk (1956) regarded them as species able to grow

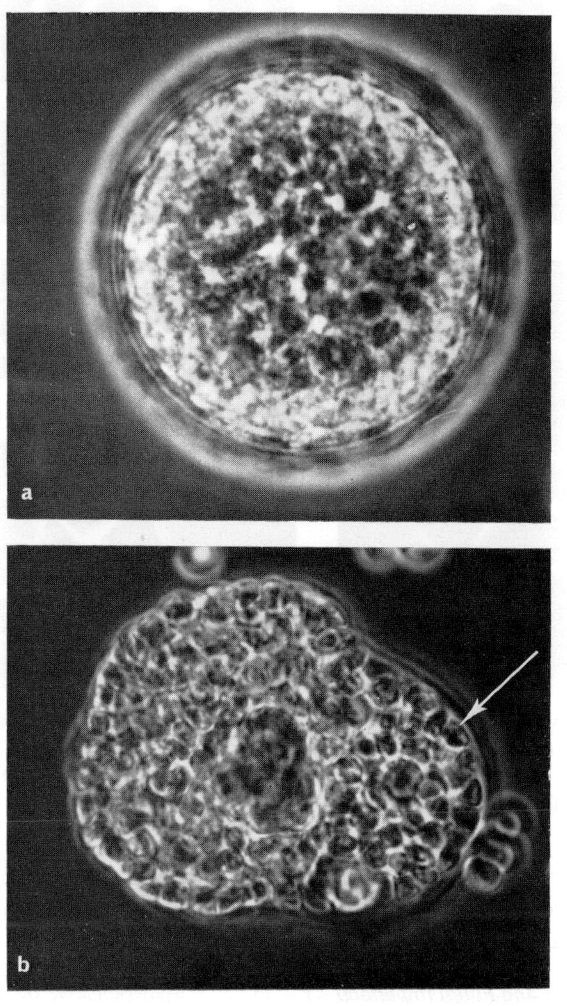

Fig. 2a. *Althornia crouchii*. Mature sporangium showing multilamellate wall and coarse cytoplasm (phase contrast) ×300. b, *Thraustochytrium striatum*. Zoospore release. Wall of sporangium indicated (phase contrast) ×600.

2. Ascomycetes

(a) *Euascomycetidae*. Ingold (1951, 1954, 1955), Ingold and Chapman (1952), and Weston (1929) have described a number of Pyrenomycetes and Discomycetes from decaying leaves and decorticated twigs submerged in lakes and rivers. Detailed studies of the colonization of timber and decorticated twigs in freshwater have been carried out by Jones and Eaton (1969); Eaton and Jones (1971*a*, *b*, *c*); Byrne (1971); Shearer (1971) and Archer and Willoughby (1973).

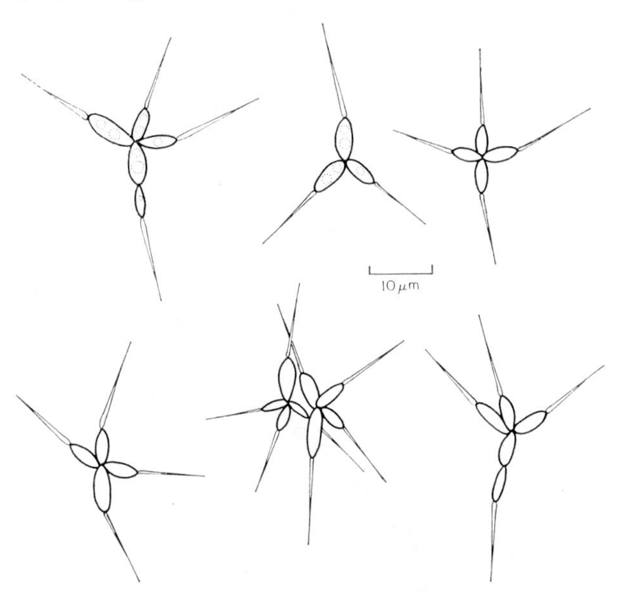

Fig. 5. *Candida aquatica*. Tetra-radiate groups of cells as collected in surface foam or scum samples at the edge of Malham Tarn.

(b) *Hemiascomycetidae*. A number of yeasts which are mainly terrestrial have been reported from freshwater (Hedrick and Soyugenc, 1967*a*, *b*; Hedrick *et al.*, 1968; Meyers *et al.*, 1970*a*). Jones and Slooff (1966) described *Candida aquatica* from water scums collected at the edge of Malham Tarn, Yorkshire. This species has tetraradiate groups of cells and is illustrated in Fig. 5.

3. Fungi Imperfecti

Nilsson (1964) listed 66 species of freshwater Hyphomycetes, growing predominantly on decaying submerged leaves, and new species have recently been described by Alasoadura (1968*a*, *c*), Ingold *et al.* (1968), Iqbal (1971, 1972) and Jones and Stewart (1972).

Sphaeropsidaceous fungi are also found growing on decaying leaves of various freshwater grasses (Cunnell, 1956, 1957, 1958). Eaton (1972), Eaton and Jones (1971*a*, *b*) and Archer and Willoughby (1973) have reported a number of dematiaceous and soil fungi growing on timber or decorticated twigs submerged in fresh water.

4. Phycomycetes

Phycomycetes are widespread and extremely common in freshwater habitats occurring as saprophytes on materials such as exuviae of insects, seeds, fruits, pollen, leaves, twigs, decorticated timber and algae. Over 300 members of all the major families of Phycomycetes are to be found in fresh water (Sparrow, 1960).

Over the past decade many new species have been described (Miller, 1962; Emerson and Weston, 1967; Elliott, 1967, 1968; Karling, 1968, 1969; Barr, 1969*a*, *b*, 1970*a–c*; Booth, 1969; Umphlett and Koch, 1969; Ulken, 1972). Distinct advances in the study of the lower fungi were photographic documentation, and the isolation and growth of so many under laboratory conditions. These have allowed the redescription of a number of taxa, and work on their physiology and ultrastructure (Fuller *et al.*, 1966; Emerson and Whisler, 1968; Emerson and Held, 1969; Kazama, 1972).

D. Polluted Waters

Fungi in polluted water and sewage were first seriously investigated by Cooke (1954, 1957, 1958) and over the last two decades he has compiled a list of over 300 fungi from such waters. He has proposed four terms to describe the distribution of fungi in relation to polluted water: lymaphobes—those not found in polluted habitats; lymaxenes—those present in few isolations or represented by only one or a few colonies; lymaphiles—those present commonly or abundantly (more than 10,000 colonies per g dry weight of sample) and lymabionts—species apparently restricted to sewage treatment plants or associated only with habitats in which faecal material appears to be the principal substrate (Cooke, 1957). Some species considered as lymaphiles are common soil fungi while examples of lymabionts are *Ascophanus carneus* and *Subbaromyces splendens*. A laboratory guide to fungi in polluted waters, sewage and sewage treatment systems has been prepared by Cooke (1963).

E. Soil Fungi

Cooke (1961, 1963) divided the fungi encountered in fresh water into two principal groups: the hydrofungi which require the presence of water to complete their life cycle and are regarded as obligate aquatic fungi, and the geofungi or typical soil fungi, who are not specifically adapted to an aquatic existence, but nevertheless they may be found in water because of an adequate supply of nutrients. These, Cooke regarded as facultative aquatic fungi or lymaxenes. Park (1972*b*) has also considered the categories of micro-organisms found in fresh water.

Terrestrial fungi have frequently been isolated from materials submerged in the sea or in estuarine habitats, but their significance is not fully understood. Are they to be considered marine or facultatively marine? (the lymaxenes of Cooke—not tolerant of such a habitat but only passing through the system). A similar problem has been found with fungi colonizing timber in cooling towers (Jones, 1972). Most of the species found growing and fruiting on the wood were not encountered when plating techniques were used. Fungi grew on the media that were not observed on the test panels, and these were sporadic in their frequency of appearance. Are these fungi active in the degradation of wood or merely present as resting spores which germinate when plated on to nutrient media?

Park (1972a) has used plating and baiting techniques to characterize fungi present in organic detritus in freshwater. Baiting techniques yielded a number of fungi not recorded by the plating method but a feature of all such work is the high number of Fungi Imperfecti isolated and the low number of Ascomycetes recorded. However, this work emphasizes the need for the use of a variety of techniques and an increase in the number of baits.

A detailed study of the role of these fungi in aquatic habitats is necessary. Particular attention should be paid to the state the fungi are in before they are isolated. Are they present as mycelium and therefore actively involved in the breakdown of materials or are they present only as spores?

III. Spore Liberation, Dispersal and Settlement

The study of dispersal of spores in aquatic fungi is in its infancy and Ingold (1971) attributes this to the lack of economic incentive to study the problem. A number of stages lead up to the settlement of spores, namely: formation, release, dispersal and settlement leading finally to penetration of the substrate. Aquatic fungi often have to accomplish these processes in turbulent and fast-moving waters.

The following types of spores are to be found amongst the aquatic fungi: (1) motile spores, e.g. Phycomycetes; (2) branched spores, e.g. aquatic Hyphomycetes and Basidiomycetes; (3) appendaged (often mucilaginous) spores, e.g. marine Ascomycetes, a few freshwater Ascomycetes and some freshwater sphaeropsidaceous fungi; and (4) non-appendaged spores.

1. Motile Spores

In recent years a number of excellent films have been prepared to show spore cleavage, release and motility in the Phycomycetes. Gay and Greenwood (1966) have followed the process of spore cleavage and release in *Saprolegnia*. They reported that at an early stage in development there is a single central vacuole present in the zoosporangium. When cleavage is initiated, furrows are pushed outwards from the tonoplast of the vacuole and join up with the plasmalemma of the sporangium. Thus the limiting membrane of each zoospore is made of a portion of the plasmalemma and a portion of the tonoplast.

Harrison (1972) has shown in *Thraustochytrium kinnei* there is no central vacuole. Cleavage is effected by membranes (formed in vesicles) which originate in the mid-central region of the sporangium, push outwards and join up with the plasmalemma (see Fig. 6).

Following cleavage in *Saprolegnia*, there is a 10% loss in volume of the zoosporangium and zoospores begin to move. A little later the tip of the apical papilla of the zoosporangium breaks down and zoospores escape. Release is quick as if the spores are squeezed out of the sporangium.

Two types of release mechanisms are to be found in the thraustochytriums. In the proliferous types, there is only a short interval between cleavage and spore release. Zoospores swim about inside the sporangium and the wall separates to allow their rapid release. In the non-proliferous forms, the spores take a long time ($\frac{1}{2}$–1 h) before they escape singly from the sporangium, as a result of their own motility.

This also occurs in the chytrids (Bremer, 1972) while in *Pythium* the protoplasm of the sporangium escapes in a mass contained in a membrane. Zoospore delimitation occurs in this vesicle, the spores start to swim, finally rupture the vesicle and escape (Webster and Dennis, 1967).

Once spores have been released, they are dispersed partly by their own motility and in part by water currents. The length of time they remain motile varies from a few hours to a few days (*Althornia crouchii*) or longer.

Large numbers of zoospores are produced. Gaertner (1968*a*) found that the number of thraustochytriums in sea water off Iceland was low (1·3–12 infective units per litre) but that fine sediments in the German Bay were rich in numbers, varying from 230–58,700 units per litre (Gaertner, 1968*b*). Willoughby (1962) showed that the number of saprolegniaceous propagules in Lake Windermere varied from less than 100 per litre during the spring and summer to almost 5000 propagules per litre at the margin of the lake in autumn.

Zoospores frequently exhibit chemotaxis. *Saprolegnia* zoospores are attracted to natural baits like insect legs (Fischer and Werner, 1958) while casein hydrolysate attracts zoospores of *Allomyces* (Machlis, 1968).

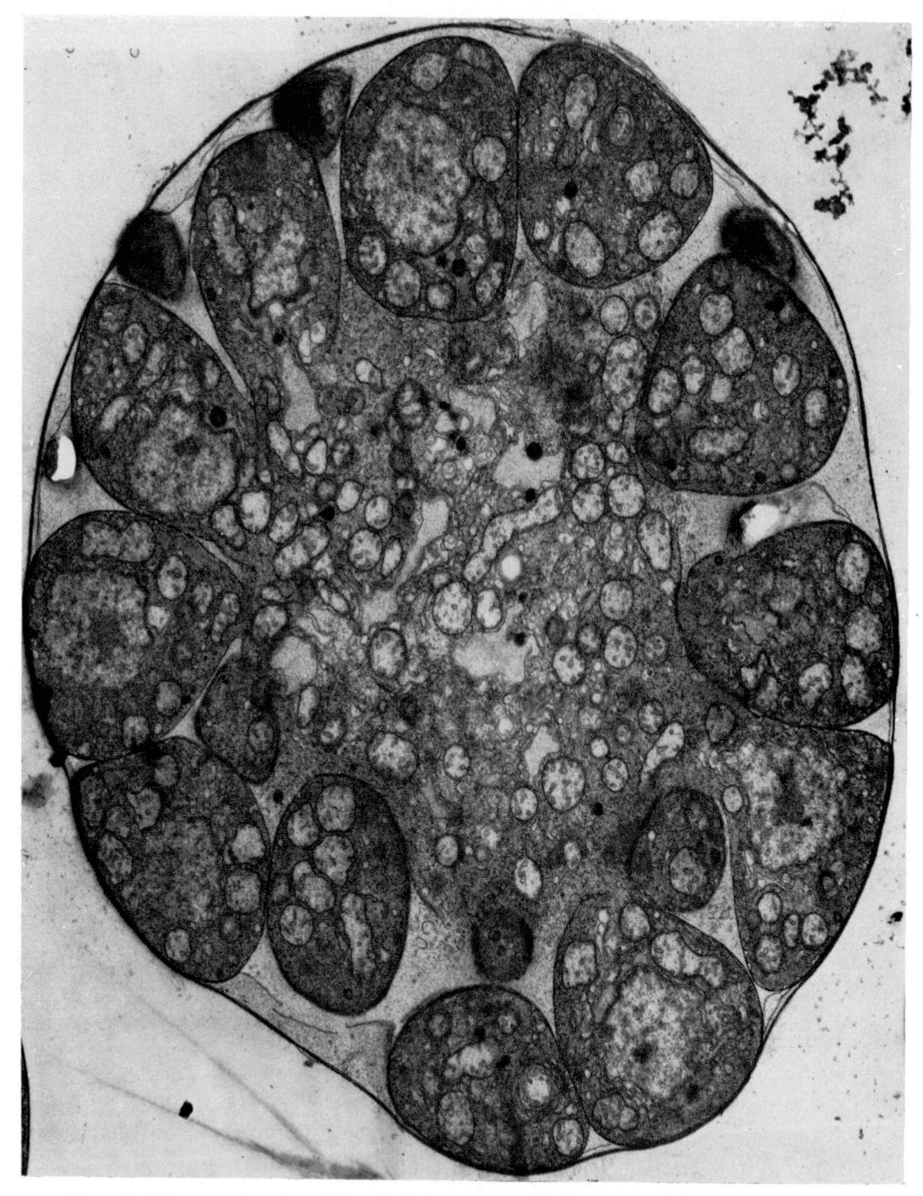

Fig. 6. *Thraustochytrium kinnei*. A late stage in cleavage of the zoosporangium. Some zoospores still retain a connection with the cytoplasm of the parent sporangium. (Electron micrograph, 5647) $\times$ 19,000.

Thraustochytriums appear to closely examine the substrate before settling (Gaertner, A., personal communication) and a similar response has been demonstrated for *Rhizophydium* (Bremer, G., personal communication).

2. *Branched Spores*

Branched or tetraradiate spores are found in a number of genera and a few are illustrated in Fig. 1. They are predominantly aleuriospores or phialospores, which are released passively.

It is interesting that in the three aquatic higher Basidiomycetes so far described all lack sterigmata and have appendaged or branched basidiospores that are released passively (Fig. 1).

Both Nilsson (1964) and Iqbal (unpublished) have presented information on the concentration of spores of freshwater Hyphomycetes in streams and

TABLE III. Concentration of Hyphomycete spores in river water

(1)	Concentration at three different sampling points		
Alatospora acuminata	220	330	270
Anguillospora longissima	1000	1870	1300
Tetracladium marchalianum	4600	4080	4800
Total number spores per ml	9120	9980	9090

(2)	Sept.	Oct.	Nov.	Jan.	May
Alatospora acuminata	60	320	750	600	200
Anguillospora longissima	260	340	750	1500	200
Flagellospora curvula	50	720	1000	2300	400
Tetracladium marchalianum	2800	5100	14100	8200	2500
T. maxilliformis	10	10	20	10	30
Total number spores per ml	3460	9490	29270	22320	4510

(3)	Oligotrophic streams		Eutrophic stream
	A	B	C
Alatospora acuminata	40	2440	360
Anguillospora longissima	240	120	160
Tetracladium marchalianum	—	80	1240
Total number spores per ml	420	4380	10360

Data from Nilsson (1964).

rivers. Nilsson collected foam samples and estimated the number of spores per ml of foam water (Table III). A number of conclusions were made: (1) the number of spores present at three different sampling points in the river at Ulva were fairly constant; (2) there was a seasonal fluctuation in the number of spores in the river with the highest number occurring in November; (3) comparison of oligotrophic streams (A–B) with a eutrophic stream (C) did not indicate a uniform preference for either.

Iqbal (unpublished) used a millipore filter to estimate the number of spores in streams and rivers. Counts of 1000–1500 spores per litre were obtained in October rising to 7000–8300 per litre in November in the River Creedy, Devon. While spore numbers were high in the autumn/winter, the summer spora was very poor.

Ingold (1966, 1971) has pointed out the advantages of the tetraradiate spore in an aquatic habitat. Undoubtedly, the arms help to keep the spore suspended in the water and help in carrying the spore along in water currents. It has been suggested (Webster, 1959) that the tetraradiate or branched nature of the spore helps to impact them on to suitable substrates submerged in the water. Little experimental work has been carried out to test these hypotheses. Webster (1959) showed that the tetraradiate spores of *Clavariopsis aquatica* were more readily impacted on to a surface than those of *Flagellospora curvula* (long curved spore) or *Dactylella aquatica* (oval spores).

3. Appendaged Spores

Some 60% of the marine Ascomycetes and many of the freshwater Ascomycetes have ascospores with mucilaginous or gelatinous appendages (see Fig. 1). The appendages of *Ceriosporopsis halima* and *Torpedospora radiata* contain a skeletal system of parallel fibrils which in the latter species appear to be continuous with the epispore (Lutley and Wilson, 1972*a*, *b*). The conidia of many freshwater sphaeropsidaceous fungi also have gelatinous appendages (Cunnell, 1958).

It has been suggested that these appendages (a) keep spores afloat by offering increased resistance to settlement; (b) entangle and attach the spores to suitable substrates; and (c) catch unorganized eddy diffusion currents. However, no quantitative experimental work has been carried out to test these ideas, or to determine the time it takes a spore to settle and attach itself to a substrate or how effective the adhesion is.

If spores are to colonize their substrate successfully, they must be capable of rapid germination and penetration. Spores have been reported as germinating within 24 h of seeding agar plates but detailed information is again lacking. Byrne (1971) has shown that temperature affects the germination of marine fungi (Table IV). Most of the fungi tested had a high

TABLE IV. Spore germination of some marine fungi at different temperatures

	°C Temperature			
	10 h (%)	15 h (%)	20 h (%)	25 h (%)
Torpedospora radiata	NG	42 (40)	18 (28)	18 (42)
Corollospora maritima	42 (100)	24 (100)	18 (85)	18 (98)
Zalerion maritimum (Linder) Anastasiou	48 (60)	24 (60)	18 (88)	18 (95)
Asteromyces cruciatus (F. & Mme Moreau) ex Hennbert	48 (55)	42 (60)	18 (70)	18 (75)
Dendryphiella salina (Suth.) Pugh et Nicot	48 (80)	24 (88)	18 (86)	18 (78)

NG = No germination after 72 h.
% germination in parentheses.
Data from Byrne (1971).

percentage germination, with the exception of *Torpedospora radiata*. These high germination rates may explain why these fungi are so successful in colonizing freshly submerged timber in the sea.

There are no figures available for the numbers of spores of marine fungi present in sea water. Millipore filtration techniques have proved disappointing as large volumes have to be filtered and even then the counts are low. This is not surprising considering the size of the oceans and the limited substrates available for colonization. However, timber placed in the sea is soon colonized by a variety of species, even at depths of 437 m (Jones and Le Campion-Alsumard, 1970).

4. Non-appendaged Spores

Many of the fungi encountered in aquatic habitats do not have branched or appendaged spores. Nothing is known of how these spores settle in fast moving waters.

Non-appendaged conidia are produced as phialospores, porospores, aleuriospores and sporangiospores. Presumably in an aquatic habitat all are released passively into the surrounding water.

Table V indicates the number of fungi present in polluted and pollution-free water samples taken from the Cache Valley, Idaho and Utah, and Logan river, Idaho. The number of fungi in spring water was low (8 per 2 l) and high where there was obvious pollution (270,000 per ml).

Fell (1967) reported 1–513 yeasts cells per litre from samples taken in the Indian Ocean but in Biscayne Bay (Fell *et al.*, 1960), the counts ranged

TABLE V. Number of fungal colonies recovered from habitat types by plate count methods. Autumn Survey (after Cooke, 1967).

Code	habitat type	Number of samples	Total species	Estimated average number of colonies per volume of sample
Cache Valley				
A	Above sources of pollution	3	44	38,000 g^{-1}
O	Obvious sources of pollution	10	44	270,000 ml^{-1}
B	Below sources of pollution	2	35	220,000 g^{-1}
V	Valley streams after absorbing pollution	12	77	209,000 g^{-1}
S	Soils (mostly agricultural, irrigated or not, sometimes with polluted water)	5	50	402,000 g^{-1}
Logan Valley				
AP	Animal pollution (pasture or grazing)	3	37	200,000 g^{-1}
RB	River bank (no domestic livestock)	3	19	34,000 g^{-1}
PT	Path and toilet area (dry)	3	23	18,000 g^{-1}
S	Spring water (Rick's Spring)	1	3	8 2/L
U	Unpolluted (apparently) by man or livestock	4	32	250,000 g^{-1}

from 1 to 5000 cells per litre. Meyers *et al.* (1970*a*) reported very high yeast populations in marshland waters of estuarine regions of Barataria Bay (Table VI). The average yeast populations in the open sea appear to be of the order 30–60 cells per litre. Various workers have observed seasonal

TABLE VI. Number of yeasts in sea water

Author	Test site	Numbers
Meyers *et al.*, 1970*b*	Lake Champlain (freshwater)	10–4100 cells l^{-1}. Average 50 cells l^{-1}.
Ahearn *et al.*, 1968*b*	South Florida	< 10–100–> 5000 cells l^{-1}.
Fell, 1967	Indian Ocean	0–513 cells l^{-1}.
Meyers *et al.*, 1970*a*	Barataria Bay, Louisiana	200–20,000 cells l^{-1}. Average 2880 cells l^{-1}.
Meyers *et al.*, 1967	North Sea	< 10–> 3000 cells l^{-1}.

variation in yeast populations. Meyers *et al.* (1970*b*) showed that large concentrations occur in July and October while in the North Sea (Meyers *et al.*, 1967) high numbers were recorded in July and September.

The salinity of the water also has an effect on yeast densities. Ahearn *et al.* (1968*b*) showed that the highest densities occur in fresh water, the number decreasing with lowered organic content, increasing salinity and distance from land.

Spore liberation in aquatic Ascomycetes (appendaged and non-appendaged spores will be considered together) depends on the unitunicate or bitunicate structure of the ascus.

Unitunicate asci and their ascospores are released from the ascocarps in a variety of ways. One method involves the active discharge of spores from asci. The ascus becomes turgid as the result of increased osmotic pressure, the wall is stretched and this leads to the rupture of the ascus in a definite way with the explosive release of the ascospores. Perhaps it is not surprising that few marine Ascomycetes (e.g. *Chaetosphaeria chaetosa*) release their spores by this method. Presumably the internal osmotic pressure of the ascus would have to be very high to absorb water from the surrounding sea water. Also pressure may affect the success of this method especially in deep sea water. Kohlmeyer (1972*b*) has shown that in the marine Pyrenomycete *Orcadia ascophylli*, the ascus has an operculum although spore discharge was not observed.

Ingold (1954) gave an account of eight Discomycetes collected on aquatic plants in the English Lake District. All these discharged their ascospores violently as in terrestrial species but the distance of discharge was small due to the viscosity of the water. Ingold (1968) showed that in *Loramyces juncicola*, the ascus dehisces by a slit to form a wide mouth, and the spores are discharged over a period of several minutes. On bursting, the stretched turgid ascus contracts to liberate the spores one by one. The ascus contains mucilage and the turgidity of the ascus is attributed to the swelling of hydrophilic mucilage rather than to osmotic intake of water.

A second mechanism involves the passive discharge of asci from ascocarps and the subsequent active release of the spores. This method is rarely found in the marine Ascomycetes but has been reported for *Gnomonia marina* (Cribb and Cribb, 1956). Ingold (1951) has shown a similar mechanism in the freshwater Pyrenomycetes *Ceriospora caudae-suis* and *Ophiobolus typhae*. In the latter, asci pass up the neck canal one at a time, emerge from the ostiole until the spore-containing part is outside the perithecium and then burst along a transverse line of weakness near the middle of the ascus releasing the eight filiform spores. In the Pyrenomycete *Sillia ferruginea*, which is frequently encountered on timber in water cooling towers (Eaton, 1972), the asci are discharged rapidly in a mass of mucilage, a few asci discharging their spores as they leave the ostiole. The majority of asci discharge their spores considerably later through a tear developing towards the base of the ascus. This is despite the fact that the asci have an apical pore (Fazzoni, K., unpublished).

The third mechanism is characteristic of many marine Ascomycetes, especially members of the Halosphaeriaceae. The ascus wall is extremely thin (Lutley and Wilson, 1972*a*, *b*) and is structurally quite different from the unitunicate asci of some terrestrial genera (Greenhalgh and Evans, 1967). The ascus wall breaks down and liberates the spores into the centrum. When mature, the tip of the neck opens, sea water may enter the ascocarp, and the mucilage around the spores or in the centrum swells. This probably helps to expel the spores up the neck and out of the ostiole into the surrounding water (Lutley and Wilson, 1972*a*; Wilson, 1954; Jones, 1962*b*, 1964). If the ascospores are appendaged, any mucilage around the spore expands and gets dissolved, and the appendages then stretch out (Fig. 1).

Amylocarpus encephaloides and *Eiona tunicata*, both cleistothecial forms, have deliquescing asci and appendaged ascospores. This suggests that the deliquescing ascus has been evolved in the marine Ascomycetes more than once. Ingold (1971) regards the non-explosive asci of *Genea*, *Chaetomium* and *Ceratocystis* as degenerate. In the sea, this seems an ideal way of releasing ascospores. All that is required is the release of the spores from ascocarps. An explosive mechanism is of limited use due to the greater viscosity of water as compared with that of air. It is therefore significant that no glycogen has been found in the epiplasm of maturing asci of *Ceriosporopsis halima* (Lutley and Wilson, 1972*a*).

The bitunicate ascus has an active method of spore discharge and is well illustrated in *Leptosphaeria discors* (Te Strake Wagner, 1965). The ascus wall consists of an outer rigid layer (ectoascus) and a very thin, inner extensile one (endoascus). Just before discharge the ectoascus ruptures apically, sometimes subapically (Johnson, 1956*a*), water is absorbed with the result that the endoascus elongates very rapidly up the neck of the pseudothecium, soon projects through the ostiole, and bursts, squirting its ascospores into the air or water. The effectiveness of this method must be very much reduced when the fungus is continuously submerged, due to the increased viscosity of water. The observations made above concerning the internal osmotic pressure of the unitunicate ascus and the effect of pressure with depth, also apply to the bitunicate asci. It is therefore interesting that in a recent study (Jones, 1968) 970 records of marine fungi were made on 266 test panels submerged for up to 120 weeks in the sea. Of these, only 12 were of bitunicate species (*Pleospora* sp. (2 records), *Leptosphaeria orae-maris* (9) and *Microthelia maritima* (1)).However, intertidal substrates, e.g. mangroves, *Spartina* culms, driftwood, are frequently rich in bitunicate fungi (Kohlmeyer and Kohlmeyer, 1971*b*). There are a number of bitunicate fungi growing on freshwater Angiosperms and Ingold (1955) has described the mechanism of spore release in *Pleospora scirpicola*, with each spore enveloped in a slug-like mucilaginous sheath.

IV. Ecology

A. Timber, Twigs and Other Cellulosic Materials

The role of marine fungi in the degradation of wood has been extensively investigated (see Jones and Eltringham, 1971), while the colonization of wood in fesh water has received little attention (Eaton and Jones, 1971*a*, *b*; Archer and Willoughby, 1973). Some 132 fungi (Ascomycetes, Fungi Imperfecti and Basidiomycetes) have been reported growing on wood submerged in the sea (Jones and Irvine, 1971). In Table VII, the frequency of occurrence of some British marine lignicolous fungi are presented. The most common species are *Zalerion maritimum*, *Ceriosporopsis halima* and *Lulworthia* spp. The observations of Poole and Price (1972) at Chatham, England, support these results, although *Lulworthia* spp. were surprisingly absent.

These appear to be commonly occurring species not only in the U.K. (Jones, 1972) but also elsewhere in the world, as the work of Jones (1971) and Jones *et al.* (1972) has shown. The latter conducted a test at 18 sites round the world and the most frequently recorded species were *Cirrenalia macrocephala* (present on 45% of the blocks), *Zalerion maritimum* (32%), *Lulworthia floridana* (25%), *L. purpurea* (16%), *Dendryphiella salina* (16%), *Ceriosporopsis halima* (15%) and *Humicola alopallonella* (13%). The frequency of occurrence of species in relation to the 18 testing sites revealed that *C. macrocephala* was found in 80% of the sites. *L. floridana* in 50%, *H. alopallonella* in 45%, *C. halima* in 40% and *D. salina* in 35%. Some of the fungi were only recorded once, e.g. *Trichocladium achraspora* (Lisbon), *Haligena viscidula* (Abidjan), *Halosphaeria appendiculata* (Langstone), *Halosphaeria maritima* (Lisbon) and *Orbimyces spectabilis* (Abidjan).

The effect of pulpmill effluents on the growth of *Zalerion maritimum* has been investigated by Churchland and McClaren (1972). They found that the caustic effluent did not stimulate growth unless nutrients were added and they suggested that the fungus was not an effective decomposer of the effluent.

Successional patterns of fungi on wood submerged in the sea have been reported by Meyers (1954), Meyers and Reynolds (1960), Jones (1963*b*, 1968) and Perkins (Chapter 22). Johnson (1967) maintained that marine fungi do not show a succession pattern, but that succession is an expression

TABLE VII. Frequency of marine fungi in British coastal waters

Fungi present	1958–60		1960–65		1969–71	
	Times present 60	%	Times present 266	%	Times present 98	%
COMMON						
Fungi Imperfecti						
Zalerion maritimum	44	73	198	74	66	65
Humicola alopallonella	—	—	48	18	17	18
Monodictys pelagica	—	—	38	14	32	33
Ascomycetes						
Lulworthia purpurea	32	53	165	62	75	76
L. rufa	—	—	55	21	—	—
L. floridana	—	—	81	31	—	—
Ceriosporopsis halima	26	43	117	44	35	36
OCCASIONAL						
Fungi Imperfecti						
Dictyosporium pelagica	—	—	15	6	9	9
Cirrenalia macrocephala	2	3	27	10	18	19
Dendryphiella salina	9	15	13	5	12	12
Sporidesmium salinum Jones	—	—	21	8	2	2
Ascomycetes						
Halosphaeria hamata	8	14	16	6	11	11
Lulworthia fucicola	—	—	10	4	—	—
Halonectria milfordensis	—	—	24	9	—	—
Ceriosporopsis cambrensis	28	47	25	9	—	—
Halosphaeria appendiculata	9	15	28	11	24	25
Ceriosporopsis circumvestita	—	—	30	11	2	2
INFREQUENT						
Fungi Imperfecti						
Trichocladium achraspora (Meyers & Moore) Dixon	—	—	2	0·8	5	5
Asteromyces cruciatus	—	—	1	0·4	2	2
Alternaria maritima Linder	7	12	1	0·4	3	3
Stemphylium maritimum Johnson	8	13	1	0·4	—	—
Ascomycetes						
Ceriosporopsis calyptrata	—	—	1	0·4	5	5
Halosphaeria torquata	—	—	2	0·8	—	—
Halosphaeria maritima	—	—	—	—	15	16
Halosphaeria mediosetigera	—	—	1	0·4	—	—
Nautosphaeria cristaminuta	—	—	6	2·3	4	4
Corollospora maritima	9	15	5	1·9	20	21

of different fruiting times, rather than replacement of one species by another. Jones (1963*b*) has shown that temperature affected the colonization of *Fagus* by *Ceriosporopsis halima*, *Lulworthia purpurea*, *L. floridana* and *L. rufa* (Table VIII). Perithecia did not develop on 6-week test blocks until the sea water temperature was about 14°C. Meyers and Reynolds

TABLE VIII. Colonization of *Fagus* test blocks submerged for 6 weeks at Langstone Harbour, Portsmouth
(Water temperature on submergence 13°C). (After Jones, 1963*b*)

Dates	Fungi on arrival	Fungi on incubation	Water temperature on withdrawal °C
21.1.58 to 2.12.58	None	*Zalerion maritimum*	8·5
2.12.58 to 13.1.59	None	None	5·2
13.1.59 to 24.2.59	None	*Z. maritimum*	6·5
24.2.59 to 7.4.59	None	*Z. maritimum*	9·8
7.4.59 to 19.5.59	None	*Z. maritimum*, *Stemphylium maritimum*, *Alternaria maritima*, *L. purpurea*, *L. floridana*	14·1
19.5.59 to 30.6.59	*Lulworthia purpurea*	*Z. maritimum*, *Dendryphiella salina*, *C. halima*	17·8
30.6.59 to 11.8.59	*Ceriosporopsis halima*, *L. purpurea*	*Z. maritimum*, *D. salina*, *S. maritimum*	20·8
11.8.59 to 22.9.59	*C. halima*	*Z. maritimum*, *A. maritima*, *L. purpurea*, *L. floridana*	17·3
22.9.59 to 3.11.59	None	*L. purpurea*, *Z. maritimum*	13·8
3.11.59 to 15.12.59	None	*Z. maritimum*	9·2
15.12.59 to 19.1.60	None	None	6·2
19.1.60 to 1.3.60	None	*Z. maritimum*	6·8

(1960*c*) have shown that *Lulworthia* species attacked test blocks at temperatures of 5°C and less, but required 100–200 days for fruiting.

Jones (1968) suggested that certain fungi show a preference for *Fagus* wood, e.g. *Halosphaeria appendiculata*, *Nautosphaeria cristaminuta* and *Halosphaeria hamata* while other fungi prefer *Pinus sylvestris Humicola alopallonella*, *Cirrenalia macrocephala* and *Ceriosporopsis circumvestita*. Johnson *et al.* (1959) found that *Ceriosporopsis* spp. and *Lulworthia* sp. colonized certain timbers (balsa, white pine and yellow poplar) while others were sparsely colonized or not at all (e.g. mahogany, black cherry, narra and prima vera).

The marine fungi colonizing mangroves are somewhat different from those colonizing submerged timber. Kohlmeyer (1966, 1969*b*) has reported 34 species from mangroves, the most common being *Lulworthia* spp. (20% of all collections), *Metasphaeria australiensis* (15%) and *Phoma* sp. (10%). Kohlmeyer (1969*b*) lists 45 species of terrestrial fungi encountered on mangroves above the water surface, though most of these were found only once.

Lignicolous freshwater fungi are quite different from those found on submerged timber in the sea. Eaton and Jones (1971*b*) recorded 42 species from *Fagus* and *Pinus* test blocks submerged in fresh water at a pumping station. Only a few were frequently present, namely *Monodictys putredinis*, *Trematosphaeria pertusa*, *Clasterosporium caricinum*, *Helicoon sessile*, *Tricladium splendens*, *Ceratosphaeria lampadophora* and *Sterigmatobotrys macrocarpa*.

Some 130 fungi have been reported from timber in freshwater. Archer and Willoughby (1973) have shown that *Fusarium* spp., *Heliscus lugdunensis*, *Anguillospora longissima* and *Clavariopsis aquatica* were the most commonly occurring Hyphomycetes on twigs exposed in fresh water at Smooth Beck. Twenty-three species of aquatic Hyphomycetes were recorded on the twigs while 41 were represented in the foam. All the common foam species were represented in the twig samples. *H. lugdunensis* and *Fusarium* spp. were early colonizers showing their best growth during the first 4 months. In contrast, *Dimorphospora foliicola* was a late colonizer, being recorded from the fifth month onwards. *Ceratostomella* sp., *Cudoniella clavus*, *Massarina* sp., *Mollisia* spp. and *Nectria lugdunensis* were frequently present on the twigs.

Timber in water cooling towers offers an ideal substrate for colonization by micro-organisms (Eaton and Jones, 1971*a*, *b*). Eaton (1972) lists some 71 microfungi observed on treated service timber and untreated test blocks removed from cooling towers. The most frequently occurring species are listed in Table IX and it can be seen that *Monodictys putredinis* was by far the most common species.

TABLE IX. The most frequently occurring fungi growing on *Fagus* and *Pinus* test blocks exposed in 16 localities. (After Eaton, 1972)

BEECH	Number of sites reported	SCOTS PINE	Number of sites reported
Ascomycetes			
Chaetomium globosum[a]	10	*Savoryella lignicola*[a]	5
Savoryella lignicola[a]	8	*Melogramma* sp.	3
Trematosphaeria pertusa	4	*Phaeonectriella lignicola*	3
Ceratosphaeria lampadophora	5	(*Halosphaeria maritima*)	2
Ascomycete 1	4		
Melogramma sp.	3		
Moniliales			
Monodictys putredinis[a]	15	*Monodictys putredinis*[a]	14
Graphium sp.[a]	9	*Fusarium* sp.[a]	8
Doratomyces microsporus[a]	7	*Septonema* sp. 1[a]	7
Fusarium sp.	5	*Graphium* sp.	5
Cephalosporium sp.	5	(*Zalerion maritimum*)	2
Torula herbarum	5		
Sphaeropsidales			
Asteromella sp.[a]	10	*Asteromella* sp.	4
Pyrenochaeta sp.[a]	6	*Pyrenochaeta* sp.	4

[a] Appearing on 25% or more of the test blocks examined.
( ) Marine fungus.

B. Leaves

The significance of fungi in the decomposition of submerged tree leaves has been reported on by Efford (1969) and Kaushik and Hynes (1971). Efford has shown that in Lake Marion, Canada, some 86% of the energy budget is derived from forest debris while Kaushik and Hynes suggest that the occurrence of certain crustaceans and insects may be limited by the availability of decomposing leaves.

Barlocher and Kendrick (1973) have shown that autumn-shed leaves, which support a rich fungal population, are an important source of food for many stream invertebrates. They also showed that mycelium of *Tricladium angulatum* proved to be a much better source of food for adult *Gammarus pseudolimnaeus* than *Acer* leaves supporting a rich microbial growth.

Nilsson (1964) has shown that *Alnus glutinosa* leaves are more quickly colonized and skeletonized than those of *Quercus robur* and *Acer platanoides*. After 3 weeks the leaves of *Corylus avellana* were slightly attacked while those of *Fagus sylvatica* remained free of fungal colonization. A more extensive study of the decay of deciduous leaves has been carried out by Newton (1971) (see Willoughby, Chapter 21).

Meyers *et al.* (1965) have investigated the fungi occurring in the leaves of the marine Angiosperm *Thalassia testudinum*. The dominant species were a *Labyrinthula*, *Lindra thalassiae*, *Hormodendron* sp., *Cephalosporium* sp. and *Dendryphiella arenaria*.

Anastasiou and Churchland (1969) have noted that a number of leaves are carried into the sea by streams and rivers, and they investigated the fungi colonizing leaves of *Prunus laurocerasus* and *Arbutus menziesii* submerged in the sea at Vancouver. The most common species recorded were *Phytophthora vesicula* (on 71% of leaves), *Zalerion maritimum* (on 59% of leaves), *Papulospora halima* (27%) and *Nowakowskiella elegans* (8%).

The intertidal Angiosperm *Spartina* supports a very diverse fungal flora. Jones (1963*a*) observed generic dissimilarities between the fungi on stems of *Spartina* and those on wood in the sea (compare Tables VII and X). Meyers *et al.* (1970*a*) suggested that yeasts may also be active in the degradation of *Spartina* as well as a number of Fungi Imperfecti including *Fusarium* spp., *Phoma* spp. and *Nigrospora* sp.

C. Algae

Sparrow (1960) has shown that a number of Phycomycetes occur as saprophytes on algae and Chesters *et al.* (1956) have observed Actinomycetes growing on marine algae. Sutherland (1916) described Fungi Imperfecti from brown algae but Johnson and Sparrow (1961) suggested that many of these may be air-borne contaminants and not true marine fungi. *Dendryphiella salina* is an exception, as this species has been frequently isolated from marine algae (Wilson, 1951; Jones, unpublished).

van Uden and Castelo-Branco (1963) found high numbers of the yeast *Metschnikowia zobellii* on decomposing *Macrocystis pyrifera* with counts of 520–39,200 propagules per g. Similar results have been reported by Suehiro (1960), Suehiro and Tomiyasu (1962) and Suehiro *et al.* (1962) for other decaying phytoplankton and macroscopic algae.

D. Animals

Kohlmeyer (1969*d*), Höhnk (1967, 1969), Cavaliere and Randall (1970) and Cavaliere and Markhart (1972) have all reported fungi from calcareous

TABLE X. Fungi on *Spartina*

Fungi	Austwick, unpublished	Lloyd, 1954	Johnson, 1956	Goodman, 1959	Jones, 1962*a*, *b* 1963*a*, unpublished	Apinis and Chesters, 1956	Total
Fungi Imperfecti							
Alternaria maritima	+	−	−	+	+	−	3
Asteromyces cruciatus	−	−	−	−	+	−	1
Cladosporium herbarum	+	+	−	+	−	−	3
Dendryphiella salina	−	−	−	−	+	−	1
Dictyosporium toruloides	−	−	−	−	+	−	1
Monodictys putredinis	−	−	−	−	+	−	1
Phoma spp.	+	+	−	+	+	−	4
Stemphylium maritimum Johnson	−	−	−	−	+	−	1
Ascomycetes							
Ceriosporopsis halima	−	−	−	−	−	+	1
Haligena elaterophora	−	−	−	−	−	+	1
H. spartinae	−	−	−	−	+	+	2
Halosphaeria hamata	−	−	−	−	+	+	2
Leptosphaeria albopunctata (West.) Sacc.	−	−	+	−	+	−	2
L. discors	+	−	+	+	+	+	5
L. halima Johnson	−	−	+	−	−	−	1
L. macrosporidium Jones	−	−	−	−	+	−	1
L. marina Ellis & Everhart	−	−	+	−	−	+	2
L. maritima (Cke. & Plowr.) Sacc.	−	−	+	−	−	−	1
L. orae-maris	−	−	+	−	−	−	1
L. pelagica Jones	−	−	−	−	+	+	2
L. typharum (Desm.) Karsten	+	−	−	+	+	+	4
Lulworthia medusa	+	+	−	+	+	+	5
Lignincola laevis	−	−	−	−	+	+	2
Gnomonia salina Jones	−	−	−	−	+	−	1
Pleospora herbarum (Fr.) Rabenhorst	+	+	−	+	+	+	5
P. pelagica Johnson	−	−	+	−	−	−	1
P. spartinea Ellis & Everhart	−	−	−	−	−	+	1
Sphaerulina pedicellata Johnson	−	−	−	−	+	−	1

This is not a complete list of species found on *Spartina*. Jones (unpublished) has collected some 80 species growing on the culms of *Spartina*.

matter of marine animals. Most of these are described from shells of Molluscs or teredinid burrows of dead animals. Recently, Kohlmeyer (1970, 1972*a*) has described fungi from hydrozoan tubes of a stony coral and keratin-like tubes of annelids (see Table XI).

TABLE XI. Marine fungi in calcareous and chitinous substances

Substrate	Fungus	Reference
1. Mollusc shells		
Ostrea edulis	*Ostracoblabe implexa*	Alderman and Jones, 1971*b*
Shell fragments	unidentified	Höhnk, 1967
Mytilidae shells and others	various—unidentified	Cavaliere and Alberte, 1970; Cavaliere and Markhart, 1972
2. Teredinid tubes	*Halosphaeria quadricornuta* *Halosphaeria salina* *Cirrenalia pygmea* *Humicola alopallonella* *Periconia prolifica*	Kohlmeyer, 1969*d*
3. Shells of balanids	*Pharcidia balani*	Kohlmeyer, 1969*d*
4. Calcareous algae		
Lithophyllum *Pseudolithophyllum*	*Lulworthia kniepii*	Kohlmeyer, 1969*d*
5. Hydrozoan tubes		
Stony coral	*Abyssomyces hydrozoicus* Kohlm.	Kohlmeyer, 1970
Hydrozoan tubes	Fungal mycelium	Kohlmeyer, 1972*a*
6. Annelid tubes		
Chaetopterus variopedatus	*Lulworthia* sp.	Kohlmeyer, 1972*a*

The occurrence of Phycomycetes on insect exuviae and baits is well documented (Sparrow, 1960; Dick, 1970).

E. Sediments

Borut and Johnson (1962) have listed 142 fungi (mainly terrestrial) from estuarine sediments. The occurrence of terrestrial fungi in marine sediments and freshwater muds and sediments has already been discussed in section IIE.

Fell *et al.* (1960) and Suehiro (1963) isolated yeasts from deep sea sediments and intertidal muds. Fell *et al.* (1960) found fewer fungi in the sediments as compared with the water samples. However, Meyers *et al.* (1970*a*) found the sediments of a *Spartina* marsh at Barataria Bay to be rich in yeasts with counts of 1800–90,000 cells per ml of sediment.

Gaertner (1967*a*) has isolated *Thraustochytrium* species from the North Sea water which contained 50–600 fungi per litre while sediments were much richer with 10,000 to 18,500 fungi per litre. Sea water collected off Iceland (Gaertner, 1968*a*) contained only 1·3–12 infective units per litre, while fine sediments in the German Bay contained from 230 to 58,700 fungi per litre (Gaertner, 1968*b*). These high counts of both yeasts and Phycomycetes suggests that these fungi may be active in the productivity of these waters.

Dick (1971) has investigated the ecology of the Saprolegniaceae in lentic and littoral muds of Marion Lake, British Colombia. He found that lentic muds (under 1 m or more water) were poor in both species and numbers while the marginal lentic muds were richer and reflect the flora of the emergent littoral muds.

F. World Distribution

Many aquatic fungi appear to be worldwide in their distribution (Nilsson, 1964; Jones, 1968, 1971) (Table XII).

Most higher marine fungi are cosmopolitan in their distribution (Jones 1971), e.g. *Cirrenalia macrocephela*, *Ceriosporopsis halima*, *Corollospora maritima*, *Lulworthia medusa*, *Torpedospora radiata* and *Zalerion maritimum*. Many workers regard *Halosphaeria quadri-cornuta* (Jones, 1968; Kohlmeyer, 1968*b*; Hughes and Chamut, 1971) and *Varicosporina ramulosa* (Kohlmeyer and Kohlmeyer, 1971a) as warm water species. Marine fungi have now been recorded from the coastal waters of over 50 countries but further information is required before meaningful discussions on their distribution can be made.

G. Distribution from Land and with Depth

Roth *et al.* (1964), Fell (1967) and van Uden and Fell (1968) have all shown that shallow coastal marine waters are richer in species and in total numbers than waters more distant from land and in deep waters. However, Roth *et al.* (1964) recovered *Aureobasidium pullulans*, *Alternaria tenuis*, *Aspergillus sydowi* and *A. niger* from a water sample taken at a depth of 4450 m, but there was no evidence to suggest that these fungi were active at these depths.

TABLE XII. World distribution of some freshwater Hyphomycetes. (After Nilsson, 1964)

I. Very common with a worldwide distribution	
A. Mainly tropical	B. Mainly temperate or cold regions
Actinospora megalospora Ingold	*Alatospora acuminata* Ingold
Campylospora chaetocladia Ranzoni	*Anguillospora crassa* Ingold
Clavatospora tentacula (Umphlett) Nilsson	*A. longissima* (Sac. & Syd.) Ingold
Flagellospora penicillioides Ingold	*Clavariopsis aquatica* De Wild
Heliscus submersus Hudson	*Flagellospora curvula* Ingold
Triscelophorus monosporus Ingold	*Tricladium splendens* Ingold
II. Common only in temperate and cold regions	
Anguillospora pseudolongissima Ranzoni	*Clavatospora stellata* (Ingold & & Cox) Nilsson
Casaresia sphagnorum Fragoso	*Culicidospora aquatica* Peterson
Dendrospora erecta Ingold	*C. gravida* Peterson
Dimorphospora foliicola Tubaki	*Heliscus lugdunensis* Sacc. & Therry
III. Species with a very limited distribution	
A. Tropical	B. Temperate or cold regions
Angulospora aquatica Nilsson	*Anguillospora gigantea* Ranzoni
Geniculospora grandis (Greathead) Nilsson	*Lemonniera cornuta* Ranzoni
Jaculispora submersa Hudson & Ingold	*Triscelophorus magnificus* Peterson

Jones and Le Campion-Alsumard (1970) recorded four fungi on wood and polyurethane panels at depths of 126–437 m. *Zalerion maritimum* was not found below 126 m, while *Corollospora maritima* and *Lulworthia purpurea* were found at 280 and 380 m respectively. *Haligena unicaudata* was found on plates submerged at 437 m. Kohlmeyer (1968a, 1969c) has also reported unidentified marine fungi growing on timber submerged at depths of 1616 and 2073 m.

V. Physiology and Biochemistry

A. Temperature

The majority of the Phycomycetes investigated have optimum growth between 20 and 30°C, as can be seen from Table XIII. None showed maximum growth below 10°C while only three species, *Rhizophydium*

TABLE XIII. Effect of temperature on some aquatic Phycomycetes

Fungus	Reference	Optimum temperatures for growth (°C)				
		below 10	10–20	20–30	above 30	killed at
Sirolpidium zoophthorum	Vishniac, 1955	—	—	+	—	36
Haliphthoros milfordensis	Vishniac, 1958	—	—	+	—	35
Thraustochytrium roseum	Goldstein, 1963*a*	—	—	+	—	40
T. motivum	Goldstein, 1963*b*	—	—	+	—	37
T. multirudimentale		—	—	+	—	37
T. aureum	Goldstein, 1963*c*	—	—	+	—	37
Dermocystidium sp.	Goldstein *et al.*, 1969	—	+	—	—	30
Rhizophydium sphaerotheca	Barr, 1969	—	+	—	—	30
R. capillaceum		—	—	—	+	40
Phlyctochytrium californicum		—	—	+	—	35
Second isolate		—	—	—	+	40
P. plurigibbosum		—	—	+	—	35
Second isolate		—	—	+	—	35
P. acuminatum		—	—	+	—	35
Second isolate		—	—	—	+	40
P. arcticum	Barr, 1970*a*	—	+	—	—	30
P. reinboldtae	Barr, 1970*b*	—	—	+	—	35
Hyphochytrium catenoides	Barr, 1970*c*	—	—	+	—	35
Entophlyctis confervae-glomeratae	Barr, 1971	—	—	+	—	30
Althornia crouchii	Alderman & Jones, 1971	—	—	+	—	35
Ostracoblabe implexa		—	—	+	—	35
Saprolegnia parasitica	Powell *et al.*, 1972	—	+	—	—	40
Phlyctochytrium mangrovii	Ulken, 1972	—	—	+	—	—

TABLE XIV. The effect of temperature on some higher aquatic fungi

Fungus	Author	Optimum temperatures for growth (°C)				
		below 10	10–20	20–30	above 30	no growth
Lulworthia salina	Barghoorn and Linder, 1944	–	–	+	–	–
Zalerion maritimum		–	–	+	–	–
Ceriosporopsis halima		–	–	+	–	–
Corollospora maritima		–	–	+	–	–
Phialophorophoma littoralis		–	–	+	–	–
Anguillospora longissima	Ranzoni, 1951	–	+	–	–	35
A. gigantea		–	–	+	–	40
Articulospora tetracladia	Thornton, 1963	–	–	+	–	–
Flagellospora penicillioides		–	–	+	–	–
F. curvula		–	+	–	–	–
Lemonniera aquatica		–	+	–	–	–
Tricladium splendens		–	+	–	–	–
T. gracilis		–	+	–	–	–
Heliscus ligdunensis		–	–	+	–	–
Tricellula aquatica		–	+ (10)	–	–	–
Candida aquatica	Jones & Sloof, 1966	–	+	–	–	37
Sterigmatomyces halophilus	Fell, 1966	–	–	+	–	–
Ceriosporopsis halima	Tubaki, 1969	–	+	–	–	–
Corollospora maritima		–	+	–	–	–
C. trifurcata		–	+	–	–	–
Leptosphaeria discors		+	+	–	–	–
Lindra thalassiae		+	+	–	–	–
Lignincola laevis		–	+	–	–	–
Remispora galerita		+	+	–	–	–
Dendryphiella salina		–	+	–	–	–
Zalerion maritimum		–	+	–	–	–
Asteromyces cruciatus	Jones & Irvine, 1972	–	–	+	–	–
Lulworthia sp.		–	+	–	–	–
Corollospora cristata		–	+	–	–	–
Tricladium varium	Jones & Stewart, 1972	–	–	+	–	–

capillaceum (35), *Phlyctochytrium californicum* (30–35) and *P. acuminatum* (30–35), showed optimum growth above 30°C. Table XIV gives the optimum growth of some higher aquatic fungi. Most of the fungi listed showed optimum growth in the range 10–20°C, only three marine Ascomycetes had optima below 10°C while none appeared to require temperatures above 30°C. The results of Barghoorn and Linder (1944) indicate that marine fungi have a high temperature requirement (25–30°C) but the results of Tubaki (1969) and Jones and Irvine (1972) do not support these observations. Meyers (1966) and Meyers and Simms (1967) have shown that *Lulworthia* sp., *Lindra thalassiae* and *Torpedospora* sp. have temperature optima of 25–30°C for fruiting and this may be higher than that required for vegetative growth.

B. Hydrogen Ion Concentration

Most aquatic fungi have pH requirements within the range 6·2–8·5, e.g. *Rhizophydium sphaerotheca* 4·95–7·20, *R. capillaceum* 4·95–6·20, *Phlyctochytrium californicum* 8·30–8·50, *P. plurigibbosum* 6·20–7·85, and *P. acuminatum* 6·20–8·50 (Barr, 1969*b*), *P. articum* 5·8–7·2 (Barr, 1970*a*), *P. reinboldtae* 8·55 (Barr, 1970*b*), *Hyphochytrium catenoides* 6·10–8·35 (Barr, 1970*c*). According to Powell *et al.* (1972) heavy growth of *Saprolegnia parasitica* was observed over the pH range 4·0–8·0, while Barr (1971) showed that *Entophlyctis confervae-glomeratae* required alkaline conditions (9·1). Barghoorn and Linder (1944) showed that some of the higher marine fungi exhibited a double pH peak for growth, e.g. *Lulworthia opaca* at 5·2 and 8·4 and *Amphisphaeria maritima* at 4·4 and 8·4. Jones and Irvine (1972) found a similar response in the fungi they tested, with one pH peak in the acidic range 6·0–6·6 and the other at 7·0–8·0.

C. Light

Alderman and Jones (1971*a*) have shown that light inhibits the growth of the two marine Phycomycetes *Althornia crouchii* and *Ostracoblabe implexa* while the growth of the pigmented marine *Thraustochytrium roseum* was stimulated by light (Goldstein, 1963*a*). Bremer (1972) has shown that while some strains of *T. roseum* were light stimulated, others appeared to have no such requirement.

Kohlmeyer (1968*a*) attributed the lack of fruiting structures of higher marine fungi at a depth of 1615 m, off the coast of California, to the constant darkness of this environment. Jones and Le Campion-Alsumard (1970) reported abundant perithecia of *Haligena unicaudata* at 437 m and Kirk (1969) stated that illumination had no apparent effect on the fruiting of marine fungi. Jones and Ward (1973) have shown septate conidia are produced in *Asteromyces cruciatus* under two main conditions, when grown on corn meal agar and under black light. *Coprinus alkalinus*, isolated from wood submerged in alkali lakes, was strongly light-sensitive and no basidiocarps were produced when cultures were kept in the dark (Anastasiou, 1967).

D. Oxygenation

Nilsson (1964) has shown that few aquatic Hyphomycetes are to be found in ditches or streams with slow-running to stagnant waters. Their absence under these conditions has been attributed to the dirty, poorly oxygenated nature of the water. Aquatic fungi are generally regarded as aerobic, although little experimental work has been carried out. Goldstein (1963*b*, *c*) frequently isolated *Thraustochytrium roseum*, *T. multirudimentale* and *T. motivum* from polluted sea water but they are obligately aerobic and do not grow under anaerobic conditions. Tabak and Cooke (1968) have shown that at least 13 geofungi they studied can survive and even grow in the absence of oxygen in a reduced medium in an atmosphere of pre-purified nitrogen gas. *Aqualinderella fermentans* is a fungus that has been isolated from stagnant pools (Emerson and Held, 1969) and shown to have a high requirement for atmospheric carbon dioxide. Good growth occurred between 5 and 20% CO_2 with some growth at 99% CO_2.

Althornia crouchii and *Ostracoblabe implexa* both required sodium bicarbonate in the medium for good growth (Alderman and Jones, 1971*a*) an observation confirmed by Bremer (1972) for a number of thraustochytrids. Alderman and Jones suggest that the effect of bicarbonate and the utilization of malate may be connected as carbon dioxide fixation is closely linked to malate metabolism in *Blastocladiella* (Cantino and Horstein, 1959).

Byrne and Jones (1972) suggest that the reduced number of fungi found on test panels submerged in Langstone harbour was due to pollution and reduced oxygenation of the water (Dunn, unpublished).

Many freshwater aquatic Hyphomycetes will not sporulate unless the cultures are submerged in water or vigorously aerated. However, there is no evidence at present to suggest this is due to increased oxygenation (Webster and Towfik, 1972).

E. Salinity Tolerances

A number of investigations of the responses of fungi to salinity have been carried out and the reader is referred to the reviews of Meyers (1971*a*, *b*), Jones (1971) and Jones *et al.* (1971). Figure 7 summarizes the results obtained by Harrison (1972) for 13 species of Phycomycetes. The Saprolegniaceae showed a decrease in tolerance with increasing salinities, the degree of intolerance depending on the species. Asexual and sexual reproduction by these organisms was more severely inhibited than vegetative growth. The two species of *Pythium* were more tolerant than the Saprolegniaceae while the chytrids showed a variable response. Finally, the

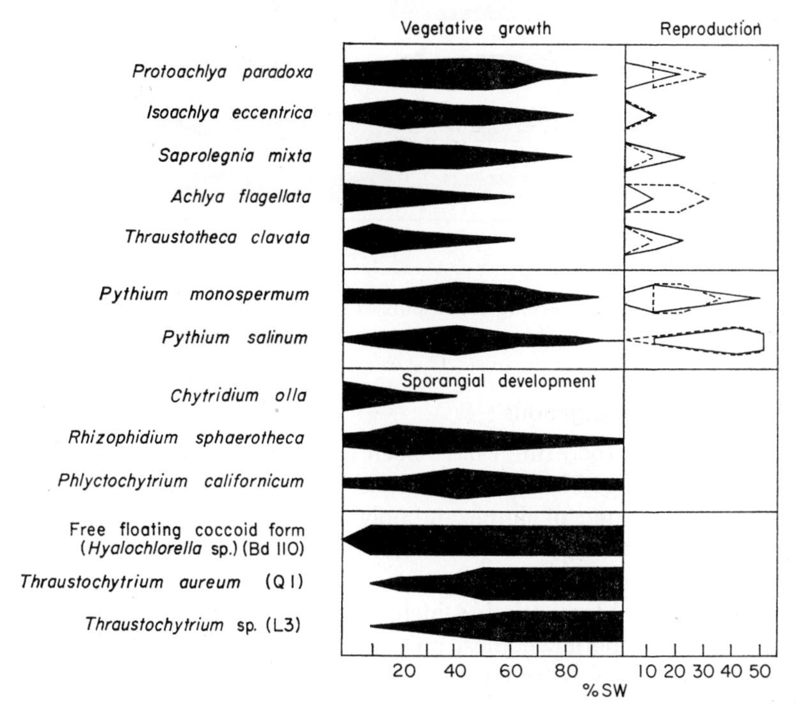

Fig. 7. Physiological responses of Phycomycetes to salinity (from Hohnk, 1953). Asexual ———, Sexual – – – – –, Medium: nutrient broth and a temperature of 20°C.

obligately marine nature of three monocentric coccoid organisms was well demonstrated. Figure 8 summarizes the results obtained by Byrne (1971) for the higher fungi. In the mucoraceous fungi, spore germination was severely inhibited and may account for their absence from the marine environment. Despite the wide tolerance of the vegetative growth of the terrestrial Ascomycetes tested, reproduction and spore germination was inhibited. A similar response was also noted for the freshwater Hyphomycetes investigated. Finally, the results show that marine Ascomycetes and Fungi Imperfecti exhibit a wide tolerance to decreasing salinities.

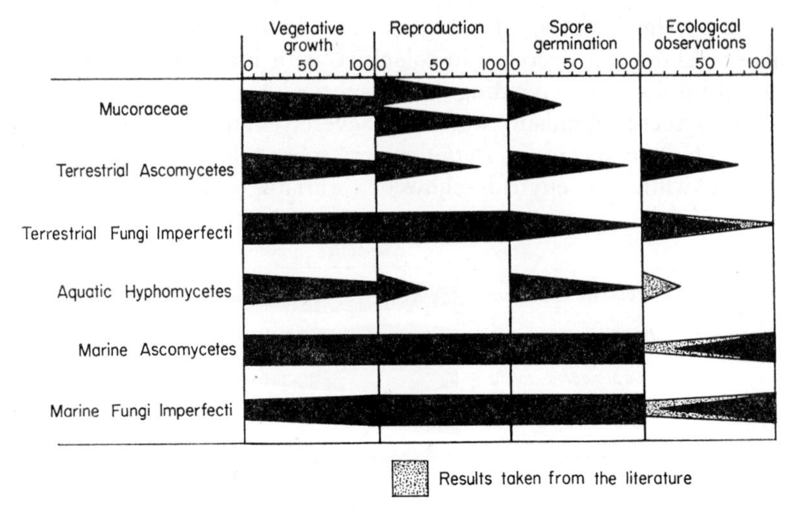

Fig. 8. Physiological and ecological responses to salinity of various fungal groups.

F. Nutritional Investigations

Approximately 1000 fungi have been isolated or observed growing on substrates submerged in aquatic habitats. Some lignicolous fungi are able to utilize a wide range of simple and polymeric carbohydrates while others are more restricted, e.g. *Oedogoniomyces* sp. (Emerson and Whisler, 1968). Table XV summarizes the ability of some aquatic fungi to utilize a range of carbohydrates. A large number, including many yeasts and Phycomycetes, are able to utilize both cellobiose and starch. The ability to degrade cellulose has been reported for a few Phycomycetes (Haskins and Weston, 1950; Barr, 1970*c*), some aquatic Hyphomycetes (Jones and Stewart, 1972) and the lignicolous Ascomycetes and Fungi Imperfecti (Jones, 1971; Eaton and Jones, 1971*a*; Meyers *et al.*, 1972). Cellulolytic activity in marine fungi has been demonstrated in a number of ways: (1) weight loss experiments (Meyers, 1968; Meyers and Scott, 1968); (2) reducing sugar method (Meyers and Reynolds, 1959*a*, *b*, 1963); (3) visual methods (Kohlmeyer, 1958*a*, *b*; Becker and Kohlmeyer, 1958); (4) loss of strength (Meyers and Reynolds, 1963); and (5) protein production (Jones and Irvine, 1972). These fungi are active not only in the breakdown of wood but also in the decomposition of marsh plants (Meyers *et al.*, 1970*b*).

Proteolytic activity has been demonstrated in 13 out of 14 marine filamentous fungi tested by Pisano *et al.* (1964). Eight of these contained measurable quantities of glutamic acid and alanine which suggested that these fungi may serve as sources of food for other micro-organisms.

TABLE XV. Effect of carbon sources on the growth of some aquatic fungi

Fungus	Reference	Xylose	Arabinose	Ribose	Rhamnose	Fructose	Sorbose	Mannose	Galactose	Glucose	Glucosamine	Mannitol	Cellobiose	Maltose	Trehalose	Sucrose	Raffinose	Starch	Cellulose	Soft rot	Lignin	Glutamate
Articulospora tetracladia	Thornton (1963)	+	—	—	—	+	—	—	—	+	—	—	+	+	—	+	—	+	o	—	—	—
Flagellospora penicillioides	Thornton (1963)	+	—	—	—	+	—	—	—	+	—	—	+	+	—	+	—	o	o	—	—	—
Tricladium angulatum	Thornton (1963)	+	—	—	—	+	—	—	—	+	—	—	+	+	—	+	—	+	o	—	—	—
T. splendens	Thornton (1963)	+	—	—	—	+	—	—	—	+	—	—	+	+	—	+	—	+	o	—	—	—
Volucrispora aurantica	Thornton (1963)	+	—	—	—	+	—	—	—	+	—	—	+	+	—	+	—	o	o	—	—	—
Pythium ultimum	Thornton (1963)	o	—	—	—	o	—	—	—	+	—	—	o	o	—	o	—	o	o	—	—	—
Dictyuchus sterile	Thornton (1963)	o	—	—	—	o	—	—	—	+	—	—	o	o	—	o	—	o	o	—	—	—
Thraustochytrium motivum	Goldstein (1963*b*)	$+^{w}$	$+^{w}$	$+^{w}$	—	$+^{w}$	—	+	+	+	—	—	+	+	$+^{w}$	+	o	+	—	—	—	+
T. multirudimentale	Goldstein (1963*b*)	$+^{w}$	$+^{w}$	$+^{w}$	—	$+^{w}$	—	+	+	+	—	—	+	+	$+^{w}$	+	$+^{w}$	+	—	—	—	+
Candida aquatica	Jones and Slooff (1966)	+	+	o	o	—	+	—	+	+	—	+	+	+	+	+	+	$+^{w}$	—	—	—	—
Sterigmatomyces halophilus	Fell (1966)	+	+	+	o	o	$+^{w}$	—	+	+	o	+	$+^{w}$	o	+	o	o	—	—	—	—	—
Corollospora maritima IMS 604	Meyers and Scott (1967)	+	$+^{w}$	—	o	+	+	—	+	+	$+^{w}$	+	+	$+^{w}$	+	+	o	+	+	+	—	—
C. maritima IMS 607	Meyers and Scott (1967)	+	+	—	o	+	+	—	+	+	+	+	+	+	+	+	o	+	+	+	—	—
Lulworthia floridana IMS 190	Meyers (1968)	+	+	o	o	+	o	+	+	+	o	o	+	+	+	o	o	+	+	+	—	—
L. floridana IMS 400	—	+	o	+	+	+	$+^{w}$	+	o	+	+	+	+	+	+	o	o	+	+	+	—	—
Pichia spartinae	Ahearn *et al.* (1970)	o	o	o	o	—	+	—	o	+	+	+	+	+	+	+	o	o	—	—	—	—
Phlyctochytrium reinboldtae	Barr (1970*b*)	o^{w}	o	o	—	+	—	o	o	+	o	—	+	+	+	o	o	+	o	—	—	—
Hyphochytrium catenoides	Barr (1970*c*)	o	o	o	—	o	—	+	o	+	$+^{w}$	—	+	+	o	o	o	+	+	—	—	—
Entophlyctis confervaeglomeratae	Barr (1971)	o	—	o	o	+	—	—	—	+	$+^{w}$	—	+	+	+	+	o	+	o	—	—	—
Althornia crouchii	Alderman and Jones (1971*a*)	o	o	o	o	o	—	o	+	o	—	—	o	o	o	o	—	+	—	—	—	—
Saprolegnia parasitica	Powell *et al.* (1972)	$+^{w}$	o	—	$+^{w}$	+	—	o	$+^{w}$	—	—	$+^{w}$	+	+	—	o	o	+	—	—	—	—

+ = positive assimilation. $+^{w}$ = assimilation weak. o = no assimilation. — = not tested.

Similar observations have been made by Thornton and Fox (1968), Thornton and McEvoy (1970) and Thornton and Griffin (1971). Ahearn *et al.* (1968b) have shown the presence of extracellular proteolytic activity among various taxa of yeasts they studied.

Chesters and Bull (1963*a*, *b*, *c*) examined a number of fungi for laminarin hydrolysing enzymes. *Dendryphiella salina*, *Halosphaeria mediosetigera*, *Culcitalna achraspora*, *Zalerion xylestrix* and *Corollospora maritima* produced laminarinase, but in low amounts. Species of *Lulworthia* were inactive.

Nematodes are frequently found on cellulosic materials undergoing fungal degradation, and Meyers *et al.* (1963, 1964), Hopper and Meyers (1967), and Meyers and Hopper (1967) have investigated the relationship between these groups. A number of marine fungi, e.g. *H. mediosetigera*, *Lulworthia floridana*, *Z. xylestrix* and *Dendryphiella arenaria*, supported animal growth while others failed to do so, e.g. *Cladosporium herbarum*, *Stachybotrys* sp. and *Aspergillus luchuensis*.

In this review, I have attempted to show that fungi are an important and active component of certain aquatic habitats, that they are present often in very large numbers, that they appear to be well adapted to such environments and that they have the ability to degrade a wide range of materials to be found in such habitats. However, further experimental work is required to determine their exact role in the productivity of natural waters.

Acknowledgements

I should like to thank Drs D. J. Alderman, J. L. Harrison (Portsmouth Polytechnic) and Dorothy E. Shaw (Papua and New Guinea) for kindly supplying photographs to illustrate this paper and to Professor W. E. Kershaw (Salford University), Drs S. H. Iqbal (Exeter University), B. Kendrick (Ontario), L. G. Willoughby and Miss J. Archer (Freshwater Biological Association, Ambleside) for allowing me to refer to unpublished observations.

References

AHEARN, D. G., MEYERS, S. P. and NICHOLS, R. P. (1968*a*). *Appl. Microbiol.* **16**, 1370–1374.

AHEARN, D. G., ROTH, F. J. and MEYERS, S. P. (1968*b*). *Marine Biology* **1**, 291–308.

AHEARN, D. G., YARROW, D. and MEYERS, S. P. (1970). *Antonie van Leeuwenhoek* **36**, 503–508.

ALASOADURA, S. O. (1968*a*). *Nova Hedwigia* **15**, 415–418.

ALASOADURA, S. O. (1968*b*). *Nova Hedwigia* **15**, 419–421.

ALDERMAN, D. J. and JONES, E. B. G. (1971*a*). *Trans. Br. mycol. Soc.* **57**, 213–225.

ALDERMAN, D. J. and JONES, E. B. G. (1971*b*). *Fishery Investigations, London, Series II* **26,** No. 8, 1–16.
ANASTASIOU, C. J. (1967). *Can. J. Bot.* **45,** 2213–2222.
ANASTASIOU, C. J. and CHURCHLAND, L. M. (1969). *Can. J. Bot.* **47,** 251–257.
APINIS, A. E. and CHESTERS, C. G. C. (1964). *Trans. Br. mycol. Soc.* **47,** 419–435.
ARCHER, J. F. and WILLOUGHBY, L. G. (1973). *Freshwater Biol.* (in press).
BAHNWEG, G. and SPARROW, F. K. (1972). *Arch. Mikrobiol.* **81,** 45–49.
BARGHOORN, E. S. and LINDER, D. H. (1944). *Farlowia* **1,** 395–467.
BARLOCHER, F. and KENDRICK, B. (1973). *Oikos* (in press).
BARR, D. J. S. (1969*a*). *Can. J. Bot.* **47,** 991–997.
BARR, D. J. S. (1969*b*). *Can. J. Bot.* **47,** 999–1005.
BARR, D. J. S. (1970*a*). *Can. J. Bot.* **48,** 2279–2283.
BARR, D. J. S. (1970*b*). *Can. J. Bot.* **48,** 479–484.
BARR, D. J. S. (1970*c*). *Mycologia* **62,** 492–503.
BARR, D. J. S. (1971). *Can. J. Bot.* **49,** 2223–2225.
BECKER, G. and KOHLMEYER, J. (1958). *J. Timber Dryers and Preservers India* **4,** 1–10.
BOOTH, T. (1969). *Syesis* **2,** 141–161.
BOOTH, T. (1971*a*). *Can. J. Bot.* **49,** 1757–1767.
BOOTH, T. (1971*b*). *Can. J. Bot.* **49,** 951–965.
BOOTH, T. and MILLER, C. E. (1968). *Mycologia* **60,** 480–495.
BORUT, S. Y. and JOHNSON, T. W. (1962). *Mycologia* **54,** 181–193.
BREMER, G. (1972). 2nd International Symposium on Marine Mycology, Bremerhaven, September, 1972 (in press).
BYRNE, P. J. (1971). Ph.D. Thesis, London University.
BYRNE, P. J. and JONES, E. B. G. (1972). 2nd International Symposium on Marine Mycology, Bremerhaven, September, 1972.
CANTINO, E. G. and HORENSTEIN, E. A. (1959). *Pl. Physiol.* **12,** 251–263.
CAVALIERE, A. R. and ALBERTE, R. S. (1970). *J. Elisha Mitchell Sci. Soc.* **83,** 203–206.
CAVALIERE, A. R. and MARKHART, A. H. (1972). *Surtsey Progress Report* **6,** 20–22.
CAVALIERE, A. R. and RANDALL, S. A. (1970). *J. Elisha Mitchell scient. Soc.* **86,** 203–206.
CHESTERS, C. G. C. and BULL, A. T. (1963*a, b, c*). *Biochem. J.* **86,** 28–31, 31–38, 38–46.
CHESTERS, C. G. C., APINIS, A. and TURNER, M. (1956). *Proc. Linn. Soc. Lond.* **166,** 87–97.
CHURCHLAND, L. M. and MCLAREN, M. (1972). *Can. J. Bot.* **50,** 1269–1273.
COOKE, W. B. (1954). *Sewage and Industrial Wastes* **26,** 539–549.
COOKE, W. B. (1957). *Sydowia, Ann. Mycol., Beiheft* **1,** 146–175.
COOKE, W. B. (1958). *Bot. Rev.* **24,** 342–375.
COOKE, W. B. (1961). *Ecology* **42,** 1–18.
COOKE, W. B. (1963). "A Laboratory Guide to Fungi in Polluted Waters Sewage and Sewage Treatment Systems." U.S. Dept. Health, Education and Welfare, Cincinnati.
COOKE, W. B. (1967). *Utah Academy Proc.* **44,** 298–315.

CRIBB, A. B. and CRIBB, J. W. (1956). *Pap. Dep. Bot. Univ. Qd* **3**, 97–105.
CUNNELL, C. J. (1956). *Trans. Br. mycol. Soc.* **39**, 21–47.
CUNNELL, C. J. (1957). *Trans. Br. mycol. Soc.* **40**, 443–455.
CUNNELL, C. J. (1958). *Trans. Br. mycol. Soc.* **41**, 405–412.
DICK, M. W. (1970). *Trans. Br. mycol. Soc.* **55**, 449–458.
DICK, M. W. (1971). *J. gen. Microbiol.* **65**, 625–637.
DOGUET, G. (1962). *C. r. hebd. Séanc. Acad. Sci., Paris*, **254**, 4336–4338.
DOGUET, G. (1967). *C. r. hebd. Séanc. Acad. Sci., Paris*, **265**, 1780–1783.
EATON, R. A. (1972). *Int. Biodetn. Bull.* **8**, 39–48.
EATON, R. A. and JONES, E. B. G. (1971*a*). *Material Organismen* **6**, 51–80.
EATON, R. A. and JONES, E. B. G. (1971*b*). *Material Organismen* **6**, 81–92.
EATON, R. A. and JONES, E. B. G. (1971*c*). *Nova Hedwigia* **19**, 779–788.
EFFORD, I. E. (1969). *Verhandlungen Internationale Vereinigung Limnologie* **17**, 104–108.
ELLIOTT, R. F. (1967). *N.Z. Jl Bot.* **5**, 418–423.
ELLIOTT, R. F. (1968). *N.Z. Jl Bot.* **6**, 94–115.
EMERSON, R. and HELD, A. A. (1969). *Am. J. Bot.* **56**, 1103–1120.
EMERSON, R. and WESTON, W. H. (1967). *Am. J. Bot.* **54**, 702–719.
EMERSON, R. and WHISLER, H. C. (1968). *Archiv. Mikrobiol.* **61**, 195–211.
FELL, J. W. (1966). *Antonie van Leeuwenhoek* **32**, 99–104.
FELL, J. W. (1967). *Bull. Mar. Sci.* **17**, 454–470.
FELL, J. W. (1970). *In* "Recent Trends in Yeast Research." (D. G. Ahearn, ed.), pp. 49–66. Georgia State University, Atlanta.
FELL, J. W. and STATZELL, A. C. (1971). *Antonie van Leeuwenhoek* **37**, 359–367.
FELL, J. W., AHEARN, D. G., MEYERS, S. P. and ROTH, F. J. (1960). *Limnol. Oceanogr.* **5**, 366–371.
FELL, J. W., STATZELL, A. C., HUNTER, I. L. and PHAFF, H. J. (1969). *Antonie van Leeuwenhoek* **35**, 433–462.
FISCHER, F. G. and WERNER, G. (1958). *Hoppe-Seyler's Z. physiol. Chem.* **310**, 65–91.
FULLER, M. S., LEWIS, B. and COOK, P. (1966). *Mycologia* **58**, 313–318.
GAERTNER, A. (1967*a*). *Veröff. Inst. Meeresforch. Bremerh.* **10**, 159–165.
GAERTNER, A. (1967*b*). *Helgolander wiss. Meeresunters* **15**, 181–192.
GAERTNER, A. (1968*a*). *Veröff. Inst. Meeresforsch. Bremerh.* **11**, 65–82.
GAERTNER, A. (1968*b*). *Veröff. Inst. Meeresforsch. Bremerh.* **11**, 105–117.
GAY, J. H. and GREENWOOD, A. D. (1966). *In* "The Fungus Spore." (M. F. Madelin, ed.), pp. 95–108. Butterworths, London.
GOLD, H. S. (1959). *J. Elisha Mitchell scient. Soc.* **75**, 25–28.
GOLDSTEIN, S. (1963*a*). *Mycologia* **55**, 799–811.
GOLDSTEIN, S. (1963*b*). *Am. J. Bot.* **50**, 271–279.
GOLDSTEIN, S. (1963*c*). *Arch. Mikrobiol.* **45**, 101–110.
GOLDSTEIN, S. and BELSKY, M. M. (1964). *Am. J. Bot.* **51**, 72–78.
GOLDSTEIN, S. and MORIBER, L. (1966). *Arch. Mikrobiol.* **53**, 1–11.
GOLDSTEIN, S., BELSKY, M. M. and CHOSAK, R. (1969). *Mycologia* **61**, 468–472.
GOODMAN, P. J. (1959). *Trans. Br. mycol. Soc.* **42**, 409–415.
GREENHALGH, G. N. and EVANS, L. V. (1967). *Trans. Br. mycol. Soc.* **50**, 183–188.

HARDER, R. and UEBELMESSER, E. (1955). *Arch. Mikrobiol.* **22,** 87–114.
HARRISON, J. L. (1972). Ph.D. Thesis, London University.
HASKINS, R. H. and WESTON, W. H. (1950). *Am. J. Bot.* **37,** 739–750.
HEDRICK, L. R. and SOYUGENC, M. (1967*a*). Pub. No. 13, pp. 69–76, Great Lakes Res. Division, Univ. Michigan.
HEDRICK, L. R. and SOYUGENC, M. (1967*b*). *Proc. 10th Conf., Great Lakes Res.* 20–30.
HEDRICK, L. R., COOK, W. and WOOLLETT, L. (1968). *Proc. 11th Conf., Great Lakes Res.* 538–543.
HÖHNK, W. (1952*a*). *Veröff. Inst. Meeresforsch. Bremerh.* **1,** 115–125.
HÖHNK, W. (1952*b*). *Veröff. Inst. Meeresforsch Bremerh.* **1,** 247–278.
HÖHNK, W. (1953). *Veröff. Inst. Meeresforsch. Bremerh.* **2,** 52–108.
HÖHNK, W. (1956). *Veröff. Inst. Meeresforsch. Bremerh.* **4,** 195–213.
HÖHNK, W. (1967). *Veröff. Inst. Meeresforsch. Bremerh.* **10,** 149–158.
HÖHNK, W. (1969). *Ber. dt. wiss Kommn Meeresforsch.* **20,** 129–140.
HOPPER, B. E. and MEYERS, S. P. (1967). *Bull. mar. Sci.* **17,** 471–517.
HUGHES, G. C. (1960). Ph.D. Thesis, Duke University.
HUGHES, G. C. and CHAMUT, P. S. (1971). *Can. J. Bot.* **49,** 1–11.
INGOLD, C. T. (1942). *Trans. Br. mycol. Soc.* **25,** 339–417.
INGOLD, C. T. (1951). *Trans. Br. mycol. Soc.* **34,** 210–215.
INGOLD, C. T. (1954). *Trans. Br. mycol. Soc.* **37,** 1–18.
INGOLD, C. T. (1955). *Trans. Br. mycol. Soc.* **38,** 157–168.
INGOLD, C. T. (1959). *Trans. Br. mycol. Soc.* **42,** 479–485.
INGOLD, C. T. (1961). *Trans. Br. mycol. Soc.* **44,** 27–30.
INGOLD, C. T. (1966). *Mycologia* **58,** 43–56.
INGOLD, C. T. (1968). *Trans. Br. mycol. Soc.* **51,** 323–325.
INGOLD, C. T. (1971). "Fungal Spores: Their Liberation and Dispersal." Clarendon Press, Oxford.
INGOLD, C. T. and CHAPMAN, B. (1952). *Trans. Br. mycol. Soc.* **35,** 268–272.
INGOLD, C. T., MCDOUGALL, P. J. and DANN, V. (1968). *Trans. Br. mycol. Soc.*, **51,** 325–328.
IQBAL, S. H. (1971). *Trans. Br. mycol. Soc.* **56,** 343–352.
IQBAL, S. H. (1972). *Trans. Br. mycol. Soc.* **59,** 301–307.
JOHNSON, T. W. (1956*a*). *Bull. mar. Sci. Gulf Caribb.* **6,** 349–358.
JOHNSON, T. W. (1956*b*). *Mycologia* **48,** 495–505.
JOHNSON, T. W. (1967). *In* "Estuaries," (A. H. Lauff, ed.), pp. 303–305. Am. Assoc. Adv. Sci. Publ. 83.
JOHNSON, T. W. and SPARROW, F. K. (1961). "Fungi in Oceans and Estuaries." J. Cramer, Weinheim.
JOHNSON, T. W., FERCHAU, H. A. and GOLD, H. S. (1959). *Phyton Internat. J. Exp. Bot.* **12,** 65–80.
JONES, E. B. G. (1962*a*). *Trans. Br. mycol. Soc.* **45,** 93–114.
JONES, E. B. G. (1962*b*). *Trans. Br. mycol. Soc.* **45,** 245–248.
JONES, E. B. G. (1963*a*). *Trans. Br. mycol. Soc.* **46,** 135–144.
JONES, E. B. G. (1963*b*). *J. Inst. Wood Sci. No.* 11, 14–23.
JONES, E. B. G. (1964). *Trans. Br. mycol. Soc.* **47,** 97–101.

JONES, E. B. G. (1968). *In* "Biodeterioration of Materials." (A. H. Walters and Elphick, eds), pp. 460–485. Elsevier, Amsterdam.
JONES, E. B. G. (1971). *In* "Marine Borers, Fungi and Fouling Organisms of Wood." (E. B. G. Jones and S. K. Eltringham, eds), pp. 237–258. O.E.C.D., Paris.
JONES, E. B. G. (1972). *In* "British Wood Preserving Association, Annual Convention," 1–18.
JONES, E. B. G. and ALDERMAN, D. J. (1971). *Nova Hedwigia* **21,** 381–399.
JONES, E. B. G. and LE CAMPION-ALSUMARD, T. (1970). *Int. Biodet. Bull.* **6,** 119–124.
JONES, E. B. G. and EATON, R. A. (1969). *Trans. Br. mycol. Soc.* **52,** 161–165.
JONES, E. B. G. and ELTRINGHAM, S. K. (1971). *In* "Marine Borers, Fungi and Fouling Organisms of Wood." O.E.C.D., Paris.
JONES, E. B. G. and IRVINE, J. (1971). *J. Inst. Wood Science* **29,** 31–40.
JONES, E. B. G. and IRVINE, J. (1972). *In* "Biodeterioration of Materials." (H. S. Walters and E. H. Hueck-van der Plas, eds), Vol. 2, pp. 422–431. Applied Science Publishers, England.
JONES, E. G. B. and JENNINGS, D. H. (1964). *Trans. Br. mycol. Soc.* **47,** 619–625.
JONES, E. B. G. and JENNINGS, D. H. (1965). *New Phytol.* **64,** 86–100.
JONES, E. B. G. and SLOOFF, W. (1966). *Antonie van Leeuwenhoek* **32,** 223–228.
JONES, E. B. G. and STEWART, R. (1972). *Trans. Br. mycol. Soc.* **59,** 163–167.
JONES, E. B. G. and WARD, A. W. (1973). *Trans. Br. mycol. Soc.* **61,** 181–186.
JONES, E. B. G., BYRNE, P. and ALDERMAN, D. J. (1971). *Vie milieu, supplement No.* 22, 265–280.
JONES, E. B. G., KÜHNE, H., TRUSSELL, P. C. and TURNER, R. D. (1972). *Material Organismen* **7,** 93–118.
KARLING, J. S. (1968). *Mycologia* **60,** 271–284.
KARLING, J. S. (1969). *Am. J. Bot.* **56,** 211–221.
KAUSHIK, W. K. and HYNES, H. B. N. (1971). *Arch. Hydrobiol.* **68,** 465–515.
KAZAMA, F. (1972). *Can. J. Bot.* **50,** 499–505.
KIRK, P. W. (1969). *Mycologia* **61,** 177–181.
KOBAYASHI, Y. and OOKYBO, M. (1953). *Bull. Nat. Sci. Mus. Tokyo* **33,** 53–65.
KOHLMEYER, J. (1958*a*). *Holz Roh-u. Werkst.* **16,** 215–220.
KOHLMEYER, J. (1958*b*). *Ber dtsch. Bot. Ges.* **71,** 98–116.
KOHLMEYER, J. (1966). *Ber. dtsch. Bot. Ges.* **79,** 27–37.
KOHLMEYER, J. (1968*a*). *J. Elisha Mitchell scient. Soc.* **84,** 239–241.
KOHLMEYER, J. (1968*b*). *Mycologia* **60,** 252–270.
KOHLMEYER, J. (1969*a*). *Can. J. Bot.* **47,** 1469–1487.
KOHLMEYER, J. (1969*b*). *Trans. Br. mycol. Soc.* **53,** 237–250.
KOHLMEYER, J. (1969*c*). *In* "Materials Performance and the Deep Sea." Special Technical Publication No. 445 American Society for Testing Materials, pp. 20–30.
KOHLMEYER, J. (1969*d*). *Am. Zool.* **9,** 741–746.
KOHLMEYER, J. (1970). *Ber. dtsch. Bot. Ges.* **83,** 505–509.
KOHLMEYER, J. (1972*a*). *Marine Biology* **12,** 277–284.

KOHLMEYER, J. (1972*b*). *In* "2nd International Symposium on Marine Mycology, Bremerhaven, September, 1972" (in press).
KOHLMEYER, J. (1972*c*). *Can. J. Bot.* **50,** 1951–1963.
KOHLMEYER, J. and KOHLMEYER, E. (1964–1969). "Icones Fungorum Maris." J. Cramer, Lehre.
KOHLMEYER, J. and KOHLMEYER, E. (1971*a*). *Mycologia* **63,** 831–861.
KOHLMEYER, J. and KOHLMEYER, E. (1971*b*). "Synoptic Plates of Higher Marine Fungi." J. Cramer, Lehre.
LLOYD, L. S. (1954). B.Sc. Thesis, University College of Wales, Aberystwyth.
LUTLEY, M. and WILSON, I. M. (1972*a*). *Trans. Br. mycol. Soc.* **59,** 219–227.
LUTLEY, M. and WILSON, I. M. (1972*b*). *Trans. Br. mycol. Soc.* **58,** 393–402.
MACHLIS, L. (1968). *Physiologia Pl.* **22,** 126–139.
MEYERS, S. P. (1954). *Bull. mar. Sci. Gulf Caribb.* **3,** 307–327.
MEYERS, S. P. (1966). *Helgol. wiss. Meeresunters* **13,** 436–443.
MEYERS, S. P. (1968). *In* "Biodeterioration of Materials." (A. H. Walters and J. J. Elphick, eds), Vol. 1, pp. 594–609. Elsevier, London.
MEYERS, S. P. (1969). *Mycologia* **61,** 486–495.
MEYERS, S. P. (1971a, b). *In* "Marine Borers, Fungi and Fouling Organisms of Wood." (E. B. G. Jones and S. K. Eltringham, eds), pp. 89–116; 217–236.
MEYERS, S. P. and HOPPER, B. E. (1967). *Helgol. wiss Meeresunters* **15,** 270–281.
MEYERS, S. P. and REYNOLDS, E. S. (1959*a*). *Can. J. Microbiol.* **5,** 493–503.
MEYERS, S. P. and REYNOLDS, E. S. (1959*b*). *Bull. mar. Sci. Gulf Caribb.* **9,** 441–455.
MEYERS, S. P. and REYNOLDS, E. S. (1960). *Can. J. Bot.* **38,** 217–226.
MEYERS, S. P. and REYNOLDS, E. S. (1963). *In* "Symposium on Marine Microbiology." (C. H. Oppenheimer, ed), pp. 315–328. Charles C. Thomas, Springfield.
MEYERS, S. P. and SCOTT, E. (1967). *Mycologia* **59,** 446–455.
MEYERS, S. P. and SCOTT, E. (1968). *Marine Biology* **2,** 41–46.
MEYERS, S. P. and SIMMS, J. (1967). *Bull. mar. Sci. Gulf Caribb.* **17,** 133–148.
MEYERS, S. P., FEDER, W. A. and TSUE, K. M. (1963). *Science N.Y.* **141,** 520–522.
MEYERS, S. P., FEDER, W. A. and TSUE, K. M. (1964). *Dev. Indust. Microbiol.* **5,** 354–364.
MEYERS, S. P., ORPURT, P. A., SIMMS, J. and BORAL, L. L. (1965). *Bull. marine Sci. Gulf Caribb.* **15,** 548–564.
MEYERS, S. P., AHEARN, D. G. and ROTH, F. J. (1967). *Bull mar. Sci. Gulf Caribb.* **17,** 576–596.
MEYERS, S. P., AHEARN, D. G. and COOK, W. L. (1970*a*). *Mycologia* **62,** 504–515.
MEYERS, S. P., NICHOLSON, M. L., RHEE, J., MILES, P. and AHEARN, D. G. (1970*b*). *Coastal Studies Bulletin* **5,** 111–124.
MEYERS, S. P., CHUNG, S. L. and AHEARN, D. G. (1972). *In* "Biodeterioration of Materials." (A. H. Walters and E. H. Hueck van der Plas, eds), Vol. 2, pp. 121–128.
MILLER, C. E. (1962). *Mycologia* **54,** 422–431.
MÜLLER, E. and VON ARX, J. A. (1962). *Beitr. Kryptogamen flora der Schweiz* **11,** 1–922.

NEWTON, J. A. (1971). Ph.D. Thesis, Salford University.

NILSSON, S. (1964). *Symb. bot. upsal.* **18,** 1–130.

PARK, D. (1972*a*). *Trans. Br. mycol. Soc.* **58,** 281–290.

PARK, D. (1972*b*). *Trans. Br. mycol. Soc.* **58,** 291–299.

PERKINS, F. (1972). *Arch. Mikrobiol.* **84,** 95–118.

PISANO, M. A., MIHALIK, J. A. and CATALANO, G. R. (1964). *Appl. Microbiol.* **12,** 470–474.

POOLE, W. J. and PRICE, P. C. (1972). *Trans. Br. mycol. Soc.* **59,** 333–335.

PORTER, D. (1969). *Protoplasma* **67,** 1–19.

POWELL, J. R., SCOTT, W. W. and KRIEG, N. R. (1972). *Mycopath. Mycologia appl.* **47,** 1–40.

POYTON, R. V. (1970). *J. gen. Microbiol.* **62,** 171–194.

RANZONI, F. V. (1951). *Mycologia* **43,** 130–141.

RITCHIE, D. (1954). *Science, N.Y.* **20,** 579.

ROTH, F. J., ORPURT, P. A. and AHEARN, D. G. (1964). *Can. J. Bot.* **42,** 375–383.

SCHAUMANN, K. (1968). *Veröff. Inst. Meeresforsch. Bremerh.* **11,** 93–118.

SCHAUMANN, K. (1972). *Veröff. Inst. Meeresforsch. Bremerh.* **14,** 23–44.

SCHMIDT, I. (1969). *Natur Naturschutz Mecklenburg* **7,** 5–14.

SCHNEIDER, J. (1967). *Kieler. Meeresforsch* **23,** 16–20.

SCHOLZ, E. (1958). *Arch. Mikrobiol.* **30,** 119–146.

SIEPMANN, R. (1959a). *Veröff. Inst. Meeresforsch. Bremerh.* **6,** 213–281.

SIEPMANN, R. (1959b). *Veröff. Inst. Meeresforsch. Bremerh.* **6,** 283–301.

SIEPMAN, R. and JOHNSON, T. W. (1960). *J. Elisha Mitchell scient. Soc.* **76,** 150–154.

SHEARER, C. A. (1971). Ph.D. Thesis, Maryland University.

SHAW, D. E. (1972). *Trans. Br. mycol. Soc.* **59,** 255–259.

SPARROW, F. K. (1936). *Biol. Bull.* **70,** 236–263.

SPARROW, F. K. (1960). "Aquatic Phycomycetes," 2nd edition. Michigan University Press, Ann Arbor.

SPARROW, F. K. (1968). *In* "The Fungi." (C. G. Ainsworth and A. S. Sussman, eds), Vol. 3, pp. 41–93. Academic Press, London and New York.

STEELE, C. W. (1967). *Pacific Science* **21,** 317–331.

SUEHIRO, S. (1960). *Sci. Bull. Fac. Agric. Kyushu Univ.* **17,** 443–449.

SUEHIRO, S. (1963). *Sci. Bull. Fac. Agric. Kyushu Univ.* **20,** 223–227.

SUEHIRO, S. and TOMIYASU, Y. (1962). *J. Fac. Agric. Kyushu Univ.* **12,** 163–169.

SUEHIRO, S., TOMIYASU, Y. and TANAKA, O. (1962). *J. Fac. Agric. Kyushu Univ.* **12,** 155–161.

SUTHERLAND, G. K. (1916). *New Phytol.* **15,** 35–48.

TABAK, H. and COOKE, W. B. (1968). *Mycologia* **60,** 115–140.

TE STRAKE, D. (1959). *Phyton, Internat. J. exp. Bot.* **12,** 147–152.

TE STRAKE WAGNER, D. (1965). *Nova Hedwigia* **9,** 45–61.

THORNTON, D. R. (1963). *J. gen. Microbiol.* **33,** 23–31.

THORNTON, D. R. and FOX, M. H. (1968). *Experientia* **24,** 393–394.

THORNTON, D. R. and MCEVOY, J. (1970). *Experienta* **26,** 24–25.

THORNTON, D. R. and GRIFFIN, M. (1971). *Experientia* **27,** 387–388.

TUBAKI, K. (1966). *Trans. mycol. Soc. Japan* **7,** 73–87.

Tubaki, K. (1969). *A. Rep. Inst. Fermentation, Osaka* **4**, 12–41.
Tubaki, K. (1971). 1st International Mycological Congress Abstracts.
Uden, N. van and Castelo-Branco, R. (1963). *Limnol. Oceanogr.* **8**, 323–329.
Uden, N. van and Fell, J. W. (1968). *In* "Advances in Microbiology." (M. R. Droop and E. J. F. Wood, eds), Vol. 1, pp. 167–201. Academic Press, London and New York.
Ulken, A. (1965). *Veröff. Inst. Meeresforsch. Bremerh.* **10**, 289–294.
Ulken, A. (1972). *Veröff. Inst. Meeresforsch. Bremerh.* **13**, 217–230.
Umphlett, C. J. and Koch, W. J. (1969). *Mycologia* **61**, 1021–1030.
Vishniac, H. S. (1955). *Mycologia* **47**, 633–645.
Vishniac, H. S. (1958). *Mycologia* **50**, 66–79.
Webster, J. (1959). *Ann. Bot.* **23**, 595–611.
Webster, J. and Dennis, E. (1967). *New Phytol.* **66**, 307–313.
Webster, J. and Towfik, F. H. (1972). *Trans. Br. mycol. Soc.* **59**, 353–364.
Weston, W. H. (1929). *Mycologia* **21**, 55–76.
Willoughby, L. G. (1962). *J. Ecol.* **50**, 733–759.
Wilson, I. M. (1951). *Trans. Br. mycol. Soc.* **34**, 540–543.
Wilson, I. M. (1954). *Trans. Br. mycol. Soc.* **37**, 272–285.

12

Protozoa

J. D. Stout

Soil Bureau
Department of Scientific and Industrial Research
Lower Hutt
New Zealand

I. Introduction

The decomposition of plant litter may be conceived as any step which initiates or completes a process that leads to the mineralization of plant tissues, whether in the living plant, in tissues eaten by herbivores or of debris falling from the living plant to its substrate. The process of decomposition may vary both with the biochemical pathways followed and with the rate of the reactions. Since the ultimate products of mineralization are simple inorganic compounds, such as water, carbon dioxide, nitrogen, ammonia, or hydrogen sulphide, the overall rate of the reaction is more important than the particular pathway followed. However, differences in metabolic pathways can influence the rate in two ways: first by determining availability of essential nutrients, such as nitrogen, phosphorus or growth factors necessary for the metabolism of the decomposing organisms, and secondly by the release of substances inhibiting further metabolic activity, such as bacteriostatic or fungistatic metabolites.

The pattern and rate of decomposition are affected by the accessibility of the plant material and by its chemical complexibility. Soluble plant residues are obviously more readily accessible to decomposition than solid tissues and simple sugars and amino acids more readily decomposed than complex polymers such as cellulose or lignin. Further some tissues are particularly resistant to decomposition and some plant residues, such as the polyphenols, tend to inhibit the activity of decomposing organisms.

The decomposition of plant litter may follow one of three pathways. The tissues may age, become senescent, and decay first by autolysis and subsequently due to the activity of saprophytes and their associated biocoenoses. Secondly, the tissues may be invaded by parasites, which initiate decay that then follows the pattern of saprophytic decomposition. Thirdly, the tissues may be eaten by vertebrate or invertebrate herbivores, and be decomposed by enzymes produced by the consumer or by its associated enteric population. Protozoa may be associated with all of these three pathways.

The role of protozoa in the decomposition of plant residues is restricted by their morphology, by their biochemical and physiological capacity, and by their distribution in relation to the morphology and life cycle of plants. Since protozoa are microscopic with no strong jaws or piercing mouthparts they play a minimal role in the physical comminution of plant tissue but their small size does ensure that they are associated with almost every stage of the plant life cycle, on growing leaves, in dead tissue, in the gut and faeces of herbivorous animals, in all kinds of water and the soil (Fig. 1). Further, they are an enormously diverse group showing extensive radiation to fill innumerable niches and remarkable convergence both amongst free-living and parasitic forms. Consequently the fauna of any particular habitat is extremely diverse and since protozoa are normally associated with an extensive microflora their exact role in the processes of decomposition is difficult to determine since only a limited range of species can be satisfactorily maintained in axenic culture. Those whose physiology has been studied in axenic culture reveal a very wide range of biochemical pathways including most of the commonly occurring pathways and others not widespread in the animal kingdom. Protozoa include strict anaerobes and strict aerobes as well as facultative species and they are found over a wide range of temperature, moisture, pH and Eh conditions.

The literature on protozoa is extensive but there are good monographs and reviews by Chen (1967*a*, 1967*b*, 1969), by Lwoff (1951) and Hutner and Lwoff (1955), Hutner (1964) and Kidder (1967). Reviews of protozoan ecology are less common but some are to be found in the volumes cited and

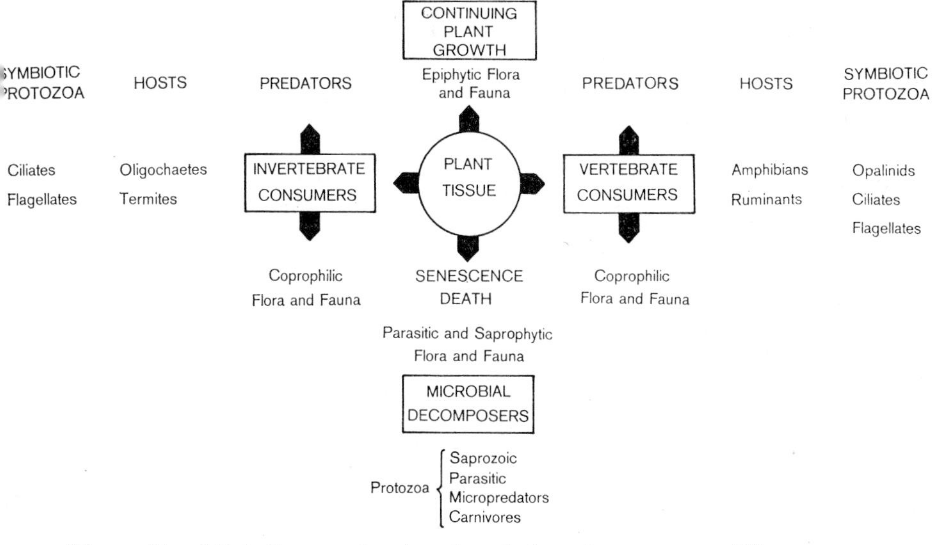

Fig. 1. Simplified diagram showing the relation of protozoa to different stages of the decomposition of plant tissue.

the biology of soil protozoa was reviewed by Stout and Heal (1967). In this chapter the relationship of protozoa to plant decomposition will be discussed primarily in regard to terrestrial and freshwater environments.

II. Nature of the Fauna

A. Biochemical Capacity and Role of Protozoa

1. Carbohydrates and Fats

Most of the metabolic pathways known in animal and bacterial cells have been identified in protozoa but the pattern varies even from species to species. A simplified diagram of carbohydrate metabolism in the protozoa is given in Fig. 2 (after Ryley, 1967). In some cases exogenous sources of sugars, including glucose, may not be metabolized due to the impermeability of the cell membrane. A wide range of polysaccharides are stored by protozoa, including starch and paramylon. Acetate and other simple fatty acids are readily used, as are alcohols. Fermentation is characteristic of most of the gut symbionts, including the termite flagellates and the rumen ciliates. Acetic acid is the principal product of the flagellates whereas the

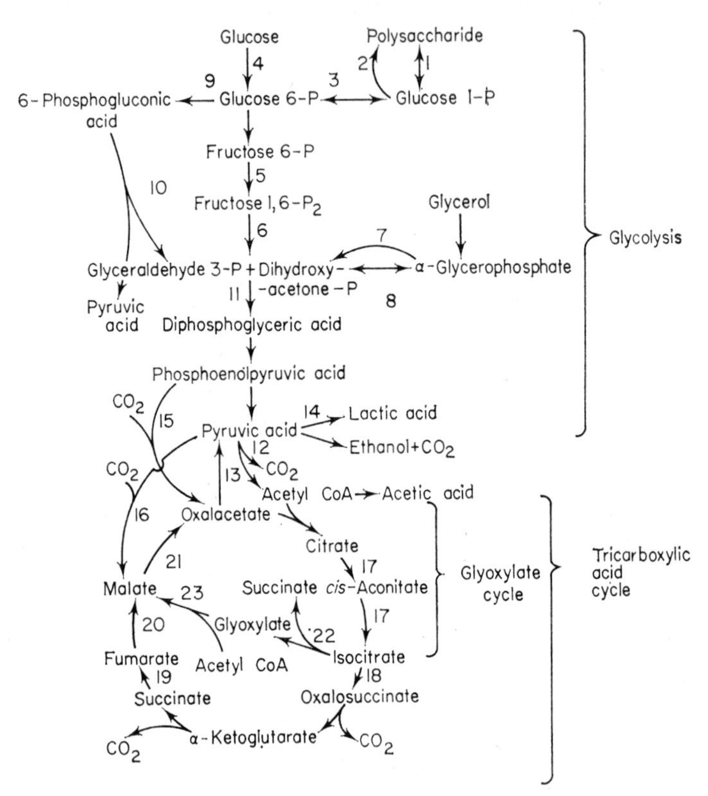

Fig. 2. Pathways of carbohydrate metabolism in the protozoa (reproduced with permission from Ryley, 1967). Metabolic pathways of carbohydrate metabolism in the protozoa. Key to enzymes used in pathways: 1, phosphorylase—degradation and perhaps synthesis of polysaccharide; 2, UDP synthesis of polysaccharide; 3, phosphoglucomutase; 4, hexokinase—uses ATP; 5, phosphofructokinase—uses ATP; 6, aldolase; 7, α-glycerophosphate oxidases—uses molecular oxygen; 8, α-glycerophosphate dehydrogenase—uses NAD; 9, glucose-6-phosphate dehydrogenase—reduces NADP; 10, Entner–Doudoroff cleavage; 11, glyceraldehyde-3-phosphate dehydrogenase—reduces NAD; 12, pyruvic oxidase; 13, oxalacetic decarboxylase; 14, lactic dehydrogenase—uses NADH; 15, phosphoenolpyruvic carboxykinase—requires inosine diphosphate and fixes carbon dioxide; 16, malic enzyme—uses NADPH and fixes carbon dioxide; 17, aconitase; 18, isocitric dehydrogenase—uses NADP; 19, succinic dehydrogenase—no coenzyme involved; 20, fumarase; 21, malic dehydrogenase—uses NAD; 22, isocitratase—liberates succinate; 23, malate synthase—uses acetyl CoA.

ciliates also produce appreciable quantities of propionic and butyric acids. Thus most of the substances comprising the mobile fraction of plant tissues are also metabolized aerobically or anaerobically by protozoa.

Cellulose, hemicellulose, cellobiose and pectin are less widely metabolized but the protozoan symbionts include species capable of attacking these substances.

2. Proteins and Amino Acids

Apart from the phytoflagellates all protozoa require at least one amino acid and most, including the ciliates, require ten. The ciliate *Tetrahymena pyriformis* also requires both purine and pyrimidine. Most protozoa excrete nitrogen as ammonia although the large amoebae can form triuret crystals. Proteins may be a source of fermentation energy for the rumen ciliates which are also remarkable for producing an unusual amino acid, 2-aminoethyl phosphonic acid. *Tetrahymena* has also been shown to have phosphatases.

3. Aromatics

Little information is available concerning the ability of protozoa to use aromatic substances, such as polyphenolics, which constitute an important part of plant residues. Apart from phytoflagellates most protozoa cannot synthesize the three aromatic amino acids invariably found in cell proteins (tryptophan, phenylalanine, and tyrosine) although they are able to catabolize them.

4. Growth Factors

There is considerable variation in the growth factors needed by different protozoan groups. All but a few flagellates require thiamine but otherwise flagellates, ciliates and amoebae show considerable variation in their requirements.

In summary, protozoa are able to break down proteins and simpler carbohydrates. A few are able to attack the larger carbohydrate polymers, such as cellulose, which are the most common rate-limiting reactions in the decomposition of plant residues and some amoebae possess chitinase. Evidence of the ability to decompose phenolics, other than the simpler phenolamino acids, is still wanting. Thus metabolically the Protozoa, with the notable exception of the gut symbionts, have a metabolic potential comparable to that of other animals and the majority of bacteria and fungi.

B. Relation to Moisture, Temperature and other Environmental Factors

Protozoa are aquatic organisms and moisture limits their distribution and activity. In saline environments, in soils or in other habitats subject to high osmotic tensions the fauna is commonly distinguished by the ability to avoid water loss, either by the presence of a test with a very narrow opening or by the ability to form cysts. Moisture relationships provide the most commonly employed parameters of ecological classification. Protozoa may be classified as aerophilic, those found in areas prone to drought; hydrophilic, those found in permanently aquatic habitats; or osmophilic, those found in saline habitats.

Temperature directly affects the activity and metabolism of protozoa but the great majority of species can tolerate a very wide range of temperatures. Species found in most hot pools are generally those found in ordinary

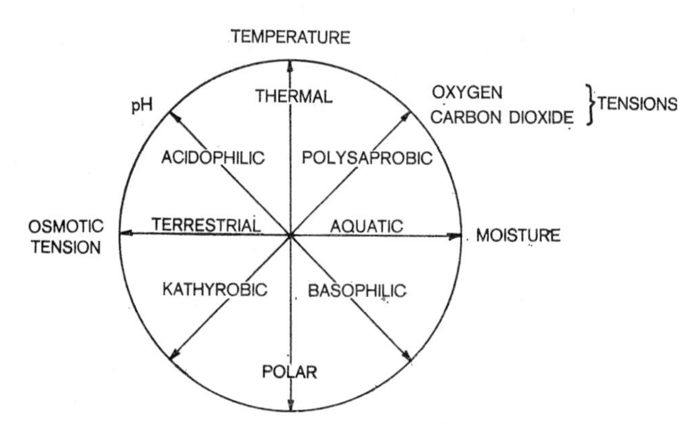

Fig. 3. Relation of protozoa to environmental factors.

waters unless the temperature is over 50°C. Although different varieties are found in separate geographical regions in only a few instances is there evidence of faunal differences being due to variation in temperature. One possible example concerns *Paramecium* and *Neobursaridium* with the former being a typical temperate ciliate and the latter a tropical species.

There is a wide range of tolerance of most other environmental factors, such as oxygen tension, carbon dioxide concentration, pH and Eh. In general there are stenotopic and eurytopic species: those found only within narrow limits and those found over a wide range of values.

The consequences of these variations of tolerance in relation to the environment is that although single species may be widespread, distinctive communities are found associated with different habitats. Since protozoa are commonly associated with bacterial populations there are significant correlations with the intensity of bacterial activity, from kathyrobic environments with little bacterial activity; through oligosaprobic, and mesosaprobic to polysaprobic environments with intense bacterial activity (Fig. 3).

C. Methods of Study

1. Field Studies

Generally protozoa are too small to be readily observed in the field, but in fresh water or terrestrial situations it is possible to use slides or capillaries to provide a means of sampling the field population. This can be useful in monitoring changes caused by pollution in a stream or lake or by variations in temperature or moisture in soil. By using a canula, the effect of changes in diet of a ruminant on the entozoic fauna can also be followed *in vivo*.

2. Direct Observation

Strictly direct observation can only be carried out in an experimental situation and not in the field but samples from the field may be examined directly or preserved and then studied. Direct observation of samples is easiest for a freshwater habitat, but is difficult for a terrestrial habitat or for a parasitic or symbiotic fauna either because it is difficult to observe the protozoa in their natural *milieu* or because the physical conditions of the environment are changed too drastically, as with the exposure of ruminant fluid to the aerobic atmosphere. Direct sampling is probably the most common method of assessing the fauna.

3. Extraction

Extraction of the protozoan fauna of natural habitats can be carried out in some cases. Thecamoeban shells and ciliates can be extracted from soil by using gas flotation and an electric current, respectively. Populations in a fluid *milieu* can be concentrated by centrifugation or selectively separated by varying heat or light gradients.

4. Culture

Culturing is the most valuable method of study since it provides material on which detailed morphological, physiological or biochemical studies can be carried out. Many protozoa, particularly symbionts, are, however, difficult to culture. Although mixed cultures are most widely used, single species cultures are ideal and particularly if it is possible to maintain the species in axenic culture, free of other living organisms. The use of antibiotics has greatly facilitated the establishment of axenic cultures but this is still difficult for slow-growing species.

D. The Free-living Fauna

The predominant groups represented in freshwater and terrestrial environments are given below. The classification follows Honigberg *et al.* (1964) and Corliss (1961).

1. The Phytoflagellates

There are ten orders, some found rarely in fresh water (Coccolithophorida), some commonly in fresh water and rarely in soil (Dinoflagellida), and a few that are relatively ubiquitous (Euglenida). Among the more common in fresh water are the Chrysomonadida with yellow to yellow-green to brown chromatophores and often amoeboid forms with a typically siliceous cyst wall, the Cryptomonadida with two chromatophores, often brown, but also red, olive-green, or blue-green and without amoeboid forms, the Dinoflagellida with their strongly grooved body with the flagella within the grooves and usually yellow or dark brown chromatophores, the Euglenida with their typical "metabolic" but not amoeboid movement and green chromatophores, and the colonial Volvocida. Phytoflagellates are uncommon in soil though they occur on its surface in many terrestrial habitats.

2. The Zooflagellates

There are four orders with free-living members. These include two small orders of sessile flagellates which are found in fresh water (Choanoflagellida and Bicosoecida), the amoeboid flagellates (Rhizomastigida) found in both fresh water and soil, and the Kinetoplastida, which include the parasitic trypanosomes and the free-living Bodonina. The last group includes the most common soil flagellates such as *Cercomonas* Dujardin, *Bodo* Ehrenberg, and *Monas* O. F. Müller, which are also found in fresh water.

3. The Rhizopods

The shelled and naked amoebae are grouped into five subclasses by the form of their pseudopodia. Each of the five subclasses includes naked amoeboid forms found in soil and generally also in fresh water. These are the Amoebida, the typical amoebae with relatively simple pseudopodia, the Aconchulinida with tapering and branching filopodia, the Athalamida, with delicate reticulopodia, the Acrasida and Eumycetozoida with a complex life cycle involving a multinucleate plasmodium and typically sexual reproduction, and the Labryrinthulida with no obvious pseudopodia but which glide on a network of mucous tracks. The shelled or testate forms found in soil and fresh water fall almost exclusively into two orders, the Arcellinida, with lobose pseudopodia, and the Gromiida, with filose pseudopodia.

4. The Actinopods

The two subclasses of Actinopodea found in soil and fresh water are the Heliozoia, with axopodia, represented by three orders, the Actinophryida, the Centrohelida, and the Desmothoracida, and the Proteomyxidia, an order with filopodia and reticulopodia which includes both parasitic and free-living forms. Only one heliozoan species, *Actinophrys sol* Ehrenberg, occurs commonly in soil, the majority being found only in fresh water. The Proteomyxidia occur in both fresh water and soil.

5. The Ciliates

Ten free-living orders are found in soil and freshwater habitats. They are grouped into four subclasses: the relatively simple structured Holotrichia, the typically sessile Peritrichia, the predatory and commonly sessile Suctoria with no cilia in the adult stage, and the morphologically elaborate Spirotrichia, commonly with reduced body ciliature but with

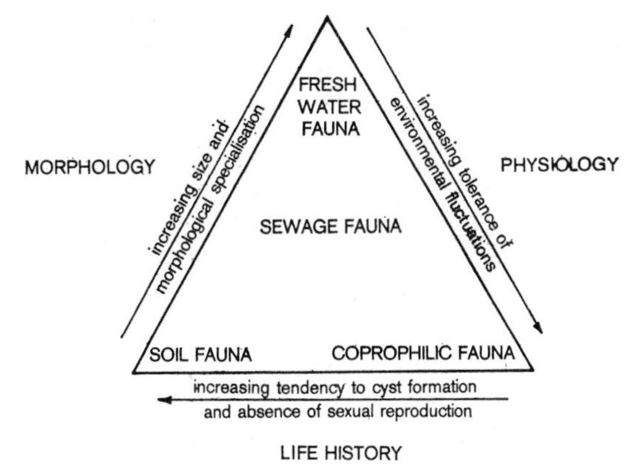

Fig. 4. Differences in the morphology, physiology and life-history of free-living protozoa in relation to their ecology.

elaborate oral ciliature. In general the soil fauna consists predominantly of holotrichs and spirotrichs while peritrichs and Suctoria are poorly represented. The freshwater fauna has more representatives of these latter groups.

Faunal differences between freshwater and terrestrial environments commonly relate to the morphology, physiology and life history of the species (Fig. 4). Species are normally able to migrate in freshwater to favourable conditions. Soil microhabitats are discrete and the moisture regime, as well as other environmental factors, fluctuates markedly so that the protozoa must be able to survive such fluctuations. In soils the microflora tends to consist predominantly of bacteria, yeasts and moulds whereas in fresh water algae may be dominant. Consequently the range of protozoa is greater in fresh water than in soil although a number of species can be found in both habitats. Terrestrial forms tend to be smaller than freshwater forms, since there is generally less living space and even within related habitats, such as the growing tip of *Sphagnum* and at its base, the populations of the same species are smaller in the more physically restricted habitat (Heal, 1962). Larger and morphologically more elaborate species are confined to large bodies of fresh water. The majority of terrestrial protozoa tend to be bacteriophagous rather than algivorous, although some are carnivorous or omnivorous. They also tend to be tolerant of fluctuating conditions of aeration, pH and Eh, and desiccation. This is commonly associated either with morphological adaptations, small size and limited oral openings for testate forms, or physiological behaviour, such as anabiosis or cyst formation. Coprophilic and sewage species also show a similar ability to tolerate fluctuating environmental conditions, large bacterial populations and often desiccation. Since terrestrial habitats commonly offer more sharply fluctuating conditions than typical freshwater habitats, the communities are less complex and the food chains simpler. However because of the importance of encystment the life-history of terrestrial species may be more complex than that of freshwater species. Comparison of the typical soil ciliate *Colpoda* with the typical fresh water ciliate *Paramecium* illustrates many of these points. *Paramecium* is typically larger, with a more elaborate oral apparatus and is able to migrate from unfavourable to favourable loci. Sexual reproduction occurs but it does not encyst. *Colpoda* is structurally simpler, smaller and reacts to adverse environmental conditions by encystment. The cysts may survive in dry soil for years but there is no record of sexual reproduction.

E. Symbionts and Commensals

1. The Zooflagellates

There are five orders with entozoic parasites in a wide range of vertebrate and invertebrate hosts. These are the Retortamonadida, comparatively simple forms with two to four flagella, and the Diplomonadida, which are bilaterally symmetrical flagellates in which the flagella, nuclei and other organelles have been duplicated, both occurring in vertebrate and invertebrate hosts, the Oxymonadida, found in termites, with an attached stage with a rostellum, the Trichomonadida, with four or more free anterior flagella and one trailing attached flagellum, occurring in a wide range of hosts, and the Hypermastigida, with numerous flagella, occurring in beetles and termites.

2. The Opalinids

These have numerous cilia-like organelles and occur in the gut of fish and amphibia. Their precise taxonomic status is uncertain.

3. The Rhizopods

These include a wide range of parasitic amoebae which occur in both vertebrate or invertebrate hosts and are classed in the order Amoebida and a parasitic order of Mycetozoia, the Plasmodiophorida, which includes plasmodial forms developing inside the host cells of both plants and animals with an elaborate life cycle involving spores and flagellate stages.

4. The Actinopods

There is one order, the Proteomyxida, which includes plant parasites, with flagellate swarmers and cysts present in some species.

5. The Ciliates

There are five ciliate orders with members parasitic in the alimentary canal of vertebrate or invertebrate hosts. There are three orders of holotrichs which include important entozoic ciliates. The Gymnostomatida, which includes the families Buetschliidae, with many entozoic species in herbivorous mammals, particularly camels and horses, and the Pycnotrichidae, also occurring in the mammalian gut; the Trichostomatida, including the entozoic families Isotrichidae, present in ungulate ruminants, the Plagiopylidae, including species known from the digestive tract of sea-urchins, the Blepharocorythidae in the digestive tract of horses and ruminants, the Balantidiidae, with the single genus *Balantidium* Claparede and Lachman known from invertebrate and vertebrate hosts, including man, and the Entorhipididae occurring in sea-urchins. The Astomatida includes ciliates in the annelid gut. Entozoic spirotrichs include two orders, the Heterotrichida, which includes the families Plagiotomidae, an entozoic group in both vertebrates and invertebrates, and Clevelandellidae, found in insects and the Entodiniomorphida, which includes many herbivore symbionts, particularly those of ruminants.

The role of symbionts and commensals in the decomposition of plant debris varies enormously (Cleveland, 1960; Coleman, 1963; Hungate, 1955, 1960, 1966). Many of the flagellates are of critical importance in the decomposition of cellulose and consequently in the nutrition of their host, but most symbionts are probably dependent, directly or indirectly, upon bacteria as the primary decomposers. Their metabolism also varies and consequently so do the end-products that become available to the host. Thus although glucose, fructose and sucrose are metabolized by rumen ciliates and largely converted to cellular starch, the holotrichs ferment part of the substrate to lactic, butyric and acetic acids whereas the entodiniomorphs produce little lactic acid and relatively more acetic and butyric acids. The importance of the rumen ciliates is difficult to assess but it has been calculated that of the rumen nitrogen 46% is bacterial, 21% protozoan,

26% plant and 7% soluble, and it has also been calculated that 20% of the fermentation acids are produced by the Protozoa. However although all this material is derived from the ingested fodder the protozoan component may well have been metabolized first by the bacterial flora. It is possible, for example, that their nitrogen requirements can be satisfied by ammonia and a few amino acids which could be produced by bacterial metabolism. The difficulty of assessing the role of protozoa arises from their very close association with bacteria which may remain metabolically active after ingestion.

The cellulolytic symbiotic flagellates act as definite primary consumers. They are associated with the major wood-eating insects, the roach *Cryptocerous* and wood-eating termites, whose high rate of metabolism is dependent upon the presence of the symbionts.

In *Cryptocerous* there are 9 families, 14 genera and over 30 species of flagellates and there are many more genera in the termites. Many of these insect flagellates are unusually large cells and they tend to pack the greatly enlarged hind-gut of the host. No other type of host has such an abundance of parasites.

However not all these species are able to digest cellulose, the cellulolytic groups being the genera *Trichonympha* Leidy and *Trichomonas* Donné.

Many rumen ciliates may ingest plant debris including cellulose but it appears that only some groups, and indeed sometimes only some species of some genera are able to digest cellulose. Like other carbohydrate substrates much of it is converted to a form of starch reserve upon which the cell depends when starved. Similarly only some of the ciliates are able to digest pectin. Thus although there is a large protozoan fauna associated with ruminants there are significant differences in the metabolic activity of different species and consequently in their role in the decomposition of plant debris.

Numbers of rumen ciliates vary with the kind and supply of food and this may also affect the specific composition of the fauna; but concentrations of up to one million ciliates per ml of rumen fluid have been recorded.

The diversity of both the flagellate and ciliate faunas indicates both radiation and convergence. There is limited data on transfaunation experiments but clearly the essential requirements are a copious supply of food in a relatively stable and typically anaerobic environment. Of the chief flagellate orders the Oxymonadida and Hypermastigida occur in both the roach and termites while the Trichomonadida occur in termites and also in vertebrates. The common rumen holotrichs belong to the family Isotrichidae, but other holotrichous ciliates have been recorded from the rumen, and although the more specialized Entodiniomorpha occur principally in ruminants they have also been recovered from primate hosts. There is therefore great variation in the nature, size and role of the protozoan symbionts of vertebrate and invertebrate herbivores (Fig. 5). Apart from the groups discussed little is known of the role of other protozoan symbionts, such as opalinids or amoebae, but protozoa are present in the gut of representatives of most animal groups and they may constitute a significant entozoic population in many cases. They undoubtedly constitute the most important group of protozoa associated with the decomposition of plant

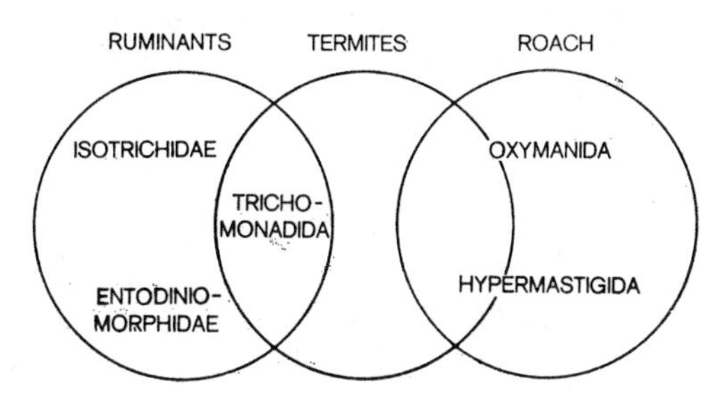

Fig. 5. Relationship of the entozoic flagellate and ciliate protozoa of the wood roach, termites and ruminants.

litter, of wood in the roach and termites, and of grass, grain, and sometimes seaweed in the case of the ruminants. There are two reasons for their effectiveness; the first is the favourable conditions offered by the habitats occupied and the second is the high degree of comminution of the plant debris. This is achieved initially by the insects during ingestion of the wood material and continually, by rumination, in the vertebrate hosts.

III. Communities Associated with Different Types of Litter

Apart from the broad zonal parameters of moisture and temperature, two aspects of plant litter determine the composition of their microbiocoenoses and their protozoan fauna. The first is differences of plant structure and in the plant community and the second is the quantity of plant litter and its rate of breakdown.

The simplest community of photosynthetic plants consists of a pure culture of unicellular algae. Where their distribution is restricted, their growth limited and environmental conditions adverse a microbiocoenose consisting of only a few species may exist. This is found in some of the very arid soils of Antarctica, in waters derived from sulphur fumaroles, or in very restricted habitats such as the reservoirs of insect-trapping plants. At the other extreme is the infinitely complex interrelationship of a tropical forest with its succession of plant strata, diversity of plant species and

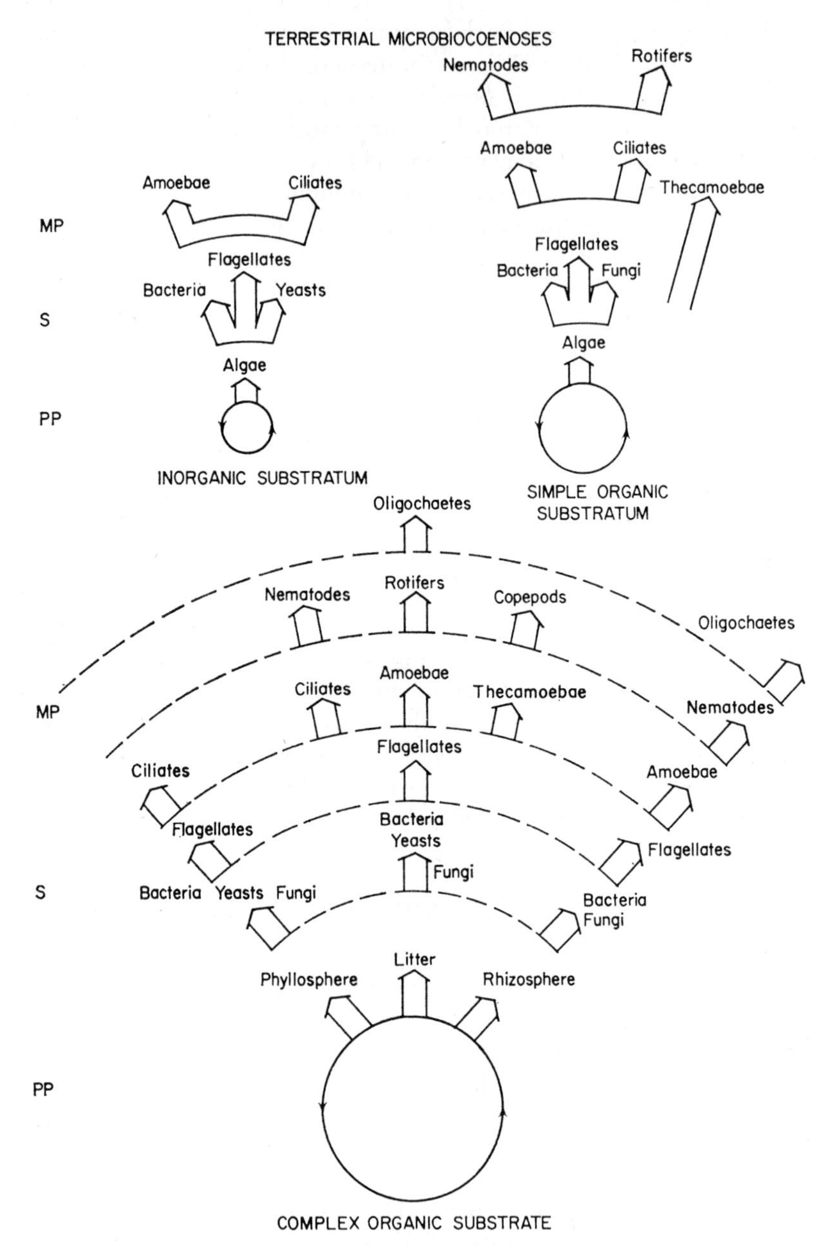

Fig. 6. Faunal composition of different terrestrial habitats in relation to the accumulation and turnover of plant debris. PP, primary producers; S, saprophytes; MP, micropredators.

abundant microbial and animal populations. Such an environment provides an enormous range of habitats for microbial life and consequently a diversity of microbiocoenoses. A square metre of the floor of a tropical forest may contain leaves from fifty species of plants and each type of leaf may have a different microflora and microfauna associated with it.

Productivity of plant debris is not a function of floral diversity: a pure stand of a single species may be much more productive than mixed vegetation. Productivity is related primarily to climate and the supply of plant nutrients. Where nutrients are not limiting, moisture and temperature determine plant growth but often where moisture and temperature are favourable there is only a limited supply of plant nutrients, as in many tropical soils and waters. Consequently productivity is often highest in cooler waters and in temperate soils. The decomposition of plant litter is commonly restricted more by moisture and temperature than by nutrient supply, since in most cases the process of decomposition itself releases nutrients for the decomposing organisms. For this reason litter tends to break down more slowly in temperate than in tropical conditions. Over-all, the balance between productivity of plant litter and its decomposition tends to distinguish polar regions with relatively low productivity and low rates of decomposition; temperate regions with moderate to high productivity and moderate to slow decomposition; and tropical regions with low to high productivity and rapid decomposition.

A. Lower Plant Litter

1. Bacteria, Actinomycetes, and Fungi

The most intimate relationships between protozoa and other organisms are with the microflora which normally provide the bulk of their food and may secrete substances which affect their activity, growth and encystment or excystment.

Because of their small size and large populations bacteria and yeasts provide the most readily available source of food for protozoa although larger protozoa may feed on other small cells such as those of flagellates, small protozoa, and more rarely, algae. Large or filamentous micro-organisms are not commonly ingested while fungal spores, even when ingested, may not be digested.

Besides their shape and size the edibility of micro-organisms may vary with their cultural condition and their taxonomic status. Capsulated forms are less edible than non-capsulated, and in some cases the production of pigment may inhibit growth and reproduction. In general fast-growing zymogenic organisms are preferred to slow-growing autochthonous microflora. This appears to be true for both bacteria and yeasts, so that nutrients which stimulate the growth of the zymogenous flora will also favour the growth of protozoa. The zymogenous flora is most commonly found in microhabitats relatively rich in nutrients, in the animal gut, in the phylloplane or in the rhizosphere, and this is where the largest protozoan populations are found. The autochthonous microflora normally forms the greatest part of the bacterial biomass of soil where micropredation by protozoa must be a major selective factor in determining the nature of the flora. The microflora may be classified in descending order of edibility (Heal and Felton, 1970):

1. Bacteria: Pseudomonadaceae and Enterobacteriaceae
   Yeasts: *Rhodotorula*, *Saccharomyces*, *Kloeckera*
2. Bacteria: *Chromobacter* and *Arthrobacter*
3. Bacteria: *Bacillus*, *Nocardia* (bacteria-like forms)
   Yeasts: *Hansenula*, *Pichia*, *Cryptococcus* and *Candida*
4. Bacteria: *Mycobacterium*, *Cytophaga*
   Fungi: a few mycelial forms
5. Bacteria: filamentous *Nocardia*
   Algae: some species
6. Actinomycetes
   Most fungal mycelium
   Most algae

In general suitable food micro-organisms either produce no exudates affecting protozoan activity or else exudates which stimulate protozoan activity (Fig. 7).

2. *Algae and Lichens*

The lower plants form the simplest plant communities, the least diversity of plant structure, and generally the smallest amounts of litter.

Algae and lichens have a very simple form, without stems, leaves or roots. Generally the algal flora of aquatic communities is more diverse than that of terrestrial communities and so too is the protozoan fauna. A very simple community, such as is found in Antarctic soils may have only algae, a few types of bacteria and yeasts, and saprozoic or bacteriophagous protozoa such as *Leptopharynx* Mermod (Flint and Stout, 1960). In an aquatic environment there is greater mobility of the motile organisms, which are able to migrate from one part of the habitat to another, and since activity is not severely restricted either by temperature or by desiccation, a more elaborate food-chain tends to be established. In particular herbivorous protozoa, such as large amoebae or ciliates such as *Nassula* Ehrenberg, and also detritus feeders, such as *Coleps*, tend to be present.

Consequently protozoa tend to play a more important role both in the direct consumption of the plant cells and in the destruction of the plant detritus than they do in Antarctic soils where they are typically microphagous, subsisting on the decomposer cycle rather than on the primary producers (Fauré-Fremiet, 1950, 1967).

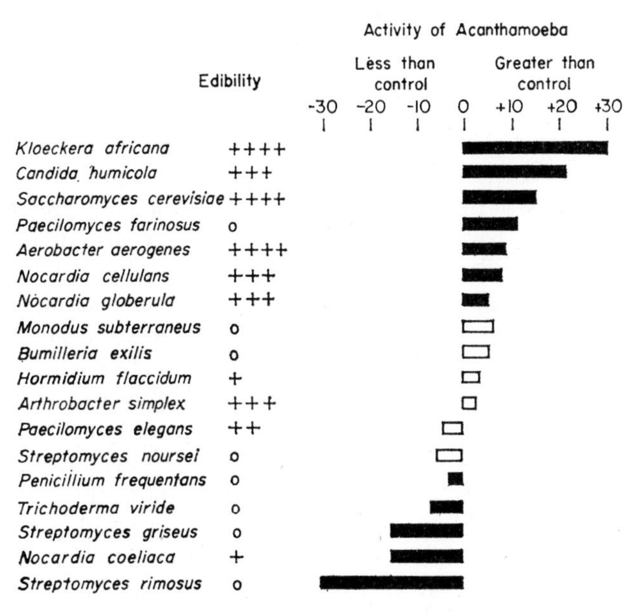

Fig. 7. Relationship between the edibility of microflora and the effect of their exudates on amoeba activity (reproduced with permission from Heal and Felton (1970)). Edibility is expressed as the reproduction of amoebae: O, none; +, very slight; ++, moderate; +++, abundant but encystment before all the food is consumed; ++++, abundant and food is completely consumed. Effect of exudate is expressed as the difference in number of active amoebae between test and control: unshaded, no significant difference; shaded, difference significant, $P < 0{\cdot}05$.

3. *Mosses and Liverworts*

Although they are only a little more complex than the algae the mosses in effect provide greater complexity of structure, primarily because of the differentiation of living, senescent and dead tissue associated with their growth form. Although algal detritus may accumulate to form peat, this is an exceptional situation, whereas *Sphagnum* has been and still is one of the major peat forming plants. The tendency to accumulate plant debris provides a much more stable habitat both physically and nutritionally for the associated flora and fauna and consequently the moss microfauna is significantly more diverse than that of algae and lichens in comparable situations (Fantham and Porter, 1945).

Thus the protozoan fauna of Antarctic moss is generally more diverse than that of Antarctic lichen or algal communities. There is a greater differentiation between the mineral soil associated with moss and that away from moss than occurs with the algal communities and there is often a greater mass of plant material than elsewhere in this terrestrial environment. The protozoan fauna is characterized both by the presence of more predatory species, including Suctoria, and by the occurrence of shelled amoebae such as *Corythion* Taranék and *Euglypha* Dujardin, otherwise only rarely encountered in Antarctica (Heal *et al.*, 1967). Where there is profuse moss growth this is often associated with warmer and wetter sites and the microflora, particularly the bacterial flora, is also different to that of other Antarctic soils (Flint and Stout, 1960).

The structure of a *Sphagnum* community is still more complex. A distinction can be made between the uppermost green leaves and stems, the older brown tissues and the decaying material. This distinction is associated with differences in microflora and microfauna. Not only does the size and diversity of the fauna increase with increasing decay but even with particular protozoa (e.g. *Nebela*) their size is less on the green leaves exposed to the air than in the dead *Sphagnum*. This is due not to the supply of food but to the available living space for the protozoa live within the thin film of moisture covering the leaf and large protozoa would be at a disadvantage when the leaves dry out. At the base of the *Sphagnum*, however, moisture is normally abundant and living space is not a limiting factor (Heal, 1962).

The second distinction between the younger tissue and the decaying tissue is in the diversity of the protozoan fauna. This is associated with a wider range of other microfauna and microflora which provide a greater diversity of nutrient substrates, and a physically more stable environment less subject to moisture and temperature fluctuations. This is the typical litter habitat which recurs throughout a wide range of environments wherever plant debris accumulates to create a predominantly organic substrate (Stout, 1968). It commonly harbours many species of thecamoebae, particularly Euglyphidae, Nebelidae and sometimes Difflugiidae.

4. Ferns

With the development of leaves, rhizomes and roots the ferns extend the complexity of plant structure and consequently of microhabitats independent of the stages of decay. Moisture, however, is more of a limiting factor than with *Sphagnum* and the fauna is more comparable to that of forest litter.

B. Herbaceous Litter

Herbaceous plants may be ephemeral, annual or perennial and they may form an open or closed cover of surface growth or a tussock. Consequently there is a great range of plant litter produced, of decomposition cycles and of available habitats. Ephemeral herbs which may be dormant for years occur in desert regions and the protozoan fauna associated with such deserts is extremely limited, with only a few species present. These are typically the small microphagous ciliates, such as *Colpoda* O. F. Müller, capable of forming highly resistant cysts. A much greater range of species is associated with annuals and perennials but where there is little standing vegetation the habitats are restricted and the nature of the fauna is closely related to the nature of the climate and soil or with aquatic vegetation to the properties of the water. Aeration, pH and Eh may be important as well as the physical structure of the habitat.

1. Meadows

Characteristically in a meadow there is a contrast between the fauna of the green leaf, of the dead grass, and of the soil. Ciliates and flagellates are dominant on the green leaf. This is because they can complete their life history within a very short space of time and since the leaf surface is subject to sharply fluctuating moisture conditions this is essential. The dominant ciliate is *Colpoda cucullus* O. F. Müller which forms strongly resistant cysts, can excyst and encyst rapidly and completes its life history within the cyst membrane (Bamforth, 1973) and *Chilodonella* Strand, a small flattened holotrichous ciliate, also occurs in the phylloplane. In the few hours following dew formation the leaf provides an adequate habitat for these ciliates. They feed on bacteria and yeasts which occur in enormous numbers on the leaf surface. Flagellates and ciliates, particularly Colpodidae include not only mirophagous species like *Colpoda cucullus* and *Colpoda steinii* Maupas but also carnivores such as *Bresslaua* Kahl (Claff *et al.*, 1941). The soil beneath such grassland offers a more stable environment and has a wider range of species. Thecamoebae, uncommon on the living vegetation are much more numerous. Some species may feed on algae or ingest dead plant debris but the majority feed either on the bacterial or yeast populations or on other protozoa.

2. Tussock

Where perennial grasses or herbs form a tussock growth, there is a differentiation of habitats such as occurs with *Sphagnum*. It is possible to distinguish between the living green tissue where the plant cells are intact and the fauna is associated with the phylloplane, the senescent leaf or stem in which damaged tissue allows entry for both microflora and microfauna and provides a nutritionally rich substrate for their growth, and a lower stratum of decaying vegetation tissue, which is both physically more stable if less nutritious. Like *Sphagnum* there is a differentiation both in the microflora and microfauna, with diversity and complexity commonly associated with increasing decay. There are fewer ciliate species associated with the tussock leaves than with the leaves of meadow grasses but thecamoebae, such as *Euglypha* Dujardin and *Assulina* Ehrenberg, occur on the older leaves. The complexity of the community is related to the climate, the size and age of the tussock and to the plant species. The most striking contrast with the protozoan community of meadows is the greater diversity of species particularly in the dead tillers and litter layers.

C. Forest Litter

Forests vary greatly in their structure, their productivity and their turnover. The physical conditions of the microclimate and consequently the microhabitats are generally important in determining the composition of the fauna but it may also be affected by the nature of the plant litter and the rapidity with which it is decomposed.

Forests, in general, offer a more stable microclimate than grass or moss. This is more pronounced in evergreen forest than deciduous forest and in dense rainforest than open xerophilic forest.

Similarly forests vary greatly in the composition of their litter: in some it is predominantly leaves, others have a higher proportion of twigs, bark branches, seeds or cones. The leaves themselves vary greatly from the modified stems of *Casuarina*, the needles of *Pinus*, the small leaves of many temperate trees and the relatively enormous leaves of many tropical plants. The leaves also differ in physical structure, the degree of siliceous hardening, and their chemical composition. The amount of litter varies and there is seasonal variation in the fall of litter and the rate at which it decays.

As a consequence populations associated with the decomposition of forest litter can vary with plant species, with season, with climate, with soil and with the structure of the forest and there is no simple way to recognize all these variables. The concept of mull and mor developed in relation to the forests of northern Europe provides one way of classifying forest soils but is not entirely satisfactory as a means of grouping the enormous range of forest habitats which are found throughout the world. It is perhaps easier to select a very restricted range of forest plants and to discuss the kinds of fauna which may be associated with a range of ecological conditions.

1. Dicotyledonous Litter

Beech occurs in the cold temperate region in both the northern (*Fagus*) and southern (*Nothofagus*) hemisphere and although there are wide specific differences there are considerable similarities in the composition and size of the beech leaves, which range up to a maximum length of 10 cm. Beech forest has been an important type of vegetation since the Cretaceous, it occurs on a wide range of soil types, and it is the classical example of the distinction between mull and mor. This distinction occurs not only between different forests but also in the same forest in different years and differences in thecamoeban fauna of these two types of forest floor were noted by P. F. Müller when he first distinguished between the two communities (Stout, 1968).

In general, the amoeboid, flagellate and ciliate fauna of beech forests is similar throughout the world but there is a distinction, especially between the northern and southern hemisphere, in the composition of the thecamoeban fauna. In European *Fagus* woods for example, the Centropyxidae are more important than the Nebelidae, whereas the reverse is commonly the case in the Australasian *Nothofagus* forests (Stout, 1969). The Euglyphidae appear to be equally represented in both hemispheres.

There are well over a hundred protozoan species that may occur in the soil and litter horizons of beech forest (Fig. 8) but the great majority of these are saprozoic, microphagous (feeding on bacteria and yeasts), mycophagous, predators or parasites. Some amoebae, thecamoebae and ciliates ingest plant detritus but although it has been suggested that thecamoebae can feed on humic substances convincing experimental data are still lacking (Schönborn, 1965). In the main the evidence strongly suggests that they lack the cellulolytic, lignolytic and even pectinolytic enzymes which would enable them to decompose directly plant residues in the litter. Only the simpler carbohydrates and amino acids are likely to be assimilated and there is intense competition for these by the microflora.

The composition of the fauna and the size of the population not only varies from site to site and from one type of ground cover to another but can also be affected by the addition of fertilizers. In one set of experiments it was found that the addition of urea reduced not only the number of thecamoeban species but also the size of the population and the biomass. The addition of potash basic slag also depressed the thecamoeban population but to a lesser extent. However the addition of both urea and potash basic slag had little affect on the number of species present and greatly increased the size of the population and the biomass. The adverse effect of urea was attributed to the high level of ammonia following hydrolysis and the physical effects of the fertilizer on water retention. The beneficial effects of the combined fertilizers was attributed to the greatly enhanced biological activity, particularly of earthworms, which made more substrate available, including silica from which tests are synthesized (Chardez *et al.*, 1972). These experiments illustrate the interrelation of physical, chemical and

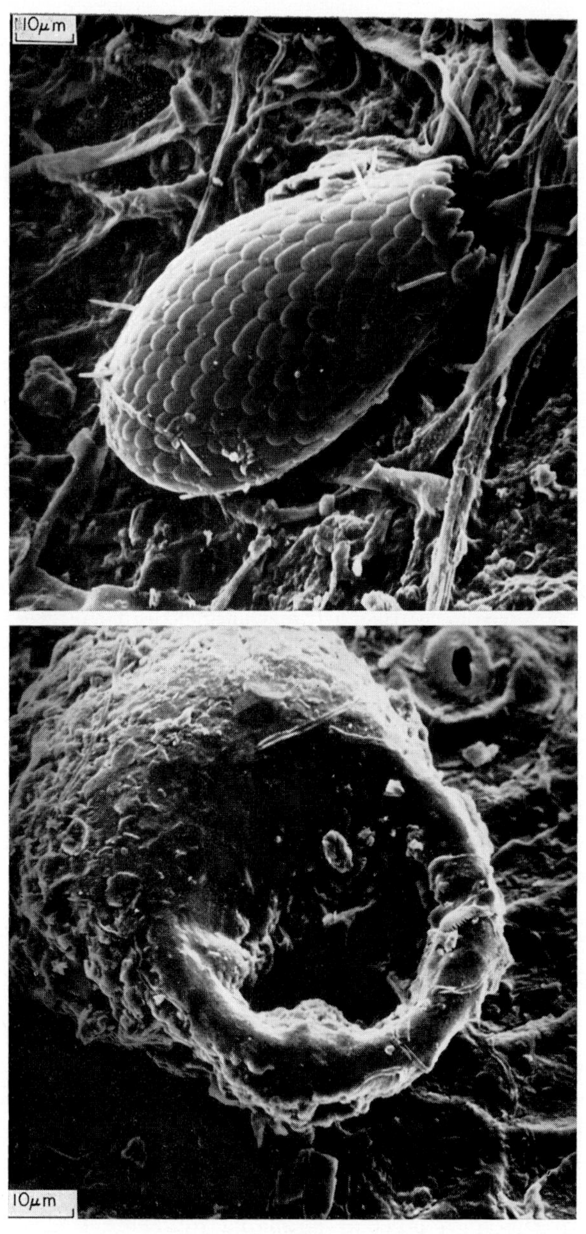

Fig. 8. Tests of two typical thecamoebae associated with decomposing beech litter. Upper, *Euglypha ciliata*, (Ehrenberg) Leidy lateral view; Lower, *Centropyxis sylvatica* (Deflandre) Thomas, ventral view, both on leaves of *Nothofagus truncata* which had been in the forest litter for one year. Note fungal mycelium in the upper photograph and the stoma in the lower photograph. *Centropyxis* Stein shows the mouth opening restricted by a diaphragm, an adaptation to terrestrial life.

biological factors in an ecosystem and how changes in any one factor can alter the activity of a population and consequently its role in plant litter decomposition.

2. *Coniferous Litter*

In the Northern Hemisphere the principal conifers are the *Pinus*, *Picea* and *Abies* forests which commonly occur at higher altitude or latitudes than the dicotyledonous forest. In the Southern Hemisphere, podocarps are amongst the more important native conifers and they occur in both tropical and temperate regions. In tropical regions there tends to be a much more rapid breakdown of the forest litter and consequently the depth of the litter is very shallow. In temperate regions both the northern and the southern conifers tend to accumulate a substantial mass of litter which is commonly sharply divided into different layers or horizons. In general the conifer litter is more fibrous, differs in chemical composition from dicotyledon litter and tends to be more acid. The protozoan fauna is normally less diverse than that of beech forests and the populations tend to be smaller but there is great variation from horizon to horizon and from soil to soil. Under natural conditions conifer forests tend to be wetter than the comparable dicotyledonous forest but both occur over a wide climatic range. The greater humidity of conifer forests is not reflected in large protozoan populations not in greater biological activity and on the whole the protozoa are less important in the decomposition of conifer litter than in that of dicotyledons.

3. Casuarina *Litter*

Casuarina occurs in Australia and throughout the Pacific region. It has nitrogen-fixing symbionts in root nodules, its leaves are modified stems and the trees tend to occur in relatively isolated clumps. The litter is physically comparable to that of conifers, being fibrous or granular. The microbial population is also similar in that there is very active fungal growth and the mycelium can matt the litter into a dense layer. However intense mycelial growth, ending with sporulation, occurs only in the wet season. During the dry season there is very limited biological activity. When the litter becomes wet again spores germinate, dormant propagules revive and a fresh cycle of mycelial growth takes place. The protozoa associated with this litter show a similar adaptation. There is a high proportion of ciliates of the family Colpodidae, capable of forming very long lived cysts. These ciliates are able to excyst, grow and encyst before the extensive mycelial matt develops. Other protozoa, whether other ciliate species or amoeboid protozoa, are poorly represented. The thecamoebae are represented by

Centropyxis and the smaller Euglyphidae. In this case therefore there is an alternation not only between the wet and dry seasons but also between the brief period of ciliate and bacterial activity and that of mycelial growth.

D. Peat

The accumulation of plant litter is normally associated with all forms of plant growth but is much more pronounced in some situations. Massive accumulations of plant litter may be found where low temperatures inhibit plant decomposition more than plant growth in polar, sub-polar and alpine regions; in many forest situations, where the nature of the plant debris retards the rate of decomposition; and in situations where a high water table limits the activity of many organisms normally involved in plant decomposition.

There is a very real difference in the microbiology of zonal peats—those formed in cold and wet situations—and intrazonal peats—those formed because of a high water table or because of the peculiarity of the plant debris—and such differences are reflected in the protozoan fauna. In a zonal peat microbial activity is limited only by temperature and a relatively high rate of respiration occurs when the temperature is raised. In intrazonal peats respiratory activities are constantly low, reflecting positive inhibition of microbial activity or a poor supply of accessible nutrients.

In zonal peats the protozoan fauna is characterized by the predominance of thecamoebae. There are generally more species of shelled and naked amoebae than of ciliates and in the southern hemisphere the Nebelidae form a conspicuous part of the thecamoebae. A similar thecamoeban fauna can be found in cold temperate forests, such as the southern beech forests, but there the ciliate fauna is much better represented. The Nebelidae are poorly represented in the lowland intrazonal peats. As the decomposition of the plant debris increases leading to the formation of a more mineral soil, ciliates and small amoebae become the more important components of the fauna and the composition of the thecamoeban fauna changes, with small Euglyphidae and Centropyxidae being predominant.

E. Wood

It seems unlikely that free-living protozoa play any specific role in the decomposition of wood in nature. They will however be associated with other microbial populations colonizing wood and also with coprophilic populations associated with wood eating animals. Mycetozoa, which can attain a relatively large biomass, may be associated with decaying wood.

F. Digested Litter—Excreta of Herbivores and Faecal Pellets

In general most coprophilic protozoa may also be found in soil or polluted water and many can pass encysted through the alimentary canal of animals and so be present in faecal material. In any event, in nature, faecal material generally comes into immediate contact with soil and can be colonized by the soil population (Watson, 1946). It is possible to classify coprophilic protozoa into three categories:

(1) Oligocoprophilic species, capable of living in faeces only after dilution. These include flagellates, such as *Amphimonas globosa* Kent; rhizopods, such as *Cochliopodium bilimbosum* (Auerbach) Hertwig and Lesser; and ciliates such as *Tetrahymena pyriformis* (Ehrenberg) Furgason.

(2) Mesocoprophilic species, capable of living in undiluted faeces, but flourishing only in diluted faeces. These also include flagellate, rhizopod and ciliate species.

(3) Polycoprophilic species, well adapted to life in undiluted faeces and capable of flourishing in this habitat. These include the flagellate, *Cercomonas crassicauda* Dujardin; the amoeba, *Acanthamoeba hyalina* (Dangeard) Volkonsky and the ciliate, *Balantiophorus minutus* Schewiakoff.

Almost all coprophilic protozoa are also found in sewage and in organically rich soils.

All these protozoa are therefore closely associated with active decomposition of faecal material which is either directly of plant origin or represents microbial cells formed in primary decomposition of plant material in the gut of the consumer. The great majority of these protozoa feed on bacteria and their role is therefore in accelerating the turnover rate of the microbial biomass rather than in directly decomposing plant residues. They are characterized by an ability to tolerate the conditions consequent upon intense bacterial activity, such as low oxygen tensions, high carbon dioxide tensions and perhaps low pH, and they have a very high rate of reproduction since food is not a limiting factor.

G. Roots

There have been numerous studies of rhizosphere populations in relation to root activity (Ghilarov *et al.*, 1968) and the earlier work on protozoa has been reviewed by Stout and Heal (1967).

Populations may be associated with the decomposition of roots after they have died or been damaged by parasites, or, more often, they are associated with the decomposition of plant exudates. The amount and nature of the exudates varies with the age and type of plant but in grassland the release of carbon dioxide from the metabolism of plant exudates may equal the release by respiration of the plant cells themselves (Warembourg and Paul, 1973). Consequently rhizosphere populations associated with the metabolism of these exudates may make a significant contribution to the decomposition of plant residues.

Protozoa populations in the rhizosphere are related to the ordinary soil population but the rhizosphere appears to stimulate the growth of some species preferentially. Although there are reports of significant faunal differences between the rhizosphere and non-rhizosphere soil (Geltzer, 1968) the most common species are flagellates belonging to the genera *Bodo* Ehrenberg, *Cercomonas* Dujardin, *Heteromita* Dujardin, and *Oikomonas* Kent, small limax amoebae, and the ciliate *Colpoda*, which are all common soil protozoa (Darbyshire, 1966; Darbyshire and Greaves, 1967; Nikoljuk, 1968).

There are considerable differences in the size of the population depending on the type of soil and the plant species. In semi-arid regions the populations are very small under natural vegetation. Larger populations are found in temperate soils but they vary with plant species.

Rhizosphere populations change with the development of the plant and the production of exudates. Generally they reach a maximum at flowering and this tends to be the case for both bacteria and protozoa (Darbyshire and Greaves, 1967; Martin, 1972). The bacterial population of the rhizosphere differs in composition from that of the non-rhizosphere soil, pigmented forms, including pseudomonads and yellow pigmented types, tending to be most active. Protozoan populations are affected by both quantitative and qualitative changes in the bacteria. On the whole the zymogenic population associated with the rhizosphere is more favourable for protozoa than the autochthonous population. Rhizosphere species inoculated into sterilized soil with different bacterial strains did not grow as well as those dominant in non-rhizosphere soil (Jarnea and Stefanic, 1965). This suggested that they may be dependent on the plant exudates directly as well as on the rhizosphere bacterial populations. However in sterile conditions the protozoa did not multiply.

In general flagellates or amoebae form the greater part of the rhizosphere population, although in some soils ciliates may also be important. Generally the protozoa are present as cysts but active as well as encysted forms can be found concentrated in the rhizoplane, in the root cap zone and around old root hairs. The greatest accumulation is associated with old, decaying roots.

The intense activity associated with roots results in higher concentrations of carbon dioxide. Protozoa found in the rhizosphere include species known to be tolerant to high carbon dioxide tensions, such as *Colpoda steinii*, and also species known to excyst in response to raised carbon dioxide tensions, such as *Naegleria gruberi* (Schardinger) Wilson (Averner and Fulton, 1966). The ability to tolerate high carbon dioxide tensions may account for the specific composition of the rhizosphere population.

Because of their close association with greatly increased bacterial activity it is difficult to assess the role of protozoa in the rhizosphere but it seems reasonable to assume that, directly or indirectly, they contribute significantly to the metabolism of plant exudates and dead plant tissues.

IV. Population Dynamics and Metabolic Activity

A. Population Fluctuations and Faunal Succession

Living space and the supply of food limit protozoan populations. Probably only the termite symbionts or the phylloplane fauna approach the physical limits of living space. Almost all other populations are limited by other environmental factors, of which food supply is the most important. As a consequence protozoan populations show very marked fluctuations. These are least for populations of stable environments, such as the rumen, and greatest for habitats with marked variations in environmental conditions, such as deserts. There is no very adequate basis for comparing protozoan populations. If values are based on a water volume unit, e.g. the millilitre, difficulty is experienced with habitats that suffer great fluctuations in moisture such as soil, or habitats, such as fresh water where the populations may be strictly localized in the immediate vicinity of growing plants or fungal hyphae. Similar difficulties are evident if mass is used. Possibly the best criterion is the proportion of nitrogen within the biomass which can be attributed to the protozoan population but there are difficulties in determining this value. The reason for using such a criterion is twofold: firstly it is the relative activity of the protozoa that is important and, secondly, this value can give a basis for comparison of very different ecosystems.

In some ecosystems with a large biomass protozoa constitute a relatively small component and this is true of most terrestrial ecosystems including that of grazing ruminants for the ciliates in the rumen are only a very small fraction of the total mass of the animal. Where there is a much smaller biomass as in some freshwater or marine ecosystems, the protozoa may constitute a much higher proportion of the biomass, as for example planktonic faunas such as foraminifera, radiolaria or tintinnida in the sea.

In many fresh water ecosystems there may be great fluctuations in the total biomass depending upon algal blooms, on varying degrees of pollution or on altering lake or river levels. Protozoa are commonly associated with such fluctuations and the tendency is for the protozoan population to increase after the initial algal or bacterial population increase and in many cases to be associated with its ultimate disappearance. In this way the protozoa tend to increase to a maximum as the substrate, whether algal or bacterial, disappears. This increase in protozoan population is associated with a change in faunal composition, some species being associated with the early stages of the sequence, some with the middle stages and some with the terminal stages.

The study of such successions has followed two main lines. The first, particularly associated with bacterial pollution, has been concerned with factors such as pH and Eh, oxygen and carbon dioxide tensions, and to a lesser extent the nature of the microflora (Fig. 9; Bick, 1971). This has

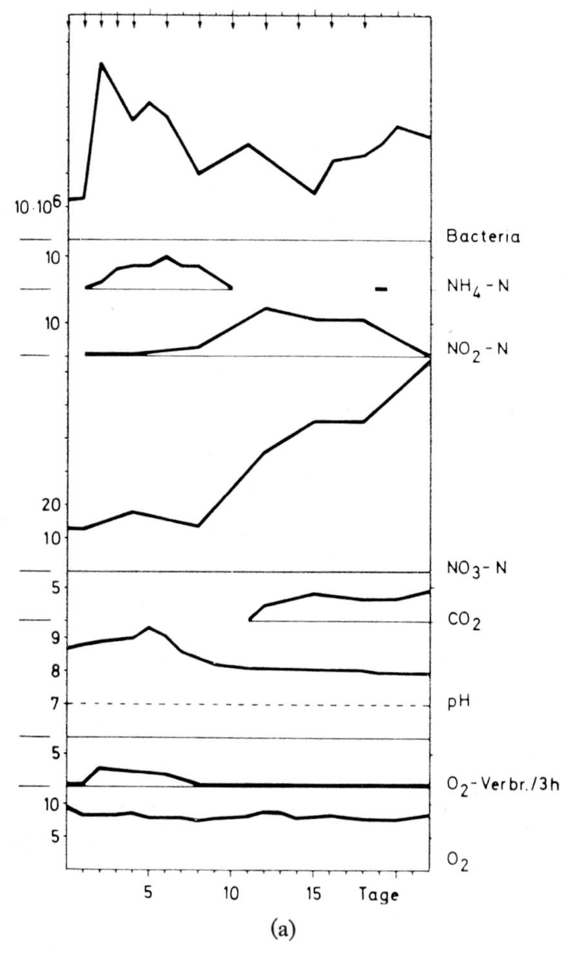

(a)

Fig. 9 (a) and (b). Sequence of protozoan populations in an aquarium in relation to bacterial population and chemical factors (reproduced with permission from Bick, 1971).

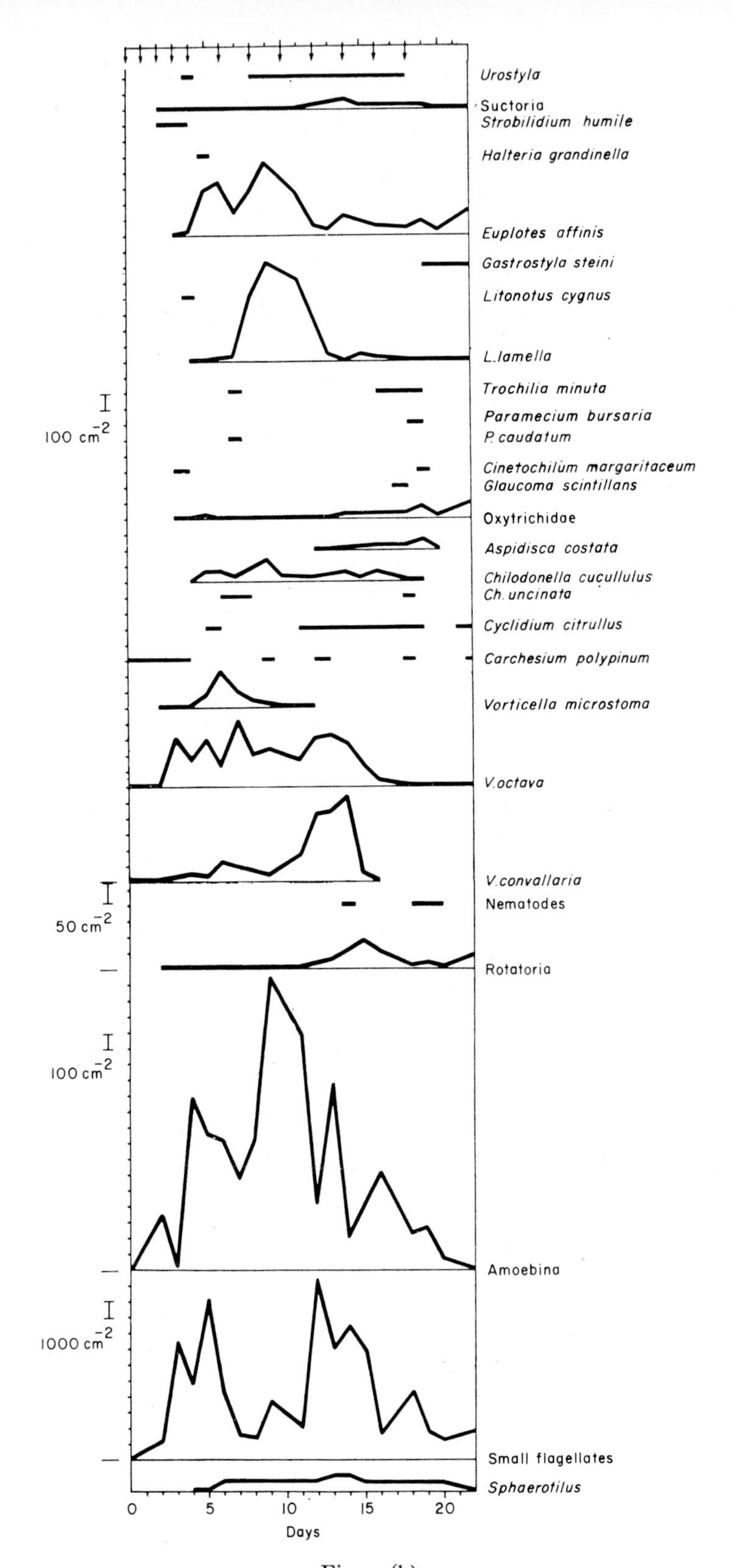
Urostyla
Suctoria
Strobilidium humile
Halteria grandinella
Euplotes affinis
Gastrostyla steini
Litonotus cygnus
L. lamella
Trochilia minuta
Paramecium bursaria
P. caudatum
Cinetochilum margaritaceum
Glaucoma scintillans
Oxytrichidae
Aspidisca costata
Chilodonella cucullulus
Ch. uncinata
Cyclidium citrullus
Carchesium polypinum
Vorticella microstoma
V. octava
V. convallaria
Nematodes
Rotatoria
Amoebina
Small flagellates
Sphaerotilus
100 cm^{-2}
50 cm^{-2}
100 cm^{-2}
1000 cm^{-2}
0
5
10
15
20
Days

Fig. 9. (b)

helped to define the ecological limits of different species and consequently classify those most useful as indicator species. In general, in fresh water, it is ciliate species, such as *Paramecium* Hill and *Vorticella* Linnaeus which provide the best index to bacterial pollution.

The second line of study has been directed to the physical properties of these biocoenoses (Fig. 10; Picken, 1937; Fauré-Fremiet, 1950). The emphasis in this case has been less on the succession of species than their interrelationship not only as a food chain but as a community intimately associated with the architecture of the algal or fungal growth which provides their substrate. In this case particular attention has also been directed to the ciliate fauna, which is commonly the most conspicuous element. The species associated with the algal filaments differ from those associated with bacteria in tending to be flattened, with cilia concentrated on the ventral surface and with the habit of moving over a surface rather than swimming freely in the ambient fluid. This mode of locomotion restricts them to the intermeshing substrate.

While the algal filaments proliferate the protozoa appear to feed on the epiphytic bacteria, detritus or unicellular algae associated with their surface. When the filaments cease to grow the protozoa may ingest the filaments and so bring about destruction of their own substrate and the end of that particular community. Similarly, the pollution of an algal community with organic matter may lead to bacterial proliferation and the multiplication of the bacteriophagous protozoa. Besides ciliate protozoa a wide range of flagellates and amoeboid protozoa are also involved in these communities.

In most protozoan communities it is possible to recognize the two basic patterns of community organization; one dependent upon bacterial proliferation and the other associated with some form of plant growth. This is true of terrestrial populations found in litter and soil. There are microphagous forms, such as *Colpoda*, which are analogous to *Paramecium* although in this case they tend not to be free-swimming but to sweep the bacteria or yeasts into their mouth while settled on the substrate; and the larger amoeboid forms which ingest various forms of plant debris. It is possible to assess from the structure of the community how far protozoa may be contributing directly to the decomposition of plant debris and how far they are acting as micropredators on primary decomposers such as bacteria and fungi.

B. Population Interactions and Metabolic Activity

The significance of the interaction of protozoa with other organisms in the decomposition of plant debris depends not only on the community

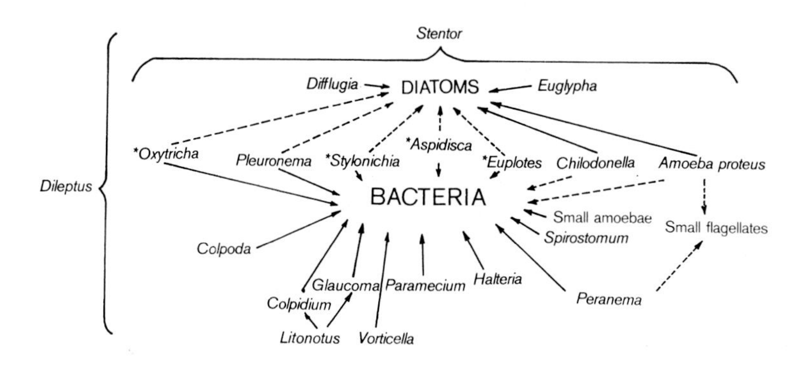

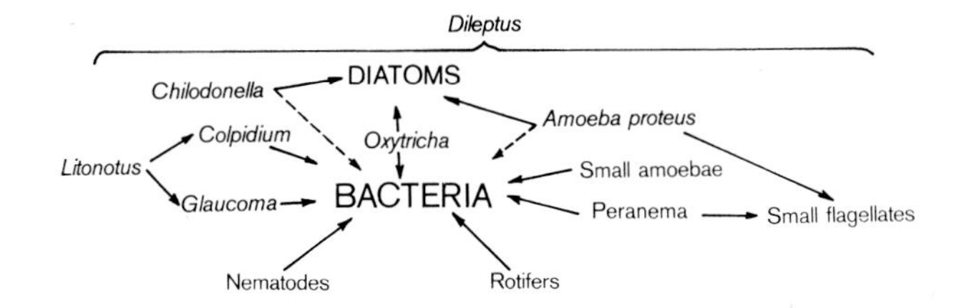

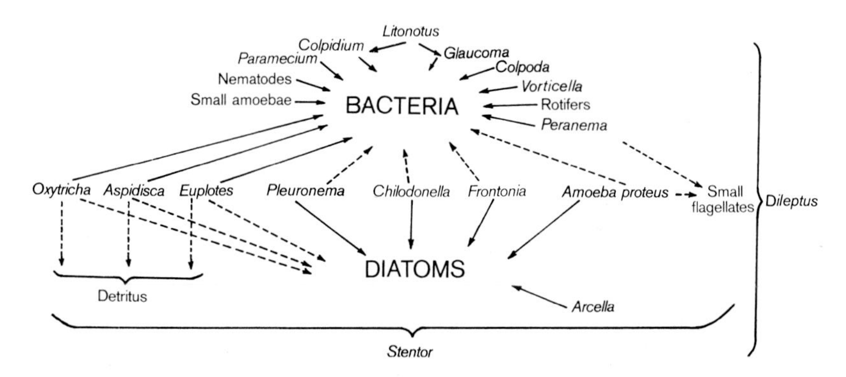

Fig. 10. Diagram of protozoan communities in relation to their substrate (reproduced with permission after Picken, 1937). The upper diagram shows a predominantly algal association and the bottom diagram a predominantly bacterial association. The middle diagram shows a transition stage from the algal to the bacterial.

structure but on the energy flow through different members of the ecosystem. This can be measured in various ways, such as the respiration of the different populations or the turnover rate of different elements, such as nitrogen, phosphorus or carbon.

A relatively simple ecosystem is provided by a *Spartina–Zostera* salt marsh. Primary production is due either to the grasses, diatoms or other algae. There are no large animal consumers. About 10–40% of energy is cycled through the grasses and through the primary decomposers of filamentous fungi, yeasts and bacteria. A mixed population of secondary decomposers is dominated by Foraminifera which feed both on the primary decomposers and on the algae (Lee and Müller, 1973). The rate of cycling of nutrients such as nitrogen and phosphorus is very rapid, a matter of hours, and this appears to be related to species diversity in the foraminiferan fauna, a rapidly changing population having a higher rate of turnover than a stable one. The essential feature, then, of this ecosystem is the rapidity of cycling compared with a forest ecosystem where nutrients may be immobilized in trees and in soil for perhaps hundreds or even thousands of years. Rapid turnover rates also occur in polluted streams, activated sludge and other ecosystems dominated by protozoan micropredation and rapidly changing microbial populations.

The essential feature of protozoan communities is therefore their short life history and rapid turnover, rather than the volume of the plant debris processed. Because of their rapid turnover rate nutrients become available for re-cycling and the overall metabolic activity of the ecosystem is enhanced. Some indication of the enhanced metabolic activity due to interaction between protozoa and primary decomposers such as bacteria and yeasts is given by respiratory experiments in which the endogenous respiration of bacteria or yeasts and the endogenous respiration of protozoa measured separately are compared with the respiration of the mixed populations. With minimal changes in population level, the respiratory rate doubles (Stout, 1973). The effect of such interactions in natural ecosystems, such as salt marshes or polluted streams, could, therefore, be enormous.

The dramatic effect of microbial population interactions depends on two principal factors: first there is the speed of metabolic activities and cell growth, and secondly there is a great contrast between the metabolic activity of a resting or encysted protozoan cell and a trophic or reproducing cell. The exponential growth of a microbial population is so much more rapid than that of larger organisms that although in terms of biomass they may constitute only a trivial fraction, in metabolic activity they may make a significant contribution to the cycling of organic nutrients in plant debris. In the case of soil ecosystems there are two examples which illustrate this point. The first is the so-called "priming effect", the enhanced mineralization of soil organic matter following the stimulation of microbial growth by the addition of fresh plant debris. The second is the partial sterilization effect, which can be achieved in nature by prolonged drought or in agricultural practice by steaming or treatment with chemicals such as methyl bromide. In the first case the effect is attributed to the enhanced activity of the natural population by the addition of fresh substrate, in the second to the release of substrate from the biomass. In each case the microbial population is stimulated and the consequence is an accelerated mineralization of nutrients such as nitrogen and phosphorus. Where plant decomposition is limited not by a rate limiting reaction, such as cellulose or lignin decomposition, but by the supply of such nutrients the enhanced rate of mineralization by the microbial population, which includes the protozoan population, is an important factor in accelerating plant decomposition.

V. Summary and Conclusions

(1) Protozoa are ubiquitous and are normally associated with every stage of plant growth and decomposition. They are found in the phylloplane and rhizosphere of growing plants associated with other microbial populations and they are equally dependent, directly or indirectly, upon plant excretions for their growth. Some species may actively parasitize growing plants. Colonization of senescent and dead tissues may start in living plants but intensifies as the dead litter falls to the substrate and the reservoir of species present in the soil or water exploits the supply of fresh nutrients.

(2) The fauna associated with decaying vegetation may vary from species to species and is affected by the range of environmental conditions which obtain in any particular habitat. There are broad distinctions between a free-living fauna and a parasitic or symbiotic fauna but there is equal diversity within fresh water or terrestrial habitats or between the symbionts of vertebrates and invertebrates. Radiation and convergence are evident in all the major protozoan groups.

(3) Like the great majority of micro-organisms, protozoa can use a very wide range of plant metabolites but only a few groups are able to hydrolyse the complex polymers, such as cellulose or lignin, which are the rate limiting reactions in many decomposition processes. The exceptions appear to be largely if not wholly confined to symbionts and of these only a relatively few groups are involved. In some habitats protozoa may constitute a significant part of the biomass acting as a reservoir of nutrients for other organisms, whether microbial or not, and promoting the rapid cycling of essential nutrients such as nitrogen, phosphorus or sulphur.

(4) The catalytic effect of protozoa on microbial metabolism can be very great, as shown in experimental situations and as known from the activated sludge process and other field situations. This is due to the very close association of protozoa with other micro-organisms, especially bacteria, as for example with the bacterial symbionts of termite flagellates or rumen ciliates, and the rapid metabolic interchange between the two, with protozoa feeding on bacteria and bacteria feeding on lysed protozoa. This very close association makes it difficult to assess the relative role of the two, as with the decomposition of cellulose by rumen ciliates. Since, however, in nature the two are invariably associated together it is more realistic to assess their joint role in the ecosystem.

(5) Judged by their metabolic activity, flagellates, small amoebae and ciliates are the most important agents of decomposition. Thecamoebae generally grow much more slowly and are therefore likely to be less important in most situations but under favourable conditions foraminifera may be the major protozoan component, as in the salt marsh ecosystem cited. Because of the stability of their habitat and the continual supply of food the termite and rumen symbionts are the most effective agents of plant litter decomposition but they represent only a small part of the total protozoan population. Further study may show a comparable role for opalinids or for the entozoic ciliate fauna of oligochaetes.

References

AVERNER, M. and FULTON, C. (1966). *J. gen. Microbiol.* **42**, 245–255.
BAMFORTH, S. S. (1973). *Am. Zool.* **13**, 171–176.
BICK, H. (1971). *Hydrobiologia* **37**, 409–446.
CHARDEZ, D., DELECOUR, F. and WEISSEN, F. (1972). *Rev. Ecol. Biol. Sol.* **9**, 185–196.
CHEN, TZE-TUAN (ed.) (1967*a*, *b*, 1969). "Research in Protozoology," Vols 1–3. Pergamon Press, Oxford.
CLAFF, C. L., DEWEY, V. C. and KIDDER, G. W. (1941). *Biol. Bull.* **81**, 221–234.
CLEVELAND, L. R. (1960). *In* "Host Influence on Parasite Physiology." (L. A. Stauber, ed.), pp. 5–10. Rutgers University Press, New Brunswick.
COLEMAN, G. S. (1963). *In* "Symbiotic Associations." (P. S. Nutman and Barbara Mosse, eds), pp. 298–324. Cambridge University Press.
CORLISS, J. O. (1961). "The Ciliated Protoza." Pergamon Press, Oxford.
DARBYSHIRE, J. F. (1966). *Can. J. Microbiol.* **12**, 1287–1289.
DARBYSHIRE, J. F. and GREAVES, M. P. (1967). *Can. J. Microbiol.* **13**, 1057–1068.
FANTHAM, H. B. and PORTER, A. (1945). *Proc. zool. Soc. Lond.* **115**, 97–174.
FAURÉ-FREMIET, E. (1950). *Endeavour* **9**, 183–187.
FAURÉ-FREMIET, E. (1967). *In* "Chemical Zoology," Vol. 1, "Protozoa." (G. W. Kidder, ed.), pp. 21–54. Academic Press, New York and London.

FLINT, E. A. and STOUT, J. D. (1960). *Nature, Lond.* **188,** 767–768.
GELTZER, J. G. (1968). *In* "Methods of Productivity Studies in Root Systems and Rhizosphere Organisms." (M. S. Ghilarov, V. A. Kovda, L. N. Novichkova-Ivanova, L. E. Rodin, V. M. Sveshnikova, eds), pp. 48–51. Nauka, Leningrad.
GHILAROV, M. S., KOVDA, V. A., NOVICHKOVA-IVANOVA, L. N., RODIN, L. E., SVESHNIKOVA, V. M. (eds) (1968). "Methods of Productivity Studies in Root Systems and Rhizosphere Organisms." Nauka, Leningrad.
HEAL, O. W. (1962). *Oikos* **13,** 35–47.
HEAL, O. W. and FELTON, M. J. (1970). *In* "Proceedings of the Tenth Symposium of the British Ecological Society." (A. Watson, ed.), pp. 145–162. Blackwell, Oxford.
HEAL, O. W., BAILEY, A. D. and LATTER, PAMELA M. (1967). *Phil. Trans. R. Soc.* 5 **252,** 191–197.
HONIGBERG, B. M., BALAMUTH, W., BOVEE, E. C., CORLISS, J. O., GOJDICS, M., HALL, R. P., KUDO, R. R., LEVINE, N. D., LOEBLICH, A. R. Jr., WEISER, J. and WENRICH, D. H. (1964). *J. Protozool.* **11,** 7–20.
HUNGATE, R. E. (1955). *In* "Biochemistry and Physiology of Protozoa." (S. H. Hutner and Andre Lwoff, eds), pp. 159–200. Academic Press, New York and London.
HUNGATE, R. E. (1960). *In* "Host Influence on Parasite Physiology." (L. A. Stauber, ed.), pp. 24–40. Rutgers University Press, New Brunswick.
HUNGATE, R. E. (1966). "The Rumen and Its Microbes." Academic Press, New York and London.
HUTNER, S. H. and LWOFF, A. (eds) (1955). "Biochemistry and Physiology of Protozoa," Vol. 2. Academic Press, New York and London.
HUTNER, S. H. (ed.) (1964). "Biochemistry and Physiology of Protozoa," Vol. 3. Academic Press, New York and London.
JARNEA, S. and STEFANIC, G. (1965). *In* "Symposium; Methods in Soil Biology," pp. 4–45. Rumanian Society for Soil Science, Bucharest.
KIDDER, G. W. (1967). "Protozoa." *In* "Chemical Zoology." (M. Florkin and B. T. Scheer, eds), Vol. 1. Academic Press, New York and London.
LEE, J. J. and MULLER, W. A. (1973). *Am. Zool.* **13,** 215–224.
LWOFF, A. (ed.) (1951). "Biochemistry and Physiology of Protozoa," Vol. 1. Academic Press, New York and London.
MARTIN, J. K. (1972). *Aust. J. biol. Sci.* **24,** 1143–1150.
NIKOLJUK, V. F. (1968). *In* "Methods of Productivity Studies in Root Systems and Rhizosphere Organisms." (M. S. Ghilarov, V. A. Kovda, L. N. Novichkova-Ivanova, L. E. Rodin, V. M. Sveshnikova, eds), pp. 126–129. Nauka, Leningrad.
PICKEN, L. E. R. (1937). *J. Ecol.* **25,** 368–384.
RYLEY, J. P. (1967). *In* "Chemical Zoology," Vol. 1. (G. W. Kidder, ed.), pp. 55–92. Academic Press, New York and London.
SCHÖNBORN, W. (1965). *Pedobiologia* **5,** 205–210.
STOUT, J. D. and HEAL, O. W. (1967). *In* "Soil Biology." (A. Burges and F. Raw, eds), pp. 149–196. Academic Press, London and New York.

STOUT, J. D. (1968). *Pedobiologia* **8,** 387–400.
STOUT, J. D. (1969). *In* "Progress in Protozoology." (A. A. Strelkov, K. Sukhanova and I. B. Raikov, eds), pp. 202–203. Nauka, Leningrad.
STOUT, J. D. (1973). *Am. Zool.* **13,** 193–202.
WAREMBOURG, F. R. and PAUL, E. A. (1972). *Pl. Soil* **38,** 331–345.
WATSON, J. M. (1946). *Biol. Rev.* **21,** 121–139.

13

Nematodes

D. C. Twinn

May and Baker Ltd.
Ongar Research Station
Essex
England

I. Introduction

Research on the Nematoda is largely divided between the study of the parasites of animals and those of plants. In contrast, work on the even greater assemblage of species which have a free-living existence is of more meagre proportions and perhaps would be negligible if it were not for the interrelations between these and the parasites. On land, both animal and plant parasites may have had their main origins in nematodes feeding on the microflora, for some animal parasites have bacterivorous stages and comparable forms could have developed into the primitive plant parasites—the fungivores (Maggenti, 1971). The ease of culture of the microflora feeders has led to their use as laboratory models in many ways, from the study of problems of nutrition and ageing to purely practical employment in the screening of soil nematicides. The presumed primitive habitat of the nematode, the sea, has tended to attract a separate discipline of nematologists despite the clear ecological links with freshwater systems, which in turn overlap with terrestrial soils in the close correspondence of much of their nematode fauna.

Parasites of green autotrophic plants and those of animals have relevance in the general concept of production and energy flow in ecosystems: the plant parasites as potential regulators of primary production and the animal parasites, whether of vertebrates or invertebrates, as regulators of consumption. Direct association of the nematodes with the decomposer cycle, of long standing if ancestral nematodes were bacterivores, resides mainly with free-living species and this review is almost entirely concerned with these.

II. Identifications

The class Nematoda forms a discrete group of pseudocoelomate Metazoa, mainly vermiform in shape and, other than some animal parasites, microscopic or nearly so in size. They rarely approach 10 mm in length but range down to 250μm, even as adults. Whilst perhaps closest to the Gastrotricha, other taxonomic relationships and evolutionary origins are obscure.

There is an evident binary division of the class into Secernentes and Adenophori (or their near equivalents Phasmidia and Aphasmidia), reflected ecologically for respectively they tend to be of terrestrial or aquatic types. However, the primary diagnostic criteria (the excretory system or the presence of phasmids) are often obscure and practical divisions are more convenient at order or lower levels (Goodey, 1963).

Most separations are made on combinations of characters, particularly of structures associated with feeding: the stoma and its armature, the form of the pharynx. Other features of importance include the head and cuticle appendages, ornamentation or markings and male copulatory structures (where males occur). The recognition of some genera or generic groups is simple but others require considerable expertise: this applies in greater force to the identification of species. The standard work on soil and freshwater genera is that of Goodey (1963); marine nematodes are less well summarized and must be sought in more scattered literature including Wieser (1959*a*) and later works by the same and other authors.

Ecological studies are complicated by the high proportion of new species in almost any previously unexamined habitat or locality. In Puget Sound, two thirds of the marine nematodes found by Wieser (1959*b*) were previously undescribed, as were most of Yeates' (1968) species from New Zealand dune sand. In an English woodland, two thirds of the population in one soil and one third of those in another differed from described forms: although only 400 m apart, the sites had very few species in common (Twinn, 1966). This is so usual that one must accept with caution the specific determinations in older works in which a high proportion of known species are recorded: the more stringent criteria of modern nematode taxonomy (Goodey, 1959) were not then applied. Many separations depend to some extent on body proportions which can vary with mounting treatment and also with the nature of the available food (Townshend, 1972).

Although accurate identifications are possible from carefully prepared material the determination of each individual present in a sample taken during an ecological study may not be easy. When a complex of species within one genus occurs, correct assignment of males to a species characterized on adult females can be troublesome and the structure of juveniles, which often form a high proportion of the population, is frequently obscure enough to prevent specific identification. Thus many comparisons must take place at generic rather than specific level despite the biological diversity which characterizes many genera.

III. Extraction Methods

Sampling strategy has an unusually large effect on the derived data, the size and number of samples often depending more on practical considerations than on statistical requirements. The design and execution of sampling is well documented for economic plant nematology where the requirement at any one time is essentially the measurement of a short series of important species relative to potential or actual plant damage. Work aimed at the determination of energy-flow patterns has the more difficult task of estimating the absolute numbers of all species to a reasonable degree of accuracy. It is inevitable that no nematode extraction and examination method is without its serious disadvantages, but the varying bias these may introduce means that in reviewing the studies of different authors one is comparing not merely data, but data and methods inextricably confounded.

Extraction methods have recently been surveyed by Oostenbrink (1971), but additional comments are opportune on some which have been used for plant litter material.

(a) *Baermann Funnel and Modifications.* Relying on nematode activity, this may give very variable results. For tree leaf litter it is often almost useless, extraction percentages varying from 80 to 20% and the apparent population structure being distorted by the extraction of surface-dwelling bacterivores but not the tissue-dwelling fungivores (Twinn, 1962). In addition, egg hatch and revival from cryptobiosis is likely; fungal infections of the *Catenaria* type may be artificially fostered.

(b) *Direct Examination.* When performed on live material (Volz, 1951) this considerably restricts the sample numbers and size; on fixed samples it requires the rather subjective discrimination of originally active or inactive specimens (Twinn, 1966). However, it has the advantage that specimens remain in their microhabitats within the litter until examined, permitting the observation of aggregation and relationships with other organisms. In particular, a means is provided of studying both nematodes and microflora *in situ* using staining methods (Minderman, 1956; Minderman and Daniëls, 1967). Nematodes are frequently noted on media following the isolation of fungi from soil and litter (Dickinson, C. H., personal communication), a further indication that joint methods might be devised. Extended culture of direct isolates reduces the diversity of the original nematode population, leading to the exaggerated development of few species (Tietjen, 1967), but is adaptable to feeding studies.

IV. Aspects of Behaviour and Physiology

A. Movement

All small organisms have a limited ability to cope with the physical forces exerted on them by their environment, which in nematodes are particularly related to dependence on free moisture and reliance on a hydrostatic skeleton maintained by internal pressure. With few exceptions, movement is based on sinusoidal waves and for efficiency requires purchase against the surface tension of a water film, the viscosity of mud or physical contact with solid objects (Crofton, 1971; Wallace, 1963). Speed of progression is related to the interaction of the dimensions of the body and those of the surroundings, such that nematodes of a given size will progress most rapidly between particles of specified dimensions or in mud of a certain viscosity (Crofton, 1971). When the fluid viscosity is low, as in clear water, swimming is inefficient and in aquatic habitats rapid movement is most likely in semi-liquid bottom deposits. Free swimming in disturbed situations may be avoided by a tendency to sink whilst swimming and by the possession of caudal cement glands which either maintain the nematode in its micro-habitat or when fixed to individual sand grains accelerate its sinking rate.

In terrestrial soils with a crumb structure, space to move in water films on the crumbs is probably rarely restricted and similarly, on plant litter, the relevant species have little difficulty in movement, provided that a continuous water film is present. A vermiform shape is well suited to passing through narrow openings and plant tissue is readily colonized; however, some restrictions apply, for the stouter microbivores only extensively occur within leaf litter after it has become porous. In the soil itself, with decreasing particle size the pore necks may become limiting, preventing free movement (Wallace, 1963). The presence of finer materials can provide a further restriction by the occlusion of pores. Bruce (1928) found 35–40% pore volume in beach sand fractions he had graded but only 20% in the original mixed sample.

Wave and current action in aquatic systems often serve to sort particles, depositing them as soils with a relatively simple size distribution. The relationship between these factors and living space (*lebensraum*) is recognized but not well understood (Tietjen, 1969). Marine nematodes may often be grouped according to their association with certain grain sizes yet Wieser (1959*b*) considered that nematodes would have little difficulty in moving in the finest sands. His data superficially question this conclusion; on the Vashon and Bainbridge Island beaches sand grains of 50–150 μm, if well packed, would provide pore necks of only 8–24 μm, much less than the body diameters of the larger species typical of these stations. Such a relationship is even more relevant in finer deposits such as mud where any species is likely to be confined to the surface semi-liquid horizon, being incapable of penetrating the stiffer material below. Records from stiff mud would indicate structuring of the soil, perhaps by the burrowing activities of larger animals. The possible concentration of the nematodes in burrow systems has many other implications in respect of their density in the micro-habitat.

B. Response to Stimuli

Nematodes are attracted by a wide variety of stimuli including carbon dioxide, ammonia, organic materials and electric potentials. Such a range indicates that a general attraction to regions of biological activity is perhaps the most usual response and, for example, in terrestrial soils this would lead to accumulations in the rhizosphere, probably equally suitable for plant parasitic, microbivorous or predatory species. Heat has also been suggested as a factor leading microbivorous nematodes such as *Turbatrix* and *Panagrellus* to centres of microbial activity (Croll, 1970).

The existence of stable gradients in such a complex system as soil over more than minute distances is improbable: interpretation of attraction as a dual pattern of kinesis and directional correction (as in insect chemoattraction) is more reasonable. In some experiments, a regularity of correction is discernible (Green, 1966): this might not be innate although asymmetry of movement with a constant deviation from straight line progression is a characteristic of some species (Croll, 1969). Providing that sufficient individuals survive, elaborate behaviour patterns are unnecessary. For instance, the colonization of leaf litter on the soil surface by fungivores does not require the supposition of responses to gravity but is easily explained by the spread of the nematodes in the train of the fungal invasion. On the other hand, more complex searching patterns might be expected in species such as bacterivores and predators, whose food has less physical continuity. These have not yet been demonstrated.

Stable chemical gradients are more conceivable in some aquatic situations, particularly in the calm of deep water or amongst submerged vegetation in which current speeds are reduced. The accumulation of *Metoncholaimus scissus* on fungal mats and senescent benthic diatom blooms in *Thalassia* beds, led Meyers *et al.* (1970) to deduce the existence of chemoattraction to regions of decomposition activity.

Many species exhibit a general response to light but a few from aquatic habitats are more efficiently equipped with light receptors. *Chromadorina* inhabits freshwater green algae and its positive response to green light has an obvious survival value in this habitat (Croll, 1970).

C. Feeding Mechanisms

In feeding, nematodes encounter the particular physical problems of taking food whilst in a fluid medium and the need to overcome the pressure of the hydrostatic skeleton to swallow it. Both are largely solved by the use of a muscular pharyngeal pump which is employed to suck in suspended particles such as bacteria, or to cause the lips to adhere to the food while other structures pierce it; the same pump forces the food into the intestine against the pressure of body fluids.

A typical system is found in *Pelodera* in which bacteria are sucked into the mouth cavity and collect within the pharynx. Retention of particles seems to be based on their physical characteristics rather than nutritive value for Indian ink is taken and the process effectively resembles a filter-feeding mechanism (Doncaster, 1962) (Fig. 1). Passing down the pharynx to the valved muscular bulb, the food is finally injected into the intestine. The accumulation of material to form a bolus indicates a means of reducing unnecessary intake of fluid although axenic culture studies indicate that in rhabditids discrete particulate food is not essential. In some bacterivorous groups, the lip region bears complicated outgrowths (*Bunonema*, some cephalobids, *Teratocephalus*, *Wilsonema*) which may serve to size-grade the food or to free it from its substrate. Sachs (1949) stated that in *Bunonema* one set of processes was used to loosen the food, another to compact it.

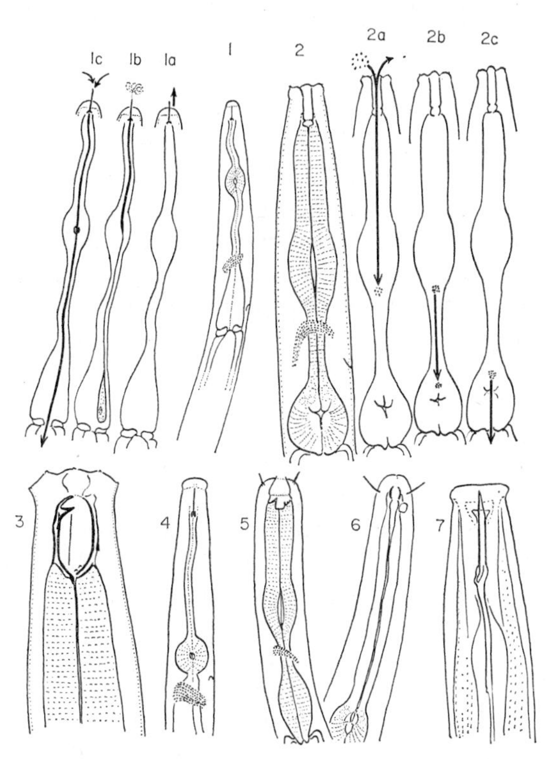

Fig. 1. Terrestrial nematodes: pharyngeal structures and feeding mechanisms.
Fungivores—plant parasites: 1, *Ditylenchus*; 4, *Aphelenchoides*.
Bacterivores: 2, *Pelodera*; 6, *Plectus*.
Facultative bacterivore: 5, *Butlerius*.
Predator: 3, *Mononchus*.
Omnivore: 7, *Eudorylaimus*.
Ditylenchus feeding sequence: 1a, stylet penetration; 1b, salivation; 1c, ingestion.
Pelodera feeding sequence: 2a, ingestion and filtration; 2b, swallowing; 2c, injection into gut.

Materials of a larger size require other feeding methods. Some species, e.g. *Tripyla* spp., whose gut may contain entire nematodes, are capable of enlarging their mouths to suck in their prey whole. Often the stoma is armed with spears or teeth which, by puncturing the food, permit liquid contents to be sucked out. A common arrangement is that of variously sized teeth arranged in opposition or facing a denticulated plate. In no case has there been a firm observation that these structures may be extruded to pierce the prey and the system of use is likely to be broadly similar to that found in the carnivore *Mononchus*. Here the food is taken into the stoma, then pierced and retained by the armature whilst contents are sucked out (Nelmes, 1972); the same is seen in *Butlerius* (Pillai and Taylor, 1968) (Fig. 1).

Spear- and stylet-bearing nematodes exemplify an adaptation which permits the piercing of food without the need first to draw it into the stoma, and this gives them the capacity to attack relatively tough-walled subjects. Whilst adhering by lip suction or by bracing of the body, repeated thrusts of the spear are used for penetration (Fig. 1) and food either passes through the spear or beside it through a salivary feeding tube, which is later left behind. Feeding in plant parasitic nematodes has been reviewed by Doncaster (1971).

D. Feeding Groups

Classification of nematodes into their trophic types is necessary before an attempt can be made to relate them to energy pathways. Purely taxonomic grouping is inappropriate when it places together species known to have different food requirements yet it has been widely used as a secondary method since definite information is meagre and relates to relatively few species.

The difficulties of fitting a given species firmly into one discrete category are considerable, for ideally a wide range of evidence should be considered: (i) the known food type of closely related species; (ii) the morphological structure of stoma and pharynx; (iii) observable gut contents; (iv) the presence or absence of possible food organisms in the habitat; (v) culture experiments to determine which food organisms promote nematode reproduction.

In practice, (i), (ii) and (iii) provide the data in most cases and a variety of factors confuse the issue. Particular species may feed in culture on a relatively wide range of organisms, perhaps chosen more for their ease of propagation than their presence in association with the nematode in nature. The nematodes may ingest materials such as sand grains (Wieser, 1953), Indian ink (Doncaster, 1962), or take apparently suitable food which cannot be digested. *Mesodiplogaster lheritieri* destroys *Chlamydomonas* but passes *Ankistrodesmus* unchanged (Leake and Jensen, 1970). Thus, as Nielsen (1949) pointed out, identifiable gut contents may represent what cannot be digested, although in predators, when the objects are nematode spears or spicules, or enchytraeid setae, they provide reasonable evidence of the organism attacked.

The studies on *Pelodera* and the spear-bearing nematodes indicate that, when using morphology and function as an indication, the complete arrangement of structures from lips to the intestine should be considered as a functional whole. Certain points require study, for there is wide variation in the form of the pharyngeal pump, which may or may not be markedly muscular or possess a valved bulb. *Rhabditis* and *Alaimus* live on similar food but possess dissimilar pharyngeal structures. Nevertheless, the mechanics of the ingestion system must determine the physical limits of the food taken and sufficient study of its operation might eliminate anomalies if taken in conjunction with the availability of food in the natural habitat.

In the definition of feeding groups, terrestrial and freshwater species are conventionally treated together and distinct from marine forms.

1. *Terrestrial Feeding Groups*

Nielsen's (1949) original classification has been modified by subsequent writers (Banage, 1963; Bassus, 1962*a*; Yeates, 1967, 1971) to suit their particular studies. The ideal scheme in which each grouping is identified by the ecological role of its food organisms (as primary producers, decomposers or various order consumers) is difficult to apply because of the variation in the feeding habits of the nematodes (either within species or within higher taxa). Nevertheless, these distinctions cannot be avoided and the grouping of algae and fungus feeders as microherbivores (Yeates, 1967) or using "plant feeders" to cover both fungus and higher green plant feeders (Banage, 1963) is confusing.

(a) *Forms with spears or stylets.* These can readily be distinguished and will include almost all Tylenchida and many Dorylaimida. Whole families of the Tylenchida have been characterized with reasonable certainty as obligate parasites of vascular plants: Heteroderidae, Hoplolaimidae, Tylenchulidae, Criconematidae. To these may be added tylenchid genera such as *Tylenchorhynchus*, *Anguina*, *Rhadinaphelenchus* and dorylaimid genera such as *Trichodorus*, *Longidorus* and *Xiphinema.* Markedly relevant to the decomposer cycle is the feeding habit of the Tylenchida on fungal hyphae (Maggenti, 1971). Some families (e.g. Neotylenchidae, Aphelenchoididae) include many species associated with bark beetles in some form of parasitism but also with the need or faculty to feed on fungi during part of their life cycle. Within these families and elsewhere in the Tylenchida are soil forms apparently subsisting on fungi alone. *Ditylenchus* and *Aphelenchoides* both contain obligate plant parasites, facultative parasites which can be cultured on fungi, and also pure fungivores. Thus specimens associated with plant litter are difficult to classify unless in the presence of fungi and the absence of living higher plants. The plant parasites of both these genera tend to attack aerial parts and those found in the rhizosphere one may suspect to be more concerned with fungi than plant roots. However, this is not a precise distinction for *D. destructor* infests potato tubers and an *Aphelenchoides* sp. has been cultured on mycorrhizal fungi (Riffle, 1967).

The widespread soil genus *Tylenchus* presents further difficulties for *T. emarginatus* and *T. costatus* attack conifer roots, *T. agricola* has been cultured on alfalfa callus tissue and *T. agricola* and *T. bryophilus* were observed to feed on the root hairs of cabbage and other plants (Gowen, 1970; Khera and Zuckerman, 1962, 1963). *T. emarginatus* would not feed on ten species of fungi (Sutherland, 1967). This evidence indicates a plant-parasitic habit, yet *T. bryophilus* itself, another closely related species and further species of the sub-genera *Filenchus*, *Lelenchus* and *Tylenchus* were found in leaf litter in the absence of plant roots where they were certainly feeding on fungi (Twinn, 1966). There are other records of *Tylenchus* from litter (Egunjobi, 1971) and this field evidence is in favour of a qualified classification of the genus as fungivore-type. Maggenti (1971) has suggested that the genus particularly exemplifies the primitive plant parasite; undoubtedly it will prove to contain the range of feeding habits found in *Aphelenchoides* and *Ditylenchus* although possibly with a more facultative bias. These and similar genera may often be designated with similar logic either as plant parasites or as fungivores; the latter has usually been preferred in this review.

Apart from the true plant parasites, the Dorylaimida equipped with a spear or spear-like tooth are generally classed as omnivores or miscellaneous feeders. Although there are few records of these being associated with the bacteria and fungi, they often form such a high proportion of the nematode numbers and biomass that they can scarcely be ignored. Nielsen (1949) counted many as algal feeders based on the frequent occurrence of chlorophyll in the gut. This conclusion has been partly confirmed for *Eudorylaimus ettersbergensis*, which was easily cultured on an alga yet also fed on protozoa, fungal conidia and other nematodes (Hollis, 1957). Other scattered evidence indicates that dorylaimids exhibit a full range of feeding habits, from algae at one extreme to pure predation (e.g. *Nygolaimus* on enchytraeids) at the other. Despite their variable feeding, information on these nematodes is enough to set them aside in a trophic group distinct from the bacterivores and fungivores and hence of indirect relevance to the decomposer cycle.

(b) *Nematodes without spears or stylets*. Of these, the Mononchidae form a discrete and recognizable group predatory on nematodes and other small animals, as shown by much published information and the close correspondence of mouth armature throughout the family. Yet a toothed buccal cavity such as they possess is not a prerequisite for predatory habits, for

Tripyla frequently feeds on other nematodes but has little more than a retaining tooth in the pharynx.

Many other genera may be arranged in an almost continuous series from those with unarmed mouths to those with teeth approaching the mononchoid type (e.g. *Butlerius*). Starting with the experiments of Nielsen (1949) there has accumulated a great deal of evidence to show that bacteria form a suitable food for species at almost any level of this series, from the unarmed *Plectus* or *Rhabditis* type to *Diplenteron* and *Butlerius* with mononchoid mouth armatures (Pillai and Taylor, 1968; Yeates, 1970). The differences in size and shape of the mouth cavity within this series suggest varying abilities to exploit different micro-organisms. In general, a not very critical choice of food, as indicated in *Rhabditis* and others, could result in the ingestion of any particle of a suitable shape and size. If available, fungus spores, yeasts and small algae are likely to be included in the normal food range of otherwise bacterivorous species.

The complex, toothed, mouth structure of many diplogasterids in the series points to a more omnivorous diet than that of bacteria alone and both *Butlerius* and *Diplenteron* will feed on protozoa. Such nematodes are often found in habitats characterized by rapid decomposition such as dung and sewage beds, compost heaps and rotting fruits, in all of which both bacteria and protozoa may be abundant: they may best be classed as facultative bacterivores.

The terrestrial feeding types fall into two series, separate in practice but not quite so taxonomically (Fig. 2). Within these groups, the sub-divisions are distinct in their typical examples but show a variable degree of overlap in transitional forms.

In fresh water Group (a) is uncommon, Group (b) forming the bulk of the population with the attendant increase in difficulty in respect of transitional forms. Algal feeding (e.g. in chromadorids) may be added as a more discrete type: in this, and in general bias, freshwater trophic groups show relationships with marine systems.

2. *Marine Feeding Groups*

Yeates (1971) has recently introduced the concept of deposit-feeders for terrestrial habitats but it is convenient to discuss this in relation to marine systems with which it is associated. The term describes more the method of feeding than the nature of the nutrient and is applied when species ingest small particles of organic detritus (deposits) or are commonly found with quantities of such material in the gut. The most interesting evidence is from the studies of Tietjen (1967) on the marine species *Monhystera filicaudata*. This genus is of particular importance since it is widely distributed in the sea, in fresh water and on land. Tietjen observed that the nematode ingested detritus particles and would survive for 9 months, but not reproduce, on non-sterile faeces of the mollusc *Aequipecten*, one of the common sources of deposits in his study area. In contrast, reproduction of the nematode was copious on marine bacteria cultures, and dinoflagellates were also accepted as food. These observations are at least as good as much of the culture evidence on terrestrial species which classes them as

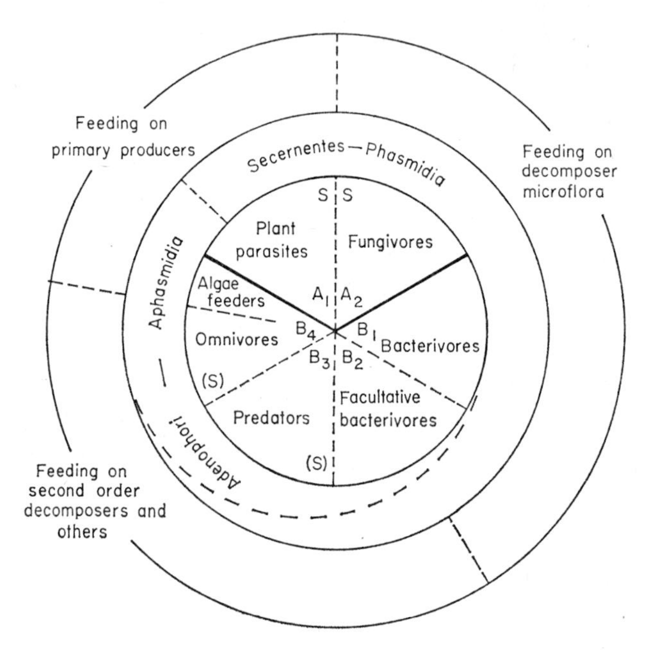

Fig. 2. Relationships between terrestrial nematode trophic groups, S, species with spears or stylets; (S), some with spears or stylets.

bacterivores. It is pertinent to compare the residual deposits of the sea with the H horizon of mor soils on land, which often consist almost entirely of invertebrate faeces. In both cases the plant material has been subjected to fragmentation, attack by micro-organisms, perhaps stripping in the digestive system of invertebrates and, certainly in the sea, to the extremes of leaching. Any suggestion that the final product is a practical food for an organism with no specially efficient digestive system is scarcely acceptable. The conclusion is inevitable that the bacteria form the food and the deposit is merely a relatively inert vehicle carrying them, as has been suggested for two marine Mollusca by Newell (1965).

However indicative the data on *M. filicaudata* may be, they cannot be applied indiscriminately to all species regarded as deposit feeders. For, much as the monhysterid would ingest dinoflagellates, so a wide variety of other food organisms may be taken, probably restricted more by size and availability than a high degree of selectivity on the part of the nematode.

Reviewing all the evidence available at that time, Wieser (1953) divided marine species into two series on the basis of whether or not the mouth cavity was armed with teeth.

1. Unarmed mouth: A. selective deposit feeders
   B. unselective deposit feeders
2. Armed mouth: A. epigrowth feeders
   B. predators

Separation on selectivity can be excluded as the essential difference between 1A and 1B is size of stoma and buccal cavity rather than observed distinctions in behaviour. Typical of 2A are the Chromadoridae, many of which are known to take diatoms and characteristically live in association with the sessile epigrowth formed by these and other small animals and plants on algae, rocks and other substrates. Class 2B consists of nematodes with heavily armed mouths and includes species known to be predatory. With further information, Wieser (1960) later altered the last group to one of omnivores.

Some form of bacteria, diatom or deposit feeding has been demonstrated or strongly suggested for members of all of Wieser's groups. Thus the classification would be of little use in relation to deposit decomposition if it were not for the fact that groups 1A and 1B are associated with situations in which fine deposits collect (Wieser, 1952). On this circumstantial evidence it may be suggested that many species in group 1 are involved in the decomposition of detritus either as consumers of bacteria themselves or at a trophic level only one or two stages higher. Variable feeding on bacteria, bacteria-laden detritus and small animals such as protozoa is probably common amongst marine nematodes and may be compared with the situation in the terrestrial diplogasterids, classed earlier as facultative bacterivores.

True fungus feeding has been described for a marine *Aphelenchoides* (Meyers and Hopper, 1967) and a few other marine examples of this typically terrestrial genus are known (Timm and Franklin, 1969). The only other tylenchid recorded from the sea, *Halenchus*, is classed as a plant parasite, the nematodes being associated with galls on *Fucus* and *Ascophyllum*. However, *Halenchus* may be found free-living (Chitwood, 1951; Twinn, unpublished) and could also be fungivorous.

E. Physiology of Particular Ecological Relevance

1. Digestion

Certain of the spear-bearing nematodes have been shown to secrete saliva into the food and digestion may be extracorporal. Otherwise, and certainly in forms without spears, digestion takes place within the body, enzymes being added from pharyngeal glands during swallowing. Jennings and Colam (1970) considered that different glands in *Pontonema* secreted, respectively, an enzyme precursor and activator, but they found no evidence of true enzyme secretion in the gut. Chitinase has been demonstrated in the plant parasite *Ditylenchus dipsaci*, possibly representing a relic of a fungivorous ancestry (Tracey, 1958), but in general enzyme spectrum is broadly related to food type.

Information that can be used to indicate the quantitative efficiency of feeding and digestion is very scarce. Crofton (1966) considered that the digestive processes were wasteful, for *Ascaris* shows a high throughput of food, the gut contents being replaced as frequently as once every 3 min. In apparent contrast, Taylor and Pillai (1967) observed the rapid disappearance of fungal hyphae contents in the gut of *Paraphelenchus acontioides* and concluded that digestion was efficient enough to keep up with rapid feeding. There is no evidence that feeding rate is repletion-regulated but on the other hand elimination is so determined, as a necessary consequence of the high body pressure, defaecation occurring before the pressure rises sufficiently to prevent the pharyngeal pump working. In *Ascaris*, most of the gut contents are voided each time (Crofton, 1966) but in *Aphelenchoides blastophthorus* the faeces volume is limited to about 9% of that of the gut by a pre-rectal valve mechanism (Seymour, M. K., personal communication). Pharyngeal secretion of digestive enzymes provides a method for intimately mixing these with the food and also a metering system related to food intake rate. Adequate digestion is probably thus assured but the extent of absorption from the gut may well be more dependent on the speed of passage and it is likely that an abundance of food would lead to considerable wastage in elimination.

2. Excretion

A swift passage of fluid through the gut is an excellent means of disposing of ammonia which forms a large proportion of the nitrogenous excretion (Crofton, 1966). However, relatively large amounts of amino acids are also formed (Myers and Krusberg, 1965) which may be a mechanism of detoxifying excess tissue ammonia: excretion of synthesized amino acids was rapid in *Ditylenchus triformis*. In this species they found no evidence of urea either as an end product or as an intermediate although a subsidiary ureotelic system exists in parallel with the ammoniotelic one in *Ascaris* (Lee, 1965). Normally, excretion pathways through the cuticle or the gut are quite adequate and the so-called excretory pore and its associated system are of enigmatic function. Secretory ability for this has been demonstrated, and in *Tylenchulus semipenetrans* females (a plant parasite) the secreting tissue is enormously enlarged, serving to produce an external gelatinous matrix. In this instance, a subsidiary excretory role is feasible, for the anus is non-functional and the gut is not available as an excretory route, unless the system is ureotelic (Maggenti, 1962). The most elaborate "excretory" canal structures are associated with the terrestrial types (Secernentes); this habitat relationship suggests a possible function under conditions of exogenous or endogenous toxin stress, quite likely to occur in thin moisture films and other situations of intense microbiological activity.

Evidence on the efficiency of overall nitrogen utilization is superficially conflicting. Crofton (1966) considered nitrogen excretion to be low, particularly in proportion to the high turnover required by egg production. In the fungivorous *Ditylenchus triformis*, the daily nitrogen excretion of starved individuals amounts to about 2–4% of the body protein (Myers and Krusberg, 1965). Wright and Smith (1972) indicated a higher rate in *Panagrellus redivivus* but much of the protein fraction might have been attributable to faeces, an inevitable confusion in recently fed individuals: in starved nematodes excretion could be abnormal. High rates have been found in some vertebrate parasites, possibly related to the amount of dietary protein taken (Lee, 1965).

3. *Food Quality Requirements*

McCoy (1929) showed that the free-living juveniles of *Ancyclostomum caninum* might be cultured on live bacteria but not on ones killed by heat; a comparable observation was made by Hirschmann (1952) for a *Rhabditis*. Similarly, whilst a marine *Aphelenchoides* thrives on the fungus *Dendryphiella arenaria*, heating to 60°C renders this food unsuitable (Meyers *et al.*, 1963, 1964). Although failure to culture on dead food may result from a different behavioural response by the nematode, Nielsen's (1949) opinion that they generally require live protoplasm is supported also by more recent biochemical studies which, for example, indicate heat-labile dietary requirements (Sayre *et al.*, 1967) and restricted sterol synthesis (Rothstein, 1968). Different bacteria vary greatly in their suitability for maintaining good growth in culture of bacterivorous nematodes (Nielsen, 1949; Tietjen *et al.*, 1970) and the same has been demonstrated for the fungivore *Ditylenchus destructor* which could be cultured on 64 fungi of 115 tested (Faulkner and Darling, 1961). The last point indicates that requirements are not always stringent and that survival, if not optimum growth, is possible on a wide range of organisms. Food that is normally suitable may become inadequate as it ages (Meyers *et al.*, 1964); fungivores tend to avoid senescent hyphae and feed for longer on young, actively growing ones (Anderson, 1964; Hechler, 1962).

On the basis of the evidence available, it seems very unlikely that nematodes could survive by feeding directly on plant litter and detritus alone, their survival undoubtedly being due to the micro-organisms actively growing on these materials. One instance may be quoted as the nearest approach to litter feeding is the observation that *Tylenchus emarginatus* would feed on recently sloughed conifer root cap cells (Sutherland, 1967).

4. Effects on Food Supply

The only extensive quantitative data on the degree to which nematodes affect their food supply concerns the spear-bearing plant parasites and those found in commercial mushroom cultures. It is difficult to relate loss of gross plant production to the biomass of nematodes, due to measurements being largely based on commercial crop yields. Pot experiments with *Rotylenchus buxophilus*, whilst ignoring turnover of plant roots, shoots and nematodes, have indicated that the production of about 15 mg nematode biomass resulted in a loss of total plant matter amounting to at least 12 g (Golden, 1956).

At the cellular level, the immediate results of feeding are very variable for whilst *Tylenchorhynchus claytoni* is a delicate feeder on plant roots, causing no obvious damage (Krusberg, 1959), *Paraphelenchus acontioides* attacks fungal hyphae voraciously, rapidly emptying the cells of their contents (Taylor and Pillai, 1967). In both root and fungus feeding, the first result is the halting of normal cyclosis of the cell contents which then stream towards the puncture site, an effect attributed to the injection of salivary juices. The damage may extend to more than one cell, for *Ditylenchus destructor* causes the death of up to five hyphal cells at each feeding site (Anderson, 1964). If a break in the continuity of the cell cytoplasm reaches the nematode's stylet, feeding may cease, thus leaving part of the cell contents uneaten. Attacked mushroom beds appear wet from the leakage of hyphal contents at the feeding sites (Hesling, 1972) and bacteria may develop extensively on these released nutrients, possibly also permitted by the limitation of the bacteriostatic action of the mycelium.

In vascular plants, one common result of nematode attack is the compensatory proliferation of the root system: a similar effect on the growth of fungus mycelium does not appear to have been observed but would seem to be possible. Nielsen (1961) suggested that bacterivorous nematodes might increase the turnover rate of bacterial colonies as a side effect of their cropping them, but in general the significance of the bacterivores and other particulate feeders will result more directly from the amount cropped and assimilated. In contrast, the plant parasites and fungivores may well spoil more than they eat.

5. *Food Utilization and Productivity*

Metabolic rate as indicated by oxygen consumption has been measured in free-living nematodes notably by Nielsen (1949) for terrestrial forms and by Wieser and Kanwisher (1961) for a range of marine species. Differences between species of the same size occur, linked in some cases to activity level by Nielsen (1949), but the expected inverse relationship to individual body weight holds good. Hourly oxygen consumption at 16°C ranges from 1440 mm^3 in *Monhystera vulgaris*, to 200 mm^3 in *Pontonema vulgare*, each per g fresh weight. Oxygen requirements were shown to vary with temperature broadly in accordance with Krogh's normal curve (Nielsen, 1949). Movement is not likely to require a large utilization of energy (von Brand, 1960) and narcotized nematodes show little change in oxygen consumption (Nielsen, 1949). The major outlet for energy is in the form of increase in body size or reproduction. Nielsen counted 209 eggs produced by *Cephalobus elongatus* in 22 days, these being equivalent to five times the female's weight. More recently, Tietjen *et al.* (1970) gave an egg production of 70–100 eggs per female for *Rhabditis marina* in culture, laid over 3–4 days and equivalent to about 53% of the female's weight per day. Even more surprising, cultured *Diplenteron potohikus* produced up to 505 eggs in 38 days (Yeates, 1970) which, from published dimensions (Yeates, 1969), seems equivalent to a total production of 52 times the female's biomass. The maximum egg laying rate noted in these last experiments averaged 15·3 eggs per day, an apparent production of over 150% in 24 h. Equivalence of specific gravity in egg and adult is assumed in this calculation; if eggs are reduced in moisture content, absorbing water after hatching, then actual production of dry matter will be correspondingly increased.

The work with *Rhabditis marina* gives data from which can be derived a quantitative relationship between food consumed and eggs produced. In the case of the most successful food (*Pseudomonas*) the nematodes ingested about ten times their own weight of bacteria daily. Thus a maximum egg production averaging at about 0·8 μg per female per day can be associated with a maximum food intake of 15 μg of bacteria in the same period. A concurrent increase in female size occurred equivalent to a length increase of 8–10% each day, although this may have resulted solely from ovarian growth and thus be confused with egg production. However, utilization of the food for maintenance, movement and reproduction together seems likely to lie in the order of one tenth to one fifteenth of the gross intake.

Such a culture situation in which food is in excess and temperature is optimum may be regarded as unnatural, although something approaching this situation may occur in blooms or flushes of micro-organisms. A steady turnover of energy in decomposing litter or deposits may actually consist of scattered bursts of localized activity, in accord with the often highly over-dispersed ("aggregated") distribution of both nematodes and microflora. A high metabolic optimum temperature (see later) would also permit the more efficient exploitation of restricted periods of higher temperatures. These are probably of regular occurrence on land, early in the year as temperatures rise but moisture remains adequate, and in the autumn on rewetting but before temperatures fall. High fecundity is one means of countering the effects of a hazardous existence, and rapid egg production rates are characteristic of terrestrial bacterivores. *Rhabditis marina* may be included in this category for even though it is found sub-tidally, it is a species characteristic of seaweed rotting on beaches (Inglis and Coles, 1961). Published data on egg production of free-living nematodes in the field are non-existent and even rarely adequate in respect of culture experiments, but many species would appear to lay eggs on a more modest scale than those noted above. Thus Gerlach (1971) quoted W. V. Thun's data on *Chromadorita tenuis* which deposited in culture 16–28 eggs, an average daily production of rather under 10% of the female's weight. The figure for *Prionchulus punctatus*, a terrestrial predator, which laid 45 eggs in less than 8 weeks (Nelmes, A., personal communication) is about 2·5% per day.

Some of the examples above indicate a very highly geared metabolic system in operation at maximum food intake, yet respiration experiments often do not specify the nutritive state of the subjects, or measurements are made on starved individuals. The topic deserves further study, for such data are widely used to estimate ecological importance in energy-flow.

6. *Temperature*

In their minimum and optimum temperatures for various activities plant parasitic nematodes tend to reflect their climate of origin (Dao, 1970), but even in temperature species such as *Ditylenchus dipsaci* the optima frequently fall between 15 and 25°C. As shown by a considerable population growth, temperate surface litter fungivores may be active well below 10°C, a temperature commonly quoted as a minimum for the activity of related species. Laboratory data are influenced by the frequent use of relatively high culture temperatures under which acclimatization may occur, sometimes with genetic involvement (Brun, 1972). Also the optima recorded are normally those at which maximum reproduction, movement, respiration or other functions are maintained and, as in other invertebrates, may differ considerably from the temperatures of maximum ecological efficiency or survival (Bursell, 1964).

In many terrestrial situations the laboratory optimum may be achieved for only relatively brief periods as the temperature rises before lack of moisture intervenes. The direct application of data from acclimatized laboratory cultures to field populations living under the widely fluctuating temperatures of, for example, surface litter or exposed beaches is thus difficult.

7. *Quiescence and Cryptobiosis*

Many nematodes have the ability to evade the lethal effects of an adverse environment by adopting a state of quiescence in which metabolism and ageing are both slowed. Possibly an extreme extension of quiescence, cryptobiosis is characterized by the apparent absence of any metabolism. When dehydration is the cause, the nematodes may survive many years; prolonged exposure to oxygen lack or to low temperatures has also resulted in a cryptobiotic state. The faculty is commonly found in species living in situations of fluctuating moisture content, such as the inhabitants of moss and leaf litter (e.g. *Plectus*, *Aphelenchoides*) or the stem and leaf parasites of plants (e.g. *Ditylenchus*, *Aphelenchoides*). The apparent absence of records of cryptobiosis among littoral marine nematodes is surprising, for there is a possible relationship between the ability to survive osmotic stress and simple dehydration (Cooper and van Gundy, 1971). Cryptobiosis provides a survival mechanism for terrestrial nematodes during temporary drought and also permits passive dispersion without a water requirement on and in dried plant litter.

V. Ecology: Introduction

Descriptions of nematode populations, in the form of relative abundance (dominance), have some value in the definition of habitats but they combine the ecologically diverse. Unless each group, genus or species can be recalculated in absolute terms (i.e. such as per unit area) comparability is poor. This is particularly true of terrestrial soils where large dorylaims ("algal feeders") are often common. In a Herning grass field, Nielsen (1949) recorded algal feeders (mainly as "*Dorylaimus obtusicaudatus*") as a low percentage (13%) of the total biomass. In a *Calluna* heath the same group (and "species") provided 93% of the biomass. This clear difference of dominance completely conceals the fact that the group occurred in both soils at about the same biomass of 1·86–2·08 g m^{-2}.

Many authors have stressed the ubiquitous occurrence of nematodes and their presence, often in immense numbers, in a wide variety of habitats sufficiently moist to support aquatic animals. They may be found in temporary Nigerian rockpools, amongst antarctic mosses, in thermal springs and the abyssal depths of the sea. Most ecological literature consists of faunistic surveys often with more emphasis on the larger omnivores and predators than on species of the decomposer cycle. These habitats are reviewed here more in proportion to the available literature than by virtue of their intrinsic importance. Few studies relate the nematode directly to identifiable plant litter itself. Thus discussion is usually limited to general relationships with the decomposer cycle, particularly in terms of numbers or biomass of bacterivores and fungivores.

VI. Ecology: Terrestrial Systems

The vertical distribution of nematodes within soil has received attention in many types of habitat. In given situations certain species may show "preferences" for particular parts of the profile but in most cases these will reflect the exogenous factors of food supply and the physical environment. High in the profile, the principal restrictions imposed will be determined by temperature and drought and below by waterlogging and poor aeration. Within the possible limits, occurrence is associated with organic content but a clear relationship can rarely be shown, because it will be with the fraction which is actively growing or being actively decomposed. Often two centres of distribution can be shown, a deeper one composed mainly of plant root parasites in the rhizosphere and one nearer the surface containing the free-living species.

Temporal cycles of population changes may or may not be shown by sequential sampling. There is no evidence of rhythmic fluctuations of reproduction and growth except when these are imposed by environmental conditions, such as temperature, moisture or food supply.

With the exception of a few faunistic surveys little information on the free-living nematodes of tropical soils has been published and most studies relate to temperate climates in the Northern Hemisphere. Forest in particular has received the most attention and, representing perhaps the most common climax vegetation, this biotope may often be related to others, most of which would normally tend to develop forest or have been derived from it.

A. Virgin Soils: Dune Sands

The successive changes in nematode population structure which follow the colonization of virgin soils by plants may be represented by the stages in the vegetational succession from beaches or inland dunes to forest or grass. Beaches were included in Nielsen's (1949) work and more detail has been provided by Yeates (1967, 1968) for a wider range of New Zealand coastal dunes in an early stage of stabilization. In these, although the highest numbers of nematodes were found within the top 20 cm of sand, an appreciable proportion of the total reached a depth of at least 90 cm. Maximum populations were of the order of $2{\cdot}1 \times 10^6\ m^{-2}$ (in the top 30 cm), on the site which had the densest cover of *Ammophila* arenaria. On most of the dunes in New Zealand half the nematodes were bacterivores but the striking facet of the data was the apparent absence or infrequency of fungivores. These were not separately listed, but in only one case did the group Tylenchida (other than certain named plant parasites) reach 9% of the total; more often the percentage was less than one. Except in this single instance, all the sands were alkaline in reaction and the virtual absence of fungivores agrees well with the studies by Brown (1958) on English dunes. She found 2–4 times as much fungus mycelium on acid dunes as on alkaline ones. In New Zealand the more alkaline sand appears to have been the most unstable and Brown associated instability with a low mycelium content. Nielsen (1949) showed a similar trend with no fungivores at the seaward station but the numbers increasing landwards to form 7–12% of the total under grass or moss cover.

Wasilewska (1970, 1971) has extended this succession series by studying a range of inland dunes with *Pinus* stands at various stages up to full forest. From site to site, in parallel with an increase in vegetation cover (particularly of the ground flora), the soil humus content and the nematode numbers increased (from $0{\cdot}5$ to $7 \times 10^6\ m^{-2}$). The critical levels for maximum species diversity appeared to be a 25% ground cover and a 1% soil humus fraction, up to which point the microflora feeders progressively enlarged their share of the populations. In older stands of forest, nematode densities were lower, $4{\cdot}4 \times 10^6\ m^{-2}$ under a *Quercus–Pinus* stand with a rich understory and ground flora and $2{\cdot}0 \times 10^6\ m^{-2}$ under *Pinus* without a ground flora. Wasilewska related this feature of older woodland to the distribution of the microflora, which Jakubczyk (1968) found to be more abundant in the dunes than in the established forest. Fungivores, excluding *Tylenchus*, produced a much smaller biomass than the bacterivores at all sites although under *Quercus–Pinus* fungivores predominated.

It would appear that virgin, unstable sands have a bacterial decomposer cycle probably fuelled by root turnover. As the sand stabilizes and its humus fraction increases a mixed fungal–bacterial population develops, which is reflected in the nematode population structure.

B. Forest: Coniferous and Deciduous

The more restricted fauna in coniferous stands found by Wasilewska (1970, 1971) in comparison with that of mixed woodland has been commonly noted. Bassus (1962*a*) quotes an average nematode biomass of

34 μg cm^{-3} for broad-leaved forest and only 11 μg cm^{-3} for coniferous woodlands. However, comparable data are rare and the nature of the humus form and the sampling methods are also involved in the differences.

Both coniferous and broad-leaved trees form some habitats with a sparse ground flora where the bulk of the litter fall consists of tree leaves. Deciduous forest, particularly *Quercus* and *Fagus*, provides both the different humus forms mull and mor under comparable canopies: it has the added advantage of a concentrated input into the decomposer system over the short period of annual leaf fall (see Jensen, Chapter 3; Millar, Chapter 4).

C. Deciduous Forest: Mor Humus

In these soils, leaf litter decomposition is concentrated in the upper, almost purely organic, horizons where the comparative absence of plant roots in the litter and F layers adds greatly to our confidence in designating the tylenchid and aphelenchoid nematodes present as purely fungal feeders.

TABLE I. Nematode frequencies: woodland mor and raw humus

Woodland type and source of data	Total nematodes per m²: Numbers 10⁶	Fresh biomass g	Microflora feeders: As % of total numbers: Fungivores	Bacterivores	As % of total biomass: Fungivores	Bacterivores
DECIDUOUS						
Fagus (Volz, 1951)	12·1	4·1	33	19	—	—
Fagus (Bassus, 1962*a*)	—	—	73	21	31	25
Quercus (Twinn, 1966)	12·3[a]	0·6[a]	—	—	31[b]	31[b]
Quercus (Bassus, 1962*a*)	—	—	49	36	12	34
Quercus,[d] mixed (Wasilewska, 1971)	4·4	0·4	38[c]	33	7	29
CONIFEROUS						
Picea (Nielsen, 1949)	1·7	4·5	12	57	2	10
Picea (Bassus, 1962*a*)	—	—	72	25	49	3
Pinus (Bassus, 1962*a*)	—	—	54	42	28	56
Pinus (Wasilewska, 1971)	2	0·2	25[c]	47	11[c]	44

[a] L + F + H horizons only
[b] L + F horizons only.
[c] Excludes *Tylenchus* spp.
[d] Humus form doubtful.

1. Vertical Distribution

Average figures for the whole soil profile (Table I) disguise changes in the balance of the two groups of microflora feeders which take place with time and also in different parts of the profile. In beech mor (Volz, 1951) the greatest nematode population and biomass occurs in the mineral soil, particularly in the upper, humus-rich portion (Table II). The H horizon

TABLE II. Deciduous woodland mor: vertical distribution of nematodes

Woodland type	Horizon	Total nematodes			Relative proportions fungivores: bacterivores
		Numbers 10^6 m^{-2}	Fresh biomass mg m^{-2}	Numbers per g dry material	
Fagus	L	0·04	15	147	1:10
(after Volz, 1951)	F	0·15	33	807	1:2·8
	H	1·68	535	615	2:1
	mineral 0–5 cm	7·42	2265	164	1:1·5
	mineral 5–10 cm	2·81	860	50	
Quercus	L	1·02	77	2000[a]	2:1
(after Twinn, 1966)	F	2·25	344	700[a]	1:1·4
	H	9·07	184	275[a]	—

[a] Maxima: mean values about half these.

contains most of the inhabitants of the more organic strata though on a unit weight basis the density of nematodes is greatest in the F horizon. This material has been derived from a larger weight of freshly fallen leaves: thus relating the nematode population to the original unit weight of leaf material (and hence to the nutrients it contains) may further alter the significance of each horizon. Volz's data suggest that by applying such a correction the highest biomass of fungivores and bacterivores per g of original leaf material occurs in the L or upper F horizons. This was also the case in an English oakwood mor (Table II), although in this site there were different proportions of the two microflora-feeding groups. Under oak, fungivores predominated in the litter but ranked about equal with bacterivores in the F horizon. In the beechwood, bacterivores, mainly *Plectus cirratus*, dominated the L populations whilst fungivores reached greater relative proportions in the lower F and H layers.

2. Colonization of Quercus Litter

After leaf fall the litter attains a high moisture content, being progressively more evenly wetted. Colonization begins with a transient bacterial flush which may develop on the leaf surface, presumably on readily available nutrients either already present or leached from the tissues. Extensive fungal colonization follows, over the surface and within the leaves, over a period of 2–3 months (Minderman and Daniëls, 1967). Only once have both

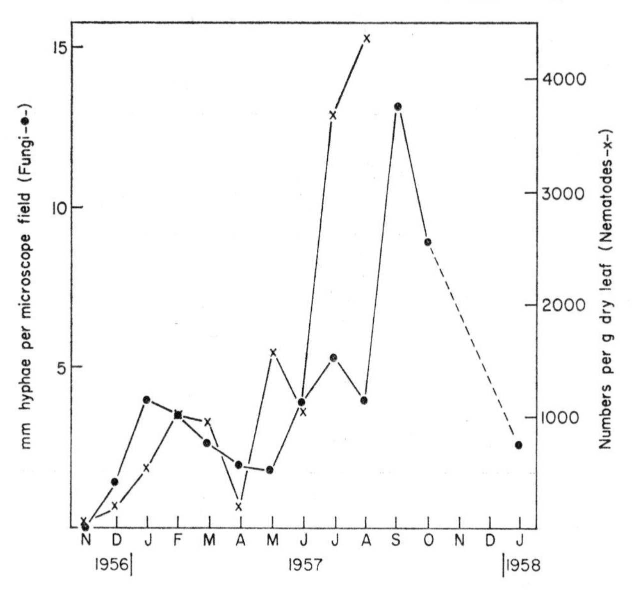

Fig. 3. Colonization by nematodes and fungi of *Quercus* litter on a mor humus form (after Twinn, 1962, and Waid, 1960).

nematodes and fungi been examined on the same material but this showed close correspondence between the colonization of the two groups (Fig. 3).

On two English sites there was no evidence of an initial bacterivore invasion, the population consisting of late juveniles and adults of fungivores *Aphelenchoides* and *Tylenchus* (Twinn, 1962, 1966). They reproduced within the leaf tissues and the resulting proliferation approached exponential form achieving a maximum between 10^3 and 2×10^4 g^{-1} dry litter. These populations were drastically reduced by drying of the leaves. Such changes occurred at the start of the spring and summer dry period, during which drying predominated in the wetting–drying cycles. Individuals were then found in the leaf tissues in tight coils and by their deranged structure they were clearly either dead or in a cryptobiotic state. From evidence on other species, those which had reached a late juvenile stage before drying would be the ones most likely to survive until rewetted in the late summer.

Fungivore resurgence occurred in autumn when the litter was more fragmented and over the subsequent months it progressively changed towards F horizon material. During this time bacterivorous species increased, in correlation with the senescence of fungus growth. Comparable events have been shown in mushroom cultures infected with both fungivorous and bacterivorous nematodes (Goodey, 1960).

3. Quercus Mor: Events in F Horizon

The new litter, the F horizon, and litter more than 7 months old followed related but different courses. Each horizon was affected by drought but the F layer maintained favourable moisture contents for longer in the spring and tended to gain moisture equilibrium earlier in the autumn. There are no data to show the full cycle from drought to drought but the information may be rearranged with enough confidence to indicate the principal events (Fig. 4).

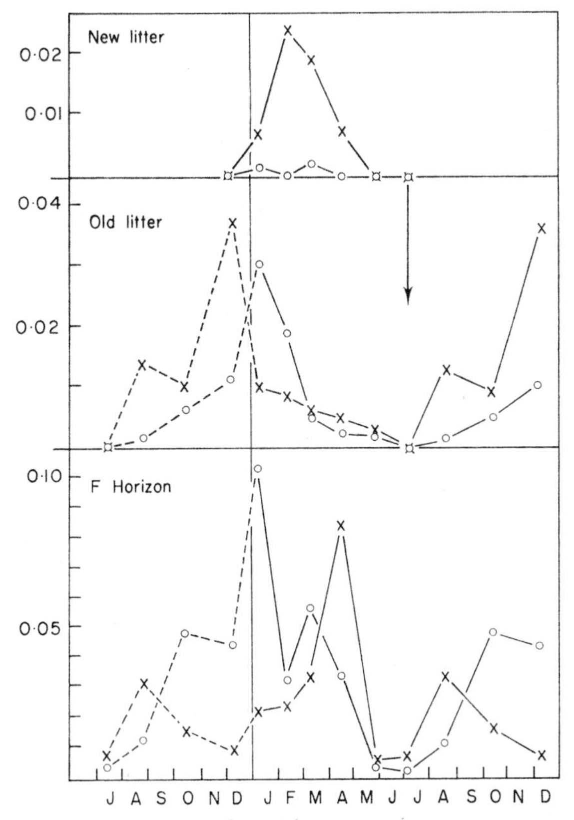

Fig. 4. Annual fluctuations of fungivorous (×) and bacterivorous (○) nematodes in *Quercus* mor litter (after Twinn, 1966). Ordinate: biomass (g m^{-2}).

Soon after rewetting, the F horizon produced a fungivore maximum, later replaced by one of bacterivores, in turn followed by a second fungivore peak. The graph suggests that there might have been a single mid-summer fungivore maximum if the moisture status had been favourable. Concurrently, in the overlying old litter, fungivores increased on autumn rewetting but did not reach their greatest biomass until after leaf-fall when, in turn, they were replaced by bacterivores.

The widest diversity of fungivore species was found in the F horizon but this included the *Aphelenchoides* sp. and *Tylenchus* (*Filenchus*) sp. which were the prime colonizers of the new litter. Thus the successive proliferations on different materials were not duplications of the preceding ones but a selective development of fewer species each time. It was not possible to determine how much of the population growth after drought in the L and F horizons was due to colonization from active specimens below and how much to the revival of cryptobiotic individuals already in the litter but presumably both occurred.

If the nematode population changes reflect the activity of the microflora, the results suggest a pattern of fungus growth upward from the F to old litter to new litter, followed by bacterial colonization. The main bacterial growth on the new litter would be delayed by drought until the subsequent autumn; the second fungus growth in the F horizon might be attributed to the leaching of nutrients as a result of microbial activity above.

4. Other Litters

The fungivore–bacterivore sequence also occurred in the colonization of rapidly decomposing *Fraxinus* litter. In this instance the process was mainly compressed into the first 6 months after leaf-fall, with a marked tendency for bacterivore invasion to occur in parallel with, but rather slower than, that of the fungivores.

The data given by Volz (1951) for *Fagus* mor differ from the events in *Quercus* mor chiefly by the presence of the large bacterivore *Plectus cirratus* which dominated the upper organic horizons. However, if this species is excluded his results also indicate alternation of fungivores and bacterivores (Fig. 5). Large *Plectus* spp. are often characteristic of upper *Fagus* litter and possibly occupy a niche different from that of other bacterivores. The difference between *Quercus* and *Fagus* sites could be explained if the *Fagus* provided a greater and longer flush of bacteria or comparable growths on superficial or leached nutrients but a slower fungal colonization of the litter itself (Saitô, 1956).

It is possible that oak-wood mor provides the optimum timing of events to demonstrate the succession. More rapid decomposition, such as in *Fraxinus* litter, contracts the sequence so that the stages are almost coincident; more extended decomposition would equally obscure events by stretching each phase so that they apparently overlapped. Summer drought assists the separation of the stages by providing a partial re-set system permitting the late summer re-growth of fungi earlier than bacteria, which tend to be less successful under dryer conditions (Hawker, 1957).

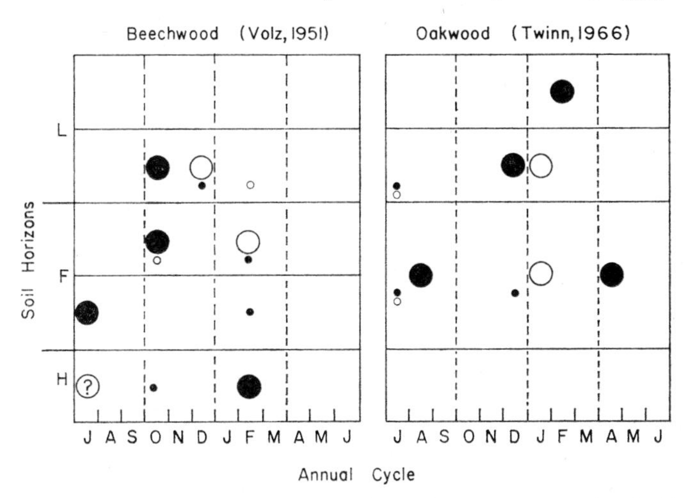

Fig. 5. Alternation of fungivorous and bacterivorous nematodes in *Fagus* and *Quercus* mor. Fungivore biomass maxima ⬤, minima ●. Bacterivor ebiomass maxima ○, minima ○ (excluding *Plectus cirratus*).

No further comparable results are available on other mor woodland soils although the extensive investigations by Bassus (1962*a*) indicate a similarity between both hardwood and softwood forests in their spectra of fungivore and bacterivore genera. He also demonstrated considerable annual fluctuations in a number of individual species of both feeding types (1962*b*). In comparing data, it should be noted that results based on numbers may differ appreciably from those of biomass, not only in the balance of feeding groups, but also in the timing of maxima.

5. *The Role of Fungivores*

With the data available from oak mor it is possible to quantify the effects of the fungivores during the first 70 days or so after leaf-fall. Over such a period, a reduction in leaf dry weight of up to 20% may be shown with little removal of particulate material (Bocock *et al.*, 1960; Twinn, 1966). Presumably the loss is caused by leaching or the action of the microflora, for larger invertebrates such as sciarid fly larvae and enchytraeids colonize the litter at a later stage. The production of nematodes in the *Quercus* leaves may be taken as equal to the biomass of juveniles, assuming negligible deaths until the end of the 70 days, which approximates to the onset of drought. In 1961 and 1962, there were total juvenile *Aphelenchoides* and *Tylenchus* productions equivalent to between 22 μg and 135 μg dry weight for each original g of litter. Minderman and Daniëls data for the fungi present after a comparable exposure period may be interpreted as a dry weight biomass of between 5 and 10 mg, using conventional conversion factors (Parkinson *et al.*, 1971). Allowing for the non-utilization of the fungus cell wall, and taking maximum figures, 55 times the nematode dry weight needs to be accounted for to equate the amounts. If spoilage at a feeding site is as high as ten times the amount taken and production less than 10% of the food intake, it is easily possible for the nematodes present to destroy this amount of fungus, particularly as death and maintenance respiration have been ignored. No estimate of the fungal turnover can be included although the fungal data might indicate the total growth if no distinction was made between hyphae with and without contents. The nematodes' actual role is probably in accelerating the turnover of fungi, making the contained nutrients more rapidly available to lytic organisms such as bacteria.

The relationships indicated above refer only to the loss of up to one fifth of the leaf dry weight, although arguably perhaps the most nutritious fraction. Within the old litter and F horizon the remaining 80% is converted, yielding eventually the H horizon, in this instance consisting almost entirely of meiofauna faeces. In one year in which both were compared the old litter and F material had an average fungivore population of about $1{\cdot}2 \times 10^{-2}$ g dry biomass m^{-2} compared with an average of 3×10^{-3} g in the new litter. With a turnover of 2–3 times in the old material, the fungivores could play a comparable role here in enhancing fungus turnover, despite the greater gross effects of the meiofauna in comminuting leaves.

Leaves form only a part of the litter falling on the forest floor. Woody litter appears to be colonized less readily by nematodes. Invasion is relatively rapid into *Fraxinus* leaflet midribs but very slow into the main petiole. However, later attack of woody substrates by Basidiomycetes, probably mainly in the F horizon, would eventually render energy available for fungivorous nematodes.

D. Deciduous Forest: Mull Humus

A typical mull soil provides little opportunity for the study of whole litter colonization owing to the rapid and selective removal of leaves by earthworms. However, both *Quercus* and *Fagus* commonly occur on mull humus and their litters usually lie longer on the surface, apparently requiring alteration by the microflora before removal. In this situation, fungal and fungivore colonization takes place as in oak mor but at a less intense level, suggesting less fungal activity in the soil below. However, fungal growth on the leaves may be comparable with that on mor (Minderman and Daniëls, 1967). Bacterivore nematodes also wander onto the litter but do not penetrate the tissues. After summer drought, the fungivore population reaches its maximum but, at this time, leaves or part of them are being removed at an increasing rate by earthworms. If the litter is confined so that such removal is hindered, bacterivores predominate in this second phase. The observations do not define a clear pattern of events but they are not inconsistent with there being a similar sequence to that described for oak mor.

Whether or not already attacked by micro-organisms, the litter is then presented for further decomposition in a comminuted state, intimately mixed with soil particles, in the form of earthworm casts above and below the soil surface. Quite fresh casts apparently contain few live nematodes (personal observation) but an extensive and relatively rapid colonization is indicated by Volz's (1951) counts on the layer of recent casts on his oak–ash mull. More nematodes per g were found in this layer than in the soil below or the leaves above. Similarly, in the top 3 cm of soil, which included casts, I also found a rich fauna, more than twice as many nematodes as in either the 3–6 or 6–9 cm strata. Although not comparable in respect of the depth examined, both these mull soils probably had populations of the same order in the uppermost few centimetres. Samples from Broadbalk Wilderness at Rothamsted (Yuen, 1966) indicated fewer nematodes in this woodland, if the conversion factor applied to the data is accurate: a similar population was found in New Zealand forest by Egunjobi (1971) and in Uganda scrub by Banage and Visser (1967) (Table III).

In temperate mull, the relative proportions of bacterivores and fungivores fall within the usual range although these groups tend to form a smaller part of the total biomass, mainly due to the presence of large dorylaimids in greater numbers than in mor. This is a particular influence on Volz's (1951) data, further emphasized by his use of few weight classes in biomass calculations. On the better documented sites, the two microflora feeding groups equate well in biomass and even between mull and mor the differences are little over one order of magnitude (Table IV). Mull and mor show considerable microflora differences. There is a much greater quantity of fungal mycelium in mor though the fungal growth rate in the two soils has been shown to be very similar (Nagel-de Boois and Jansen, 1967). As fungivorous nematodes tend to attack actively growing hyphae, some correspondence is likely between the humus forms in their fungivore biomass. On the other hand, the two types are often compared

TABLE III. Nematode frequencies: woodland mull and soils other than mor

Woodland type and source of data	Total nematodes per m²		Microflora feeders					
			Numbers 10^6 m^{-2}		As % of total numbers		As % of total biomass	
	Numbers 10^6	Fresh biomass (g)	Fungivores	Bacterivores	Fungivores	Bacterivores	Fungivores	Bacterivores
Quercus (Twinn, 1966)	4[c]	0·63[c]	2·3[c]	1·1[c]	59	26	15	10
Quercus-Fraxinus (Volz, 1951)	29·0[a]	15·2[a]	2·95[b]	2·68[b]	27[b]	24[b]	3·7[b]	4[b]
Quercus-Fraxinus (Yuen, 1966)	2·4[d]	—	0·7	1·1	31	47	—	—
Fagus (Bassus, 1962*a*)	—	—	—	—	60	25	23	23
Fagus (Bassus, 1962*a*)	—	—	—	—	58	31	18	23
Fagus, mixed (Bassus, 1962*a*)	—	—	—	—	67	26	25	30
New Zealand forest (Egunjobi, 1971)	2·4 to 1·6[d]		*Tylenchus* spp. 2–31% in soil; 0–47% in litter					
Uganda scrub (Banage and Visser, 1967)	3·0[d]	—	—	—	—	—	—	—

[a] To 20 cm including litter.
[b] Upper soil horizons only.
[c] Mean, estimated to 9 cm depth.
[d] Arbitary conversion, 10 cm² × 1 m deep = 100 kg soil.

TABLE IV. Beech and oak woodlands: mean biomass of nematode microflora feeders

Site	Sample	Fresh biomass of nematodes, g m^{-2}	
		Fungivores	Bacterivores
Fagus mull (after Volz, 1951)	to 5 cm deep	0·44	0·39
Quercus mull (after Twinn, 1966)	to 3 cm deep	0·05	0·033
Quercus mor (after Twinn, 1966)	L + F horizons	0·13	0·13

both in microbiological and nematological studies by sampling the mineral soils. These differ considerably in their relationship to the decomposer processes: in mor decomposition takes place above the mineral soil, in mull mainly within it, owing to earthworm activity. Strictly, the soils should be compared on the basis of their total activity.

At present, the role of nematode microflora feeders in temperate mull cannot be further defined. Studies based on serial soil sampling are unlikely to clarify this situation unless linked with work on the fate of the litter material. Comparison should be made between the nematode colonization of litter on the soil surface or within the mouths of earthworm burrows, and the succession on earthworm casts themselves.

E. Grassland, Moor and Cultivated Land

Within Nielsen's wide range of Danish grasslands the poorest growths usually, but not invariably, showed the lowest nematode populations whereas the best lush growth yielded up to 20×10^6 m^{-2}. For the most part, fungivores were very low throughout but the bacterivores tended to vary more with the overall population level, commonly forming some 50% of all species. In this characteristic, the grass sites resembled those of the beach under moss or grass, both of which also had low fungivore populations but, by percentage, high bacterivore ones. This contrasts with woodland which often has a fungivore bias of numbers if not of biomass. Studies in England, restricted to only three grass sites, showed a balance of the two trophic groups different from the Danish examples. Yuen (1966) in Rothamsted Broadbalk grassland found a relatively low nematode population but one in which fungivores predominated over bacterivores. Similarly, in upland *Nardus* and limestone grassland, Banage's (1963) somewhat higher numbers also contained more Tylenchida: he did not separately list fungivores.

Raw humus and peats with various types of vegetation have low overall nematode populations, culminating in only some tens of thousands per m^2 in bare peat (Banage, 1963). Again, the same distinction is shown between the results of Nielsen and Banage, the former's data from *Calluna* heath and high moor peat being biased towards bacterivores, the latter's from *Calluna* and *Juncus* moors towards Tylenchida, mainly fungivores.

The variation in nematode numbers in grassland and the differences in the balance of the decomposer feeding groups cannot at present be explained. One contributory factor may prove to be the spacing of the vegetation, of particular relevance on sites with open communities. It has been suggested that almost all the litter decomposition of *Eriophorum* takes place within the tussocks themselves (Goodman, 1963): thus one may predict that the selection of samples in relation to individual plants may well have a considerable effect on the results obtained.

TABLE V. Nematode frequencies: cultivated land

Crop	Numbers 10^6 m^{-2}		
	Total population	Fungivores	Bacterivores
Rye field (Nielsen, 1949)	2·5	0·34	1·48
Potato field (Witkowski, 1958)	1·8 (2·12[a])	0·21	0·59
Spring wheat (Witkowski, 1958)	2·3 (3·8[a])	0·21	1·25
Winter wheat (Witkowski, 1958)	1·8	0·13	1·04

[a] maxima.

Data on cultivated land are commonly restricted to the important species of plant parasites. When all groups are listed, the low numbers of free-living nematodes found in these situations are notable (Table V), the greatest biomass recorded being one of 3 g m^{-2} for a rye field (Nielsen, 1949). The population structure resembles that of the grass fields in a low fungivore: bacterivore ratio.

VII. Ecology: Aquatic Systems

Despite their taxonomically different nematode faunas, freshwater and marine habitats may conveniently be considered together. Deep water sampling of both involves considerable technical difficulties and even faunistic surveys are very restricted. As in work on soil nematodes, most studies have been concentrated on temperate regions.

A. Marine

McIntyre (1969) has comprehensively reviewed the frequency of marine benthic nematodes, pointing out the decline in density which occurs towards deeper waters. Depth, or some correlated factor, can be shown from the work of Thiel (1971) and other authors to be apparently limiting the maximum numbers found (Fig. 6). The same relationship is shown for

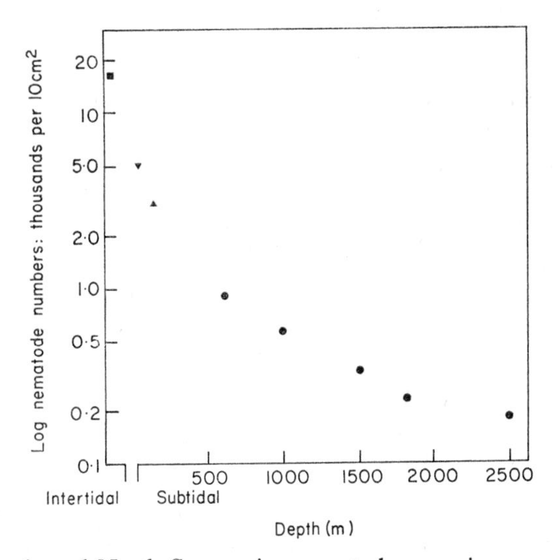

Fig. 6. Atlantic and North Sea marine nematodes: maximum numbers related to depth (data from McIntyre, 1964 ▲; Teal and Wieser, 1966 ■; Thiel, 1971 ●; Tietjen, 1969 ▼).

biomass, which infrequently exceeds 3 g m^{-2} at depths of several metres or more but may reach over 20 g m^{-2} in tidal or immediately subtidal zones (Table VI). The largest figure quoted so far for a single species is 28 g fresh weight m^{-2} for *Metoncholaimus scissus* but the species was aggregated and such a high density may not have covered a large area (Meyers *et al.*, 1970).

Water serves not only to modify the benthic habitats themselves by the sorting of mineral particles but it also carries organic detritus, depositing it selectively. Areas of fine mineral and organic deposits tend to carry the highest nematode populations (McIntyre, 1969) and the deposit feeders—Wieser's Group 1—are particularly prevalent in such situations (Wieser, 1960). Even where relatively dense vegetation occurs in tidal or subtidal

TABLE VI. Marine nematodes: selected population biomass data

Locality	Original source	Fresh biomass g m^2
Beach sand, Loch Ewe	McIntyre *et al.*, 1970	<0·35
Salt marsh, Massachusetts	Teal and Wieser, 1966	0·2–4·9[a] (2·76[b])
Salt marsh, Georgia	Wieser and Kanwisher, 1961	8·7–18·4[a]
Estuary, Rhode Island, to 1 m depth (*Zostera* beds)	Tietjen, 1969	14·9–21·6[c] (62·5[d])
Estuary, Florida, to 1 m depth (*Thalassia* beds)	Meyers *et al.*, 1970	23·6[d]
12–18 m, Buzzards Bay	Wieser, 1960	0·4–1·9
45 m, Plymouth	Mare, 1942	0·35[a]
80 m, Northumberland	Warwick and Buchanan, 1970	1·1–2·8
101 m, Loch Nevis	McIntyre, 1964	0·4–1·1[a]
146 m, Fladen, North Sea	McIntyre, 1964	1·1–2·6[a]
290 m, Iceland–Faroe Ridge	Thiel, 1971	0·78–1·74
600 m, Iceland–Faroe Ridge	Thiel, 1971	7·14[d]
2500 m, Iceland–Faroe Ridge	Thiel, 1971	1·26[d]

[a] As given by Tietjen, 1969.
[b] Mean value given by Teal, 1966.
[c] Range of means.
[d] Maxima.

zones, much of the litter is likely to be transported from the site of production (den Hartog, 1967; Teal, 1962). When it reaches the benthic nematodes it will be macerated and probably already extensively attacked by microorganisms. In contrast to some terrestrial systems, the amount of detritus or litter input is not easy to estimate (see Perkins, Chapter 22). Measures of marine soil organic matter are difficult to relate to nematode numbers for not only are decomposable and refractory portions inseparable, but an appreciable amount of animal material may be included (Tietjen, 1969).

Water movement also has effects on the local distribution of nematodes. In the relatively calm conditions of deep water, or within vegetation growths where current velocity may be reduced, active movement is possible. This is indicated by the rapid colonization of experimentally provided substrates, such as mats of fungus growth (Hopper and Meyers, 1966). Differential wandering of age classes may occur: these fungus traps accumulated gravid females of *Metoncholaimus scissus* and Ott (1967) found *Enoplus* juveniles at the tips of the alga *Cystoseira* but adults mainly at the base.

The effects of rapid water action are mitigated by the possession of caudal cement glands which help the nematodes to maintain their stations. Even so, if they live in the superficial horizons of sandy beaches, tidal movements passively transport them. There is evidence that some species react by moving vertically up and down the sand profile in response to tides, some moving downwards as the sea recedes, others travelling in the reverse direction (Rieger and Ott, 1969). Such voluntary or involuntary migrations have considerable implications in the definition of micro-habitats. For example, if extensive and selective interchange between algae and substrate occurred, algal samples alone would not adequately describe the community as a whole.

The continual or intermittent presence of surface water permits the colonization of aquatic vegetation to a much greater extent than on land. However, plant parasites as such appear to be uncommon in that the habit has only been recorded for *Halenchus* on *Fucus* and *Ascophyllum*. Other nematodes are mainly surface dwellers and the vegetation is to be regarded more as a substrate than a direct source of food. The extent of colonization is determined in part by factors which reduce the scouring effects of water movement. Relatively smooth algae disturbed by water currents carry low populations, but these are increased by the presence of superficial growths of other animals or plants (i.e. epigrowth, *aufwuchs*) which give protection for the nematodes. Thus Wieser (1952) counted between 7·7 and 200 nematodes per 100 g fresh weight of *Fucus serratus* without epigrowth, but up to 1590 when the epigrowth was considerable.

Density may vary within the algal growth related to the degree of exposure: Ott (1967) found basal populations on *Cystoseira spicata* and *C. abrotanifolia* which were higher than those in the crowns. Again, on *C. adriatica* in an exposed situation, dense epiphytic growths of the red alga *Jania rubens* permitted nematode numbers to rise to a higher level. Algal growth habit itself has a considerable effect on population density, not only by virtue of the protection afforded but also because dense growths retain fine deposit material. This was clearly shown by Wieser (1952) whose high populations were associated with heavily silted *Gelidium corneum*. Unfortunately, published data do not define any differences in population structure between silty and silt-free growths: one might expect a higher proportion of deposit feeders in the former, associated with the decomposing organic fraction.

More typically, the algal inhabitants exhibit a characteristic bias towards epigrowth feeders, as exemplified by chromadorids, many of which may be inferred to feed on diatoms. This family also forms an appreciable proportion of the populations on coarse sandy substrates and also on relatively firm mud which often has a considerable surface bloom of diatoms.

Seasonal cycles of primary production and input into the decomposer cycle are likely to lead to nematode population peaks. Annual fluctuations have, however, rarely been demonstrated: Warwick and Buchanan (1971) could find no evidence of any at 80 m depth off the Northumberland coast and most of the species in their samples appeared to breed throughout the year. Annual changes may be discerned rather more readily in much shallower water. Tietjen (1969) suggested a correlation between benthic microflora growth and the increased abundance of epigrowth feeders during spring and summer on four estuarine sites. Deposit feeders tended to have the reverse cycle, being more frequent in autumn and winter when total benthic organic matter increased. This was only particularly marked at one station which was perhaps the poorest in deposit material, being subject to relatively strong currents and with a low gain of detritus from the shore. Conceivably, this annual cycle was emphasized by the winter input of detritus being more limiting at this station than the others. However, particulate deposits are not the sole source of organic carbon and may be insufficient to meet the energy requirements of the benthic community, the deficit apparently being made up by "soluble" organic matter (McIntyre *et al.*, 1970).

In estuarine conditions, where much of the primary production and hence much of the detritus is provided by benthic macro- and microflora or from terrestrial origins, Odum (1970) postulated a recurring cycle of bacterial degradation of the litter and protozoan cropping of the bacteria. Such a mixture of bacteria and protozoa suggests an appropriate food source for nematodes of rather varied feeding habits, particularly for facultative bacterivores. The possible relationships of this system are indicated in Fig. 7.

In deeper water, much of the deposit matter will have been extensively decomposed before it reaches the bottom (Finenko and Zaika, 1970) but input is sufficient to support high populations of benthic bacteria (Zatzepin, 1970). Thus the ultimate dependence of many nematodes on bacteria is probable, either directly or no more than one or two stages removed, i.e. via protozoa.

Fungus–nematode associations also occur in the sea as demonstrated by the discovery of an *Aphelenchoides* sp. in shallow water *Thalassia* beds and its successful culture on a range of marine fungi (Meyers *et al.*, 1964). It is possible that stylet-bearing fungivores have often been overlooked in marine studies, owing to their small size. A number of the fungi used in culture are implicated in the degradation of vegetable material but the role of the nematode in the fungus bionomics of the *Thalassia* beds is not yet known.

A particular problem in relation to benthic nematodes in the presence of considerable bacterial action is that of low oxygen tensions. Such situations are usual at a depth of one or two centimetres in intertidal or deep

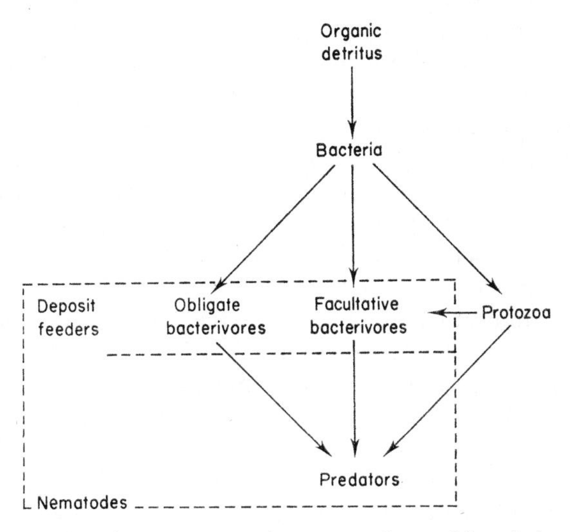

Fig. 7. Marine decomposer cycle: nematode trophic relationships.

water muddy substrates (Wieser and Kanwisher, 1961). Many species are able to maintain activity under almost completely anaerobic conditions and *Enoplus brevis* possesses haemoglobin which may assist respiration at low oxygen tensions (Atkinson *et al.*, 1972). The efficiency of food utilization under such circumstances is unknown. Teal (1962) assumed that comparable movement activity under aerobic and anaerobic conditions indicated equivalent rates of energy dissipation, but if movement accounts for only a small fraction of total energy utilization (von Brand, 1960), this is not necessarily justified.

B. Freshwater

Much of the work concerning these habitats was reviewed by Winslow (1960). The main conclusion of earlier workers such as Micoletsky (1925) was that there existed no clear boundary between true soil and true freshwater species, the gradation between the two biotopes being much more gradual than that between the land or fresh water and the sea. No doubt in the latter situation the widely fluctuating salinity has the effect of sharpening the distinction between the faunas. However, freshwater systems have much in common with the sea in the presence of water as a transport medium, particularly for detritus, much of which is of allochthonous origin as in estuarine and inshore areas. Fresh water also has a recognizable but more restricted fauna analogous to marine epigrowth feeders and associated with the periphyton or with benthic diatoms. Comparable extremes of anoxia occur in muddy situations in both biotopes.

The periphyton of *Phragmites* reeds has been studied by Pieczyńska (1964) who followed earlier authors in considering it as an independent unit containing all three major trophic levels: producers, consumers and reducers. However, the system is not in balance for much of the periphyton falls as litter during winter to the benthic environment below. The periphyton population consisted largely of the chromadorid algal feeders *Prochromadorella bioculata* and *Punctodora ratzeburgensis*, other species generally forming less than 10% of the total. Many of the others were of benthic types, indicating a degree of interchange between the periphyton and the benthos, which was not studied. The data of Filipjev, compiled by Winslow (1960), illustrate that both the periphyton and bottom sand were inhabited by appreciable proportions of chromadorid algal feeders. The sand and mud benthos were similar in the presence of *Trilobus* spp. and *Ironus* sp. As in the sea, spear-bearing fungivores appear to be rare except in habitats with extensive vegetation where the water and land biotopes overlap: bacterivores or facultative bacterivores are much more common.

Hirschmann (1952) examined a range of pond and river banks, finding amongst the bacterivores, species of *Monhystera* and *Plectus* to be constantly present in large numbers. At the edge of still water, the greatest variety of other species of the same feeding group was found when vegetation was richly developed. The stream series included three graded in respect of pollution type and it was notable that an increase in diplogasterids and rhabditids was associated with greater contamination. Similarly, these groups were common in the most nutrient-rich and polluted of the still waters (Table VII). This aspect of the population structure shows a clear relationship to that found in sewage treatment beds (Pillai and Taylor, 1968).

Few quantitative data are available on benthic freshwater nematodes. Winslow (1960) quotes Filipjev's figures of 200 to 18,000 m^{-2} for profundal regions and Por (1968) recorded "tens of thousand/m^2" of *Eudorylaimus andrassyi* at 30–43 m depth in Lake Tiberius. This species lives and reproduces in association with ciliates and oligochaetes in a habitat which is anaerobic for some eight months of the year. The scarcity of data on population dynamics is surprising, since freshwater systems would appear to be well suited to the study of nematodes in relation to annual cycles of primary production.

TABLE VII. Freshwater banks: bacterivorous nematodes

Habitat	Frequency index[a]				
	Mon-hystera	*Plectus*	Rhab-ditids	Diplo-gasterids	Others
Still water					
old river bed pond	19	4		4	1
polluted: rich vegetation	13	6	11	16	4
small, temporary	13	2			
small, permanent	13	5	3		18
woodland *Sphagnum* pool	4	3			8
Running water					
fast flow	17				
slow flow	13	6		5	1
weakly polluted	11	3		11	1
strongly polluted	13	7	6	13	4
very strongly polluted	8	5	20	32	

[a] Data after Hirschmann, 1952.
Scored on abundance: 5 (very abundant) to 1 (infrequent); index is the sum of the individual species scores.

VII. Ecological Role in Decomposer Cycles

In considering the marine benthic meiofauna, which often contains a high proportion of nematodes, Gerlach (1971) concluded that although they are only equivalent to about 3% of the benthic macrofauna biomass, their role in terms of energy flow was likely to be about five times greater than this suggests. He pointed out that such calculations are based on inadequate information, in particular on assumed rates of turnover for both meio- and macrobenthos. The respiration data employed are derived from laboratory determinations and the efficiency of energy utilization under anaerobic benthic conditions is unknown. Even greater proportional importance might be inferred in some cases, for freshwater microbenthos may equal 50–60% of the macrobenthos (Kayak and Rybak, 1966).

Using data from a grassland habitat rich in nematodes, Nielsen (1961) calculated, from respiration data, that protozoa, nematodes and enchytraeids consumed bacteria in the proportion 20:2:1. In one year they consumed quantities of the same order as the standing stock of bacteria which implied a relatively minor role in cropping bacteria, which must have a turnover of many times annually. However, most early figures of both bacterial and fungal frequencies include a high proportion of inactive propagules and do not fairly measure the active standing stock (Clark, 1967; Harley, 1971).

In his comprehensive study of a Georgia salt marsh, Teal (1962) apportioned to animal consumers, which included nematodes as a small fraction, less than 8% of the energy of net production in comparison with 47% dissipated by bacteria. More recently, Wasilewska (1971) used Nielsen's (1949) nematode respiration measurements to compute the energy requirements of the nematodes of afforested dunes. Maximal figures, omitting corrections for drought and low temperatures, were estimated to be about 13·8 kcal m^{-2} year, of which the fungivores, bacterivores and predators required 9·2 kcal. Given a leaf fall of about 250 g m^{-2} with the calorific value of oak leaves (Cummins, 1967), this part of the input into the decomposer system is of the order of 250 $\times$ 4·8 kcal m^{-2} year. Nematodes therefore use only a fraction of the energy and are even less important if allowance is made for other litter.

Work in which standard species biomasses were used for calculation may have considerably over-estimated total nematode fresh weights. On the other hand, if conditions of optimum food supply and temperature obtain, both respiration and production of eggs might have been grossly underrated. Eggs are rarely counted in work on free-living nematodes and a high death rate at this stage under variable environmental conditions could represent a considerable level of production. Nevertheless, the data available are consistent with the view that nematodes generally account for a low percentage of the energy of net production in an ecosystem, despite their often immense numbers and locally considerable biomass. When viewed in proportion to the energy taken by other animals, the nematode fraction might well be much more appreciable. However, Teal's work indicates that these levels can be of the same order as or less than the minimum error one can expect in ecological work. It is difficult to see how the role of nematodes can be further defined in terms of energy transfer without much more research on small defined biotopes.

Although themselves a minor route for energy, animals may possess important functions in facilitating and directing the energy dissipation of the bacteria and fungi. Nematodes ("deposit feeders"), which take in refractory organic matter with their bacterial food, might alter the nature of the food but the already finely divided nature of the particles suggests that this is not an important function. The effects of the nematodes are more directly on bacteria and fungi or, within the next consumer order, on organisms such as protozoa.

The efficiency of their digestive processes is of considerable importance in this respect in that a high intake rate coupled with that of low assimilation would indicate a much greater effect on the food than could be measured by respiration and production. In addition spoilage due to incomplete feeding would switch energy from fungi to other organisms such as bacteria. The net effect of such activities is likely to be an enhanced substrate use and a more rapid transfer of nitrogen and trace elements. In the case of oak-wood mor soils, fungivorous nematodes in the upper litter may well foster the process of differential removal of nutrients carried out by fungi: this is one facet of the selective accumulation of the more refractory portion of the litter in lower horizons. Similar nematodes in a mull mineral soil, in attacking microflora growing on more comminuted materials, might more directly accelerate the return of nutrients to the primary producers.

In terrestrial systems, such nutrient cycling may occur within a relatively small area, e.g. within *Eriophorum* tussocks, whereas in aquatic environments export and import is often considerable. Thus, in the latter, outside the euphotic zones, the whole benthic nematode population is likely to be ultimately dependent on the detritus input whether of a particulate or "soluble" nature. On land such a complete relationship would appear to be less likely, owing to the presence of vascular plant parasites and frequently high populations of omnivorous dorylaimids, mainly regarded as algal feeders. Yet, in the field, these omnivores might feed less on autotrophic algae than on those in a heterotrophic phase or on other organisms. In this case only the higher plant parasites could be regarded as primary consumers, the bulk of the nematode biomass being then related more to decomposition processes. Even predatory forms feeding on enchytraeids and other decomposer fauna may be so classed.

At whatever level, nematodes would seem to be subservient to the microflora in decomposition processes: further definition of their roles requires the parallel study of both nematodes and microflora on the same materials.

Acknowledgement

The sections on woodland soils have been derived in part from research fulfilling the requirements of the Ph.D. degree of London University: this work was aided by a grant from the University Central Research Fund.

References

ANDERSON, R. V. (1964). *Phytopathology* **54**, 1121–1126.

ATKINSON, H. J., ELLENBY, C. and SMITH, L. (1972). *XIth Int. Symp. Nematol., Reading*, 1972, p. 1 (abstr.).

BANAGE, W. B. (1963). *J. Anim. Ecol.* **32**, 133–140.

BANAGE, W. B. and VISSER, S. A. (1967). *In* "Progress in Soil Biology." (O. Graff and J. E. Satchell, eds), pp. 93–100. North-Holland, Amsterdam.

BASSUS, W. (1962*a*). *Wiss. Z. Humboldt-Univ. Berlin, Math-Nat. R.* **11**, 145–177.

BASSUS, W. (1962*b*). *Nematologica* **7**, 281–293.

BOCOCK, K. L., GILBERT, O., CAPSTICK, C. K., TWINN, D. C., WAID, J. S. and WOODMAN, M. J. (1960). *J. Soil Sci.* **11**, 1–9.

VON BRAND, T. (1960). *In* "Nematology." (J. N. Sasser and W. R. Jenkins, eds), pp. 257–266. University of North Carolina Press, Chapel Hill.

BROWN, J. C. (1958). *Trans. Br. mycol. Soc.* **41**, 81–88.

BRUCE, J. R. (1928). *J. mar. biol. Ass. U.K.* **15**, 535–552.

BRUN, J. (1972). *XIth Int. Symp. Nematol., Reading*, 1972, p. 4 (abstr.).

BURSELL, E. (1964). *In* "The Physiology of Insecta." (M. Rothstein, ed.), Vol. 1, pp. 283–321. Academic Press, New York and London.

CHITWOOD, B. G. (1951). *Tex. J. Sci.* **3**, 617–672.

CLARK, F. E. (1967). *In* "Soil Biology." (A. Burges and F. Raw, eds), pp. 15–49. Academic Press, New York and London.

COOPER, A. F. and VAN GUNDY, S. D. (1971). *In* "Plant Parasitic Nematodes." (B. M. Zuckerman, W. F. Mai and R. A. Rohde, eds), Vol. II, pp. 297–318. Academic Press, New York and London.

CROFTON, H. D. (1966). "Nematodes." Hutchinson, London.

CROFTON, H. D. (1971). *In* "Plant Parasitic Nematodes." (B. M. Zuckerman, W. F. Mai and R. A. Rohde, eds), Vol. 1, pp. 88–113. Academic Press, New York and London.

CROLL, N. A. (1969). *Nematologica* **15**, 389–394.

CROLL, N. A. (1970). "The Behaviour of Nematodes." Edward Arnold, London.

CUMMINS, K. W. (1967). "Calorific Equivalents for Studies in Ecology and Energetics," 2nd edition. Pymatuning Laboratory of Ecology, University of Pittsburgh.

DAO, F. (1970). *Meded. LandbHoogesch. Wageningen* 70–72.

DONCASTER, C. C. (1962). *Nematologica* **8**, 313–320.

DONCASTER, C. C. (1971). *In* "Plant Parasitic Nematodes." (B. M. Zuckerman, W. F. Mai and R. A. Rohde, eds), Vol. II, pp. 137–157. Academic Press, New York and London.

EGUNJOBI, O. A. (1971). *N.Z. Jl Sci.* **14**, 568–579.

FAULKNER, L. R. and DARLING, H. M. (1961). *Phytopathology* **51**, 778–786.

FINENKO, Z. Z. and ZAIKA, V. E. (1970). *In* "Marine Food Chains." (J. H. Steele, ed.), pp. 32–44. Oliver and Boyd, Edinburgh.

GERLACH, S. A. (1971). *Oecologia* (*Berl.*) **6**, 176–190.

GOLDEN, A. M. (1956). *Univ. Maryland Ag. Exp. Stat. Bull.* A.85.

GOODEY, J. B. (1959). *Nematologica* **4**, 211–216.

GOODEY, J. B. (1960). *Ann. appl. Biol.* **48**, 655–664.

GOODEY, T. (1963). "Soil and Freshwater Nematodes," 2nd edition, revised by J. B. Goodey. John Wiley, New York.
GOODMAN, G. T. (1963). *J. Ecol.* **51**, 205–220.
GOWEN, S. R. (1970). *Nematologica* **16**, 267–272.
GREEN, C. D. (1966). *Ann. appl. Biol.* **58**, 327–339.
HARLEY, J. L. (1971). *J. appl. Ecol.* **8**, 627–642.
DEN HARTOG, C. (1967). *Helgoländer wiss. Meeresunters.* **15**, 648–658.
HAWKER, L. E. (1957). *In* "Microbial Ecology." (R. E. O. Williams and C. C. Spicer, eds), Seventh Symposium Soc. gen. Microbiol. London, pp. 238–258. Cambridge University Press.
HECHLER, H. C. (1962). *Proc. helminth. Soc. Wash.* **29**, 19–27.
HESLING, J. J. (1972). *In* "Economic Nematology." (J. M. Webster, ed.), pp. 435–468. Academic Press, London and New York.
HIRSCHMANN, H. (1952). *Zool. J. (Syst.)* **81**, 313–436.
HOLLIS, J. P. (1957). *Phytopathology* **47**, 468–473.
HOPPER, B. E. and MEYERS, S. P. (1966). *Nature, Lond.* **209**, 899–900.
INGLIS, W. G. and COLES, J. W. (1961). *Bull. Brit. Mus. (Nat. Hist.)* **7**, 320–333.
JAKUBCZYK, H. (1968). *Ekol. pol. B* **14**, 325–328.
JENNINGS, J. B. and COLAM, J. B. (1970). *J. Zool., Lond.* **161**, 211–221.
KAYAK, Z. and RYBAK, J. I. (1966). *Verh. Int. Verein. Limnol.* **15**, 441–451.
KHERA, S. and ZUCKERMAN, B. M. (1962). *Nematologica* **8**, 272–274.
KHERA, S. and ZUCKERMAN, B. M. (1963). *Nematologica* **9**, 1–6.
KRUSBERG, L. R. (1959). *Nematologica* **4**, 187–197.
LEAKE, P. A. and JENSEN, H. J. (1970). *J. Nematol.* **2**, 351–354.
LEE, D. L. (1965). "The Physiology of Nematodes." Oliver and Boyd, Edinburgh.
MAGGENTI, A. R. (1962). *Proc. helminth. Soc. Wash.* **29**, 139–144.
MAGGENTI, A. R. (1971). *In* "Plant Parasitic Nematodes." (B. M. Zuckerman, W. F. Mai and R. A. Rohde, eds), Vol. 1, pp. 65–81. Academic Press, New York and London.
MARE, M. F. (1942). *J. mar. biol. Ass. U.K.* **25**, 517–554.
MEYERS, S. P. and HOPPER, B. E. (1967). *Helgoländer wiss. Meeresunters.* **15**, 270–281.
MEYERS, S. P., FEDER, W. A. and TSUE, K. M. (1963). *Science, N.Y.* **141**, 520–522.
MEYERS, S. P., FEDER, W. A. and TSUE, K. M. (1964). *Dev. in indust. Microbiol.* **5**, 354–364.
MEYERS, S. P., HOPPER, B. E. and CEFALU, R. (1970). *Mar. Biol.* **6**, 43–47.
McCOY, O. R. (1929). *Am. J. Hyg.* **10**, 140–156.
McINTYRE, A. D. (1964). *J. mar. biol. Ass. U.K.* **44**, 665–674.
McINTYRE, A. D. (1969). *Biol. Rev.* **44**, 245–290.
McINTYRE, A. D., MUNRO, A. L. S. and STEELE, J. H. (1970). *In* "Marine Food Chains." (J. H. Steele, ed.), pp. 19–31. Oliver and Boyd, Edinburgh.
MICOLETSKY, H. (1925). *K. danske. Vidensk. Selsk. Skr. ser.* 8, **10**, 57–310.
MINDERMAN, G. (1956). *Nematologica* **1**, 216–226.
MINDERMAN, G. and DANIËLS, L. (1967). *In* "Progress in Soil Biology." (O. Graff and J. E. Satchell, eds), pp. 3–7. North-Holland, Amsterdam.

MYERS, R. F. and KRUSBERG, L. R. (1965). *Phytopathology* **55**, 429–437.
NAGEL-DE BOOIS, H. M. and JANSEN, E. (1967). *In* "Progress in Soil Biology." (O. Graff and J. E. Satchell, eds), pp. 27–35. North-Holland, Amsterdam.
NELMES, A. J. (1972). *XIth Int. Symp. Nematol. Reading*, 1972, p. 47 (abstr.).
NEWELL, R. (1965). *Proc. zool. Soc. Lond.* **144**, 25–45.
NIELSEN, C. O. (1949). *Nat. Jutl.* **2**, 1–131.
NIELSEN, C. O. (1961). *Oikos* **12**, 17–35.
ODUM, W. E. (1970). *In* "Marine Food Chains." (J. H. Steele, ed.), pp. 222–240. Oliver and Boyd, Edinburgh.
OOSTENBRINK, M. (1971). *In* "Methods of Study in Quantitative Soil Ecology." IBP Handbook No. 18, (J. Phillipson, ed.), pp. 72–82. Blackwell, Oxford and Edinburgh.
OTT, J. (1967). *Helgoländer wiss. Meeresunters.* **15**, 412–428.
PARKINSON, D., GRAY, T. R. G. and WILLIAMS, S. T. (1971). "Methods for Studying the Ecology of Soil Micro-organisms." IBP Handbook No. 19. Blackwell Scientific Publications, Oxford and Edinburgh.
PIECZYŃSKA, E. (1964). *Ekol. pol. A* **12**, 185–234.
PILLAI, J. K. and TAYLOR, D. P. (1968). *Nematologica* **14**, 89–93.
POR, F. D. (1968). *Israel J. Zool.* **17**, 51–79.
RIEGER, R. and OTT, J. (1969). *IIIe Symp. européen de Biologie marine*, 1968, *Arcachon* (abstr.).
RIFFLE, J. W. (1967). *Phytopathology* **57**, 541–544.
ROTHSTEIN, M. (1968). *Comp. Biochem. Physiol.* **27**, 309–317.
SACHS, H. (1949). *Zool. Jb. (Syst.)* **78**, 323–470.
SAITÔ, T. (1956). *Ecol. Rev.* **14**, 141–147.
SAYRE, F. W., LEE, R. T., SANDMAN, R. P. and PEREZ-MENDEZ, G. (1967). *Archs Biochem. Biophys.* **118**, 58–72.
SUTHERLAND, J. R. (1967). *Nematologica* **13**, 177–324.
TAYLOR, D. P. and PILLAI, J. K. (1967). *Proc. helminth. Soc. Wash.* **34**, 51–54.
TEAL, J. M. (1962). *Ecology* **43**, 614–624.
TEAL, J. M. and WIESER, W. (1966). *Limnol. Oceanogr.* **11**, 217–222.
THIEL, H. (1971). *Ber. dt. wiss. Kommn. Meeresforsch.* **22**, 99–128.
TIETJEN, J. H. (1967). *Trans. Am. microsc. Soc.* **86**, 304–470.
TIETJEN, J. H. (1969). *Oecologia (Berl.)* **2**, 251–291.
TIETJEN, J. H., LEE, J. J., RULLMAN, J., GREENGART, A. and TROMPETER, J. (1970). *Limnol. Oceanogr.* **15**, 535–543.
TIMM, R. W. and FRANKLIN, M. T. (1969). *Nematologica* **15**, 370–375.
TOWNSHEND, J. L. (1972). *XIth Int. Symp. Nematol. Reading*, 1972, pp. 76–77 (abstr.).
TRACEY, M. V. (1958). *Nematologica* **3**, 179–183.
TWINN, D. C. (1962). *In* "Progress in Soil Zoology." (P. W. Murphy, ed.), pp. 261–267. Butterworths, London.
TWINN, D. C. (1966). Ph.D. Thesis, University of London.
VOLZ, P. (1951). *Zool. J. (Syst.)* **9**, 514–566.
WAID, J. S. (1960). *In* "The Ecology of Soil Fungi." (D. Parkinson and J. S. Waid, eds), pp. 55–75. Liverpool University Press.

WALLACE, H. R. (1963). "The Biology of Plant Parasitic Nematodes." Edward Arnold, London.

WARWICK, R. M. and BUCHANAN, J. B. (1970). *J. mar. biol. Ass. U.K.* **50,** 129–146.

WARWICK, R. M. and BUCHANAN, J. B. (1971). *J. mar. biol. Ass. U.K.* **51,** 355–362.

WASILEWSKA, L. (1970). *Ekol. pol. A* **18,** 429–443.

WASILEWSKA, L. (1971). *Ekol. pol. A* **19,** 651–688.

WIESER, W. (1952). *J. mar. biol. Ass. U.K.* **31,** 145–176.

WIESER, W. (1953). *Ark. Zool.* **4,** 439–484.

WIESER, W. (1959*a*). *Lund. Univ. Årsskrift N.F. avd.* 2 **55,** 111 pp.

WIESER, W. (1959*b*). "Free-living Nematodes and Other Small Invertebrates of Puget Sound Beaches." University of Washington Press, Seattle.

WIESER, W. (1960). *Limnol. Oceanogr.* **5,** 121–137.

WIESER, W. and KANWISHER, J. (1961). *Limnol. Oceanogr.* **6,** 262–270.

WINSLOW, R. D. (1960). *In* "Nematology." (J. N. Sasser and W. R. Jenkins, eds), pp. 341–415. University of North Carolina Press, Chapel Hill.

WITKOWSKI, T. (1958). *Zesyty Nankowe Uniwersytetu Mikolaja Kopernika w Toruniu. Nauki Matematyczno-Przyrodnicze,* No. 3 (Biologia), pp. 61–101.

WRIGHT, D. J. and SMITH, L. (1972). *XIth Int. Symp. Nematol. Reading,* 1972, p. 81 (abstr.).

YEATES, G. W. (1967). *N.Z. Jl Sci.* **10,** 927–948.

YEATES, G. W. (1968). *Pedobiologia* **8,** 173–207.

YEATES, G. W. (1969). *Nematologica* **15,** 115–121.

YEATES, G. W. (1970). *J. nat. Hist.* **4,** 119–136.

YEATES, G. W. (1971). *Pedobiologia* **11,** 173–179.

YUEN, P.-H. (1966). *Nematologica* **12,** 195–214.

ZATZEPIN, V. I. (1970). *In* "Marine Food Chains." (J. H. Steele, ed.), pp. 207–221. Oliver and Boyd, Edinburgh.

14

Oligochaetes

J. R. Lofty

Rothamsted Experimental Station
Harpenden
Hertfordshire
England

I. Introduction

As long ago as 1789, Gilbert White of Selborne appreciated that earthworms contribute considerably to the fertility of soil by their mode of burrowing and feeding. Just under a century later Charles Darwin (1881) in his book, "The Formation of Vegetable Mould through the Action of Worms", showed that they play an important role in the disappearance of surface plant litter, its breakdown and incorporation within the soil. Since Darwin's time, especially in recent years, much work has been done to evaluate more precisely the part played by oligochaetes, particularly earthworms, in plant litter breakdown. Today, the role of oligochaetes in plant litter breakdown is seen as a complex one, with many factors involved. There is still much work to be done before the subject is completely understood.

II. Types of Oligochaetes Involved

A. Enchytraeidae

These are very small, pale-coloured worms, commonly known as pot worms. They are small compared with the Lumbricidae, with which this chapter is mainly concerned. The largest species grow up to 5 cm long, but the majority of species are 1 cm or less in length. They are found in soils in large numbers, especially in acid soils with a high organic matter content. O'Connor (1957) has recorded up to 250,000 m^{-2} in a coniferous forest soil, and Peachey (1963) 300,000 m^{-2} in a *Juncus* moorland soil. Despite the large numbers that can occur in some soils, the biomass is considerably less than most lumbricid populations. For example, in Peachey's *Juncus* moorland site, the biomass was only 53 g m^{-2}.

B. Lumbricid Earthworms

Lumbricid earthworms are the most important group of terrestrial oligochaetes involved in litter breakdown and turnover, particularly in temperate soils, as it is within the temperate zones that they are most widely distributed. They are the dominant oligochaete family, especially in those areas cultivated by man. It is in temperate soils, therefore, that earthworms contribute most to litter breakdown and turnover. Edwards and Heath (1963) considered that they may be responsible for up to 78% of the total litter disintegration.

Lumbricid earthworm species can be subdivided into 3 groups according to their feeding and burrowing activities: (a) surface feeding, (b) surface casting, and (c) other species that do not, to any extent, do either.

1. Surface Feeders

Lumbricus terrestris (Linn.) is undoubtedly the most important species in this group. Although a certain amount of nutrient is obtained from soil passing through the gut during burrowing activities, most food material is obtained from surface litter. The food material is foraged from an area around the mouth of the burrow, and is pulled into the burrow so as to form a plug at the mouth. Feeding then takes place upon the stored material. The material pulled down depends upon what is available, but, generally, only material that has become detached from growing plants is used.

L. terrestris is not the only surface feeder in temperate areas, but other species make more limited forays, and their contribution to the removal of surface litter is small.

2. *Surface Casting Species*

Some lumbricid species void their excreta onto the soil surface, rather than within the soil structure or onto the walls of their burrows. Species responsible for much of the surface casting include *Allolobophora longa* (Ude), *A. nocturna* Evans, and *A. caliginosa* (Sav.), and several others also do so to some extent. *L. terrestris* also casts on the surface on occasion. Casts are produced in greatest numbers during spring and autumn months.

Many authors have calculated the amounts of cast material brought to the surface by such species during one year. Darwin (1881) estimated that up to 40 t ha^{-1} were produced annually in English pasture, Guild (1951) 27 t ha^{-1} also in England, and Stockli (1949) 100 t ha^{-1} in Zurich, Switzerland.

3. *Underground Feeding and Casting Species*

These constitute the majority of lumbricid species. Evans (1948) assumed that, in English pasture, the weight of cast material produced by species within the soil was (weight for weight) at least equal to that voided by casting species on the surface. He calculated that in pasture fields at Rothamsted, up to 82 t ha^{-1} of total cast material was ejected by earthworms per annum. He also calculated that the top 10 cm of soil of a densely populated pasture field would pass through the guts of the earthworms in $11\frac{1}{2}$ years, although in an arable field with a low earthworm population it would take up to 80 years.

C. Other Terrestrial Oligochaetes

Although members of the Lumbricidae dominate the temperate areas of the globe, in tropical and sub-tropical regions the terrestrial oligochaetes are represented by other families, such as the Eudrilidae in tropical and sub-tropical Africa; the Glossoscolecidae, dominant in Central and South America and also represented in North Africa and Southern Europe; and the Megascolecidae, found particularly in East and South-east Asia, and Australasia. Some species of this last family, particularly those of the genus

Pheretima, are widely distributed throughout the world, often competing with the Lumbricidae. Many members of these tropical and sub-tropical families are surface casting species producing very large casts. Lumbricids rarely produce casts weighing more than 100 g, but, in Africa, *Dichogaster jaculatrix* (Octochaetidae) produce casts 10–12 cm high and 4 cm in diameter, and the very large *Notoscolex* earthworms in Burma produce casts up to 25 cm high, one of which Gates (1961) found to weigh 1·6 kg.

III. Populations in Different Habitats

Earthworm populations vary greatly in numbers according to their habitats (Table I). Unfortunately, no method of sampling gives a completely satisfactory quantitative result. Hand-sorting gives a very good estimate of those species living near the surface and among the litter layers and vegetation root system. It is, however, very inaccurate for *L. terrestris*; this species can be extracted very efficiently by the application of dilute formalin to the soil surface, but this method is less suitable for other species. However, the figures given in Table I provide some indication of relative populations. There are many fewer earthworms in moorland and mor than in fertile soils, and the build-up of organic matter on the surface of the former habitats is partly due to the lack of earthworms. Fallow soils also have comparatively few earthworms.

Earthworm populations in arable land are very variable. Arable soils with a high organic content can support quite large populations (Reynoldson *et al.*, 1955). The biomass is low, but this is probably due to the lack of *L. terrestris* in the samples, because of the sampling method used. The sites that support the largest populations are those that provide earthworms with an abundant food supply, such as orchards under grass, pasture lands, and where local aggregations of organic matter occur (e.g. under pig litter). Woodlands on mull soils usually support rather lower populations than pastures, but the biomass may be higher as there are often many *L. terrestris* individuals feeding on the abundant surface litter.

There is comparatively little information on the populations of earthworms in tropical habitats. Block and Banage (1968), using formalin, sampled a range of habitats, including swamp forest, bush, coffee and banana plantations and arable fields in Uganda, and recorded mean numbers of worms (families Acanthodrilidae and Eudrilidae) ranging from 101·8 m^{-2} (banana plantation) to 7·4 m^{-2} (swamp forest). Madge (1969) has recorded 33 m^{-2} in Nigerian grassland, and El-Duweini and Ghabbour (1965) reported mixed populations of lumbricids and non-lumbricids ranging from 8 to 788 m^{-2} from a variety of cultivated or semi-cultivated habitats in Egypt.

Table I. Populations of earthworms in different habitats

	Site	Extraction method	No. m^{-2}	g m^{-2}	Reference
Arable land	Bardsey Island, U.K.	Handsorting	287	76	Reynoldson *et al.* (1955)
Arable land	Herts., U.K.	Formalin	18	1·6	Raw and Lofty (unpublished)
Arable land with dung	Herts., U.K.	Formalin	79	39·9	Raw and Lofty (unpublished)
Fallow soil	U.S.S.R.	Handsorting	18·5–33·5	4·6–8·4	Džangaliev and Belousova (1969)
Orchard with grass	Cambs., U.K.	Formalin and Handsorting	848	287	Raw (1959)
Orchard with grass	Holland	Handsorting	300–500	75–122	van Rhee and Nathans (1961)
Pasture	N. Wales	Handsorting	481–524	112–120	Reynoldson (1955)
Pasture	Westmorland, U.K.	Handsorting	389–470	52–110	Svendsen (1957)
Under pig litter	U.S.A.	Formalin	960	272	Edwards (unpublished)
Pseudotsuga mor	N. Wales	Handsorting	14·0	4·7	Reynoldson (1955)
Mixed woodland	N. Wales	Handsorting	157	40	Reynoldson (1955)
Quercus woodland	Hants., U.K.	Formalin	184	68	Edwards and Heath (unpublished)
Pinus woodland	Hants., U.K.	Formalin	40	17	Edwards and Heath (unpublished)

It seems likely that populations of non-lumbricid earthworms in the tropics and sub-tropics are smaller than lumbricid populations in temperate regions.

Enchytraeid worms attain their greatest density in acid soils with a high organic content, such as under coniferous forests and in moorland soils. In these habitats, where there are few lumbricid earthworms, enchytraeids play an increased role in litter fragmentation and breakdown.

IV. Role in Fragmentation and Incorporation

When litter is deposited on the soil surface, the first stage in its breakdown is fragmentation. For many types of deciduous forest litter, a period of "weathering" is required before fragmentation commences. In temperate regions, litter is fragmented mainly by earthworms, diplopods, isopods, dipterous larvae, Collembola and oribatid mites. Fragmentation is a very important stage in the process of decomposition, for if it is slowed down experimentally, then the whole process of decomposition is retarded. Witkamp and Crossley (1966) found that 45% of litter treated with napthalene to exclude most invertebrates broke down after one year, whereas in the same period 60% of litter decomposed in untreated soil.

Earthworms are important in this initial process of organic matter cycling, because certain lumbricids, in particular *L. terrestris*, seem to be responsible for the majority of litter fragmentation in woodlands and other tree-stand sites. Raw (1962) has shown that *L. terrestris* may remove more than 90% of the autumn leaf-fall in an apple orchard. In two woodland sites, earthworms consumed more *Fagus* (beech) and *Quercus* (oak) litter leaves than the rest of the soil invertebrates together (Edwards and Heath, 1963).

Soils with a low earthworm population often have a layer of undecomposed litter on the surface, with a sharp demarcation from the underlying soil, which usually has a poor crumb structure.

Figure 1 shows the weight of leaf material which became buried in relation to the weight of *L. terrestris* in *Malus sylvestris* Mill. (apple) orchards. The leaves were confined beneath wire mesh cages, each containing 100 or 200 leaves, and the *L. terrestris* population beneath each was estimated, using formalin. For a given weight of *L. terrestris*, more leaves were buried in arable than in grass orchards, probably because leaves lie directly on the bare soil in arable orchards, and are thus more directly available, and also because the leaves are the only form of litter present.

Edwards and Heath (1963) buried oak and beech leaf disks (cut from freshly fallen leaves) in nylon mesh bags in woodland and old pasture soil. They used four different size meshes, of which only the largest, of 7 mm aperture, would allow earthworms entry. After one year, none of the oak disks remained intact, 92% of the tota oak leaf material and 70% of the beech had been removed from the bags. Among other workers using similar techniques, von Perel *et al.* (1966) compared the disappearance of litter in 1 mm mesh nylon bags with that held under nylon net, but exposed

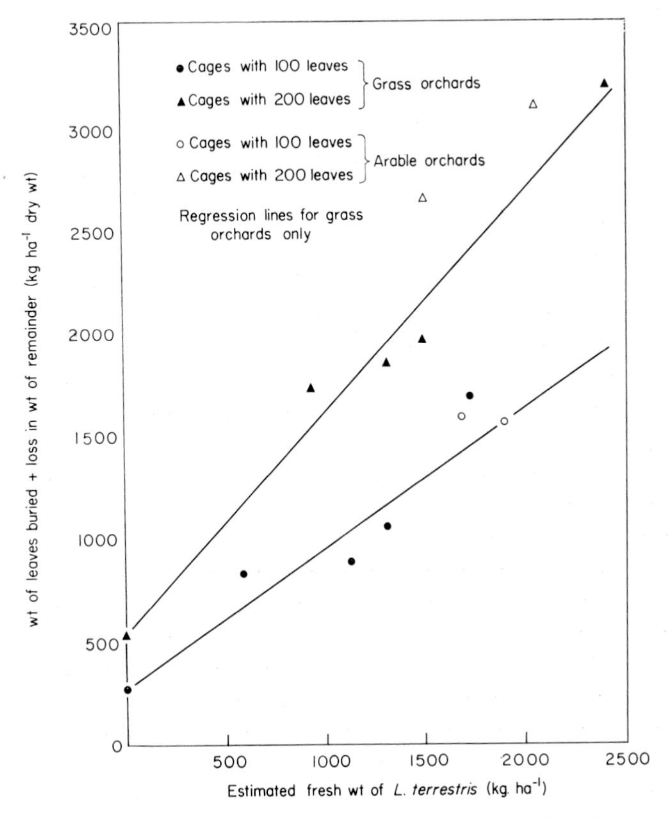

Fig. 1. Weight of leaf material buried in apple orchards in relation to the populations of *Lumbricus terrestris* in the soil.

beneath. They found that litter under the net disappeared 2–3 times faster than the litter in nylon bags and they concluded that earthworms were mainly responsible for this difference.

A striking demonstration of the effectiveness of *L. terrestris* in fragmentation and incorporation of leaf litter has been provided by a comparison between the soil profile of a grass orchard (A) with a typical *L. terrestris* population, and an adjacent orchard (B) where *L. terrestris* was absent, and other species rare, due to long-term heavy spraying with a copper-based fungicide. The soil profile of orchard (A) had no litter accumulating on the surface, and the soil beneath possessed a well developed crumb structure, whereas that of orchard (B) had a surface mat, 1–4 cm thick, consisting of leaf litter in various stages of decomposition with the boundary between it and the soil beneath quite sharply defined (Fig. 2).

In New South Wales, pasture land deficient in earthworms also accumulated a similar layer up to 4 cm thick. When earthworms were introduced experimentally, this mat gradually disappeared. Grass plots on Park Grass, Rothamsted, that have very few earthworms because of regular treatment with sulphate of ammonia also have similar accumulations.

It is not only in deciduous woodlands and tree stands that earthworms are active. They also accelerate litter decomposition in coniferous woods. Heungens (1969) put individuals of *Dendrobaena octaedra*, *D. attemsi*, *D. rubida*, *Bimastos eiseni* and *Allolobophora chlorotica* (Sav.) into cultures of pine litter containing many needles. Initial fragmentation by the worms was indicated by a rapid decrease in volume of the culture material, as the size of the constituents decreased. After 3 months, the litter had been broken down into still smaller particles, and had decreased in volume by 13% in cultures with 2 worms per litre of litter, and by 26% when there were twelve worms per litre. The litter had also become mineralized to a certain extent. Further humification and mineralization processes were accelerated by the initial "macro-decomposition".

In contrast to the limited period of leaf fall in temperate woodlands, that of tropical and subtropical forests is continuous or nearly so. Where seasonal climatic changes occur, leaf fall is usually higher in the dry season. During the wet season, litter is fragmented and broken down by a variety of soil organisms, and decomposes very quickly, about 1% per day in West African forests (Madge, 1969), hence little or no litter accumulates (in temperate woodlands the rate is from 0·1 to 0·3% per day). There is little information on the ability of tropical earthworms to fragment and incorporate surface litter, and there seems no counterpart to *L. terrestris* in tropical regions. Many tropical earthworms are surface casting species, and as the amount of cast material is sometimes very great, these species may indirectly contribute to litter incorporation when the casts crumble in the dry season, burying litter in the process. Nye (1955) has reported that the material from casts decomposing during each dry season (mainly belonging to the eudrilid species *Hippopera nigeriae*) was sufficient to form a 1·3–2·5 cm layer of fine sandy loam over the soil surface.

Although enchytraeid worms are found in large numbers of plant litter, it is not easy to define their part in litter fragmentation. O'Connor (1957) suggested that many of the species present are in fact feeding on

(a) (b)

Fig. 2. Soil profile of a grass orchard. (a) No *Lumbricus terrestris*, showing surface matt of debris, (b) with a typical *L. terrestris* population.

fungi present in the litter layer. Nielson (1962) has found that at least some species do not possess the enzymes that would enable them to break down the complex higher plant polysaccharides which form the bulk of the litter layer. However, enchytraeids must be responsible for a certain amount of initial fragmentation, as they pass vegetable fragments through their guts.

Clark (1949) has reported that, in Australian forests, Enchytraeidae ingest fragmented plant material together with fungal mycelium, and Zachariae (1963) has stated that in both deciduous and coniferous woodlands, Enchytraeidae living in the litter layer consume loose particles of leaf material, together with Collembola frass.

V. Palatability of Different Types of Litter

Earthworms do not take different kinds of litter indiscriminately. Darwin and later workers have stated that they show a preference for leaves of different shapes. Many species of earthworms, including *Lumbricus terrestris*, *L. rubellus*, *Allolobophora caliginosa*, *A. rosea* (Sav.), *Eisenia foetida* and *Eiseniella tetaedra* (Sav.), are able to distinguish between different kinds of leaf litter (Lindquist, 1941). Bornebusch (1953) offered *L. terrestris* individuals a selection of leaves from a number of different plant species, including some herbaceous plants such as *Mercurialis perennis* L. (dog's mercury), deciduous trees, and some conifers. In general, the herbaceous plant leaves were preferred to deciduous tree leaves, and the latter to coniferous needles.

Many workers have shown that when earthworms are offered a choice of leaves all cut to a uniform size and shape, such as a disk, they still show a definite preference with the order of selection being little changed from that when the leaves are offered whole. The choice depends ultimately on the species to which the leaves belong, and the prior degree of weathering. Edwards and Heath (1963) buried 2·5 cm diameter leaf disks, taken from trees just before leaf fall, in nylon mesh bags with 7·0 mm apertures, in woodland soil. They found that *Quercus* leaves were taken in preference to *Fagus*. Heath *et al.* (1966), in a similar experiment using a range of leaves of different species, found that earthworms fed preferentially on *Ulmus* (elm), *Zea mays* (maize), *Betula* (birch), *Fraxinus* (ash), *Tilia* (lime), *Quercus* and *Fagus*, in that order (Fig. 3).

Many leaves become more palatable after a period of weathering on the ground, and the order of preference for weathered leaves may be different from that for leaves fresh from the tree. Thus, *L. terrestris* prefers fresh *Fagus* leaves to fresh *Malus sylvestris* leaves, but if the leaves have weathered for a few weeks, the preference is reversed.

The preference for some species of leaf is an indication of their relative palatibility. Although Gast (1937) attributed differences in palatability to variations in mineral content, this has not been subsequently confirmed. Sugar content, which has been correlated with protein content (Laverack, 1960) may be of importance, as protein-deficient litter is less readily acceptable than protein-rich litter. Leaves of *Mercurialis perennis*, *Urtica* sp. (nettle), *Sambucus nigra* (elderberry), *Fraxinus* and *Ulmus glabra* Huds.

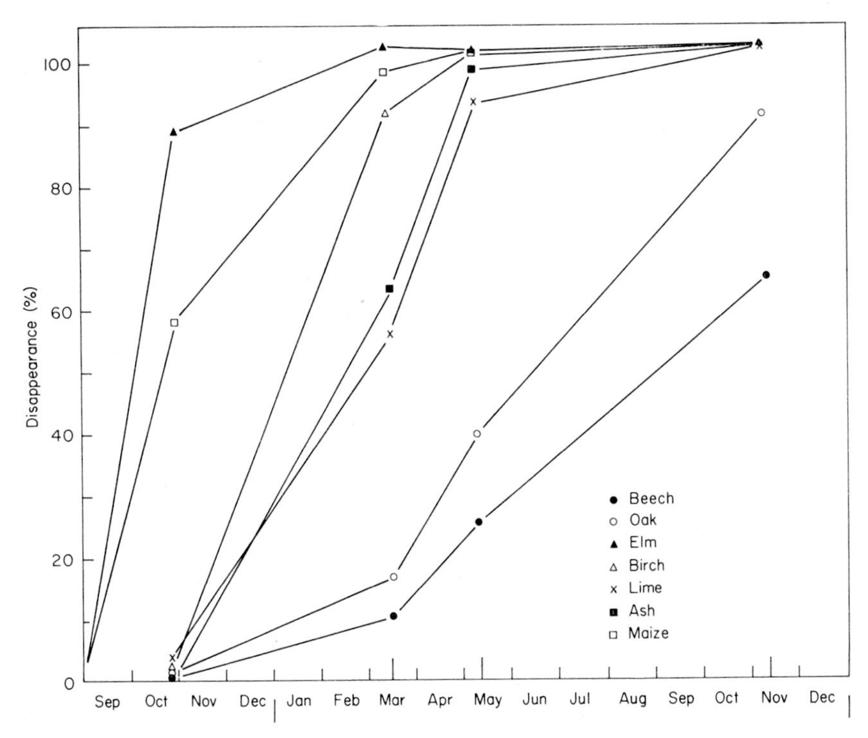

Fig. 3. Disappearance of leaf discs in 7 mm mesh bags in woodland soil (after Heath *et al.*, 1966).

(wych elm), which are all particularly palatable to earthworms, contain no condensed tannins, whereas leaves of *Larix* (larch), *Picea* (spruce), *Quercus* and *Fagus*, all of which do, are very much less palatable (Brown *et al.*, 1963). King and Heath (1967) found that litter containing a high proportion of water-soluble polyphenols was consumed more slowly than litter with only a smaller proportion, and that, generally, litter became more palatable after some weeks of weathering, when much of the water soluble polyphenolic content had been leached out (see Williams and Gray, Chapter 19).

In a comprehensive series of experiments Satchell and Lowe (1967) and Satchell (1967) found that amongst the factors that contribute towards litter palatability must be included nitrogen and carbohydrate content, and the presence or absence of a number of polyphenolic substances, particularly tannins.

VI. Role in Organic Matter Cycling

A. Consumption

Earthworms pass a mixture of organic and inorganic matter through their guts, the proportion depending upon the feeding and burrowing behaviour of different species. *Lumbricus terrestris*, a surface litter feeder, passes a high proportion of organic matter. Small species, such as *L. castaneus* (Sav.) and *Eisenia foetida*, living within the litter layer in woodlands, produce casts that often consist almost entirely of fragmented plant material. Species such as *Allolobophora longa*, *A. nocturna* and *A. caliginosa*, which live and, to a large extent, feed within the soil layers of woodlands and pastures, consume large quantities of mineral material when feeding. The casts of these species therefore contain a much smaller proportion of organic matter.

Satchell (1967) calculated that *L. terrestris* individuals consume 10% to 30% of their live weight (or about 100–120 mg) of plant litter per day, which compares with Needham's (1957) figure of 80 mg of *Ulmus* leaves per gram fresh weight of the species (8% of life weight) per day.

Many of the smaller species, e.g. *A. caliginosa*, *A. chlorotica* and *A. rosea*, can consume, weight for weight, similar quantities of litter material. However, the disappearance of litter due to these species is usually slower, as they tend to feed on the softer leaf tissues only, leaving the more lignified vascular tissues (van Rhee, 1963).

B. Turnover

Guild (1955) fed three species of lumbricid earthworms, *A. caliginosa*, *A. longa*, and *L. rubellus*, on cow dung for two years. During this period, individuals consumed 35–40 g, 20–24 g and 10–20 g respectively, equivalent to an annual consumption of 17–20 t of dung per ha by a field population of 120,000 adults per ha. Raw (1962) reported that, in an arable orchard, a *L. terrestris* population consumed over 93% of the annual leaf-fall (2000 kg of leaf litter per ha) between October and the end of the following February.

These observations show that earthworms are capable of consuming (or removing from the surface) a considerable proportion of the plant debris that is produced annually in various ecosystems.

The weight of annual leaf-fall in woodlands and forests varies considerably depending on the climatic zone within which they occur. Thus temperate deciduous forests produce 2·5–3·5 t ha^{-1} year, tropical forests 5·5–15 t, and arctic forests as little as 0·5 t ha^{-1} year. Satchell (1967) has calculated that the earthworm population could consume the entire leaf-fall of a temperate deciduous woodland, assuming an annual leaf-fall of 3 t ha^{-1}, and the consumption of 27 mg g^{-1} of leaf litter per day by the earthworms.

One of the major factors in litter destruction by earthworms is the amount of organic material available to them. In an experiment where mesh cages containing leaves were put down in apple orchards, *L. terrestris* removed more leaves from cages with 200 leaves than from those with half this number (Raw, 1962) (Fig. 1). It seems likely that, provided the soil conditions are optimum, numbers of earthworms will increase until the amount of food material available becomes the limiting factor.

C. Humification

After fragmentation and incorporation within the soil, the final process of plant litter breakdown is humification. This is the final breakdown into a complex stable product consisting of an amorphous colloidal substance containing a number of phenolic substances.

Much of the final decomposition is brought about by the direct action of microflora in the lower soil layers. This activity increases during the autumn in temperate areas. Other small organisms that contribute to humification are the small arthropods such as mites and springtails. The role of the earthworm is not clear. Some contribution must come from the action of microflora in the earthworm's gut, and the process is continued in cast material. Earthworm casts have been shown to contain larger populations of micro-organisms than the surrounding soil (Stöckli, 1928; Ponomareva, 1962; Zrazhevskii, 1957). Another contribution is that of combining fragmented organic matter with soil taken in with food—thus completing incorporation of the organic matter with soil, and thereby enhancing the effect of microfloral action.

D. Energetics

Satchell (1967) has calculated the oxygen consumption of a *L. terrestris* population of a *Fraxinus–Quercus* wood, with a biomass of 140 g m^{-2} to be about 22·9 l m^{-2}, an energy of respiration equivalent to 110 kcal m^{-2} per annum, and about 8% of the total energy content of the annual leaf fall.

Barley and Kleinig (1964), working from the oxygen consumption of *A. caliginosa* individuals in a respirometer, calculated that a population of this species in a pasture, with a biomass of 80 g m^{-2}, would contribute only about 4% of the total annual energy of decay of the organic matter present.

O'Connor (1963) estimated the energy of respiration of an enchytraeid population in a douglas fir plantation to be 150 kcal m^{-2}, 11% of the total energy content of the annual litter fall.

There is no information on the direct contribution to soil metabolism by tropical and sub-tropical oligochaetes; however, there is no reason to suppose that it is any greater than that of temperate oligochaetes.

The direct contribution to soil metabolism by the Oligochaeta may therefore be considered to be much less than the contribution by other soil animal groups taken together.

VII. Nitrogen Mineralization

Up to 72% of the dry weight of earthworms is protein (Lawrence and Millar, 1945) which indicates that on their death earthworms are a valuable source of nitrogen which is being returned continuously to the soil. Satchell (1967) calculated the weight of earthworm tissue produced by a *L. terrestris* population in a *Fraxinus–Quercus* wood at about 364 g m^{-2}. Working on the basis that the nitrogen content of *L. terrestris* is about 1·75% of fresh weight, he estimated the amount of nitrogen returned to the soil annually from worms dying in this population to be 6–7 g m^{-2}.

After death, earthworms decay rapidly in most soils, disappearing completely after 2–3 weeks. About 25% of the nitrogen from decomposing worms is in the form of nitrate and 45% as ammonia, with the remainder as soluble organic matter in components such as setae and cuticle.

Mineralizable nitrogen is also produced by living earthworms in their excreta, since they consume large quantities of plant material, which itself is a considerable source of nitrogen. Of the total nitrogen excreted by worms about half is secreted by gland cells in the epidermis as mucoproteins, and half excreted from the nephridiopores as ammonia, urea, allantoin and uric acid.

Barley and Jennings (1959) found that when non-available nitrogen, in the form of fragmented plant litter, was fed to immature individuals of *A. caliginosa*, about 6% was excreted as available nitrogen. Needham (1957) has provided data on the rate of nitrogen excretion by earthworms. His figure for *L. terrestris* was 269 μg g^{-1} day at 23°C. Needham considered that very little nitrogen is excreted in the faeces, although other workers, notably Lunt and Jacobsen (1944) and Graff (1972), have found that earthworm casts contain considerably more nitrogen than the surrounding soil. Using Needham's figures Satchell (1967) estimated the amount of nitrogen excreted annually by earthworms in the oak/ash wood previously referred to, and calculated the total annual yield of nitrogen both from the excreta of the population and from dead worms to be about 100 kg ha^{-1}.

VIII. Effects on the C:N Ratio

Unless the ratio of carbon to nitrogen (C:N ratio) in organic matter is less than 20:1, plants are unable to assimilate the unchanged nitrogen. Freshly fallen leaves have a much higher ratio than this; *Ulmus* leaves 24·9:1, *Fraxinus* 27·6:1, *Tilia* 38·2:1, *Quercus* 42·0:1, *Betula* 43·5:1, *Sorbus aucuparia* L. 54·0:1 and *Pinus sylvestris* L. 90·6:1 (Wittich, 1953). Often the C:N ratio of plant material is correlated with the degree of palatability to earthworms. Soft, succulent leaf tissues frequently have comparatively low C:N ratios, whereas the more initially unpalatable, tough tree leaves that have a high percentage of cellulose and lignin often have high C:N ratios. This means that large amounts of nitrogen in litter remain unavailable to plants until such time as the C:N ratio is reduced to below 20:1. Many of the tougher, less palatable tree leaves (with the higher C:N ratios) have to undergo a period of weathering before they become acceptable to earthworms. Once earthworms begin to feed on plant litter, they gradually reduce its C:N ratio. The extent to which they lower the C:N ratio of their food material can be judged by the amount of carbon assimilated, which in turn can be assessed approximately by measuring their respiration. This is not a particularly accurate method, as earthworm respiration can be affected by many factors. However, Barley and Kleinig (1964) calculated that a population of *A. caliginosa* accounted for only 4% of the total carbon consumption, and Satchell (1967) calculated a comparable figure of 8% for *L. terrestris* in two English woodlands. These rates of carbon consumption are too small to reduce the C:N ratio of most litter to a level at which nitrogen is available to plants. The earthworms were inactive when the measurements were made, so the estimates are probably too low, and Satchell considered that a more accurate figure for the carbon respired by earthworms would be 12% of that consumed. Satchell estimated that in one of his woodland sites, the C:N ratio of the litter was 385. Even using his revised carbon respiration estimate this would only lower the C:N ratio to about 33·5. Satchell suggested, therefore, that other groups of animals, such as nematodes, have a smaller biomass but have a higher metabolic rate and may play a more important role in affecting the C:N ratio of litter.

IX. Effect on Available Mineral Nutrients

Many workers have analysed earthworm casts, and soil with and without earthworms, for available mineral elements, and have reported that earthworm casts have higher concentrations of these than associated soils. This has been demonstrated by Lunt and Jacobson (1944) (Table II), Nijhawan and Kanwar (1952) in India, and Nye (1955) in Nigeria. Soils containing a large population of earthworms are also reported to have more exchangeable calcium, magnesium and potassium and available phosphorus than soils with few or no earthworms (Stöckli, 1949; Ponomareva, 1950; Nye, 1955). It has been reported from New Zealand that when European earthworm species were introduced into pastures that were poor in available molybdenum, although rich in total molybdenum, they greatly increased the supply of the element available to plants. Graff (1972) has reported that earthworms were responsible for a large increase in amount of available nitrogen, phosphorus and potassium in brown earth soil.

TABLE II. Comparison between available mineral elements in earthworm casts and ploughed soil

In ppm	Earthworm casts	0–15 cm soil layer
Nitrogen (nitrate)	21·9	4·7
Calcium (exchangeable)	2793	1993
Magnesium (exchangeable)	492	162
Phosphorus (available)	150	20·8
Potassium (available)	358	32·0

Adapted from Lunt and Jacobson (1944).

It seems to be generally agreed from all the evidence available that the presence of earthworms and their faecal material are responsible for increasing the proportions of soil minerals which are available to plants.

X. Effects of Earthworms on Soil Micro-organisms

The alimentary canal of earthworms contains many species of micro-organisms. Most workers are agreed that these species are the same as those found in the soil from which the worms have been taken. Bassalik (1913) was the first to isolate micro-organisms from earthworms, and he found more than 50 species of bacteria in the gut of *Lumbricus terrestris*, which were also present in the soil in which the worms lived. This finding has been confirmed by Parle (1963) and Satchell (1967), who also concurred with this view, and concluded that earthworms are unlikely to possess an indigenous microfloral population in their guts.

There is, however, much evidence to show that numbers of micro-organisms can greatly increase in earthworm guts compared with the numbers in surrounding soil (Stöckli, 1928; Ponomareva, 1953). The greatest numbers are found in the more posterior parts of the gut (Parle, 1959) (Table III).

It is therefore to be expected that earthworm casts will contain many more micro-organisms than will the associated soil. This has been confirmed by Ponomareva (1962), who found 13 times more bacteria in earthworm casts than in the surrounding soil. Ghilarov (1963), Zrazhevskii (1957), Went (1963), Dawson (1948) and Teotia (1950) have all reported larger microfloral populations in cast material than in surrounding soil. The casts are therefore foci from which soil micro-organisms can spread into surrounding soil. Koylovskaya and Zhdannikova (1961) compared the microfloral populations of the gut contents of two lumbricid species living at different soil depths. They found that *L. rubellus*, which lives in the top 5 cm of soil, had more than ten times the numbers of bacteria in its gut compared with the soil in which it lives, whereas *Octolasium lacteum* which lives at between 10 and 40 cm, had a bacterial population in its gut which was little, if at all, greater than in the soil. However, the casts of both

TABLE III. Numbers of micro-organisms in the fore-, mid- and hind-gut of *L. terrestris* ($\times 10^6$)

	Fore gut	Mid-gut	Hind-gut
Actinomycetes	26	358	15,000
Bacteria	475	32,900	440,700

Data from Parle (1959).

species contained proportionately more fungi, actinomycetes and cellulose-decomposing and butyric acid-forming bacteria than the surrounding soil. Organic matter decomposed much faster in the casts of these two species than in the soil.

Differences in microbial populations in earthworm casts and surrounding soil are due either to environmental changes within the earthworm's gut or to the food material ingested which provides a rich substrate, thereby increasing microfloral activity. Which of these is the major factor is debatable; certainly the type and quality of ingested material helps determine the size of microfloral populations in casts, but, as Parle (1959) showed, numbers of bacteria and actinomycetes in ingested material increased by a factor of 1000 while passing through the earthworm's intestine, and microbial activity in the casts was still much greater than in the surrounding soil 50 days after their excretion.

Tenney and Waksman (1929) and Harmsen and van Schreven (1955) have reported that when simple nitrogenous compounds are added to soil, then the decomposition of the organic matter it contains is accelerated (particularly if the organic matter is already poor in nitrogen). Therefore, since earthworm casts are usually rich in nitrogenous compounds, the presence of earthworms may help to stimulate microbial decomposition of organic matter in the soil. Barley and Jennings (1959) added grass, clover litter and dung pellets to soil cultures containing populations of *Allolobophora caliginosa* equivalent to those found in the field, and to soil cultures without earthworms. They found that the rate of litter decomposition (estimated by the accumulation of nitrate and ammonia) was 17–20% greater in the cultures with earthworms. They estimated that half of the increased rate of decomposition was due directly to the earthworms, and half to the increase in microbial activity induced by the worm's excretory products.

XI. Influence of Environmental Factors on Litter Breakdown by Earthworms

A. Soil Type

Numbers of earthworms and the population composition are much influenced by soil type (Table IV). Light and medium loams have a higher population than heavy clay soils and alluvial soils. Acid peaty soils have a much reduced population, and also fewer species. In England *L. terrestris*

TABLE IV. Relation of soil type to earthworm populations

Soil type	Population	
	Number per ha	Number of species
Light sandy	57	10
Gravelly loam	36	9
Light loam	63	8
Medium loam	56	9
Clay	40	9
Alluvium	44	9
Peaty acid soil	14	6
Shallow acid peat	6	5

From Guild, 1951.

is most abundant in light loamy soils. *A. caliginosa* is the commonest species in all but the acid peaty soils, in which *Bimastos eiseni* and *Dendrobaena octaedra* (both small species feeding in the surface mat) are most commonly found. The greatest diversity of species in temperate areas is found in mull soils. Satchell (1967) reported eight characteristic species, including *L. terrestris*, *A. longa* and *A. caliginosa*, from mull sites in the English Lake District. In moder sites, there were usually only four species of small size, of which the largest was *L. rubellus* a humus and surface mat feeding species, which was much less abundant than in the mull sites. In the most extreme acid mor sites, only two species, *D. octaedra* and *B. eiseni*, were found. However, although mor soils contain few species of small biomass, whose contribution towards litter incorporation is comparatively small, other faunal groups, including considerable populations of many species of enchytraeids, are found in these sites. Layers of unincorporated organic matter persist in these sites despite the presence of earthworms, enchytraeids, microarthropods and the associated microflora. Handley (1954) has suggested that organic nitrogenous material in mor sites is not so easily digestible as is that in mull.

B. Effects of Temperature

Most species of earthworms in the temperate zones are most active during the spring and autumn months. During the summer and winter months their activity is much reduced or ceases temporarily, depending on the species. During the summer months some, such as *Allolobophora longa*, go into an obligatory diapause when soils become dryer. Others, such as *L. terrestris*, retreat deeper into their burrows, and feed less. During winter, as the soil becomes colder, most species retreat deeper into the soil; feeding and other activities become minimal. *L. terrestris* ceases to feed on the surface when the temperature of the soil surface approaches 0°C. Table V gives the results obtained when cultures of *L. terrestris*, provided with apple leaves as food, were exposed to four different temperatures for a period of four months.

Although worms kept at 15°C buried most leaves during the first month, the rate was much reduced during the remainder of the period. Over the whole period, most leaves were buried by worms kept at 10°C. Kollmannsperger (1955) noted a positive correlation between the number of earthworms on the soil surface at night and temperature; 10·5°C being the optimum temperature for activity. Satchell (1967) concluded that earthworms were most active on the soil surface, when temperatures at night were not less than 10·5°C, provided the grass–air temperature was above 2°C, and the soil surface was moist from recent rain.

TABLE V. Effects of temperature on leaf burial. Number of leaves buried by 2 *L. terrestris* at different temperatures in four successive months.

	0°C	5°C	10°C	15°C
After 1 month	0	25	35	67
After 2 months	0	33	31	30
After 3 months	0	53	63	33
After 4 months	0	67	75	44
	0	178	204	174

From Lofty unpublished.

XII. Conclusion

The major contributions by oligochaetes to plant litter breakdown can be summarized by the following stages:

1. *The removal of leaf and other litter material from the soil surface.* This is achieved by the lumbricid earthworm *Lumbricus terrestris* in temperate regions, particularly in deciduous tree stands on mull soils, where a population is capable of removing up to 90% of a seasons leaf fall annually, and is of great importance in initiating the process of plant litter decay.

2. *Litter burial.* Many species of earthworms cast on the surface, and this cast material contributes to litter burial, bringing the litter into greater contact with other plant material decomposers. In tropical and subtropical soils, where earthworms otherwise play a much smaller part in litter decomposition than they do in temperate regions, some species produce surface casts to such an extent that, in certain areas, they will form, on breaking down, a layer up to 4 cm deep over the surface.

3. *Fragmentation.* When feeding on plant remains, oligochaetes, particularly *L. terrestris* and those earthworms and Enchytraeidae living in litter layers, fragment the material, greatly increasing the surface area on which other decomposers may feed.

4. *Incorporation.* Of considerable importance is the action of earthworms in incorporating fragmented and decomposed plant material throughout the soil horizons within which they live, by their burrowing activities. Cast material is either deposited freely in the soil, or forms a lining of the burrows. Mor type soils, with few earthworms, have a mat of slowly decomposing plant material on the surface, showing no incorporation with the soil layers below.

5. *Effect on the C:N ratio.* During its passage through the guts of earthworms, the ratio of carbon to nitrogen in the ingested plant material is

reduced towards the level at which the material can be directly assimulated by plants.

6. *Effects on the soil microflora.* Earthworm faeces (cast material) are much richer in numbers of microflora than the surrounding soil, many of which are involved in the final humification of plant material in the soil.

References

BARLEY, K. P. and JENNINGS, A. C. (1959). *Aust. J. agric. Res.* **10,** 364–370.
BARLEY, K. P. and KLEINIG, C. R. (1964). *Aust. J. Sci.* **26,** 290.
BASSALIK, K. (1913). *Z. Garungsphyiol* **2,** 1–32.
BLOCK, W. and BANAGE, W. V. (1968). *Rev. Ecol. Biol. Sol.* **5,** 515–521.
BORNEBUSCH, C. H. (1953). *Dansk. Skovforen. Tidskr.* **38,** 557–579.
BROWN, B. R., LOWE, C. W. and HANDLEY, W. R. C. (1963). *Rep. For. Res. London*, Part 3, 90–93.
CLARK, D. P. (1949). Unpublished thesis, quoted in Birch, L. C. and Clark, D. P. 1953), *Q. Rev. Biol.* **28,** 13–36.
DARWIN, C. (1881). *In* "The Formation of Vegetable Mould through the Action of Worms." John Murray, London.
DAWSON, R. C. (1948). *Proc. Soil Sci. Soc. Am.* **12,** 512–516.
DŽANGALIEV, A. D. and BELOUSOVA, N. K. (1969). *Pedobiologia* **9,** 103–105.
EDWARDS, C. A. and HEATH, G. W. (1963). *In* "Soil Organisms." (J. Doeksen and J. van der Drift, eds), pp. 76–80. North-Holland, Amsterdam.
EL-DUWEINI, A. K. and GHABBOUR, S. I. (1965). *J. appl. Ecol.* **2,** 271–287.
EVANS, A. C. (1948). *Ann. appl. Biol.* **35,** 1–13.
GAST, J. (1937). *J. For.* **35,** 11–16.
GATES, G. E. (1961). *Am. Midl. Nat.* **66,** 61–86.
GHILAROV, M. S. (1963). *In* "Soil Organisms." (J. Doeksen and J. van der Drift, eds), pp. 255–259. North-Holland, Amsterdam.
GRAFF, O. (1972). *In* "Proc. IV Coll. Pedobiologiae, Dijon 1970."
GUILD, W. J. McL. (1951). *J. Anim. Ecol.* **20,** 88–97.
GUILD, W. J. McL. (1955). *In* "Soil Zoology." (D. K. McE. Kevan, ed), pp. 83–98. Butterworths, London.
HANDLEY, W. R. C. (1954). *Bull. For. Comm. Lond.* No. 23.
HARMSEN, G. and VAN SCHREVEN, D. (1955). *Adv. Agron.* **7,** 299–398.
HEATH, G. W., ARNOLD, M. K. and EDWARDS, C. A. (1966). *Pedobiologia* **6,** 1–12.
HEUNGENS, A. (1969). *Pl. Soil* **31,** 22–30.
KING, H. G. C. and HEATH, G. W. (1967). *Pedobiologia* **7,** 192–197.
KOLLMANNSPERGER, F. (1955). *Decheniana* **180,** 81–92.
KOYLOVSKAYA, L. S. and ZHDANNIKOVA, E. N. (1961). *Dokl. Akad. Nauk. S.S.S.R.* **139,** 470–473.
LAVERACK, M. S. (1960). *Comp. Biochem. Physiol.* **1,** 155–163.
LAWRENCE, R. D. and MILLAR, H. R. (1945). *Nature, Lond.* **155,** 517.
LINDQUIST, B. (1941). *Svenska SkogsvFör. Tidskr.* **39,** 179–242.
LUNT, H. A. and JACOBSEN, G. M. (1944). *Soil Sci.* **58,** 367.

MADGE, D. S. (1969). *Pedobiologia* **9,** 188–214.
NEEDHAM, A. E. (1957). *J. exp. Biol.* **34,** 425–446.
NIELSEN, C. O. (1962). *Oikos* **13,** 200–215.
NIJHAWAN, S. D. and KANWAR, J. S. (1952). *Indian J. agric. Sci.* **22,** 357–373.
NYE, P. H. (1955). *J. Soil Sci.* **6,** 73–79.
O'CONNOR, F. B. (1957). *Oikos* **8,** 161–199.
O'CONNOR, F. B. (1963). *In* "Soil Organisms." (J. Doeksen and J. van der Drift, eds), pp. 32–48. North-Holland, Amsterdam.
PARLE, J. N. (1959). Ph.D. Thesis, University of London.
PARLE, J. N. (1963). *J. gen. Microbiol.* **31,** 1–12.
PEACHEY, J. E. (1963). *Pedobiologia* **2,** 81–95.
PEREL, T. S. VON, KARPACHEVSKIJ, L. O. and YEGOROVA, S. V. (1966). *Pedobiologia* **6,** 269–276.
PONOMAREVA, S. I. (1950). *Pochvovedenie* 476–486.
PONOMAREVA, S. I. (1953). *Trudy pochv. Inst. Dokuchaeva* **41,** 304–378.
PONOMAREVA, S. I. (1962). *Vtoraya Zoologischeskaya Konferenciya Litovskoi S.S.R.* 97–99.
RAW, F. (1959). *Nature, Lond.* **184,** 1661.
RAW, F. (1962). *Ann. appl. Biol.* **50,** 389–404.
REYNOLDSON, T. B. (1955). *North Wales Nat.* **3,** 291–304.
REYNOLDSON, T. B., O'CONNOR, F. B. and KELLY, W. A. (1955). *Bardsey Obs. Rep.* **9.**
RHEE, J. A. VAN (1963). *In* "Soil Organisms." (J. Doeksen and J. van der Drift, eds), pp. 55–59. North-Holland, Amsterdam.
RHEE, J. A. VAN and NATHANS, S. (1961). *Neth. J. agric. Sci.* **9,** 94–100.
SATCHELL, J. E. (1967). *In* "Soil Biology." (A. Burges and F. Raw, eds), pp. 259–322. Academic Press, London and New York.
SATCHELL, J. E. and LOWE, D. G. (1967). *In* "Progress in Soil Biology." (O. Graff and J. E. Satchell, eds), pp. 102–119. North-Holland, Amsterdam.
STÖCKLI, A. (1928). *Landw. Jb. Schweiz* **42.**
STÖCKLI, A. (1949). *Z. PflErnähr. Düng.* **45,** 41–53.
SVENDSEN, J. A. (1957). *J. Anim. Ecol.* **26,** 409.
TENNEY, F. G. and WAKSMAN, S. A. (1929). *Soil Sci.* **28,** 55–84.
TEOTIA, S. P. S. (1950). University of Nebraska Doctorial abstracts 16–19.
WENT, J. C. (1963). *In* "Soil Organisms." (J. Doeksen and J. van der Drift, eds), pp. 260–265. North-Holland, Amsterdam.
WITKAMP, M. and CROSSLEY, D. A. (1966). *Pedobiologia* **9,** 293–303.
WITTICH, W. (1953). *Schrift Reihe forstl Fab Univ. Göttingen* **9,** 7–33.
WHITE, G. (1789). *In* "Natural History of Antiquities of Selborne in the County of Southampton."
ZACHARIAE, G. (1963). *In* "Soil Organisms." (J. Doeksen and J. van der Drift, eds), pp. 109–124. North-Holland, Amsterdam.
ZRAZHEVSKII, A. I. (1957). "Dozhdevye chervi kak faktor plodorodiya lesnykh pochv." Kiev. 133 pp.

15

Microarthropods

D. J. L. Harding and R. A. Stuttard

Department of Biological Sciences
Wolverhampton Polytechnic
England

I. Introduction

This chapter is concerned mainly with the feeding habits of Collembola, or springtails (apterygote, "primitive" insects, with basically mandibulate mouthparts), and of Acari, or mites, which have basically chelate jaws known as chelicerae. These are usually the most abundant animals in dry-funnel extracts of plant litter, and the term microarthropods is often used in terrestrial studies to refer to these two groups together. Other groups which may also be encompassed within this term include Diplura, Thysanura and Protura (apterygotes), Symphyla and Pauropoda (small myriapods), and Tardigrada ("degenerate" arthropods).

Many species of microarthropod are zoophagous, while certain phytophages consume living tissues of higher plants. We shall deal with the utilization of higher-plant litter and lower plants, which are the major food sources of phytophagous litter-dwelling species. Particularly with lower plants, it is often difficult to distinguish between the ingestion of living and of dead tissues (e.g. in gut contents), so that some of the examples in this chapter may not be true litter feeders.

Most Collembola crush their food between mandibular plates, although some possess styliform, piercing and sucking mouthparts. Styliform chelicerae occur in various families of the acarine order Prostigmata, but in

Fig. 1. Mouthparts of a phthiracaroid mite, with chelate-dentate chelicerae. (Reproduced by permission of the Trustees of the British Museum (Natural History).)

many of these cases it is not known whether the food consists of animal or of plant tissues. Most litter-dwelling Astigmata and the vast majority of Cryptostigmata, or oribatids, have chelate-dentate chelicerae (Fig. 1) and are phytophagous, as are certain species of the predominantly zoophagous order Mesostigmata, notably among the suborder Uropodina. In most sites Acari are more abundant than Collembola, with Cryptostigmata (or, occasionally, Prostigmata) numerically dominant among the mites.

The ecology of microarthropods in litter and soil has been discussed by Kühnelt (1961) and Kevan (1962) and, more recently, by Wallwork (1970), while quantitative methods in soil ecology are considered by Phillipson (1971). Review articles include those by Christiansen (1964) and Hale (1967) on Collembola, Wallwork (1967) on Acari, Edwards *et al.* (1970) on invertebrates, and Butcher *et al.* (1971) on Collembola and Acari. In an important paper by Schuster (1956), plant-feeding oribatid species were classified as *macrophytophagous* (generally feeding solely on higher-plant remains, the gut contents including at most only insignificant amounts of, for example, fungal fragments) or *microphytophagous* (with gut contents consisting entirely of lower-plant remains, with only occasional traces of phanerogam material) or *unspecialized* (gut contents either include a mixture of cryptogam and phanerogam remains, or are mainly of one type in some individuals and of a different type in others of the same species). Luxton (1972), in his review of oribatid nutritional biology, uses the term *panphytophagous* instead of unspecialized.

II. Types of Food and Feeding Habits

A. Cryptogams

Despite numerous population and community studies of micro-arthropods associated with cryptogams (see Frankland, Chapter 1), there is relatively little information on their feeding habits. For example, mosses and lichens have long been recognized as likely habitats for high population densities of animals such as tardigrades, mites and Collembola, while oribatids are commonly known as moss mites (Jacot, 1930) or Moosmilben (Willmann, 1931), but Gerson (1969) points out that the exact nature of the relationships between oribatids and mosses is largely unknown.

1. Fungi

Acari and Collembola are frequently found in the sporophores of polypores (Graves, 1960; Pielou and Matthewman, 1966), and some of these inhabitants are thought to be mycophagous (Graves, 1960). Hingley (1971) concluded from gut-content studies that ten species of micro-arthropod, including *Thyreophagus* sp. (Astigmata), *Cepheus latus* C.L.K. (Cryptostigmata) and *Isostoma arborea* (L.) (Collembola), were feeding predominantly on remains of dead stromata of the ascomycete *Daldinia concentrica*.

Most studies of mycophagy among litter microarthropods have involved the microfungi on plant debris. Ever since Michael (1884) found that mouldy cheese provided a satisfactory diet for various oribatids, numerous microarthropod species have been cultured either on fungi or on organic matter undergoing fungal decay (Woodring, 1963; Butcher *et al.*, 1971). Observations of feeding in culture, coupled with the dominance of fungal remains in the visible gut contents of certain species, led many to conclude that fungi were among the most important components of the food of many Acari and Collembola (Forsslund, 1939; Macnamara, 1924). More extensive surveys of gut contents and the offer of a wider range of foods have shown that some species are indeed strictly mycophagous, but that others also feed on other cryptogams or are panphytophagous. Further details of mycophagy will be found in section IIB.

2. *Lichens*

Lichens were noted by Michael (1884) as a favourite resort of oribatids, and several species have been cultured on lichens, including *Camisia segnis* (Hermann), *Platynothrus peltifer* (C.L.K.) and *Nothrus* spp. (Grandjean, 1950), *Scheloribates laevigatus* (C.L.K.) and *Oppia nova* (Oudemans) (Woodring and Cook, 1962). Two oribatid species of the genera *Scapheremaeus* and *Cryptoribatula* which dwell in and feed on crustose lichens were described by Jacot (1934). In his studies of saxicolous and arboreal oribatids Travé (1963, 1969) described certain species, such as *C. segnis* and *Pirnodus detectidens* Grandjean, as strictly lichenophagous, and also recorded the presence of Collembola, Thysanura and Prostigmata in lichens. The oribatid *Maudheimia petronia* Wallwork forms feeding cavities in *Usnea antarctica* (Gressitt and Shoup, 1967), while *Halozetes belgicae* (Mich.) probably feeds on foliose lichens (Strong, 1967). Schuster (1956) recorded fragments, apparently of lichens, in the gut contents of most microphytophagous and "unspecialized" oribatids which he examined.

3. *Algae*

Certain microarthropods of littoral and salt-marsh sites are directly or indirectly dependent on algae for food. Mites in the intertidal zone and those associated with early stages of *Fucus* (wrack) decomposition are mainly predatory, but intertidal species of *Ameronothrus* (Cryptostigmata) are algivorous (Evans *et al.*, 1961). Oribatids and Astigmata on wrack in the later stages of decay are probably largely mycophagous (Schuster, 1966), and although Moeller (1967) concluded that Collembola have no visible effect on wrack decomposition, Zachariae (in the discussion of Moeller's paper) suggested the likely importance of fungi to these wrack inhabitants. *Hygroribates schneideri* (Ouds.) is an example of an algivorous oribatid from salt marshes which may also ingest fungi (Luxton, 1966).

Lund (1967) points out that scarcely any study has been made of grazing on soil algae, and the same applies to most terrestrial habitats. Limited experiments with antarctic microarthropods suggest that algae may be an important food source for certain species such as the collembolan *Cryptopygus caecus* Wahlgren (Tilbrook, 1967) and the prostigmatid *Nanorchestes antarcticus* Strandtmann (Gressitt and Shoup, 1967). Various microarthropods have been successfully cultured on *Protococcus* or *Pleurococcus* including Collembola (Healey, 1971), the oribatids *Nanhermannia nana* (Nicolet), *P. peltifer*, *Nothrus silvestris* Nicolet, *Camisia* spp., *Damaeus* spp. and immatures of *Galumna* spp. (Woodring, 1963; Littlewood, 1969). Tarman (1968) observed that *Pleurococcus* cells were broken open by the chelicerae of oribatids, while Schuster (1956) and Tarras-Wahlberg (1961) recorded that oribatids such as *Xenillus tegeocranus* (Hermann) and *Nothrus pratensis* Sellnick prefer these algae to phanerogam litter. Epiphytic algae on bark are often plentiful in forest and heathland litter, but there is very little evidence of the importance of these or of soil algae as food for microarthropods. Algal remains were detected among the gut contents of 14 oribatid species, including *X. tegeocranus* and *Belba* spp., by Schuster (1956) and also in *P. peltifer* and the Collembola *Folsomia quadrioculata* (Tullberg) and *Onychiurus quadriocellatus* Gisin (H. Faasch, personal communication).

4. Bryophytes

Very high microarthropod densities are indicated by seasonal means of *ca.* 28 cm^{-2} for Acari plus Collembola under mosses on limestone boulders (Wood, 1967), and by a mean of 27·5 cm^{-2} of moss for the collembolan *Cryptopygus antarcticus* Willem, which is the dominant arthropod of maritime Antarctica (Tilbrook, 1967). Gerson (1969) commented that arthropods could find shelter, as well as food, among mosses, and Strong (1967) suggested that microclimatic requirements of large species such as *C. antarcticus* might be more significant than the food value of the mosses. Nevertheless, various oribatids, such as adults of *Galumna nervosa* (Berlese) (Sengbusch, 1954), and Collembola, including *Isotoma klovstadi* Carpenter (Pryor, 1962), have been cultured on mosses, while certain tardigrades (Kühnelt, 1961) and Prostigmata (Gerson, 1972) are known to feed on living mosses.

Identifiable remains of moss tissues, among other plant material in gut contents, were recorded in four species of oribatid (including *Liacarus* sp.) out of 40 examined by Schuster (1956), and in *Tomocerus* spp. (Collembola) by McMillan and Healey (1971). The occurrence of faecal pellets of isotomids beneath *Polytrichum* led Drift (1964) to conclude that these Collembola were the only mesofauna of importance in the breakdown of this moss in an inland dune site.

5. Pteridophytes

Elton (1966) found a flourishing fauna, including microarthropods, in *Pteridium* (bracken) litter, and further evidence from *Pteridium*-dominated sites is provided by maximum densities of 50,000 m^{-2} for the oribatid *P. peltifer* (Harding, 1973), and of 13,100 m^{-2} for *Onychiurus procampatus* Gisin, a species which comprised *ca.* 70% of the collembolan population biomass in Healey's (1967) site. Healey (1967) conducted quantitative feeding experiments with this species on fungi which decompose *Pteridium* (Frankland, Chapter 1), while preliminary, unpublished tests by the present authors indicate that a range of oribatid species will not feed on *Dryopteris* prothalli, but much remains to be investigated concerning the role of microarthropods in the decomposition of litter of pteridophytes and other cryptogams.

B. Trees, Shrubs and Other Phanerogams

1. *Laboratory Evidence*

Most microarthropods cannot ingest freshly fallen leaves of trees, particularly if this litter is dry. Among Collembola, Schaller (1950) recorded high mortalities in five out of seven species cultured on fresh litter of various trees, ranging from *Alnus glutinosa* (alder) to *Quercus pedunculata* (oak), while Dunger (1956) observed traces of feeding by *Folsomia fimetaria* (Tullberg) on fresh litter of only *Alnus* and *Sambucus nigra* (elder). Hartenstein (1962*a*) found no evidence of feeding on fresh litter by any of 20 species of Cryptostigmata, but, according to Riha (1951), *Pelops* sp. prefers to feed on litter which is dry, rather than partially decomposed, and *Achipteria* sp. possibly shows a similar preference for the epidermis of dry *Tsuga canadensis* (hemlock) needles (Wallwork, 1967). Kühnelt (1961) commented that machilids (Thysanura), when abundant, may be of significance among the initiators of decomposition, feeding, for example, on fresh *Fagus* litter.

There is virtually no information on the longevity or fecundity of those species that can feed on fresh litter, but most litter-feeders seem to prefer to utilize moist, partially decomposed leaves (Kühnelt, 1961). Phthiracaroids, for example ("box mites" or "armadillo mites"; Cryptostigmata), feed readily in culture on fresh *Quercus* leaves which have been immersed in water for as little as two days (Harding, 1966), but sustained culturing of microarthropods usually necessitates providing material which is somewhat more decomposed. Attack normally starts on the abaxial surface of the leaf, where oribatids tend to remove the superficial, non-lignified tissues of the midrib and main veins, before proceeding to feed on intervein tissues (Schuster, 1956; Führer, 1961). Removal of the lower epidermis and mesophyll from the regions between the finest veins produces a pattern which is common to several groups of animals, particularly when feeding on leaves of *Quercus* and *Fagus*. This "Fensterfrass" (Schaller, 1962) was described by Noordam and Vlieger (1943) as the "normal mastication pattern" of Collembola such as *Onychiurus armatus* and *Tomocerus minor* (Lubbock), and of various oribatids (e.g. *Hermannia gibba* (C.L.K.), *N. silvestris* and *P. peltifer*) on *Quercus*. Complete perforation of intervein regions, resulting in "Lochfrass" (Brauns, 1954) or "Skelettfrass" (Dunger, 1964) can be effected by Collembola on leaves of *Sambucus*, *Alnus* or *Carpinus betulus* (hornbeam) (Schaller, 1950; Dunger, 1956), but similar patterns on *Quercus* or *Fagus* are usually associated with more robust animals such as oribatids of the genus *Liacarus* (Schuster, 1956) or of the superfamily Phthiracaroidea; the latter may also consume some of the finer veins, forming circular feeding areas (Riha, 1951; Murphy, 1953).

Feeding within petioles and midribs of broad-leaved species has been recorded mainly among immature oribatids, especially phthiracaroids, e.g. *Steganacarus diaphanum* Jacot in *Fagus grandifolia* petioles (Hartenstein, 1962*e*) and *Steganacarus* sp. and *Phthiracarus anonymum* Grandjean in *Quercus* and *Fagus* (Harding, unpublished). Endophagous development of *Adoristes ovatus* (C.L.K.) and of *Steganacarus* and *Phthiracarus* spp. in conifer needles was described by Jacot (1936, 1939; see also Murphy, 1953; Hartenstein, 1962*e*); eggs were laid on needles which had been partially decomposed by fungi, and the immature stages fed within, chewing an exit hole when mature. Adult phthiracaroids in culture have been seen to consume all but the vascular tissues of needles (Murphy, 1953; Führer, 1961; Hartenstein, 1962*e*), but attacks by sminthurid Collembola are largely ectophagous (Zachariae, 1963).

Among the few published observations of oribatids feeding in culture on stems and roots are Michael's (1884) record of burrowing by *Minunthozetes semirufus* (C.L.K.) in grass stems, feeding on *Calluna vulgaris* stems by *Steganacarus magnus* (Nicolet) (Webb and Elmes, 1972), and the preference of *Rostrozetes flavus* Woodring for the decomposing outer tissues of roots (Woodring, 1965). Among Collembola, Schaller (1950) described feeding by *O. armatus* within roots, while Kooistra (1964) found that *F. quadrioculata* showed a preference for decaying roots of *Trifolium repens*.

2. *Field Evidence*

(*a*) *Leaves*. Noordam and Vlieger (1943) attributed normal mastication patterns (see above) on litter in a *Quercus* site to feeding by microarthropods. However, Zachariae (1963, 1965) concluded that these patterns could also be produced by other animals or by physical processes, and that microarthropods were relatively unimportant in comminuting *Quercus* or *Fagus* litter; Collembola and non-predatory mites, and possibly also pauropods, symphylids and proturans, were thought to be feeding on the microflora, or, at most, on decomposed litter fragments, including macroarthropod pellets. Sminthurids and certain oribatids ingested less decomposed leaves, but they were considered to play a subsidiary role to macroarthropods and lumbricids in Zachariae's sites. Where damp *Fagus* leaves accumulated in places inaccessible to larger animals, e.g. under logs and stones, ideal conditions were provided for the development of leaf-feeding phthiracaroids. On the basis of performance in culture, microarthropods should be able to feed more readily on leaves with a higher nitrogen content than *Quercus*

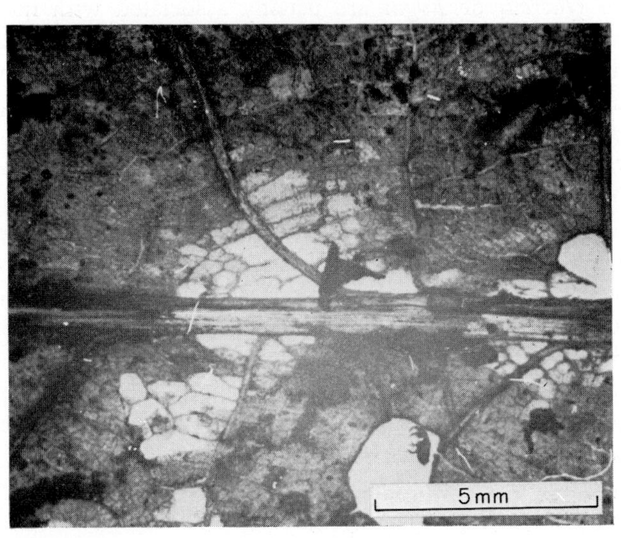

Fig. 2. Part of a *Quercus* leaf from the forest floor, showing stripped midrib and veins, Fensterfrass and Lochfrass, and an adult *Platynothrus peltifer*. (Photograph by P. W. Murphy.)

or *Fagus*, but Zachariae considered that these richer food sources were normally almost monopolized by the macrofauna.

In a mull-like moder site under *Quercus*, *Fagus* and *Sambucus*, where lumbricids were virtually absent and molluscs and macroarthropods scarce, Harding (1966) observed exploitation of *Sambucus* remains by scatopsid larvae (Diptera), but there were also numerous examples of leaflets being perforated by adults and immatures of *Platynothrus peltifer*. *Fagus* and *Quercus* leaves exhibited stripped midribs and veins, Fensterfrass and fine perforations of the lamina; this damage was often accompanied by faecal pellets similar to those of oribatids in culture, and occasionally adults of *Phthiracarus nitens* (Nicolet), *Rhysotritia duplicata* (Grandjean) and *P. peltifer* were found on the damaged areas (Fig. 2). Microphytophagous mites such as *Galumna lanceata* (Oudemans) (Fig. 3) and *Ceratoppia bipilis* Hermann were sometimes seen on newly fallen *Quercus* litter,

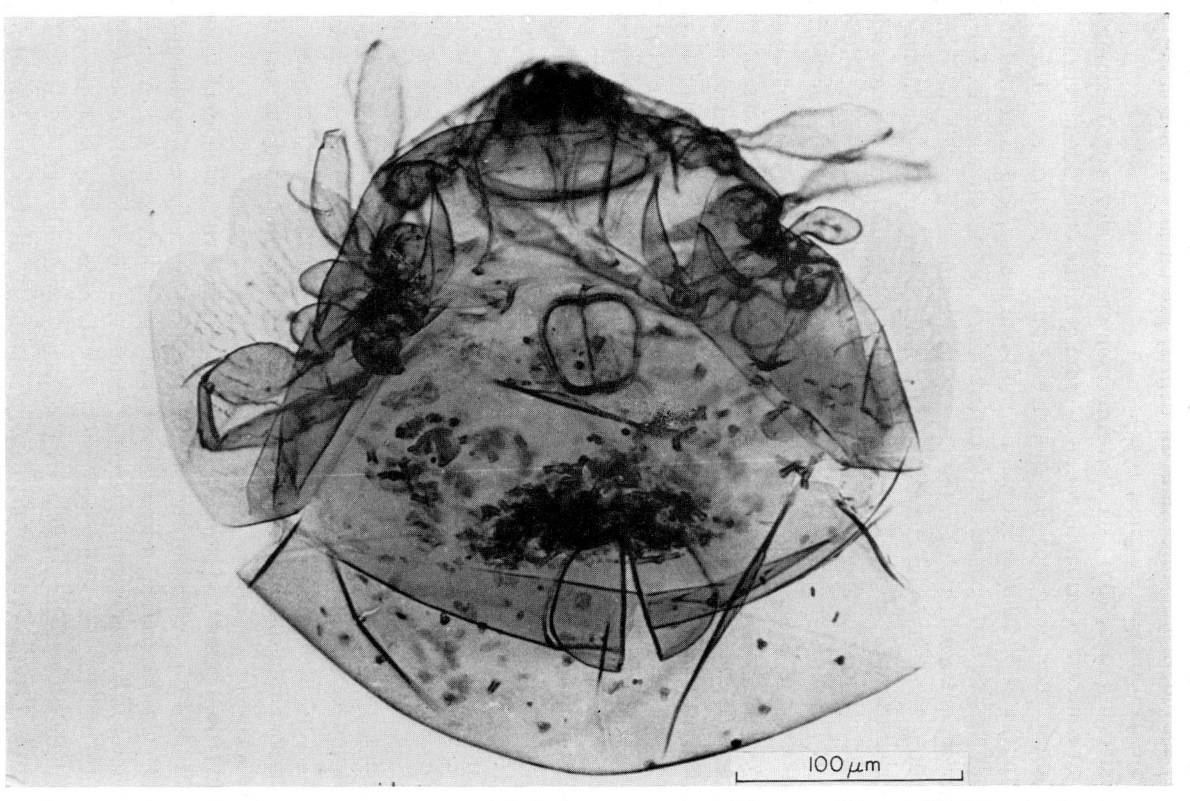

Fig. 3. *Galumna lanceata* adult, with fungal gut contents. (Photograph by P. W. Murphy.)

whereas the most abundant oribatid, *Tectocepheus velatus* (Mich.), occurred mainly among fine detritus in the A_F sub-horizon, and anoetids (Astigmata) moved in moisture films on debris, presumably filter-feeding on micro-organisms (see Hughes, 1953).

Anderson (1970, 1971; see also Anderson and Healey, 1970) examined the gut contents of oribatids removed from gelatine-embedded sections of the forest floor of a mor-like moder site under *Castanea sativa* (sweet chestnut) and of a *Fagus* site with a mull-like moder, and discerned that certain species had characteristic distributions and feeding habits. *Carabodes labyrinthicus* (Mich.) in the upper sub-horizons fed mainly on fungi, including those associated with freshly fallen litter and with humified material transported by lumbricids. The panphytophage *Hermanniella granulata* (Nicolet) ranged between the A_0 and A_{F2}, feeding primarily on *Castanea* leaves in various stages of decomposition, but also on lower plants. Most adults of *Steganacarus magnus* inhabited the A_{F1}, feeding mainly on relatively undecomposed mesophyll. The characteristic species of the A_{F2} was *Rhysotritia ardua* (C.L.K.), which as adults fed on raw humus and mycorrhizal fungi; a few were also found in deep pockets of *Fagus* litter in the A_{00}, resembling the situation described by Zachariae (1965; see p. 532). Collembola were not preserved well in these sections, so specimens of three species were collected from the *Castanea* site, their gut contents dispersed on gridded Millipore filters, and an assessment made of the relative abundance of various types and sizes of particles (Anderson and Healey, 1972; see also McMillan and Healey, 1971). The proportion of individuals with empty guts, presumably associated with moulting, ranged among the species from 38 to 54%. The relative amounts of leaf and fungal material varied seasonally, but on average over 90% of the particles in *Orchesella flavescens* (Bourlet) were of higher plant origin, signifying feeding on superficial, relatively undecomposed litter, whereas a fungal content of *ca.* 40% in *Tomocerus* spp. suggested that the diet of deeper dwelling species consisted largely of partially humified leaf material (Figs 4 and 5).

Bal (1968) compared laboratory feeding patterns with evidence of faunal activity in thin sections of the humus profile in two adjacent moder sites. Under *Quercus rubra*, microbially conditioned leaves in the A_F were skeletonized by mycetophilid larvae (Diptera) and by *Nothrus silvestris*, the oribatid pellets containing cell walls and hyphae, while midribs were mined by *Rhysotritia minima* (Berlese). Endophagous activity was also evident in *Pseudotsuga taxifolia* needles, mainly by all stages of *Rhysotritia* spp. and *Steganacarus striculus* (C.L.K.) in the A_{F2}.

Sectioning of *Pinus sylvestris* needles from a mor site (Kendrick and Burges, 1962) revealed that the formation of pellet-filled cavities in the

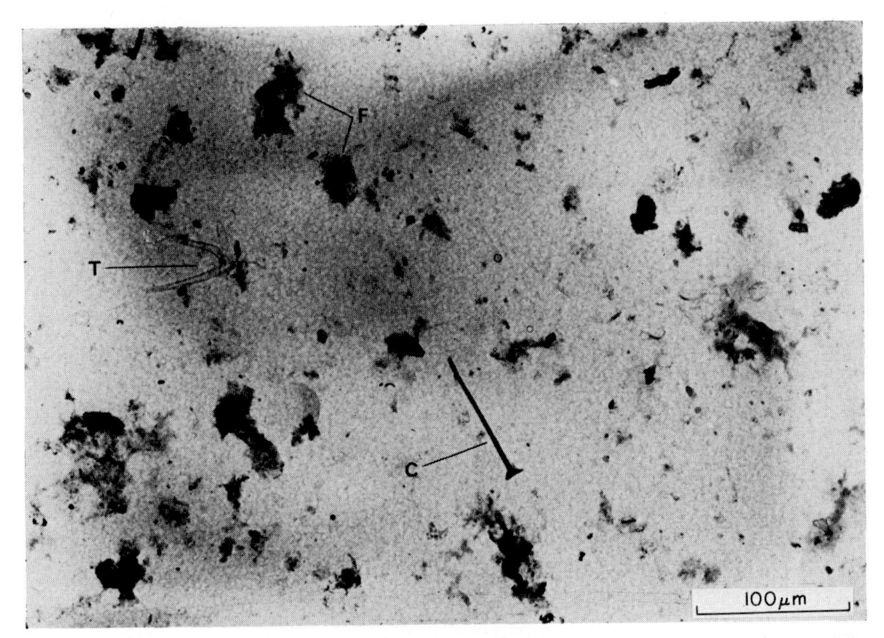

Fig. 4. Gut contents of *Orchesella flavescens*, including leaf fragments (F), trichome (T) and conidiophore (C). (Reproduced with permission from Anderson and Healey, 1972.)

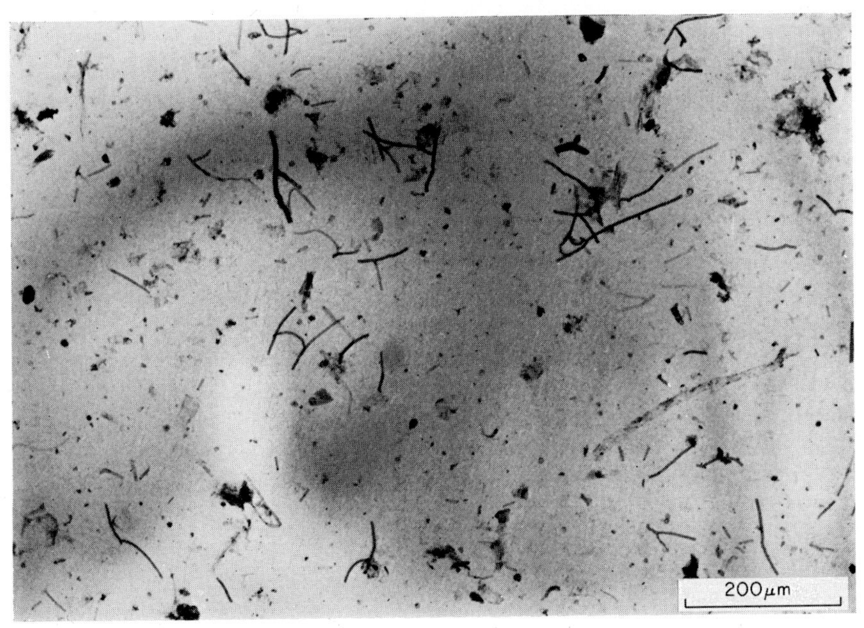

Fig. 5. Gut contents of *Tomocerus minor*. (Reproduced with permission from Anderson and Healey, 1972.)

mesophyll of needles in the A_{F1} was apparently dependent on a preliminary, internal attack by fungi. From the species list for this site, which does not include phthiracaroids, it seems likely that these cavities were produced by *Adoristes ovatus*, an oribatid recorded by Jacot (1939) as developing within conifer needles (see p. 495) and by Michael (1884) from needles of *Ulex* (furze). Faecal evidence suggested that mites and Collembola fed on hyphae and conidiophores on the needle surface.

Truly ectophagous behaviour by Collembola, involving ingestion of needle material, was thought by Zachariae (1963) to be of minimal importance, except when sminthurids were locally abundant. Seasonal variations in the gut-content composition of Collembola in a *Pseudotsuga* site were described by Poole (1959), who concluded that in general larger species such as *Tomocerus longicornis* (Muller) were feeding mainly on fungi, with decaying mesophyll as an alternative, while smaller species (e.g. *Tullbergia krausbaueri* Borner) possibly subsisted on humus fragments or arthropod pellets.

Animals responsible for the breakdown of plant material in particular sites may also be studied by using radioactively labelled material. For example, Gifford (1967) recorded relatively high levels of ^{14}C in oribatids such as *P. peltifer* exposed to tagged *Pinus sylvestris* needles, and Coleman and McGinnis (1970) found that mites belonging to four orders fed on ^{65}Zn-labelled mycelium of *Geotrichum* in field soil, but that Collembola were not labelled, indicating selective feeding in relation to this fungus. Various criteria of ^{137}Cs uptake from *Liriodendron tulipifera* (tulip tree) leaves, including accumulation rate, were employed by McBrayer and Reichle (1971) to distinguish between members of different trophic levels; fungivores included Onychiuridae, Carabodidae and Rhodacaridae (the latter, mesostigmatid family is normally considered predatory), while Isotomidae, Phthiracaridae and Uropodina were classified as litter-ingesting saprophages, and soil-dwelling Symphyla as feeding on translocated organic matter.

(*b*) *Pollen*. Pollen grains make up a large proportion of the diet of certain surface-dwelling Collembola (Christiansen, 1964; Scott and Stojanovich, 1963), as well as being utilized by certain Prostigmata (Evans *et al.*, 1961). Pollen is important to rock dwellers, such as oribatids of the genus *Saxicolestes* (Travé, 1963), but may also be a component of the gut contents of forest-floor Collembola and Acari, including microphytophages (Schuster, 1956; Zachariae, 1965) and panphytophages, such as *Xenillus tegeocranus* (Schuster, 1956); Hammer (1972) found that during the spring certain panphytophagous oribatids fed almost exclusively on *Alnus* pollen.

(*c*) *Fruits and cones*. Microarthropods played a minor role in the decomposition of *Quercus rubra* acorns (Winston, 1956); shell material was ingested by certain Collembola and oribatids, whereas astigmatids and prostigmatids were mycophagous. Harding (unpublished data) found up to 30 immature phthiracaroids burrowing into the tissues of a single cupule of *Fagus sylvatica*; examination of younger cupules revealed deutova ("eggs") embedded in the surface of the pedicel, the valves and the base

Fig. 6. Phthiracaroid deutovum embedded in valve of *Fagus sylvatica* cupule. (Photograph by R. Turner, Rothamsted Experimental Station.)

of the locule. Larvae burrowed from these sites into the pedicel or valves, developing via nymphs to adults of *Steganacarus* sp. or *Phthiracarus anonymum*, the latter continuing to burrow when adult (Figs 6 and 7). Adults of *Phthiracarus japonicus* Aoki were abundant in *Alnus hirsuta* cones (Aoki, 1967), but although various Collembola were recorded from *Picea excelsa* cones by Arend (1967), the only oribatid was the microphytophage, *Belba gracilipes* Kulcz. In striking contrast, N. R. Webb (personal communication) has obtained up to 50 immatures of *S. magnus* from within a single cone of *Pinus sylvestris*.

C. Coprophagy

Vertebrate dung may be colonized by various microarthropods, but most of these, such as macrochelids (Mesostigmata) are predatory.

Feeding on the excreta of litter-ingesting invertebrates is said to be common among Collembola, particularly in small species (Christiansen, 1964), and also among oribatids, mainly as immatures (Wallwork, 1967). However, conclusions based on laboratory experiments should be accepted with caution, and, as Anderson and Healey (1972) pointed out, it is extremely difficult to demonstrate the significance of coprophagy in natural populations of soil animals.

Schuster (1956) recorded that macrophytophagous and unspecialized oribatids fed for several weeks on fresh pellets of isopods, but that pellets obtained from the forest floor were ignored. Collembola such as *Tomocerus flavescens* Tullberg, *F. fimetaria* and *Sinella coeca* (Schött) will comminute the excreta of certain macroarthropods and lumbricids (Schaller, 1950; Dunger, 1956, 1958*a*), but Dunger (1956) found that these species preferred to feed on litter which had not been processed by other animals. On the other hand Wallwork (1958) observed that the oribatids *Galumna formicarius* (Berl.) and *Oppia* spp., found inside fallen *Betula* twigs, could not ingest woody tissue, but they were able to feed on the pellets of endophagous phthiracaroids. According to Wallwork (1967), intraspecific coprophagy is common among immatures of oribatids which are xylophagous as adults, and Rohde (1955) assumed that larvae of *Rhysotritia* sp. in culture were feeding on adults' pellets. However, coprophagy was not observed by Walker (1965) in larvae of *Plesiotritia megale* Walker, where all stages were apparently xylophagous, and is unlikely to occur, at least between adults and larvae, in species where the larvae hatch from embedded deutova (e.g. in beech cupules, see p. 501) and burrow into woody tissues independently of adult feeding activity.

Reports of evidence of coprophagy under field conditions are rare. Nicholson *et al.* (1966) found that pellets of the millipede *Glomeris marginata*, placed on the soil surface in nets, were often visited by Acari, Collembola and enchytraeids and converted into smaller pellets. Bal (1968) reported evidence from soil sections of feeding by *O. quadriocellatus* on pellets of *Quercus*-feeding mycetophilid larvae, but Zachariae (1963) believes that Collembola found among decomposing excreta are microbivorous.

D. Artificial Litter

Harding (1966) used the cellulose-film technique devised by Tribe (1957) to provide microbially conditioned material for cultures of litter-feeding oribatids, including phthiracaroids and the panphytophagous *P. peltifer*, and also for microphytophagous Acari. Evidence of feeding activity in a *Quercus-Fagus* site was obtained by burying cellophane inserts in the forest floor (Harding, 1967). Some of the pellets on these inserts (Figs. 9 and 10) were similar to those of cultured oribatids, including *R. duplicata*, a species observed by Went and de Jong (1966) on cellophane recovered from a forest soil. Small pellets are often difficult to detect in soil sections (Drift, 1964), but very large numbers of pellets of *ca.* 30 μm breadth were produced on inserts in the late summer, possibly by immatures of *T. velatus* and *P. peltifer*.

The role of Collembola in the decomposition of filter paper is discussed on p. 509.

E. Successions

Some of the temporal changes in the species composition of populations associated with decomposing litter have been correlated with physical and chemical properties of the litter, e.g. of *Fucus* (Strenzke, 1963). Cernova (1971) suggested that certain microarthropods could be used as indicator species for particular stages in the decomposition of manure. However, Moeller (1965) concluded that variations in the abundance of wrack-dwelling species were largely seasonal, and did not constitute a true, ecological succession. Curry (1969*b*) also distinguished between these two types of temporal variation when discussing faunal changes accompanying the decomposition of grassland herbage. During the decomposition of lucerne and rye straw in soil Naglitsch (1966) recognized a bacterial phase, succeeded by a phase which included Acari (Anoetids and pyemotids), and finally a humification phase, involving Collembola, uropodids, *R. ardua* and other oribatids, pauropods and diplurans. In many papers on succession there are few if any details of the role of the dominant organisms, but it appears that successions of Collembola on wrack (Zachariae, in discussion of Moeller, 1967) and on compost (Gisin, 1952) may be related to microbial successions, and that the animals have little direct effect on litter breakdown. Crossley and Hoglund (1962) and Crossley and Witkamp (1964) recorded an initial invasion of tree litter in mesh bags by a few species of microarthropod, many probably mycophagous, and only later was there an increase in species diversity.

Anderson (1970) found no evidence of a single group of pioneer fauna on *Fagus* or *Castanea* leaves sampled one month after being placed in the forest floor. Insufficient animals were recovered from the litter in mesh bags to ascertain whether there was a faunal succession corresponding to the known microbial succession on these leaf species; Anderson expected the

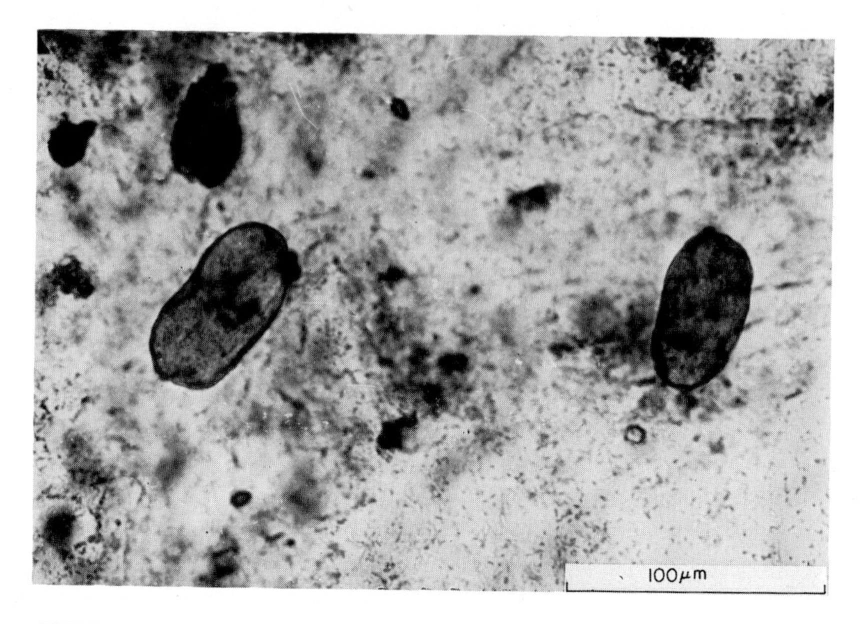

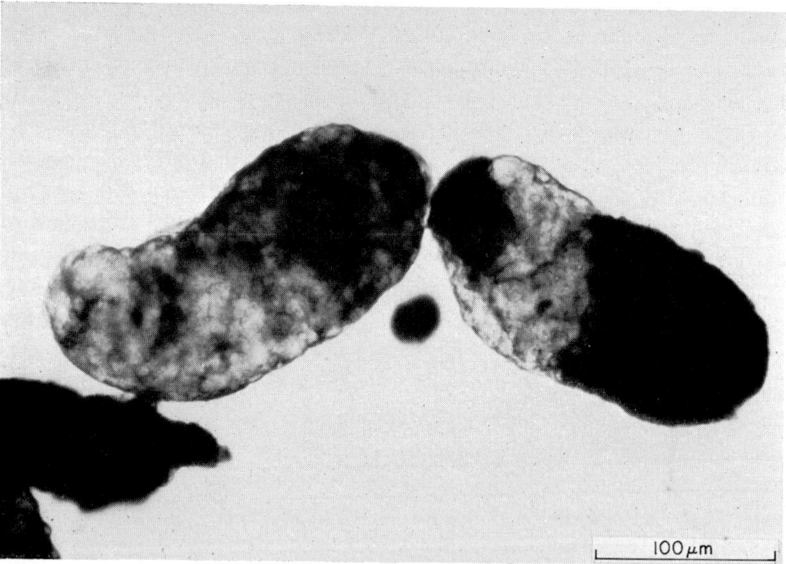

Figs 9 and 10. Faecal pellets on cellophane inserts recovered from anoak -beech forest floor. (Reproduced with permission from Harding, 1967.)

faunal species composition to change as microhabitats and exploitable substrates, including micro-organisms and their exometabolites, increased in diversity, but more data are needed to support this hypothesis.

F. Preference and Feeding Specificity

Food preference may be assessed in various ways: by measuring ingestion or defaecation rates on a range of foods when offered in combination ("choice" experiments) or separately; by noting which foods are consumed before others; or by determining which food has the most animals associated with it. The precise relationships between these various criteria are in many cases unknown (Hayes, 1963).

Dunger (1956) found that if certain species of trees and shrubs were arranged in series according to the volume of litter ingested by *F. fimetaria*, then this series agreed closely with that of Wittich (1943) for natural decomposition rates, ranging from *Sambucus* to *Fagus*. Berthet (1964) observed that *N. palustris* fed on *Carpinus* leaves, but not on *Corylus avellana* (hazel), *Fagus* or *Quercus*, whereas *S. magnus* fed on each species in turn, starting with *Carpinus*, on which ingestion rates were considerably higher than for *Quercus* or *Fagus* (see p. 513). Schuster (1956) and Hayes (1963) were unable to detect preferences among oribatids for particular litter species, but Hayes observed larger numbers of phthiracaroids and their pellets on conifer needles in later stages of decay, and Führer (1961) recorded that *R. ardua* adults were about 20 times more abundant around dead roots of *Artemisia campestris* than around healthy ones. H. Faasch (personal communication) observed feeding by *P. peltifer* on shade leaves of *Fagus*, but virtually none on sun leaves (*cf.* Heath and Arnold, 1966).

It is rarely possible to determine defaecation rates on different foods from data published on preference experiments, but by culturing oribatids such as *R. duplicata* on each of the foods which were simultaneously being used in choice experiments, Harding (1966) found that in most cases there were appreciably higher rates of pellet production on those litter species or stages of decomposition (of litter or cellophane) which attracted most mites, and that pellets were significantly larger, as well as more abundant, on more decomposed cellophane.

Hartenstein (1962*a*) gives several examples of preference for particular species of fungi among strictly fungivorous oribatids, e.g. *Trichoderma koningi* by belbids; *Aspergillus niger* repelled all species except *Oppia nova*. The behaviour of various Collembola and Acari (including oribatids, Mesostigmata and Astigmata) when offered *Trichoderma viride* or *Rhizopus nigricans* was investigated by Farahat (1966), who found that certain species in all groups ingested hyphae and spores, whereas *Achipteria* *coleoptrata* (L.) fed only on hyphae, and *Oppia nitens* C.L.K. only on spores. Other species of *Oppia*, sminthurids and phthiracaroids failed to feed at all, and populations declined in soils containing these fungi. Singh (1969) presented gut-content microflora to Collembola, and observed slightly different preferences between *T. longicornis* and *O. armatus*, both of which fed on hyphae and spores, whereas only spores were seen in *Neanura muscorum* (Templeton). Mignolet (1971) concluded that fungi cultured from faeces could be used to demonstrate feeding specificity in the microphytophagous *Damaeus onustus* C.L.K., but not in the panphytophagous *Euzetes globulus* Nicolet, nor in the macrophytophagous *Nothrus palustris* C.L.K., the latter failing to feed on any of the fungi. Preference experiments with *Nothrus biciliatus* C.L.K. led Saichuae *et al.* (1972) to the conclusion that the suitability of particular fungi as food may alter as these age, possibly due to production of toxins, and that pellet production may be an unreliable parameter for suitability, since some fungi were consumed but failed to support development of immatures. Luxton (1972) has recorded relative amounts of feeding by various oribatids from a *Fagus* woodland when offered a wide range of fungi, yeasts and bacteria from the same site. Two species of *Steganacarus* consumed none of this microbial food, whereas *Phoma* sp. was selected by *Adoristes ovatus* and by adults of *Damaeus clavipes* (Hermann), immatures of the latter preferring *Trichoderma viride. Belba corynopus* (Hermann) was the only species which preferred *Penicillium*, while *Gustavia microcephala* (Nicolet) specialized on bacteria, and *Hypochthonius rufulus* C.L.K. included bacteria among a wide range of foods.

Dunger (1962) concluded that the preference order for various species of tree litter was remarkably similar in members of different faunal groups, e.g. Diplopoda and Collembola. Earlier (Dunger, 1958*b*), possible factors had been suggested as being responsible for food selection by macrofauna, and these were also discussed, with reference to *Lumbricus terrestris*, by Satchell and Lowe (1967), who confirmed Dunger's view that texture is of little significance in determining palatability, but that the latter is broadly correlated with nitrogen and soluble carbohydrate levels; the increased palatability which often accompanies weathering was thought to be due partially to microbial degradation of various distasteful substances, such as tannins (see Williams and Gray, Chapter 19). Heath and Arnold (1966) considered that palatability of litter was more important than digestibility in determining the preferences of soil fauna. The preference of many litter-feeding microarthropods for decaying vegetation, sometimes at a particular stage, also suggests the importance of micro-organisms, but these could be providing food, either in the form of microbial tissues or as litter breakdown products (exometabolites), rather than merely decomposing tannins.

The presence of living micro-organisms is apparently essential for the feeding of certain species. Woodring and Cook (1962) showed that larvae of the oribatid *Ceratozetes cisalpinus* Woodring and Cook, derived from surface-sterilized eggs, were unable to consume aseptic lichen until they had ingested fungal hyphae. Decayed leaves, which are the preferred food of *Protoribates lophotrichus* Berlese, were rendered unpalatable by heating, sulphanilamide treatment or prolonged use in culture (Hartenstein, 1962*d*); this suggested that living micro-organisms were important, either multiplying in the gut and being digested or aiding digestion as gut symbionts. Various bacteria and other bodies have been demonstrated in the gut of this and other oribatid species, e.g. by Woodring (1963) in *Galumna* adults, by Rohde (1955) in *Rhysotritia*, and by N. Haarløv (personal communication) in the proventricular caeca of *P. peltifer*, but it is not known whether these are symbiotic. Finally, Törne (1967, 1968) has stressed the importance of extraintestinal cellulolytic attack in Collembola associated with decomposing filter paper; interspecific variation in the ability to utilize this substrate is probably a consequence of characteristic gut floras, the microbes being spread mainly via the faeces.

Luxton (1972) concedes that microbes may be essential in the diet of *Protoribates* and other panphytophages, but suggests that they are not utilized in macrophytophages, since he was unable to demonstrate in *Steganacarus* spp. the presence of enzymes capable of attacking trehalose, an important reserve carbohydrate in certain fungi. Fungal remains are said by Luxton to be rare in the gut contents of macrophytophages (but see p. 510), and he suggests that the greater rates of pellet production on later stages of litter decomposition, e.g. by phthiracaroids on conifer needles, are a consequence of reduced availability of nutrients in litter tissue, due to microbial activity. It seems to us, however, that microbial protoplasm or exometabolites could be utilized by these macrophytophages, thus explaining the fact that these mites prefer to feed on microbially conditioned foods, when offered a choice, congregating there as well as producing pellets at a greater rate. Attraction by living bacteria or their metabolic products was demonstrated by Führer (1961) using an arrangement of tissue extracts from *Artemisia* roots which precluded actual feeding by *R. ardua*. If these microbes are not actually providing nourishment, and if texture is unimportant in determining palatability, then the only remaining way in which they could be indispensable is in enabling the animals to recognize the plant material as food; this was one possible suggestion made by Littlewood (1969) to account for the cessation of feeding by surface-sterilized microphytophagous and panphytophagous oribatids on aseptic algae, a diet which is unlikely to require symbiotic assistance except, perhaps, to deal with mucilage.

It is unwise to consider the results of a choice experiment as representing an immutable order of preference of a particular species, since choice may vary with time and pre-test conditions (Drift, 1965). Healey is quoted by Hale (1967) as suggesting that the results of preference tests with Collembola such as *O. procampatus* are of little significance, because of the limited sensory powers of these species, but Coleman and Macfadyen (1966) found greatest numbers of *O. procampatus* recolonizing irradiated soil cores containing species of fungi which Macfadyen (1969*b*) described as being selected in the laboratory. If individuals in nature are able to exercise some choice over food, then there are indications of more rapid development on certain foods, e.g. *Belba kingi* Hartenstein on *Trichoderma koningi* (Hartenstein, 1962*b*).

Certain species, such as those found in particular kinds of lichen, are possibly as fastidious in their natural feeding as they appear to be in culture. The difficulties which are encountered in trying to rear certain oribatids in culture suggest that the immatures of these species are more food-specific than the adults (Arlian and Woolley, 1970). Nevertheless, studies of feeding habits, particularly those based on examination of gut contents of animals collected from different sites or at different seasons, generally indicate that a given species ingests a variety of foods. Schuster (1956) recorded a predominance of fungal remains, with some algal and lichen fragments, in the microphytophagous *Damaeus auritus* (C.L.K.), whereas panphytophages such as *Liacarus* and *Notaspis* spp. contained a wider range of dietary components. Many Collembola are panphytophagous (Dunger, 1956), but the potential choice as indicated in culture may be restricted in the field by availability (Wallwork, 1970), so that the gut-content composition of a particular species may vary markedly, either between sites (Gilmore and Raffensperger, 1970), or, within one site, at one time or seasonally (Poole, 1959; Anderson and Healey, 1972); intraspecific variation between sites may be more marked than interspecific differences within one site (Anderson and Healey, 1972). Schuster (1956) was unable to discern seasonal variation of this kind in oribatids, but Anderson (personal communication) found considerable intraspecific variation, both within and between samples of oribatids from *Fagus* and *Castanea* sites (see p. 498). *Adoristes ovatus* was a typically opportunistic species, feeding predominantly on fungi in the autumn and phanerogam material in the summer, while certain individuals of species in three genera of Phthiracaroidea (usually considered to be the macrophytophagous group *par excellence*) contained nothing but fungal material (cf. Hartenstein, 1962*a*, who recorded occasional selection of fungi in phthiracaroid cultures). The use of different culture techniques and foods may also be responsible for some of the conflicting reports in the literature. For example, *Nothrus palustris* is described by Lebrun (1970 and personal communication) as a typical macrophytophage, feeding on microbially conditioned leaves, preferably of species with a low C:N ratio, such as *Alnus*; Luxton (1972), however, found that this species consumed only small amounts of *Fagus* leaves, but fed readily on a wide range of fungi, and he therefore classified it as panphytophagous.

Because of this variability, rigid categorization of feeding habits of particular microarthropods should be avoided until more data are available, but a selection of examples belonging to the main feeding types is included here.

Out of 14 collembolan species from a mixed woodland, Dunger (1956) found *Lepidocyrtus* spp. and *Entomobrya muscorum* Nicolet to be predominantly mycophagous, the gut contents of the remainder (panphytophagous species of *Tomocerus*, *Orchesella*, *Folsomia* and *Onychiurus*) including at least 50% leaf material. Half of the oribatids described by Schuster (1956) as microphytophagous, such as *Amerus troisii* (Berlese) and *Oppia subpectinata* (Oudemans), seemed to prefer fungi. *Ceratoppia bipilis* and *Belba kingi* were among several oribatid species designated as strictly fungivorous by Hartenstein (1962*a*), the latter species belonging to the superfamily Damaeoidea which, together with the Oppioidea and possibly the Eremaeoidea, Luxton (1972) suggests are wholly microphytophagous groups. *Cepheus latus*, *Hermanniella granulata* and phthiracaroids are classified by Luxton as macrophytophagous, leaving the majority of oribatids, such as species of *Camisia*, *Platynothrus*, *Carabodes*, *Ceratozetes*, *Hermannia*, *Adoristes*, *Liacarus* and *Nothrus*, as panphytophagous.

Hartenstein (1962*a*, 1962*c*) described certain panphytophages, such as *P. peltifer*, as primarily fungivorous while also being able to consume decomposing litter (Fig. 11), and Luxton (1972) has pointed out that this plasticity is of ecological advantage, enabling certain species to exploit a variety of habitats.

The vertical distribution of types of litter or micro-organisms suitable for particular microarthropod species or their immatures may be largely responsible for species distribution patterns in the field (Anderson, 1971; Luxton, 1972). Anderson and Healey (1972), in trying to explain the combination of high species diversity with a comparatively low degree of feeding specificity in detritivore communities, such as Collembola, suggested that interspecific competition is possibly reduced not only by different vertical distributions, but also by the horizontal compartmentalization of resources. Other suggestions included the presence of excess food for decomposers, possibly aided by non-feeding phases associated with moulting, and undetected differences in the digestive abilities of the various species, with further sharing of resources through coprophagy.

III. Microarthropod Energetics

From the point of view of litter breakdown, we need to know the amount of litter which is processed by microarthropods, that is, transformed

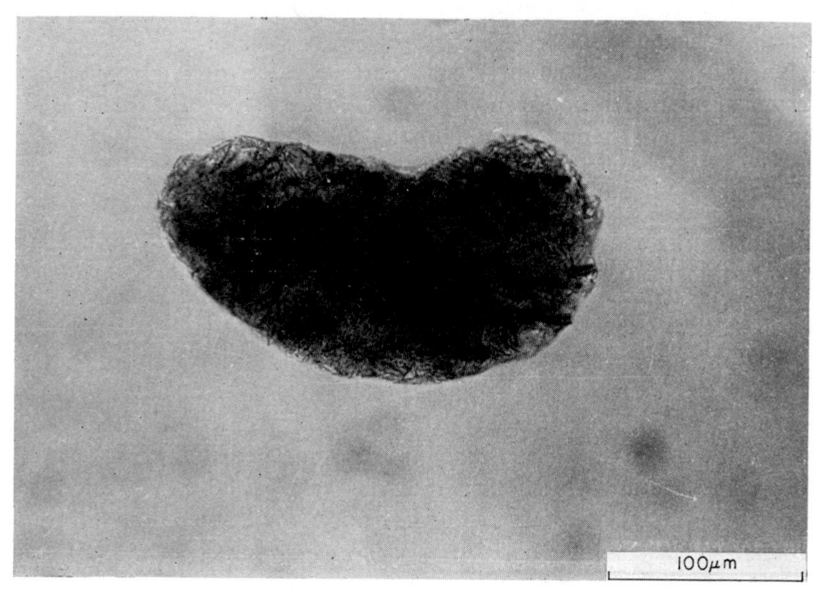

Fig. 11. Faecal pellet produced by *Platynothrus peltifer* adult cultured on decomposing *Quercus* leaf. (Photograph by P. W. Murphy.)

chemically or physically, mainly as a result of being ingested. Some of this food is passed out as faeces, while the remainder is assimilated and metabolized, the associated energy being represented in production (growth, exuviae and reproductive products) and respiration (maintenance metabolism). The energy relationships may be represented by two equations:

$$C = A + FU$$

and

$$A = P + R$$

where C = consumption (ingestion); A = assimilation; FU = rejection (normally considered in microarthropod studies as undigested faecal material); P = production; R = respiration.

These litter feeders effect some chemical decomposition of ingested material as it passes through the gut, but litter decomposition is probably mainly dependent on the subsequent utilization of their faeces by other organisms, breakdown of their body tissues by predators and decomposers being of indirect significance. We are, therefore, primarily interested in the amounts of litter ingested, assimilated and defaecated by microarthropods. Recently, much information has been published about energy flow in soil ecosystems, but this refers mainly to maintenance metabolism or, occasionally, to production, and generally there are only very approximate estimates of ingestion and defaecation rates for particular populations.

A. Amounts of Litter Processed

Gravimetric determinations of litter consumption by microarthropods are complicated by inaccuracies arising from the small amounts ingested and from estimates of the oven-dry weight of the actual food. Thus, Noordam and Vlieger (1943), Spencer (1951) and Hayes (1963) all recorded instances of controls losing more weight than litter exposed to animals, so that Spencer doubted the validity of his few positive results, one of which is given in Table I.

TABLE I. Ingestion (C) and defaecation (F) rates of Acari[a]

Taxon	C	F	Food	°C	Author
Steganacarus magnus	3·5	1·7	*Fraxinus*	4	Spencer (1951)
Steganacarus magnus	—	1·9	*Quercus* (fresh)	20	Madge (1964)
Steganacarus magnus	—	2·5	*Quercus* (old)	20	Madge (1964)
Steganacarus magnus	5·2–5·5	—	*Quercus, Fagus*	18	Berthet (1964)
Steganacarus magnus	8·3–8·7	—	*Corylus, Carpinus*	18	Berthet (1964)
Steganacarus magnus	—	11	*Corylus*	22	Berthet (1964)
S. magnus (young and old ♂♂, ♀♀)	6·3–15·1	3·2–6·3	^{32}P-*Calluna*	18	Webb and Elmes (1972)
Cultroribula juncta	1·06	—	^{45}Ca-*Pinus* mor	20	Kowal (1969)
Oribatids (mean for 7 species)	2	—	*Quercus*	Summer	Witkamp (1960)
Oribatids (mean for 7 species)	0·7	—	*Quercus*	Winter	Witkamp (1960)
Damaeus clavipes (nymphs, adults)	0·4–5·8	0·1–2·2	^{32}P-fungus	15	Luxton (1972)
Nothrus biciliatus	65[b]	—	Yeast	28	Saichuae *et al.* (1972)
Astigmata (4 species)	10[b]–50[b]	—	*Neurospora*	27	Pimentel *et al.* (1960)

[a] μg oven-dry weight ingested or defaecated per mite per day.
[b] μg fresh weight ingested per mite per day.

It should be easier to assess weights of defaecated material, but relevant data are scarce. Gravimetric estimates in Table I refer mainly to *S. magnus*, for which a comparison of rates at 10°C and 20°C (Madge, 1964) indicates a Q_{10} of *ca.* 1·6; we have obtained similar results (unpublished) with *R. duplicata* on *Fagus* leaves.

Radiotracers have been used in attempting to quantify feeding on phanerogams and fungi, but the results need to be interpreted carefully (Reichle and Crossley, 1965; Kowal and Crossley, 1971). Various methods and formulae can be used, but it is usually assumed that both radiotracer and food are assimilated in similar proportions, and that the accumulated material occupies a single body-pool. Engelmann (1961) fed yeast containing ^{14}C-labelled glycine to an unnamed oribatid and estimated that the ingestion rate per day was equivalent to 40% of dry body-weight and the assimilation rate to 8%, giving a value for the assimilation/consumption ratio (as a percentage) of 20%; these values were then applied to an old-field mite population, with an average biomass of 54 mg m^{-2}, to obtain figures of 10,248 cal ingested and 2058 cal assimilated m^{-2} year^{-1}.

Using Kowal's (1969) data for the oribatid *Cultroribula juncta*, and assuming an exponential relationship between temperature and ^{45}Ca elimination, Kowal and Crossley (1971) estimated weight-specific ingestion rates for various groups feeding on *Pinus echinata* mor, i.e. forest-floor material (Table II). Their values are lower than some given in Table II, possibly because of the values assigned to the elimination rates; however, when applied to a field population of Collembola and oribatids with a biomass of 200 mg m^{-2}, a rate of 0·046 is equivalent to an ingestion of 9·1 mg m^{-2} day^{-1} or, with approximate temperature corrections, to an annual consumption of 0·5% of the total detritus input of the *Pinus* forest floor. The data in Table II suggest that ingestion rates per day range from 1 to 40% of dry body-weight.

Accurate measurements can be made of the area of food consumed, particularly when leaves are perforated, and these values can be converted to volume or weight. Dunger (1956) gives an average ingestion rate of 0·049 mm^3 day^{-1} at 18°C for *F. fimetaria* on leaves of various trees, while *O. armatus* at similar temperatures was estimated by Witkamp (1960) to consume 0·07 mm^3 day^{-1} of *Mortierella pusilla* mycelium.

Berthet (1964) used planimetry to assess the ingestion of leaf litter by *S. magnus* at 18°C, noting that it was difficult to take partial perforation into account. The ingestion rates quoted in Table I were derived by using the area/weight ratio of the litter, but no allowance was made for differential feeding on vascular and intervein tissues. Defaecation rates were determined at 22°C, the mean dry weight of 1·7 μg per pellet being identical to Madge's (1964) value for this species. Berthet then applied a respiration/

temperature regression to estimate that a mite weighing 360 μg would use 1·5 μl of O_2 per day at 22°C, equivalent to the complete combustion of 1·86 μg of carbohydrate. Adding this value to 11 μg of faeces, he concluded that 12·9 μg of litter had been ingested, giving an A/C ratio of 14%. A similar approach was used by Berthet to calculate the amount of litter which

TABLE II. Weight-specific ingestion rates[a] of microarthropods

Taxon	Ingestion rate	Food	°C	Author
One oribatid species	0·40	^{14}C-yeast	?	Engelmann (1961)
C. juncta	0·25	^{45}Ca-*Pinus* mor	20	Kowal (1969)
Oribatids (adult)	0·125	^{45}Ca-*Pinus* mor	Field (*ca.* 20)	Kowal and Crossley (1971)
Oribatids (immature)	0·059	^{45}Ca-*Pinus* mor	Field (*ca.* 20)	Kowal and Crossley (1971)
Collembola	0·018	^{45}Ca-*Pinus* mor	Field (*ca.* 20)	Kowal and Crossley (1971)
Oribatids and Collembola	0·046	^{45}Ca-*Pinus* mor	Field (*ca.* 20)	Kowal and Crossley (1971)
S. magnus (young and old, ♂♂, ♀♀)	0·03[b]–0·08[b]	^{32}P-*Calluna*	18	Webb and Elmes (1972)
S. magnus	0·02[b]–0·04[b]	*Fagus, Corylus*	18	Berthet (1964)
Phthiracaroids	0·01	^{137}Cs-*Liriodendron*	20	McBrayer and Reichle (1971)
Fungivorous microarthropods	0·04	^{137}Cs-*Liriodendron*	20	McBrayer and Reichle (1971)
D. clavipes (nymphs and adults)	0·06–0·09	^{32}P-fungus	15	Luxton (1972)
Onychiurus procampatus	0·25[c]–0·38[c]	Fungi	15	Healey (1967)

[a] μg ingested/μg body weight/day; both weights oven-dry.
Dry body weight assumed to be [b]60% or [c]45% of fresh weight.

would have to be ingested to account for the annual oxygen consumption of the oribatid population of a *Quercus* forest floor (Berthet, 1963). Adult respiration rates for 16 species were measured in the laboratory, Q_{10} values ranging from 2·6–5·6 (average *ca.* 4) for 5–15°C. The relationship between body weight and respiration rate was studied in *S. magnus*, and a common regression of mean body weight on respiration was established for the 16 species. Using this information, the oxygen requirements of each of 46 species were calculated with reference to monthly abundance figures and mean temperatures of the forest floor, yielding an annual adult consumption of 4·5 l O_2 m^{-2}. The formula used to calculate adult respiration rates from body weight and temperature was applied to abundance data for all developmental stages in a *Pinus* forest, and it was concluded that 70% of the total oribatid respiration rate was accounted for by immatures. The annual total for the *Quercus* population was therefore increased to 10–15 l m^{-2}, corresponding to the combustion of 12–19 g carbohydrate. Taking 10% as a value for A/C in the oribatids, while admitting that the only data available were for *S. magnus*, Berthet estimated that *ca.* 150 g m^{-2}, half the annual leaf-fall, were ingested by oribatids in this site. Berthet (1967) later calculated annual respiration rates for adult oribatids in the litter of a number of woodland sites; these ranged from 33 to 123 ml O_2 l^{-1} of litter. Allowing for the activities of immatures and humus dwellers, he obtained an average value for energy flow of 30 kcal m^{-2} for forest oribatids, which, with an A/C value of 14%, gave an annual ingestion equivalent to *ca.* 20% of the annual litter fall.

Despite the uncertainties inherent in extrapolating from laboratory determinations to natural populations (Macfadyen, 1967; Berthet, 1971), Berthet's respiration data are the most comprehensive so far, since they include specific, seasonal fluctuations, as well as indicating the relative significance of various species in terms of annual oxygen consumption. Berthet (1967) stressed that in each site at least half of the total respiratory activity was accounted for by only four species, usually large mites such as *S. magnus*, *P. peltifer* and *N. palustris*, but with *T. velatus* compensating for small size by great abundance in a coniferous site.

However, Berthet's estimates of assimilation and ingestion rates may have to be reconsidered in the light of more recent results. Berthet (1964) quoted Engelmann's (1961) value of 20% for the A/C ratio to support his suggestion that this ratio might be higher in mycophages than the range of 5–14% for macrophytophages such as *S. magnus*, diplopods and isopods. Considerably higher values have in fact been obtained for mycophages by Healey (1967) and Luxton (1972). Healey gives values of 40–70% for *O. procampatus*, depending on fungal species, while Luxton's values for *D. clavipes*, based on ^{32}P uptake, range from 47% in tritonymphs to *ca.* 62% for the other nymphal stages and adults. Luxton pointed out that the use of phosphate possibly gives a false picture of general food requirements, especially in immatures, but nevertheless he suggested that A/C ratios for microphytophages in general might be 50–65%. High values were possibly achieved with *D. clavipes* protonymphs and adults by slow movement of relatively rich food through the gut, defaecation rates of *ca.* 0·75 pellets per mite per day being contrasted with a mean of 6·5 for *S. magnus* on *Corylus* (Berthet, 1964). Luxton tentatively suggested A/C ratios of 40–50% for panphytophages and 10–15% for macrophytophages. However, Webb and Elmes (1972), also using ^{32}P, calculated that young adults of *S. magnus* assimilated 58% of ingested *Calluna* litter, the ratio falling to 52% and

19% in mature females and males, respectively; when calculated in terms of calorific values of food and faeces, values ranged from 41 to 69%. Ingestion rates were comparable to those given by Berthet (1964; see Table I) and although defaecation rates were about half those quoted by Berthet for *Corylus* (Table I), they were about twice the rates for *Quercus* (Madge, 1964). Since defaecation was assessed gravimetrically by Webb and Elmes, major errors seem more likely to have arisen from non-uniform distribution of ^{32}P in the food, thus leading to inaccurate estimates of ingestion. Nevertheless, these results indicate the need for more studies before we can make valid generalizations about assimilation/consumption ratios.

Berthet (1967) assumed that only a small part of the energy assimilated by soil organisms is used for biosynthesis, and in fact his conversions of oxygen consumption to ingestion rates make no allowance for production. Admittedly, Engelmann (1961) had estimated that 95% of the energy assimilated by an old-field mite population was respired, and Macfadyen (1969*a*) considered this to be a realistic value in sluggish mites which presumably feed on relatively non-nutritious foods. However, both Berthet (1964) and Macfadyen (1969*a*) realized that Engelmann's numerical values were unreliable in several respects; for example, estimates of production were based on rather atypical generation times for one species. Webb and Elmes (1972) calculated the amount of energy available for production in *S. magnus* by finding the difference between assimilation and respiration, when feeding on *Calluna*, and obtained P/A percentages of 96, 74, and 88 for young adults, mature females and males, respectively. Respiration rates were similar to those given by Berthet (1964), and these high values for production may be based on inaccurate ingestion and assimilation data. Webb and Elmes, however, consider that their estimates of production could be accounted for by cuticular thickening in young adults, and reproduction in mature adults. Thomas (1972) gives a P/A percentage of *ca.* 27, and an A/C percentage of *ca.* 32, with an annual ingestion of *ca.* 18 kcal m^{-2}, for the oribatid population of the I.B.P. deciduous woodland site at Meathop.

Michael (1884) commented on the special vigour of immature oribatids' feeding, but there have been very few quantitative assessments of this vigour. Saichuae *et al.* (1972) recorded similar ingestion rates for tritonymphs and adults of *N. biciliatus* on yeast, whereas Luxton (1972) found that adults of *D. clavipes* ingested *ca.* 4, 8 and 15 times as much fungus as tritonymphs, deutonymphs and protonymphs, respectively. Luxton's figures have not yet been applied to field populations, as Webb (1970*a*) did for respiration rates of the various stages of *N. silvestris*. Laboratory estimates of these rates (Webb, 1969; see Table III) were applied to a heathland population, with temperature corrections as determined for

DRIFT, J. VAN DER (1970). *In* "Methods of Study in Soil Ecology." (J. Phillipson, ed.), pp. 295–300. UNESCO, Paris.
DUDICH, E., BALOGH, J. and LOKSA, J. (1952). *Acta biol. hung.* **3,** 295–317.
DUNGER, W. (1956). *Zool. Jb. (Syst.)* **84,** 75–98.
DUNGER, W. (1958*a*). *Z. PflErnähr. Düng. Bodenk.* **82,** 174–193.
DUNGER, W. (1958*b*). *Zool. Jb. (Syst.)* **86,** 138–180.
DUNGER, W. (1962). *Abh. Ber. NaturkMus.-ForschStelle, Görlitz* **37,** 143–162.
DUNGER, W. (1964). "Tiere im Boden." Ziemsen, Wittenberg.
DUNGER, W. (1968). *Abh. Ber. NaturkMus.-ForschStelle, Görlitz* **43,** 1–256.
EDWARDS, C. A. (1961). *Entomologia exp. appl.* **4,** 239–256.
EDWARDS, C. A. (1965). *In* "Ecology and the Industrial Society." (G. T. Goodman, R. W. Edwards and J. M. Lambert, eds), pp. 239-261. Blackwell, Oxford.
EDWARDS, C. A. (1967). *In* "Progress in Soil Biology." (O. Graff and J. E. Satchell, eds), pp. 585–594. Vieweg, Braunschweig.
EDWARDS, C. A. and FLETCHER, K. E. (1971). *In* "Methods of Study in Quantitative Soil Ecology." (J. Phillipson, ed.), pp. 150–185. Blackwell, Oxford.
EDWARDS, C. A. and HEATH, G. W. (1963). *In* "Soil Organisms." (J. Doeksen and J. van der Drift, eds), pp. 76–84. North-Holland, Amsterdam.
EDWARDS, C. A., REICHLE, D. E. and CROSSLEY, D. A. (1970). *In* "Analysis of Temperate Forest Ecosystems." (D. E. Reichle, ed.), pp. 147–172. Springer-Verlag, New York.
ELTON, C. S. (1966). "The Pattern of Animal Communities." Methuen, London.
ENGELMANN, M. D. (1961). *Ecol. Monogr.* **31,** 221–238.
ENGELMANN, M. D. (1966). *Adv. ecol. Res.* **3,** 73–115.
EVANS, G. O., SHEALS, J. G. and MACFARLANE, D. (1961). "The Terrestrial Acari of the British Isles." Vol. 1. British Museum, London.
FARAHAT, A. Z. (1966). *Pedobiologia* **6,** 258–268.
FORSSLUND, K. H. (1939). *In* "7th Int. Congr. Ent., Berlin (1938)" **3,** 1950–1957.
FOURMAN, K. L. (1938). *Mitt. Forstw. Forstwiss.* **9,** 144–169.
FÜHRER, E. (1961). *Pedobiologia* **1,** 99–112.
GASDORF, E. C. and GOODNIGHT, C. J. (1963). *Ecology* **44,** 261–268.
GERSON, U. (1969). *Bryologist* **72,** 495–500.
GERSON, U. (1972). *Acarologia* **13,** 319–343.
GHILAROV, M. S. (1971). *In* "Productivity of Forest Ecosystems." (P. Duvigneaud, ed.), pp. 433–442. UNESCO, Paris.
GIFFORD, D. R. (1967). *In* "Secondary Productivity of Terrestrial Ecosystems." (K. Petrusewicz, ed.), Vol. II, pp. 687–693. Polish Academy of Sciences, Warsaw.
GILMORE, S. K. and RAFFENSPERGER, E. M. (1970). *Pedobiologia* **10,** 135–140.
GISIN, G. (1952). *Revue suisse Zool.* **59,** 543–578.
GRANDJEAN, F. (1950). *Bull. Mus. natn. Hist. nat., Paris* (2) **22,** 224–231.
GRAVES, R. C. (1960). *Ann. ent. Soc. Am.* **53,** 61–78.
GREENSLADE, P. J. M. and GREENSLADE, P. (1968). *Pedobiologia* **7,** 362–370.
GRESSITT, J. L. and SHOUP, J. (1967). *In* "Entomology of Antarctica." (J. L. Gressitt, ed.), pp. 307–320. American Geophysical Union, Washington.

Grossbard, E. (1969). *J. Soil Sci.* **20**, 38–51.
Haarløv, N. (1960). *Oikos* Suppl. **3**, pp. 1–176.
Hale, W. G. (1966). *Pedobiologia* **6**, 65–99.
Hale, W. G. (1967). *In* "Soil Biology." (A. Burges and F. Raw, eds), pp. 397–411. Academic Press, London and New York.
Hammer, M. (1972). *Pedobiologia* **12**, 412–423.
Harding, D. J. L. (1966). Ph.D. Thesis, University of Nottingham.
Harding, D. J. L. (1967). *In* "Progress in Soil Biology." (O. Graff and J. E. Satchells, eds), pp. 10–20. Vieweg, Braunschweig.
Harding, D. J. L. (1968). *Rep. E. Malling Res. Stn. for* 1967 (1968), 169–172.
Harding, D. J. L. (1973). *Proc. 3rd Int. Congr. Acarology* (1971), 79–83.
Hartenstein, R. (1962*a*). *Ann. ent. Soc. Am.* **55**, 202–206.
Hartenstein, R. (1962*b*). *Ann. ent. Soc. Am.* **55**, 357–361.
Hartenstein, R. (1962*c*). *Ann. ent. Soc. Am.* **55**, 709–713.
Hartenstein, R. (1962*d*). *Ann. ent. Soc. Am.* **55**, 587–591.
Hartenstein, R. (1962*e*). *Ann. ent. Soc. Am.* **55**, 713–716.
Hayes, A. J. (1963). *Entomologia exp. appl.* **6**, 241–256.
Healey, I. N. (1967). *In* "Secondary Productivity of Terrestrial Ecosystems." (K. Petrusewicz, ed.), Vol. II, pp. 695–708. Polish Academy of Sciences, Warsaw.
Healey, I. N. (1970). *In* "Methods of Study in Soil Ecology." (J. Phillipson, ed.), pp. 175–182. UNESCO, Paris.
Healey, I. N. (1971). *In* "Methods of Study in Quantitative Soil Ecology." (J. Phillipson, ed.), pp. 209–232. Blackwell, Oxford.
Heath, G. W. and Arnold, M. K. (1966). *Pedobiologia* **6**, 238–243.
Heath, G. W., Arnold, M. K. and Edwards, C. A. (1966). *Pedobiologia* **6**, 1–12.
Hingley, M. R. (1971). *J. Anim. Ecol.* **40**, 17–32.
Hughes, T. E. (1953). *Proc. Acad. Sci. Amst.* **56C**, 278–287.
Jacks, G. V. (1965). *In* "Experimental Pedology." (E. G. Hallsworth and D. V. Crawford, eds), pp. 219–226. Butterworths, London.
Jacot, A. P. (1930). *Mycologia* **22**, 94–95.
Jacot, A. P. (1934). *Ann. ent. Soc. Am.* **27**, 462–467.
Jacot, A. P. (1936). *Can. Ent.* **68**, 31.
Jacot, A. P. (1939). *J. For.* **37**, 858–860.
Jalil, M. (1963). M.Sc. Thesis, University of Nottingham.
Jongerius, A. (1963). *In* "Soil Organisms." (J. Doeksen and J. van der Drift, eds), pp. 137–148. North-Holland, Amsterdam.
Kendrick, W. B. and Burges, A. (1962). *Nova Hedwigia* **4**, 313–344.
Kevan, D. K. McE. (1962). "Soil Animals." Witherby, London.
Kooistra, G. (1964). *Neth. J. Pl. Path.* **70**, 136–141.
Kowal, N. E. (1969). *Am. Midl. Nat.* **81**, 595–598.
Kowal, N. E. and Crossley, D. A. (1971). *Ecology* **52**, 444–452.
Kubiena, W. L. (1953). "The Soils of Europe. Illustrated Diagnosis and Systematics." Murby, London.
Kubiena, W. L. (1955). *In* "Soil Zoology." (D. K. McE. Kevan, ed.), pp. 73–82. Butterworths, London.

KUBIENA, W. L. (1964). *In* "Soil Micromorphology." (A. Jongerius, ed.), pp. 1–13. Elsevier, London.
KÜHNELT, W. (1961). "Soil Biology, with Special Reference to the Animal Kingdom." Faber and Faber, London.
KÜHNELT, W. (1963). *In* "Soil Organisms." (J. Doeksen and J. van der Drift, eds), pp. 333–341. North-Holland, Amsterdam.
KURCHEVA, G. F. (1960). *Soviet Soil Sci.* 1960, No. 4, 360–365.
LEBRUN, P. (1970). *Acarologia* **12,** 193–207.
LEBRUN, P. (1971). *Mém. Inst. r. Sci. nat. Belg.* **165,** 1–203.
LITTLEWOOD, C. F. (1969). *Proc. 2nd int. Congr. Acarology* (1967), 53–56.
LUND, J. W. G. (1967). *In* "Soil Biology." (A. Burges and F. Raw, eds), pp. 129–147. Academic Press, London and New York.
LUXTON, M. (1966). *Acarologia* **8,** 163–175.
LUXTON, M. (1972). *Pedobiologia* **12,** 434–463.
MACFADYEN, A. (1961). *Ann. appl. Biol.* **49,** 216–219.
MACFADYEN, A. (1963). *In* "Soil Organisms." (J. Doeksen and J. van der Drift, eds), pp. 3–17. North-Holland, Amsterdam.
MACFADYEN, A. (1964). *C. r. 1er Congr. Int. d'Acarologie* (1963); *Acarologia* **6** (h.s.), 147–149.
MACFADYEN, A. (1967). *In* "Secondary Productivity of Terrestrial Ecosystems." (K. Petrusewicz, ed.), Vol. II, pp. 383–412. Polish Academy of Sciences, Warsaw.
MACFADYEN, A. (1969*a*). *Proc. 2nd int. Congr. Acarology* (1967), 565–571.
MACFADYEN, A. (1969*b*). *In* "The Soil Ecosystem." (J. G. Sheals, ed.), pp. 191–197. The Systematics Association, London.
MACNAMARA, C. (1924). *Can. Ent.* **56,** 99–105.
MADGE, D. S. (1964). *Acarologia* **6,** 199–223.
MADGE, D. S. (1965). *Pedobiologia* **5,** 273–288.
MCBRAYER, J. F. and REICHLE, D. E. (1971). *Oikos* **22,** 381–388.
MCMILLAN, J. H. (1969). *Pedobiologia* **9,** 372–404.
MCMILLAN, J. H. and HEALEY, I. N. (1971). *Revue Écol. Biol. & Sol.* **8,** 295–300.
MCNEILL, S. and LAWTON, J. H. (1970). *Nature, Lond.* **225,** 472–474.
MICHAEL, A. D. (1884). "British Oribatidae," Vol. I. Ray Society, London.
MICHELBACHER, A. E. (1938). *Hilgardia* **11,** 55–148.
MIGNOLET, R. (1971). *In* "IV Colloquium pedobiologiae, Dijon (1970)," pp. 153–162.
MOELLER, J. (1965). *Z. Morph. Ökol. Tiere* **55,** 530–586.
MOELLER, J. (1967). *In* "Progress in Soil Biology." (O. Graff and J. E. Satchell, eds), pp. 148–155. Vieweg, Braunschweig.
MURPHY, P. W. (1953). *J. Soil Sci.* **4,** 155–193.
NAGLITSCH, F. (1966). *Pedobiologia* **6,** 178–194.
NEF, L. (1957). *Agricultura, Louvain* **5** (2. Ser.), 245–316.
NICHOLSON, P. B., BOCOCK, K. L. and HEAL, O. W. (1966). *J. Ecol.* **54,** 755–766.
NIELSEN, C. O. (1949). *Natura jutl.* **2,** 1–131.
NIELSEN, C. O. (1962). *Oikos* **13,** 200–215.

NOORDAM, D. and VLIEGER, S. H. VAN DER VAART DE (1943). *Ned. BoschbTijdschr.* **16**, 470–492.
NOSEK, J. (1967). *In* "Progress in Soil Biology." (O. Graff and J. E. Satchell, eds), pp. 141–147. Vieweg, Braunschweig.
PHILLIPSON, J. (1963). *Oikos* **14**, 212–223.
PHILLIPSON, J. (ed.) (1971). "Methods of Study in Quantitative Soil Ecology: Population, Production and Energy Flow." I.B.P. Handbook No. 18. Blackwell, Oxford.
PIELOU, D. P. and MATTHEWMAN, W. G. (1966). *Can. Ent.* **98**, 1308–1312.
PIMENTEL, D., RUMSEY, M. W. and STREAMS, F. A. (1960). *Ann. ent. Soc. Am.* **53**, 549.
POOLE, T. B. (1959). *Proc. zool. Soc. Lond.* **132**, 71–82.
PRYOR, M. E. (1962). *Pacif. Insects* **4**, 681–728.
REICHLE, D. E. (1971). *In* "Productivity of Forest Ecosystems." (P. Duvigneaud, ed.), pp. 465–477. UNESCO, Paris.
REICHLE, D. E. and CROSSLEY, D. A. (1965). *Hlth Phys.* **11**, 1375–1384.
RIHA, G. (1951). *Zool. Jb. (Syst.)* **80**, 407–450.
ROGERS, W. S. (1939). *J. Pomol.* **17**, 99–130.
ROHDE, C. J. (1955). Doctoral dissertation, Northwestern University, Illinois.
SAICHUAE, P., GERSON, U. and HENIS, Y. (1972). *Soil Biol. Biochem.* **4**, 155–164.
SATCHELL, J. E. (1967). *In* "Soil Biology." (A. Burges and F. Raw, eds), pp. 259–322. Academic Press, London and New York.
SATCHELL, J. E. and LOWE, D. G. (1967). *In* "Progress in Soil Biology." (O. Graff and J. E. Satchell, eds), pp. 102–119. Vieweg, Braunschweig.
SCHALLER, F. (1950). *Zool. Jb. (Syst.)* **78**, 506–525.
SCHALLER, F. (1962). "Die Unterwelt des Tierreiches." Springer-Verlag, Berlin.
SCHUSTER, R. (1956). *Z. Morph. Ökol. Tiere* **45**, 1–33.
SCHUSTER, R. (1966). *Veröff. Inst. Meeresforsch. Bremerh.* **2**, 319–328.
SCOTT, H. G. and STOJANOVICH, C. J. (1963). *Fla. Ent.* **46**, 189–191.
SENGBUSCH, H. G. (1954). *Ann. ent. Soc. Am.* **47**, 646–667.
SINGH, S. B. (1969). *Revue Écol. & Biol. Sol.* **6**, 461–467.
SPENCER, J. A. (1951). Special subject report, University of Oxford.
STRENZKE, K. (1963). *Pedobiologia* **3**, 95–141.
STRONG, J. (1967). *In* "Entomology of Antarctica." (J. L. Gressitt, ed), pp. 357–371. American Geophysical Union, Washington.
STUTTARD, R. A. (1972). Ph.D. Thesis, C.N.A.A.
TARMAN, K. (1968). *Biol. Věst.* **16**, 67–76.
TARRAS-WAHLBERG, N. (1961). *Oikos*, Suppl. **4**, 1–56.
THOMAS, J. O. M. (1972). Ph.D. Thesis, C.N.A.A.
TILBROOK, P. J. (1967). *In* "Entomology of Antarctica." (J. L. Gressitt, ed.), pp. 331–356. American Geophysical Union, Washington.
TILBROOK, P. J. (1970). *In* "Antarctic Ecology." (M. W. Holdgate, ed.), Vol. 2, pp. 886–896. Academic Press, London and New York.
TÖRNE, E. VON (1967). *Pedobiologia* **7**, 220–227.
TÖRNE, E. VON (1968). *Pedobiologia* **8**, 526–535.
TRAVÉ, J. (1963). *Vie Milieu*, Suppl. **14**, 1–267.

Travé, J. (1969). *Revue Écol. & Biol. Sol.* **6,** 239–248.
Tribe, H. T. (1957). *In* "Microbial Ecology." (R. E. O. Williams and C. C. Spicer, eds), pp. 287–298. Cambridge University Press.
Walker, N. A. (1965). *Fort Hays Studies,* n.s., science series no. 3.
Wallwork, J. A. (1958). *Oikos* **9,** 260–271.
Wallwork, J. A. (1967). *In* "Soil Biology." (A. Burges and F. Raw, eds), pp. 363–395. Academic Press, London and New York.
Wallwork, J. A. (1970). "Ecology of Soil Animals." McGraw-Hill, London.
Wallwork, J. A. (1972). *J. Anim. Ecol.* **41,** 291–310.
Webb, N. R. (1969). *Oikos* **20,** 294–299.
Webb, N. R. (1970*a*). *Oikos* **21,** 155–159.
Webb, N. R. (1970*b*). *Pedobiologia* **10,** 447–456.
Webb, N. R. and Elmes, G. W. (1972). *Oikos* **23,** 359–365.
Went, J. C. and de Jong, F. (1966). *Antonie van Leeuwenhoek* **32,** 39–56.
Willmann, C. (1931). *Tierwelt Dtl.* **22,** 79–200.
Winston, P. W. (1956). *Ecology* **37,** 120–132.
Witkamp, M. (1960). *Meded. Inst. toegep. biol. Onderz. Nat.* **46,** 1–51.
Witkamp, M. (1971). *A. Rev. Ecol. Systematics* **2,** 85–110.
Wittich, W. (1943). *Forstarchiv* **19,** 1–18.
Wood, T. G. (1967). *Oikos* **18,** 102–117.
Woodring, J. P. (1963). *Adv. Acarol.* **1,** 89–111.
Woodring, J. P. (1965). *Acarologia* **7,** 564–576.
Woodring, J. P. and Cook, E. F. (1962). *Acarologia* **4,** 101–137.
Zachariae, G. (1963). *In* "Soil Organisms." (J. Doeksen and J. van der Drift, eds), pp. 109–124. North-Holland, Amsterdam.
Zachariae, G. (1965). *Forstwiss. Forsch.* **20,** 1–68.
Zinkler, D. (1966). *Z. vergl. Physiol.* **52,** 99–144.
Zinkler, D. (1971). *In* "IV Colloquium pedobiologiae, Dijon (1970)," pp. 329–334.
Zlotin, R. I. (1971). *In* "IV Colloquium pedobiologiae, Dijon (1970)," pp. 455–462.

16

Macroarthropods

C. A. Edwards

Rothamsted Experimental Station
Harpenden
Hertfordshire
England

I. Introduction

Plant litter ranges from soft and fragile deciduous leaf tissue to tough, undecomposed wood. Soft and delicate plant tissues are usually rapidly decomposed by microbial action, but some of the tougher tree leaves may take several years to break down, and some woody tissues considerably longer than this. The decomposition of such tissues is a complex process involving both the soil fauna and microflora, each acting to facilitate further breakdown by the other.

Soil animals aid in the breakdown of plant organic matter in several ways:

1. By physically disintegrating tissues and increasing the surface area available for bacterial and fungal action.
2. By selectively decomposing material such as sugar, cellulose and even lignin.
3. By transforming plant residues into humic materials.
4. By mixing decomposed organic matter into the upper layer of soil.
5. By forming complex aggregates between organic matter and the mineral fractions of soil.

For several weeks after fresh deciduous litter reaches the forest floor, very little of it is eaten by the fauna, although it becomes invaded by some micro-organisms, especially fungi. During this period the litter darkens, becomes weathered, and water-soluble substances, particularly sugars, organic acids and polyphenols, are leached out. Edwards and Heath (1963) reported that *Fagus* (beech) and *Quercus* (oak) leaves that did not darken appreciably after reaching the soil, began to be fragmented by litter animals much sooner than darker leaves. These darker leaves had a much greater percentage of both polyphenols and sugars than did "shade" leaves (Heath and King, 1964). As the water-soluble polyphenols were removed by weathering, the litter became more palatable to arthropods (Edwards and Heath, 1974). It was suggested that the increase in polyphenolic materials precipitated protein complexes in leaves and made them less digestible by the microfauna (Edwards and Heath, 1963) and King and Heath (1967) confirmed that the amount of polyphenols in beech litter was inversely proportional to the rate at which arthropods consumed the litter (Williams and Gray, Chapter 19).

Satchell and Lowe (1967) studied the palatability of many kinds of leaf litter to the earthworm *Lumbricus terrestris* and reached similar conclusions.

The groups of animals having the greatest influence on the rate of decomposition of leaf litter are earthworms, termites, millipedes, woodlice and larvae of various insects, especially Diptera. The role of earthworms is considered in Chapter 14, and here we are concerned only with the role of macroarthropods. The importance of some of these animals in fragmenting litter has been demonstrated experimentally. When naphthalene was used to minimize the numbers of arthropods in *Quercus* litter, about one twentieth of the litter broke down in 140 days, compared with one tenth when the soil animals were present (Kurcheva, 1960). In similar experiments in U.S.A., 60% of the *Quercus* litter in untreated soil broke down in one year, compared with only 45% in soil treated with naphthalene. When the soil fauna was totally excluded from *Quercus* and *Fagus* litter, by confining it in fine (0·003 mm aperture) mesh bags there was little breakdown, even after 3 years (Edwards and Heath, 1963).

Macroarthropods can also chemically decompose plant material, either by various digestive enzymes or with the aid of symbiotic protozoa and bacteria that live in their intestines. The combined effects of macroarthropod feeding and microbial action are gradually to produce dark, stable humic substances and to decrease the ratio of carbon to nitrogen in the organic matter.

Finally, as most of the organic matter becomes humified, the soil animals mix this material with the mineral soil, and when this process is well advanced in a rich mull soil there are no obvious horizontal strata in the soil profile; instead there is a homogeneous mixture of soil and organic matter.

It is extremely difficult to assess the relative contributions of bacteria and fungi on the one hand, and the micro-, meso- and macrofauna on the other, to the breakdown of plant litter. It seems probable that microbes make a major contribution, and in their absence soil invertebrates would be relatively unimportant. Of the soil animals, the macrofauna seem to play a much more important part than some of the smaller animals. However, all are complementary and have a more or less essential role in breaking down the tougher plant materials.

II. Groups of Macroarthropods Involved in Litter Breakdown

The main groups of macroarthropods that contribute to the breakdown of plant litter are the Isopoda (woodlice), Symphyla (symphylids), Diplopoda (millipedes), Isoptera (termites), Diptera larvae (flies) and Coleoptera larvae and adults (beetles). Small contributions may be made by Psocoptera (booklice), Thysanoptera (soil-inhabiting thrips) and Hemiptera (root-aphids) and even occasionally Trichoptera (terrestrial caddis fly larvae). Of these animals there is little doubt that in temperate countries, millipedes and fly larvae are most important and in tropical areas Isoptera make a major contribution. Each of these groups of animals will be considered in turn.

A. Isopoda—Woodlice

Woodlice are familiar animals usually restricted to sheltered, moist habitats such as plant litter, decaying wood or under stones; they are nocturnal, only emerging from their hiding places at night. They are a terrestrial group of Crustacea, most of which are aquatic, and this relationship means that they are very susceptible to loss of water by evaporation. Woodlice do not hibernate in winter but they remain quiescent for most of this period. They are comparatively long-lived, surviving for as long as 4 years in ideal conditions. Species common in Great Britain in deciduous woodlands are *Trichoniscus pusillus*, *Philoscia muscorum*, *Oniscus asellus* and *Porcellio scaber*.

Woodlice are omnivorous, but feed mainly on dead and decaying plant material, fragmenting this and making it more susceptible to attack by other invertebrates and micro-organisms. They also feed on their own faeces and on those of other animals. Damp leaves and wood are usually preferred, and it is not certain to what extent the leaf tissue must be decomposed by micro-organisms before it becomes acceptable to woodlice. When young, they eat only the softer parts of leaves, but adults of the more robust species eat the whole leaf including veins. As with the soil invertebrates, such as earthworms, woodlice feed much more readily on some species of plant litter than others, and one of the main factors that influence palatability seems to be the sort and amounts of polyphenols in the litter. Leaves that contain large amounts of certain polyphenols, e.g. *Fagus*, are not eaten until the leaves have been weathered and these chemicals broken down or leached out; by contrast, leaves with a low polyphenol content such as *Fraxinus* (ash) and *Acer* (sycamore) are readily eaten soon after they reach the litter layer. Woodlice also seem to feed on decaying wood, but there is no information on the way in which they digest this or how much they break it down. Kühnelt (1961) considered that woodlice play a considerable role in the breakdown of litter and wood residues, but Kevan (1962) concluded that where woodlice are abundant they may be important in the comminution of dead vegetable matter, but cannot be considered to play a major general role in litter breakdown.

B. Symphyla—Symphylids

Symphylids are often considered to be rare animals, but this is more due to their small size and habit of penetrating deep into the subsoil than their actual abundance. They are small, white, active animals with twelve pairs of legs when fully grown. Although they are found in many soils, they are very common in leaf litter and, due to their large numbers and voracious appetites, there is little doubt that they contribute appreciably to the breakdown of plant material. Symphylids are very long-lived, surviving for as long as 10 years in the laboratory. Although all symphylids are polyphagous, eating all sorts of living and dead plant material and micro-organisms, the common species, from the two suborders Scutigerellidae and Scolopendrellidae, differ in their habits. Species from the Scutigerellidae are larger and more active, showing a distinct preference for living plant tissues, whereas those belonging to the Scolopendrellidae tend to feed on decaying plant and animal material or dead micro-organisms. It seems likely that, because of the very rapid throughput of food, e.g. one symphylid can consume 15 times its own weight in a day (Edwards, 1961); there is very little digestion of the food, and their main contribution to litter breakdown is in fragmentation. They seem to prefer the softer leaf tissues, being unable to eat some of the tougher leaves, veins and woody tissues. A fair conclusion would be that although symphylids contribute appreciably to litter breakdown, their role is a relatively minor one.

C. Diplopoda—Millipedes

Millipedes are probably some of the most common of the larger arthropods encountered in forest litter. They are almost exclusively vegetarian or saprophagous, feeding on a great variety of living or dead plant organic matter, although they will also feed on decomposing animal tissues.

They are nocturnal, usually hiding by day and becoming active by night; this is probably due to their susceptibility to desiccation. The litter-inhabiting species include the pill millipedes which roll into a ball when disturbed, the long cylindrical iulid forms which are sometimes called snake millipedes and can actively tunnel into soil, and the flat-back millipedes which are dorso-ventrally flattened and are restricted to the loose litter layers. Millipedes are quite long-lived, some species surviving for several years, and they breed all the year round. They can feed on leaves which have not been otherwise disintegrated (Dunger, 1958; Zachariae, 1965), but usually they feed more readily on weathered leaves (van der Drift, 1951). They eat large quantities of leaf litter, some species of leaves being much more palatable than others. Lyford (1943) reported that the order of decreasing palatability of different species to millipedes, was: *Betula lutea* Mickx. (yellow birch), *Fraxinus americana* L. (ash), *Populus grandidentata* Michx. f. (aspen), *Acer rubrum* (red maple), *Acer saccharum* (sugar maple), *Betula papyrifera* Marsh (paper birch), *Quercus borealis* Michx. f. (red oak) and *Fagus grandifolia* Ehrh. (beech). He believed palatability to be correlated with a high calcium content of the leaves, and Marcuzzi (1970) also agreed that there was some correlation between calcium and palatability. However, it seems much more likely that this order of palatability is a feature of the polyphenol content of the litter, in the same way as is the palatability of leaves to earthworms and woodlice. This idea is supported by observations that *Iulus* and *Cylindroiulus* eat many more beech leaves that have been weathered for a year than those that are newly fallen (van der Drift, 1951), and that oak is preferred to beech (Cloudsley-Thompson, 1951). Striganova (1971) reported that the order of decreasing palatability of four species of leaf litter to millipedes was: *Carpinus*, *Acer*, *Quercus* and *Fagus*. Dunger (1962) gave the order of decreasing preference in the species he studied as: *Tilia* (lime), *Fraxinus* (ash), *Alnus* sp. (black alder), *Ulmus* (elm), *Acer* (maple), *Fagus* (beech), *Quercus* (oak) and red beech, and he also confirmed that weathered leaves were preferred to freshly fallen ones. Although there are marked food preferences, van der Drift (1951) considered that they may not always be clearcut because of a need for a variety of food. It seems probable that availability of food may also influence preferences.

Millipedes can also feed on decaying wood and although it is not clear how they manage to digest this material, they probably do so with the aid of a symbiotic intestinal flora.

There is considerable controversy over the ability of millipedes to break down their food chemically. In experiments on litter from a Dutch beech wood, van der Drift (1951) concluded that there was little difference in chemical composition between the food eaten by *Glomeris marginata* (Villers) and its faeces, and Dunger (1958) reach a similar conclusion. However, Bocock (1963) showed that *Glomeris marginata* could assimilate some holocellulose, soluble carbohydrates and fats. During passage of *Fraxinus* leaf litter through the intestine of this millipede, the percentage of holocellulose fell from 30% to 27%, that of soluble carbohydrates from 4·6 to 4·0% and that of crude fat from 3·6 to 1·7%. These amounts are small, and much more breakdown was reported by Striganova (1971), who found that more than half the cellulose in leaf litter was digested during passage through the intestines of millipedes. Specifically, millipedes consumed 40% of the cellulose in oak and hornbeam litter, 70% of that in beech litter, with an overall daily consumption of 3–5 mg per individual.

Franz and Leitenberger (1948) claimed that the C:N ratio of millipede faeces was greater than that of the litter on which they fed, so that considerable humification occurred during passage through the millipedes' intestine. Dunger (1958) also reported that some humification occurred when millipedes were fed on several kinds of fresh litter that was rich in nitrogen. He obtained similar results with weathered litter of *Fraxinus excelsior* and *Alnus glutinosa*, if it still contained considerable nitrogen. Humification did not occur when millipedes were fed with litter of *Quercus robur*, *Acer platanoides* and *Ulmus carpinifolia*. Bocock (1963) also reported more nitrogen in millipede faeces than in their food (Table I).

In a very careful investigation, Marcuzzi (1970) studied changes in the C:N ratio during digestion by millipedes. The criticism that van der Drift (1951) made of Franz and Leitenberger's work was that humification

TABLE I. Changes in the nitrogen content of *Corylus* litter eaten by *Glomeris*

	Amino Nitrogen ($mg\ g^{-1}$)	Ammonia Nitrogen ($mg\ g^{-1}$)	Total Nitrogen ($mg\ g^{-1}$)
Food	193	175	13,600
Faeces	103	1385	15,700

Adapted from Bocock (1963).

may have occurred microbially, after the faeces had been excreted. Marcuzzi took trouble to eliminate this as a source of error, and found quite marked decreases in the C:N ratio in his experiments (Table II).

TABLE II. Changes in the carbon:nitrogen ratio after passage through the intestines of *Glomeris* spp.

		C:N ratio	
Millipede sp.	Leaf species	Of leaves	Of faeces
(a) *G. euganeorum*	*Castanea sativa*	31·44	20·25
(b) *G. undulata*	*Castanea sativa*	34·48	27·55
(c) *G. conspersa*	*Castanea sativa*	29·82	26·31
(a) *G. euganeorum*	*Fagus sylvatica*	28·50	25·56
(b) *G. undulata*	*Fagus sylvatica*	30·50	26·80
(c) *G. conspersa*	*Fagus sylvatica*	29·86	27·20
(a) *G. euganeorum*	*Quercus ilex*	31·86	27·95
(b) *G. undulata*	*Quercus ilex*	32·10	26·90
(c) *G. conspersa*	*Quercus ilex*	38·79	25·61

Adapted from Marcuzzi (1970).

To summarize the findings of these various workers, it seems that although there is chemical breakdown of the litter and some humification by millipede feeding, by far the most important role millipedes play is in fragmentation and disintegration of large quantities of litter.

Kubiena (1955) emphasized the importance of millipedes in organic matter turnover, when he stated that mull-like moder soil consisted mainly of the droppings of iulid and glomerid millipedes, small earthworms and insect larvae. In an almost pure *Fagus* stand in northern England, Blower (1956) found that the millipedes *Cylindroiulus punctatus* (Leach) and *Proteroiulus fuscus* (Am Stein) were the dominant species of invertebrates, and the soil humus consisted mainly of cylindrical millipede faeces. Similar observations were made in a Dutch beech wood by van der Drift (1951). Eaton (1943) described the formation of a mull soil mainly by a millipede, *Apheloria*, and Jacot (1940) suggested that stocking new forest plantations with millipedes would promote mull formation. Blower (1956), however, concluded that, although mull humus can support large populations of millipedes, there is no evidence that millipedes can produce the clay-humus complex characteristic of mull soils. Lyford (1943) believed that the importance of millipedes in breakdown of litter has been over-emphasized, and he concluded that they consumed only 1–5% of the total leaf fall. Bocock (1963) considered that a large population of *Glomeris* would consume 1·7–10% of the annual deciduous litter fall. In complete contrast, Striganova (1971) stated that diplopod activity in some broad-leaved forests of the U.S.S.R. seemed to be the most important factor of leaf litter decomposition. There seems little doubt that millipedes are one of the important elements of the soil fauna in promoting litter breakdown, particularly in temperate regions.

D. Isoptera—Termites

Most species of termites are tropical or sub-tropical, although they sometimes occur in temperate areas. They differ considerably in their habits and food, but all are social insects living in nests or colonies of varying degrees of complexity. They feed almost exclusively on decaying vegetation or wood. In the process of feeding and building their nests, they not only disintegrate plant organic matter, but intimately mix it into the soil. In much the same way as earthworms (see Chapter 14) and millipedes, termites alter the carbon : nitrogen ratio in soil, the ratio being much greater in termite mounts or subterranean colonies than in the surrounding soil.

Thus, termites are very important animals in the breakdown of plant litter and, together with earthworms, they probably contribute more to leaf litter breakdown than all the other soil-inhabiting invertebrates. Termites are one of the few groups of animals that are able to degrade chemically large quantities of cellulose and they rely on cellulose as their main source of energy. Termites have three main types of feeding habit—feeding on leaf litter or humus, feeding on wood or feeding on fungi.

The termite species that are most important in breaking down leaf litter are called foraging termites, and most of these belong to the family Termitidae. They collect freshly fallen or old leaf litter and carry it to their nests, where it is either eaten immediately or stored until required. Although there are distinct preferences by different termite species for particular sorts of plant litter, we have little information on these preferences. There is also some evidence that a particular kind of litter may become either more or less palatable to certain species of termites as it decomposes, just as it does to earthworms. Other species of termites (Hodotermitenae, some Amitermitinae and Nasutiterminae) are called harvesting termites. These tend to prefer to harvest fragments of living plants, especially grasses, which they carry to their nests as food.

Many termites depend upon wood as their source of energy. A few species feed on living wood (Kalotermitidae and Mastotermitidae), but more commonly, they feed on decaying or rotten wood (Kalotermitidae, Rhinotermitidae, Hodotermitidae and some Termitidae) and most prefer wood in advanced stages of decomposition. It is likely that many of the fungi growing on the wood are an important component of the food of termites.

Different species of termites seem to have distinct preferences for decaying wood from particular species of trees, and the preferences are sometimes complex, with certain parts of a tree being preferred to others. It is impossible to assess the relative importance of termites in breaking down particular species of trees. In fact, many species of termites are polyphagous, and eat not only many different kinds of wood, but can also consume plant debris, animal dung, stored products and many other materials.

Termites obtain most of their energy from the digestion of polysaccharides including cellulose and hemicelluloses, and some species can also digest lignin. Four to five times as much cellulose as lignin is decomposed. However, most of this digestion is made possible by symbiotic protozoa and bacteria. Most important are the protozoa found mostly in the hind-gut of termites that can digest not only decaying but also sound wood. Most cellulose digestion is anaerobic, and there is some evidence that the initial product is glucose, which in turn is broken down to carbon dioxide, hydrogen and acetic acid (Hungate, 1943). It is not known how lignin is broken down, but it is probably an aerobic process, although it is not clear how this and anaerobic cellulose breakdown can occur together. Nevertheless, there is good evidence that the rate of decomposition of cellulose and lignin can be quite rapid; in experiments *Kalotermes flavicollis* (Fabricius) decomposed 94–95% of the cellulose, 60–70% of the hemicelluloses and 3–4% of the lignin in the wood upon which it fed (Seifert, 1962).

Some species of termites actually consume large quantities of mineral soil, particularly soil containing much humus (Kalshoven, 1941; Adamson,

1943). Those species that have adopted this habit seem to have modifications of their mouthparts, the grinding ridges of the molar plates being adapted for crushing particles. There is little evidence on how much the humus used as food is decomposed by the termites.

Thus, there is little doubt of the importance of termites in facilitating the breakdown of plant litter. In fact, they are so efficient in this process that they can sometimes decrease the organic carbon content of soil, in situations where the vegetation is comparatively sparse. Harris (1955) considered that termite digestion is so complete that they make only small contributions to soil fertility.

E. Diptera—Fly Larvae

The larvae of Diptera are extremely common in leaf litter and decaying wood. Many different species are found, with a great variety of food preferences. Some feed directly on freshly fallen leaf litter, but the majority of species important in litter breakdown are saprophagous, and feed after there has been some microbial breakdown. Perel *et al.* (1971) stated that in Russia, dipteran larvae are encountered in great numbers in areas ranging from the overmoist taiga forests to dry forests in the forest steppe zone. They studied the role of larvae of Tipulidae in litter breakdown, and found that these larvae would feed on fresh litter and increase its reducibility and ash content. Tipulid larvae can also feed on decomposing litter; they can disintegrate considerable quantities of material. Tipulids are usually much more common in deciduous woodland litter than in that from coniferous forests, and Bornebusch (1930) considered tipulid larvae to contribute the greatest biomass of all Diptera. Chironomid larvae are also very abundant in deciduous leaf litter, particularly in *Fagus* woodlands. Other larvae common in litter are those of the Lonchopteridae and Platypezidae and some Stratiomyidae, all of which are adapted for this habit by their flattened form. Bibionidae larvae are also common in leaf litter and, with the aid of a symbiotic gut flora, can feed on only slightly decomposed leaves as well as well decayed ones. Szabo *et al.* (1966) estimated that the bibionid population in a mixed woodland in Hungary could consume 15% of the total annual leaf fall. Karpachevsky *et al.* (1968) reported that they found evidence of a small degree of humification of litter during passage through the gut of larvae of *Bibio marci*, and d'Aguilar and Bessard (1963) reported a decrease in the C:N ratio of leaves of *Castanea* and *Quercus*, indicating a tendency for humification to occur. Sciarid larvae can also feed on litter in various stages of decay.

Many dipterous larvae feed mainly on fungi; one important group with this habit is the Mycetophilidae and these larvae are very common in decaying wood. Some species such as the larvae of the Sapromyzidae mine in leaf litter. Many of the muscoid fly larvae feed in rotting vegetation, but tend to prefer litter in an advanced stage of decay. Other families which have larvae that feed on various stages of decomposing litter, are the Sciaridae, Borboridae, Trichoceridae, Anisopodidae, Psychodidae and Lonchopteridae.

It seems that fly larvae make a significant contribution to litter breakdown, although less than termites, earthworms and millipedes, but we need much more information on the relative importance of the many different species.

F. Coleoptera—Beetles

A great variety of beetles and their larvae live in forest litter and decaying wood. Many of these, such as the larvae of Clambidae, Corylophidae and Trichopterygidae, feed on decomposing plant material. Some Scaphididae, Cryptophazidae and Lathridiidae feed on micro-organisms and fungi. It is certain that many species of beetle larvae, such as those of Elateridae, live on decomposing leaf litter, but the details of their feeding habits are poorly known, although elaterids are one of the dominant groups of beetle larvae in forest litter. Another very common group of larvae in litter is the Tenebrionidae, many species of which feed on plant remains and living or dead roots. The larvae of some species of Staphylinidae eat decomposing litter as do those of some species of Scarabaeidae.

Many beetles and larvae are wood-feeders but usually, as in the Scarabaeidae and Cerambycidae, they depend on a symbiotic gut flora to digest the cellulose. Not all wood-feeding beetles can digest cellulose; for instance, larvae of the Bostrichidae and Scolytidae and some Staphylinidae merely fragment the tissues. Some species of Elateridae, Ipidae, Tenebrionidae, Buprestidae, Cerophytidae, Throscidae and Eucnemidae also feed on decaying wood, and members of the Scaphidiidae live on fungi that grow on decaying wood.

To summarize, beetles are much less important than fly larvae in the breakdown of leaf litter, but probably more important in the decomposition of decaying wood.

G. Other Groups

Several other groups of macroarthropods contribute to breakdown of plant material. In the Hymenoptera, some species of ants in the Myrmicinae, Dolichoderinae and Formicinae are vegetarian and break down leaf litter. In tropical countries, leaf-cutting ants (Attini) carry large amounts of leaf material to their nests, where it is fragmented and used as a substrate for fungi. However, ants as a group break down a great deal of leaf material.

A few terrestrial species of Trichoptera larvae, e.g. *Enoicyla pusilla* (Burmeister) feed on leaf litter, particularly the underside of *Quercus* leaves.

III. Populations of Macroarthropods

Although populations of macroarthropods vary very considerably between habitats, soil types, species of woodland and climatic region, some groups are clearly more numerous and contribute a greater biomass than others. Figure 1 summarizes some of the data in the classical survey of Danish woodlands by Bornebusch (1930). Diptera larvae were generally the most common but they included a great diversity of types and habits. In functional terms, the Diplopoda are probably much more important.

A. Isopoda

Woodlice aggregate very much in response to environmental conditions, dispersing much more at night. This makes it very difficult to make accurate estimates of populations other than by mark and recapture methods which are relatively inaccurate. As with many other soil arthropods, populations tend to be greater in alkaline conditions than in acid soils (Kühnelt, 1961). This is not surprising when the need for calcium to build their exoskeleton is considered. Woodlice seem to be most characteristic of mull soils, only *Trichoniscus pusillus*, *Oniscus asellus* and *Porcellio scaber* occurring in mor soils. Not all workers agree with this conclusion, and van der Drift (1951) found more woodlice in acid mull (moder) woodland sites than on calcareous mulls in Holland, although he reported that *Armadillidium vulgare* was more common on the alkaline sites. Populations of 250 m^{-2} for *Philoscia* in Britain, 1000 m^{-2} for *Trichoniscus pusillus*, 4000 m^{-2} for woodlice in an Australian forest, 5000 m^{-2} for *Armadillidium vulgare* in California and as many as 7900 m^{-2} for *T. pusillus* in scrub grassland on the Chiltern Hills in Britain have been reported (Sutton, 1972) and large numbers in a British woodland (Table III). These figures are larger than those reported for other woodland species (van der Drift, 1951; Dunger, 1958).

Such populations are rather lower than comparable ones for millipedes and it is generally accepted that isopods are usually less important than millipedes in litter breakdown.

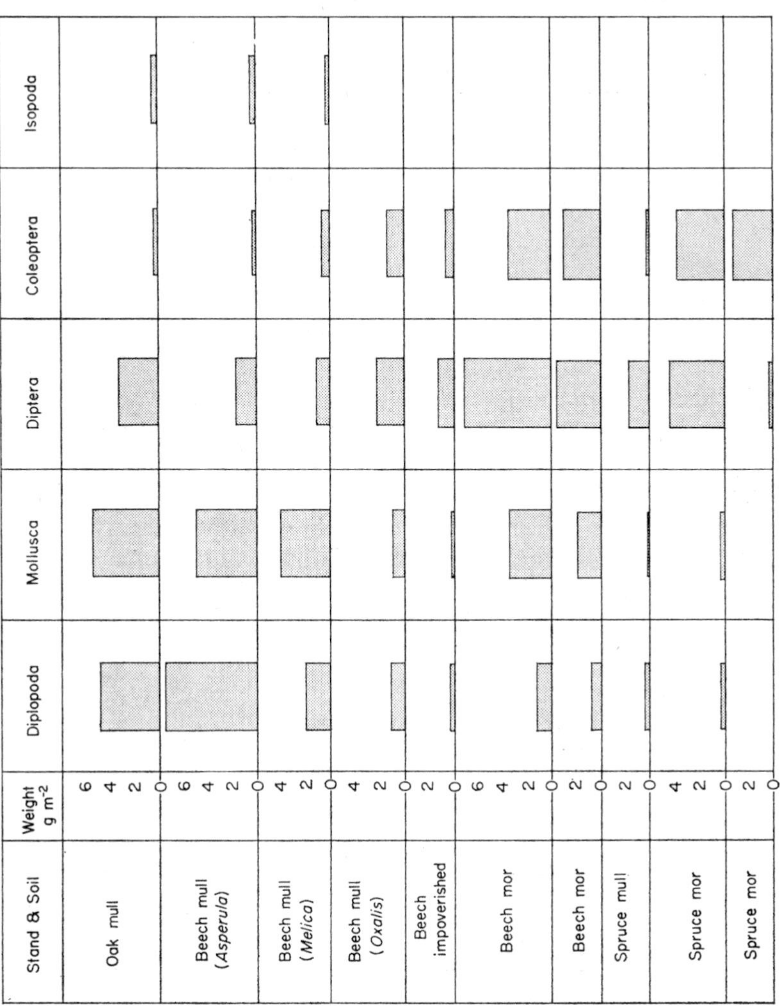

Fig. 1. Biomass of macroarthropods in 10 Danish woodlands (From data by Bornebusch, 1930.)

TABLE III. Numbers and weights of woodlice in grassland in a British wood in March

	Biomass ($g\ m^{-2}$ live wt)	Numbers (no. m^{-2})
Trichoniscus pusillus	1·19	1490
Philoscia muscorum	0·70	103
Armadillidium vulgare	0·21	6
Total woodlice	2·137	1820

Adapted from Sutton (1972).

B. Symphyla

Symphylids are often considered to be rare but in a survey of 415 sites in southern England, Edwards (1958) found that 54% had symphylid populations, and symphylids were found in 26% of all woodland sites.

Because symphylids cannot actively tunnel through soil, but rather move through existing cracks and crevices, they are rather sensitive to the type of soil they inhabit. The soil types that favour build-up of symphylid populations are loams or clay loams, although neither heavy clays nor sandy soils are suitable (Table IV) (Edwards, 1958).

TABLE IV. Populations of symphylids in different soil types (based on 33 sites)

Soil type	Mean number per m^2
Clay	2,100
Clay loam	43,700
Sandy clay loam	59,100
Loam	111,100
Sandy loam	29,600
Sandy	4,800

Data from Edwards (1958).

The largest numbers of symphylids were found in cultivated soils (mean population of 89,700 per m^2), but forest litter also supported large numbers (mean population of 51,520 per m^2) (Edwards, 1958).

C. Diplopoda

Millipedes, like woodlice, are very aggregated, and this makes estimates of populations very difficult. Although soil sampling can be used in grassland, populations in woodlands have to be estimated by methods that depend on activity, such as mark and recapture, pitfall traps or baits. Nevertheless, it is clear that very large populations of millipedes exist in many woodlands.

Millipedes seem to be much more abundant in calcareous soils than in acid ones, and one species, *Cydroiulinlus londiniensis* (Leach), is very sensitive to acidity (Blower, 1956).

Thiele (1956) reported that numbers of *Glomeris marginata* in German woodlands were small, average values ranging from 2·3 to 7·5 per m^2 and Shaw (1968) found populations of *Narceus annularis* (Raf.) of 18·4 per m^2 in a mixed woodland in New York, but Blower (1956) reported at least 70 per m^2 in a sandy woodland soil, van der Drift (1951) estimated populations of 80 individuals of *Cylindroiulus punctatus* per m^2 in a *Fagus* wood, Nef (1957) calculated 110 per m^2 in *Quercus* woodland and Bornebusch (1930) up to 177 per m^2 in Swedish woodlands, and much larger numbers with peak populations of 200–300 per m^2 were estimated by Dunger (1958).

D. Isoptera

Little data are available on termite populations, and there is no exact estimate of the termite population for a particular habitat. Most estimates of the abundance of termites have been made on the basis of the numbers of termite mounds, which are easily counted. Since the numbers of individuals in a colony may range from a few hundred to several million individuals (Lee and Wood, 1971), this is a relatively poor index of abundance. Woodlands tend to have larger numbers of small mounds, rather than fewer large mounds.

There is even less information on the numbers of those species of subterranean termites that do not construct obvious mounds yet are often very abundant. A few of the population estimates that have been made for these species are given in Table V. Some of these estimates are of dubious value, but they do show that in tropical forests termites constitute an appreciable proportion of the soil fauna, and must make a considerable contribution to litter breakdown.

E. Diptera and Coleoptera

The species represented in these two families vary so much in form and habits that overall population estimates are relatively meaningless. Bornebusch's survey of Danish woodlands indicated populations of Diptera ranging from 232 to 1076 per m² and of Coleoptera from 44 to 426 per m². In view of his methods of extraction these can be considered as under-estimates and it would not be unreasonable to expect insect populations of the order of 1000 per m². Nef (1957) estimated populations of saprophagous insects as 400–800 per m², which agrees with this figure.

TABLE V. Abundance of subterranean termites in various habitats

Habitat	Numbers per m²	Percentage of total macro-arthropods	References
Rainforests (Panama)	12	0·5	Williams, 1941
Savanna woodland	50	3·0	Lee and Wood, 1971
Semi-arid woodland	69	11·0	Lee and Wood, 1971
Forest (Congo)	500	—	Maldague, 1964
Forests (Trinidad)	2790	34·0	Strickland, 1945
Rainforests (Trinidad)	4450	43·0	Strickland, 1944

Adapted from Lee and Wood (1971).

IV. The Role of Macroarthropods in Litter Decomposition

When data are available for an ecosystem on the energetics and biomass of these groups of animals, it is possible to construct energy budgets and to estimate how much of the total litter fall the animal is consuming and how much it is assimilating. Edwards *et al.* (1970) calculated that large decomposers (macroarthropods, molluscs and earthworms) represented 27–82% of the total invertebrate biomass in soil and contributed 3–33% of the total respiration.

The total litter fall in a stable temperate deciduous forest is about 3·0 t ha⁻¹ (ray and Gorham, 1964), which is equivalent to 3000 g m⁻². It has been calculated that, for many tree species, earthworms can consume most of this if there is no competition from other animals. If we look at the groups of macroarthropods in turn we can make comparative estimates of their relative contributions to litter breakdown.

A. Isopoda

Woodlice consume a lot more food than they assimilate, although the amount of material assimilated is very variable, ranging from 10 to 70% of the food consumed. It seems probable that much of the variability can be explained on the basis of abundance of food. When food is plentiful it passes through the animal much faster, and much less is assimilated, whereas when there is a shortage it is retained in the gut much longer. When food is abundant, woodlice usually consume much larger amounts (Hubbell *et al.*, 1965). Much of the energy assimilated is used up in metabolism. For instance, *Ligidium japonicum* has been estimated to use more than 80% of the energy it assimilates in this way (Saito, 1965). Saito (1970) reported that the assimilation efficiency of *Armadillidium vulgare* was 0·48, and Phillipson (1966) that the equivalent figure for *Oniscus asellus* was 0·41. Using a figure of 0·45 for *Ligidium japonicum*, Saito calculated that the food consumption of the population in a Japanese evergreen broad-leaf forest was 5·8 kcal m^{-2} year, and such a forest may produce more than 500 kcal m^{-2} year of energy in litter fall.

The total biomass of woodlice in litter is quite small compared with that of some of the other macroarthropods. For instance, Bornebusch (1930) calculated it to be 0–0·28 g m^{-2} in 10 Danish woodlands. With a figure for the average weight per individual of 0·7 mg (Edwards, 1967), and if populations of about 1000 per m^2 are considered to be about average for woodlands, the biomass of woodlice might be expected to be about 7 g m^{-2}; this may be an underestimate, however, because the figure of 0·7 mg was based on small species of woodlice. Hubbell *et al.* (1965) calculated that woodlice eat up to 3% of their body weight per day, so 7 g m^{-2} of woodlice would consume 0·2 g of litter per day or 73 g per year. This is only a very small proportion of the total litter fall.

B. Symphyla

No data for assimilation efficiency of symphylids are available, but Edwards (1961) reported that they could eat up to 15 times their own weight per day. Such a large throughput indicates a very poor assimilation efficiency, particularly since symphylids grow very slowly, often hardly changing in weight for several years.

If an average population estimate for woodlands is 3260 per m^2 (Edwards, 1958), and a figure for the average weight per individual is 0·082 mg (Edwards, 1967), the average biomass would be only about 0·25 g m^{-2}. The rate of food consumption varies greatly with age and moulting phase, but an estimate of 0·05 mg per day per individual is not unreasonable. Using these figures, an average population would consume 0·326 mg of litter per day or 60 g m^{-2} year. This is a rather larger consumption than might be expected, but is probably an overestimate, because symphylids feed on materials other than litter, and at any time a considerable part of the population would be in deeper soil. Obviously they are not of major importance in litter breakdown.

C. Diplopoda

There is considerable variability in the estimates of the amounts of food assimilated by millipedes. van der Drift (1951) claimed that *Glomeris marginata* assimilated only about 6% of the food ingested, Gere (1957) reported assimilations of 3·8–12·4%, and Bocock (1963) of 7·5–10·5%. O'Neill (1968) and Shaw (1970) presented data that showed an assimilation efficiency of about 15%. Striganova (1971) reported that millipedes assimilated 39·4% of the *Carpinus* (hornbeam) litter and 37·6% of the *Quercus* litter that they ate. Franz and Leitenberger (1948) claimed that millipedes could assimilate about 50% of the food they ingest. Gere (1957) reported that millipedes converted 1·7–7·4% of the food they consumed into body tissue, but Bocock (1963) stated that *Glomeris marginata* converted only 0·29–0·45% of the food it consumed into body tissue. His feeding tests showed that this species utilized 6·0–10·5% of the dry matter, 43·2% of the crude fat, 28·4% of the holocellulose, 28·7% of the soluble carbohydrates and 0·3–0·4% of the nitrogen in the *Fraxinus* leaf litter that they ate.

There have been various estimates of the amounts of litter consumed by millipedes. Gere (1957) stated that they consumed 0·5–4·0% of their live weight daily. Striganova (1971) estimated that they ate 3–5 mg of food per individual per day, Gere (1957) found that they ate 0·33–4·1 mg per individual per day with an average of about 2 mg, and Bocock (1963) reported daily intakes of 3·75–11·77 mg per individual.

Populations of millipedes in woodlands differ considerably, but 100 individuals per m^2 can be considered a reasonable average, and with an average biomass of 2·8 mg per individual (Edwards, 1967), the total biomass would be about 0·3 g m^{-2}. This would be valid only for British woodlands, because some species from the U.S.A. and the tropics are very large and weigh several grams when fully grown. Shaw (1968) calculated that the biomass of the population of one of these larger species, *Narceus annularis*, in a mixed woodland in New York State was 2·86 g m^{-2}. If the average daily consumption of litter is assumed to be about 4 mg per individual, a population of 100 millipedes per m^2 would consume about 400 mg or 146 g per year. Striganova (1971) made similar calculations and reached the conclusions that millipedes could consume up to 4800 kg ha^{-1} (480 g m^{-2}) per year of *Quercus*, *Acer* or *Carpinus*. Such a turnover rate would deal with the whole of the annual litter fall at the site quoted, and is almost certainly an over-estimate. Bocock (1963) made more modest claims, estimating that *Glomeris* populations would consume 1·7–10% of the annual deciduous litter fall.

Clearly, millipedes consume large amounts of litter and are one of the most important groups involved in litter breakdown in temperate woodlands.

D. Isoptera

Termite populations per unit area are so poorly known that it is difficult to give an accurate figure for the biomass in tropical woodlands. In a *Brachystegia* forest in the Congo, it was estimated that there was a biomass of 11 g m^{-2} (Maldague, 1964), and in a *Eucalyptus* forest in S. Australia a biomass of 6·0 g m^{-2} was estimated (Lee and Wood, 1971). It seems likely that the range in biomass is of the order of 5–50 g m^{-2} (Lee and Wood, 1971).

The average amounts of food consumed by termites is only poorly known. In feeding experiments, *Kalotermes flavicollis* consumed 2–3% of its body weight (*ca.* 0·2 mg) daily and decomposed 61–62% of this (Seifert, 1962). Lee and Wood (1971) reported that *Nasutitermes exitiosus* (Hill) consumed 0·06 mg per individual per day. They calculated that this represented 16·6% of the total fall of wood or 4·9% of the total litter fall, but the litter fall was small in this woodland.

Maldague (1964) estimated that a termite population in the Congo could consume 570 g plant material m^{-2} year, which represented almost half of the total litter fall, but Lee and Wood (1971) believed this to be an overestimate and suggested that 15% of the total litter fall would be a more realistic figure. Nevertheless, this was only a moderate population and a large population may contribute much more to total litter turnover. Clearly, termites have an important role in the metabolism of woodland ecosystems and in breakdown of plant litter in tropical ecosystems.

E. Diptera and Coleoptera

The populations, habits and food of these animals vary so greatly that it is impossible to make a really valid estimate of their contribution to plant litter breakdown. An average population would consist of about 1000 of these animals per m^2. With an average weight of about 1·5 mg per individual this represents a total biomass of 1·5 g m^{-2}, which is probably an underestimate. If each individual consumes about 0·5 mg of litter per day the amount of litter consumed would be 0·5 g m^{-2} day or a total of 180 g m^{-2} year. This estimate involves a considerable potential error as it takes no account of the variation in habits, size, and type of food of these animals.

V. Relative Importance of Macroarthropod Groups in Litter Breakdown

An estimate of the relative importance of macroarthropods must be extremely approximate in view of variations in numbers and lack of knowledge of the ecology and physiology of these animals (see Käärik, Chapter 5).

Some of the estimates of turnover made in section III are summarized in Table VI.

This summary shows that these animals may consume only about one third of the total litter fall; other groups such as earthworms, bacteria and fungi must therefore account for the rest of the turnover. These data do not account for consumption of faeces by other animals, a common habit of many litter-inhabiting arthropods. We need much more information on food chains, and interrelationships between soil organisms, before we can understand the overall contribution of macroarthropods to litter breakdown.

TABLE VI. Approximate consumption of litter by macroarthropods (g m^{-2})

Average deciduous litter fall	3000
Isopoda	73
Symphyla	60
Diplopoda	146
Isoptera	570[a]
Diptera and Coleoptera	180

[a] Maldague (1964).

References

ADAMSON, A. M. (1943). *Trop. Agric., Trin.* **20,** 107–112.

D'AGUILAR, J. and BESSARD, A. (1963). *In* "Soil Organisms." (J. Doeksen and J. van der Drift, eds), pp. 103–108. North-Holland, Amsterdam.

BLOWER, J. G. (1956). "*6th Congr. Sci. du Sol. Paris,*" 169–175.

BOCOCK, K. L. (1963). *In* "Soil Organisms." (J. Doeksen and J. van der Drift, eds), pp. 85–91. North-Holland, Amsterdam.

BORNEBUSCH, C. H. (1930). *Forstl. Forsogsvaesens Meddel. Bereta, Copenhagen* **96,** 1–225.

BRAY, J. R. and GORHAM, E. (1964). *Adv. ecol. Res.* **2,** 101–157.

CLOUDSLEY-THOMPSON, J. L. (1951). *Proc. Zool. Soc. Lond.* **121,** 253–277.

DRIFT, J. VAN DER (1951). *In* "Experimental Pedology." (E. G. Hallsworth and D. V. Crawford, eds), pp. 227–235.

DUNGER, W. (1958). *Zool. Jb. (Syst.)* **86,** 139–180.

DUNGER, M. (1962). *Proc. 11th Int. Congr. Ent.* Wien 1960 **3,** 169.

EATON, T. H. Jr. (1943). *Am. Midl. Nat.* **29,** 713–723.

EDWARDS, C. A. (1958). *Ent. exp. appl.* **1,** 308–319.

EDWARDS, C. A. (1961). *Ent. exp. appl.* **4,** 239–256.

EDWARDS, C. A. (1967). *In* "Progress in Soil Biology." (O. Graff and J. E. Satchell, eds), pp. 585–594. North-Holland, Amsterdam.

EDWARDS, C. A. and HEATH, G. W. (1963). *In* "Soil Organisms." (J. Doeksen and J. van der Drift, eds), pp. 76–84. North-Holland, Amsterdam. (1974). *Pedobiologia*, in press.

EDWARDS, C. A., REICHLE, D. E. and CROSSLEY, D. A. (1970). *In* "Ecological Studies, Analysis and Synthesis." (D. E. Reichle, ed.), Vol. 1, pp. 147–172. Springer-Verlag, Berlin.
FRANZ, H. and LEITENBERGER, L. (1948). *Öst. zool. Z.* **1**, 498–518.
GERE, G. (1957). *Acta Biol.* (Budapest) **7**, 257–271.
HARRIS, W. V. (1955). *In* "Soil Zoology." (D. K. McE. Kevan, ed.), pp. 62–72. Butterworths, London.
HEATH, G. W. and KING, H. G. C. (1964). *Proc. 8th Int. Cong. Soil Sci.* Bucharest, 779–986.
HUNGATE, R. E. (1943). *Ann. ent. Soc. Am.* **36**, 730–739.
HUBBELL, S. P., SIKOVA, A. and PARIS, O. H. (1965). *Health Physics* **11**, 1485–1501.
JACOT, A. P. (1940). *Q. Rev. Biol.* **15**, 28–58.
KALSHOVEN, L. G. E. (1941). *Tectona* **34**, 568–582.
KARPACHEVSKY, L. O., PEREL, T. S. and BARTSEVICH, V. V. (1968). *Pedobiologia* **8**, 146–149.
KEVAN, D. K. McE. (1962). "Soil Animals." H. F. and G. Witherby, London.
KING, H. G. C. and HEATH, G. W. (1967). *Pedobiologia* **7**, 192–197.
KUBIENA, W. L. (1955). *In* "Soil Zoology." (D. K. McE. Kevan, ed.), pp. 73–82. Butterworths, London.
KÜHNELT, W. (1961). "Soil Biology." Faber and Faber, London.
KURCHEVA, G. F. (1960). *Pochvovedenie* **4**, 16–23.
LEE, K. E. and WOOD, T. G. (1971). "Termites and Soils." Academic Press, London and New York.
LYFORD, W. H. Jr. (1943). *Ecology* **24**, 252–261.
MALDAGUE, M. E. (1964). *Proc. 8th Int. Cong. Soil Sci.* Bucharest, Vol. 3, 743–755.
MARCUZZI, G. (1970). *Pedobiologia* **10**, 401–406.
NEF, L. (1957). *Agricultura* (Louvain) **5**, 245–366.
NIELSEN, C. O. (1962). *Oikos* **13**, 200–215.
O'NEILL, R. V. (1968). *Ecology* **49**, 803–809.
PEREL, T. S., KARPACHEVSKY, L. O. and YEGOROVA, E. V. (1971). *Pedobiologia* **11**, 66–70.
PHILLIPSON, J. (1966). *Proc. Symp. on Terrestrial Secondary Production*, Warsaw, 459–473.
SAITO, S. (1965). *Jap. J. Ecol.* **15**, 47–55.
SAITO, S. (1970). *In* "Methods of Study in Soil Ecology." (J. Phillipson, ed.), pp. 215–223. UNESCO, Paris.
SATCHELL, J. E. and LOWE, D. G. (1967). *In* "Progress in Soil Biology." (O. Graff and J. E. Satchell, eds), pp. 102–120. North-Holland, Amsterdam.
SEIFERT, K. (1962). *Holzforschung* **19**, 105–111.
SHAW, G. G. (1968). *Ecology* **49**, 1163–1166.
SHAW, G. G. (1970). *Pedobiologia* **10**, 389–400.
STRICKLAND, A. H. (1944). *Trop. Agric., Trin.* **21**, 107–114.
STRICKLAND, A. H. (1945). *J. Anim. Ecol.* **14**, 1–11.

Striganova, B. R. (1971). *In* "Organismes du Sol et Production Primaire 4th Colloquium Pedobiologiae Dijon." Institut National de la Recherche Agronomique, Paris.

Sutton, S. L. (1972). "Woodlice." Ginn and Company, Limited, London.

Szabo, J., Bartfay, T. and Marton, M. (1966). *In* "Progress in Soil Biology." (O. Graff and J. E. Satchell, eds), pp. 475–489. North-Holland, Amsterdam.

Thiele, H. U. (1956). *Z. angew. Ent.* **39,** 316–367.

Williams, E. C. (1941). *Bull. Chicago Acad. Sci.* **6,** 63–124.

Zachariae, G. (1965). *Forst. wiss Forschungen Hamburg* **20,** 1–68.

17

Mollusca

C. F. Mason

School of Biological Sciences
University of East Anglia
Norwich
England

I. Introduction

The phylum Mollusca is second only to the Arthropoda in terms of numbers of species, with some 80,000 known. The phylum is structurally more diverse than the arthropods, ranging from chitons and limpets which cling to rocks and feed on algae, through burrowing bivalves, which feed on suspensions and deposits, to the highly active and carnivorous cephalopods. The majority of species are marine, where diversity of form is also greatest.

A large number also occur in freshwater habitats and many gastropods (but no other class) are found on land, probably originally colonizing via fresh water.

Most molluscs are generalized feeders, taking a wide variety of living and dead food, and hence do not fit simply into a functional classification based on trophic levels. For this reason the present review will consider those molluscs which feed on plant material, whether living or dead. Carnivores will be ignored. Concepts formulated from herbivorous molluscs will probably be largely applicable to litter-feeding molluscs, the literature on which is very slim indeed.

The effect of molluscs on the vegetation of a community will be a function of the size and biomass of the molluscan populations, their diets, their consumption rates and their assimilation efficiencies.

II. Population, Biomass and Production

A. Marine and Estuarine

The majority of studies of populations of molluscs in marine and estuarine habitats have been conducted in the intertidal zone, where the influence of tides, currents, substrate, etc., cause a marked zonation. Zonation and its causes have been amply reviewed by Lewis (1964), Southward (1965) and Newell (1970). The present section will deal largely with the ranges of populations and biomasses to be expected in marine and estuarine habitats, with some examples of biotic interactions.

There have been a number of studies of estuarine molluscan faunas (Spooner and Moore, 1940; Anderson, 1972; Bloom *et al.*, 1972). In a study of Morecambe Bay, England, Anderson (1972) observed populations of 2000–4000 m^{-2} of the bivalve *Macoma balthica* (L.), with up to 56,000 m^{-2} if spat was included. *Cerastoderma edule* (L.) had a lower population of 250 m^{-2}, while there were 8525 m^{-2} of the prosobranch *Hydrobia ulvae* (Pennant). Very high populations of *Hydrobia* are usual, for example, densities of 5000–9000 m^{-2} in the Clyde estuary (Hunter and Hunter, 1962), 15,000–46,300 m^{-2} at Skallig, Denmark (Thamdrup, 1935) and an average of 50,000 m^{-2} (but up to 100,000 m^{-2}) in the Ythan estuary (Anderson, 1971).

In a saltmarsh habitat an average standing crop of 700 *Littorina irrorata* per m^2 had an annual production of 29 kcal m^{-2} (Odum and Smalley, 1959). The population of the mussel *Modiolus demissus* Dillwyn in a Georgia saltmarsh averaged 6·66 m^{-2} (Kuenzler, 1961*a*), with a mean biomass of 4·11 g m^{-2} and an annual production of 16·7 kcal m^{-2} year.

Nixon *et al.* (1971) studied the populations and biomasses of animals in a mussel bed. The dominant mussel was *Mytilus edulis* L., with a population of 2139 m^{-2} (biomass 10,962 g dry wt m^{-2}). The other bivalves present were 507 *Mya arenaria* L. m^{-2} (1275 g m^{-2}) and 10 *Mercenaria mercenaria* L. m^{-2} (257 g m^{-2}). The commonest gastropod was *Littorina littorea* L., with a population of 2805 m^{-2} (1006 g dry wt m^{-2}), and also present were 736 *Crepidula fornicata* L., m^{-2} (329 g m^{-2}), 21 *Nassarius trivittatus* (Say.) m^2 (9 g m^{-2}) and 5 *Eupleura caudata* (Say.) m^2 (0·5 g m^{-2}). The total dry weight of molluscs was 13,821 g m^{-2} out of a total animal biomass of 14,179 g m^{-2}. Most of the molluscan biomass was however shell, the total weight of meat being only 1712 g m^{-2}.

In an estuarine mud-flat in Wales, Hughes (1970*a*, *b*) found densities of 0–140 *Scrobicularia plana* (da Costa) m^{-2} along a series of transects. The spat which settled in 1966 had a 50% mortality, whilst mortality amongst large size-groups was 10–34%. The standing crop, measured monthly, varied from 163–286·5 kcal m^{-2} in the lower shore and 12·3–29·6 kcal m^{-2} on the upper shore. At both sites the population was lowest in March and highest in July. The average production was 70·8 kcal m^{-2} $year^{-1}$.

On rocky shores Hughes (1971*a*) recorded a population of 45·7 m^{-2} of the key-hole limpet *Fissurella barbadensis* Gmelin on intertidal bedrock on Barbados, with a production of 50·9 kcal m^{-2} $year^{-1}$, and densities ranging from 2·4–40 m^{-2} of five species of *Nerita* in Aldabra Atoll (Hughes, 1971*b*). Jones (1948) recorded the highest densities of *Patella vulgata* L. (240 m^{-2}) on rocks which also contained barnacles, and the mean size of the limpets in the population varied inversely with population density. Rao and Ganapata (1971) found densities of 15–150 m^{-2} of the tropical limpet *Cellana radiata* (Born), being most common on exposed surfaces.

Very high densities can however be reached on rocky shores. For instance, Hadfield (1970) records up to 100,000 m^{-2} of the vermetid prosobranch *Petaloconchus montereyensis* in California and Hadfield *et al.* (1972) noted densities as high as 60,000 m^{-2} (in Hawaii) of *Dendropoma gregaria* Hadfield and Kay, these occurring on water-levelled beaches.

Low densities of molluscs seem to occur on coral reefs. Of ten species occurring on the Great Barrier Reef, densities ranged from 0·01 to 16 per 100 m^2 (Frank, 1969). The biomass of the abyssal benthos also seems low. For instance Filatova (1959) recorded biomasses of bivalves not exceeding 2–5 g m^{-2}, and usually less, in the Pacific. Deposit feeders predominated.

A rock dwelling population of *Tegula funebralis* (Adams) had an average density of 800 m^{-2} in the upper intertidal and 412 m^{-2} lower down (Paine, 1971). These are similar to the figures of Glynn (1965) and Wara and Wright (1964). The absence of small *Tegula* in the lower intertidal suggested that this subpopulation was maintained by immigration. The biomass was 36·8 g m^{-2} (upper shore), and 86·9 g m^{-2} (lower shore). Thus, although the population in the lower intertidal was smaller, it consisted of larger individuals. The composite production for the two areas was 178 kcal m^{-2} $year^{-1}$.

Sutherland (1970) studied the dynamics of high shore and low shore populations of the limpet *Acmaea scabra* (Gould) in California. The populations were considered separate because the homing behaviour of the limpets precluded migration between the levels. Recruitment was lower in the high zone and the densities were lower, but the annual growth rate was higher, so that the limpets were larger and the biomass was greater than the lower zone. The interaction of tides and climate in the high zone resulted in a seasonal pattern of food availability and a seasonal pattern of growth and reproduction in *A. scabra*, whereas in the low zone growth and reproduction was less variable. The distribution of *Acmaea* was contagious in the high zone and random in the low zone.

The work of Paine and Sutherland shows that populations of molluscs that are spatially very close in the intertidal can differ very markedly in their densities, age and size structures and pattern.

McIntyre (1970) has studied populations of the bivalve *Tellina tenuis* (da Costa) in detail on the west coast of Scotland. In one bay in Loch Ewe, the population in early spring 1965 had a mean density of 768 m^{-2} By the following spring an absence of significant recruitment, together with mortality, had caused an overall reduction in the population of 41%. Failure in recruitment continued and the population had fallen to 192 m^{-2} by 1968. The populations of seven beaches in north-west Scotland were then examined and the age analysis is shown in Fig. 1. Quite local differences in recruitment were occurring, so that it seemed unlikely that unusually favourable or unfavourable environmental conditions were the main factors involved. Beaches in south-west Scotland showed a more regular pattern of annual recruitment.

Three phases must be completed for successful recruitment into a bivalve population, viz. spawning, the planktonic phase and the settlement phase (McIntyre, 1970). With reference to *Tellina tenuis* Trevallion *et al.* (1970) suggested that siphon regeneration occurred at the expense of gonad development after the grazing of siphons by young flat-fishes. This predation ceased after the *Tellina* population dropped below a critical level. Thus, the *Tellina* population may have been controlled locally by the inhibition of spawning due to high populations of flat-fishes.

In the planktonic phase (discussed by Thorson, 1966) the larvae must especially face the effects of dispersal by adverse currents, and on isolated pocket beaches on the west coast of Scotland the probability of successful recruitment by *Tellina* spat must be very low indeed. Even when the spawning and planktonic phases are successful and spat are available it

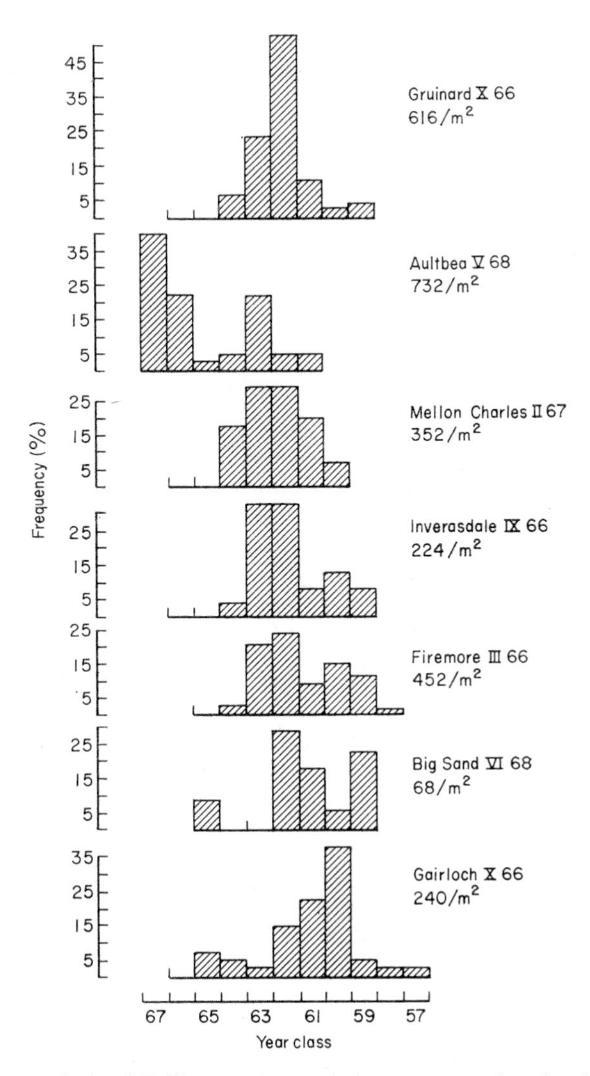

Fig. 1. Age analysis of *Tellina tenuis* populations on seven beaches in north-west Scotland. Numbers below each name show density per m^2. (Reproduced, with permission, from McIntyre, 1970)

seems that the rigorous physical conditions in exposed areas may either inhibit settlement or survival of *Tellina* spat on the beach (McIntyre, 1970).

Segerstråle (1965) also gives evidence of the biotic control of a bivalve population. In the Baltic Sea *Macoma balthica* suffered total failure of recruitment over several successive years below a depth of 20 m. During these periods there were especially large numbers of the amphipod *Pontoporeia affinis*, which was presumed to eat the spat of *Macoma* along with the bottom sediment, its major food.

On the Dogger Bank densities of *Mactra stultorum* (L.) may reach 200,000 m^{-2} at spatfall (Birkett, 1970). During the first summer over 70% of *Mactra* is taken by predators, mainly by fish. As the shells thicken the importance of predatory fish declines. During the second summer about 14% of the remaining *Mactra* are taken by predators, this time mainly by asteroids and the prosobranch *Natica*.

Estuaries carry large populations of wading birds and these feed extensively on molluscs. Drinnan (1957) calculated that the predation of *Cerastroderma* (*Cardium*) *edule* by oystercatchers, *Himantopus ostralegus* L., in Morecambe Bay amounted to 21·9% of the total population in the winter of 1954–55. The total cockle mortality was 73·8%, and flat-fishes may have been responsible for a proportion of the rest.

Also in Morecambe Bay it was estimated that knots, *Calidris canutus* L., whose diet consisted of 80% by volume of *Macoma balthica*, removed only between 0·9 and 3·6% of the *Macoma* population per annum, despite the presence of a wintering population of some 70,000 birds (Prater, 1972) Again Hughes (1970*a*) records a mortality of 5–6% of *Scrobicularia plana* due to oystercatchers in his study area.

The mortality of bivalves due to predators can in some cases be very significant, though no generalizations can be made. More work on the cumulative effect of complementary predators is needed.

B. Freshwater

A number of factors influence the abundance of molluscs in freshwaters (Soska, 1968; Combrinck and Eeden, 1969; Foin and Stiven, 1970; Williams, 1970; Gale, 1971; Harman, 1972). Calcium is the most important chemical factor and waters in Britain with a calcium concentration of less than 3 mg l^{-1} support only two species of snail and five species of bivalve, whereas waters of 8–10 mg Ca l^{-1} may have up to 17 species of molluscs (Boycott, 1936; Hunter, 1957, 1964). In Africa the highest population densities of snails occurred in waters of medium hardness, from 5 to 40 mg l^{-1} of calcium (Williams, 1970). The planorbid *Promenetus exacuous* (Say) inhabited areas where the sulphate ion rarely exceeded 100 ppm (Pip and Paulishyn, 1971).

Soska (1968) found in a brackish lake that molluscs were scarcest in areas with a sandy bottom, with no vegetation and much wave action. They were most abundant on a muddy bottom with abundant vegetation and little wave action. A sandy substrate was suggested as a reason for the low densities of molluscs in a Welsh river (Learner *et al.*, 1971). Substrate preference tests showed that the bivalve *Sphaerium transversum* (Say) preferred a mud substrate to a sand or sandy mud (Gale, 1971). Korinkova (1971) stressed the importance of submerged vegetation to molluscs, though Cvancara (1972) found lower populations of mussels where aquatic plant cover was greater. Population densities tended to decrease with depth (Cvancara, 1972) and to increase with organic enrichment (Hynes, 1960; Learner *et al.*, 1971), though Morgan (1970) has shown that the number of species of gastropods in Loch Leven has declined from 12 to 7 during a period of nutrient enrichment and macrophyte loss.

Densities of molluscs can often be high in freshwaters. *Potamopyrgus jenkinsi* (Smith), which is often associated with enriched and unstable conditions, has been recorded in densities up to 74,000 m^{-2} by Heywood and Edwards (1962). Figures for other localities range from 1700 to 30,000 m^{-2} (Whitehead, 1935; Adam, 1942; Macan, 1950; Crisp and Gledhill, 1970). It was noted earlier that a close relative of *P. jenkinsi*, *Hydrobia ulvae*, reached very high densities in estuaries.

Freshwater limpets regularly have high densities in flowing waters, for instance an average of 2000 *Ancylus fluviatile* (Müll.) per m^2 (Maitland, 1965), and 1276 *Ferrissia rivularis* (Say) per m^2 (Burky, 1971). Three species of the bivalve *Pisidium* had a mean density of 1480 m^{-2} (Ladle and Baron, 1969), while in the upper reaches of an estuary the commonest of 13 species of *Pisidium* was *P. subtruncatum* Malm. with a maximum density of 777 m^{-2} (Kuiper and Wolff, 1970). The average standing crop of mussels (four species) in the River Thames was 2922 kg ha^{-1}, with an average body tissue production of 205·1 kg ha^{-1} $year^{-1}$ (Negus, 1966). Mussels accounted for 90% of the energy content of the bottom fauna of the River Thames (Mann, 1964).

Gillespie (1969) studied the populations of four species of molluscs in Madison River over two seasons. The peak populations in numbers per m^2 (biomass in g ash-free dry wt m^{-2} in parentheses) were 2460 (1373) for *Physa gyrina* Say, 15,378 (1342) for *Gyraulus deflectus* (Say), 7686 (2671) for *Valvata humeralis* Say and 6040 (1270) for *Pisidium compressum* Prime. The total net annual production was 5·2 g m^{-2}. The peak populations for three species occurred in late summer (August–October), *Valvata humeralis* had peaks in spring and late summer. The peak in biomass did not necessarily occur at the same time as the population peak, which consisted of large numbers of newly hatched young. Crisp and Gledhill (1970) also found late summer peaks in populations of snails in southern England.

There is less population data available for lakes and ponds. Clampitt (1970) had peaks for *Physa integra* and *P. gyrina* of 3340 m^{-2} and 740 m^{-2} respectively. Within a shallow lake densities of *Valvata piscinalis* (Müller) reached 1600 m^{-2}, *By thinia tentaculata* (L.) 111 m^{-2}, *Planorbis albus* (Müll.) 259 m^{-2} and Sphaeridae 2200 m^{-2} (Mason, unpublished). Densities of molluscs in a reed bed reached 7087 m^{-2} (Dvorak, 1970), the biomass being 9·98 g m^{-2}. The standing crop of *Anodonta anatina* (L.) in Lake Borrevann, Norway, was 2593 kg ha^{-1}, representing 96% of the bottom fauna (Okland, 1963). Three populations of *Viviparus malleatus* (Reeve) had a mean biomass of 236·3 g m^{-2} and an annual production of 89·2 g m^{-2} $year^{-1}$ (Stanczykowska *et al.*, 1971).

The majority of freshwater snails have an annual life cycle or one lasting less than a year (Boycott, 1936; De Wit, 1955; Duncan, 1959; Hunter, 1961; Maitland, 1965; Ladle and Baron, 1969; Jobin, 1970; Burky, 1971; Heppleston, 1972), though Berrie (1965) found a biennial life-cycle in *Limnaea stagnalis* (L.) with some individuals breeding in both years. Variations in breeding of snails are dependent on both environmental conditions, especially temperature, and on endogenous factors.

C. Terrestrial

Molluscs have a skin which is permeable to water and hence the maintainance of internal water in an atmosphere of variable humidity is a prerequisite for terrestrial life. Few groups of molluscs are successful in such conditions, namely the order Pulmonata and some largely tropical prosobranch genera (Hunter, 1955). The water relations of snails were reviewed by Hunter (1964). Many terrestrial molluscs are cryptozoic or tend to be active at night or in dewfall or rainfall. Some species however, notably in the genus *Helicella*, are xerophilic and they have regular periods of aestivation when they cluster together in large numbers on vegetation (Bigot, 1967; Pomeroy, 1968). Pomeroy showed that by climbing off the ground to aestivate *Helicella virgata* (da Costa) avoids the extremes of temperature and was less often wakened by dew-falls and light rains. In the Negev Desert of Israel *Trochoidea seetzeni* Pfeiffer aestivates by attaching to shrubs, while *Sphincterochila boissieri* Charpentier buries itself (Yom-Tov, 1971).

The environmental factors controlling the activity of slugs have been investigated by a number of workers (Barnes and Weil, 1944, 1945; Dainton, 1954; Kucera, 1954; White, 1959; Webley, 1962, 1964; Crawford-Sidebotham, 1972). Crawford-Sidebotham (1972) studied the activity of slugs in relation to temperature, relative humidity, vapour pressure and vapour pressure deficit. The activity of four species increased with increase in temperature and decreased with increasing vapour pressure deficit. Relative humidity and vapour pressure were not important.

The distribution of shelled gastropods is often limited by calcium. In

Britain 9 species out of 95 are obligate calcicoles while 23 prefer calcareous localities and only one species is a calcifuge (Boycott, 1934). Both species diversity and abundance will increase with an increase in soil and litter pH and calcium ion concentration (Atkins and Lebour, 1923; Agocsy, 1968; Valovirta, 1968; Wäreborn, 1969, 1970). The addition of calcium citrate increased the percentage of *Cochlicopa lubrica* and *Discus rotundatus* hatching in laboratory experiments (Wäreborn, 1970). Feeding behaviour may allow snails to exist in areas of low calcium; for instance *Achatina fulica* Bowdich was common in tea plantations in Ceylon (pH 4·5), where it fed on the fallen leaves of *Camillia sinensis* (tea plant) whose ash consisted of 40% calcium (Mead, 1961), and the calciphilic *Cepaea hortensis* (Müller) on an acidic island in Nova Scotia obtained calcium from sea-shells which had been dropped and broken by gulls (Bleakney, 1966).

There are relatively few studies on populations of land snails and the situation is aggravated by the difficulties involved in sampling (reviewed in Mason, 1970*b*; Newell, 1971). Differences in sampling technique make comparisons difficult. Mason (1970*b*) sampled the snails in the litter of a *Fagus sylvatica* wood over 13 months; the mean number per m^2 and the mean biomass of the 21 species recorded are shown in Table I. The average population was 489 m^{-2}, the most numerous species being *Carychium tridentatum* (Risso). On a less exposed beech wood site a density of 1085 snails m^{-2} was recorded (Mason, 1970*b*). Lower densities were recorded by Dzieczkowski (1971) for a Polish *Fagus* woodland. Okland (1929) found a density of 250 snails m^{-2} in Norwegian woodlands and Mörzer-Bruijns *et al.* (1959) had averages of 100–400 snails m^{-2} in Dutch *Quercus* and *Fagus* woodlands. In Hungarian oakwoods Agocsy (1968) had mean densities of snails of 320 m^{-2} on limestone and 28 m^{-2} on sandstone substrates. Valovirta (1968) found densities of 123–225 snails l^{-1} of litter in Finnish woodlands. On exposed limestone cliffs in Malaya, Berry (1966) recorded 3197 snails m^{-2} on mossy rocks and 1597 m^{-2} on moss-free rocks.

Population densities of single species have been determined by Foster (1937), Strandine (1941), Goodhart (1962), Blinn (1963), Owen (1965), Baker (1968), Grime and Blythe (1969), Pomeroy (1969), Wolda (1969) and Yom-Tov (1972). In relatively open grassland and shrubland sites, densities generally ranged from 1–20 snails m^{-2} but Baker (1968) recorded up to 232 *Helicella caperata* Mont. m^{-2} in English sand-dunes and Pomeroy (1969) found a mean density of 158 m^{-2} *H. virgata* in south Australian shrublands.

Slugs are usually considered separately from snails because different sampling and extraction techniques are required (reviewed by South, 1964; Hunter, 1968; Newell, 1971). Estimates of slug densities in agricultural land have been given by Gimmingham and Newton (1937), McC. Calan

(1940), Thomas (1944, 1948) and Newell (1967). They ranged from 3 to 150 m^{-2}, the latter being in a *Triticum* crop badly damaged by slugs (Thomas, 1944). An average figure is about 50 slugs m^{-2} in farmlands. Drift (1951) recorded an average density of 14 *Arion subrufus* m^{-2} in *Fagus* woodland.

There is rather less information on the biomasses and production of terrestrial molluscs. The shell, consisting largely of metabolically inactive mineral material, accounts for much of the biomass, as much as 80% in *Carychium tridentatum* (Mason, 1970*b*). The population biomasses of twenty-one species of snails from a *Fagus* woodland (Mason, 1970*b*) are given in Table I. The mean total of all species was 699 mg dry weight m^{-2} (278 mg ash-free dry wt m^{-2}). The total production was 4·4 kcal m^{-2} year (Mason, 1971). The biomasses of snails in Danish forests (4–146 mg live wt) determined by Bornebusch (1930) are considered very low, probably because of inadequate sampling techniques (Birch and Clark, 1953). Bigot (1965), in an invertebrate study of the sansouire of the Rhone delta, determined the biomass of molluscs (chiefly *Leucochroa candidissima* Drap. and *Euparypha pisana* Müll.) in three vegetation types in this open and saline habitat. The peak biomass of about 8·2 g m^{-2} was in the Salicornietum fruticosae, but the populations varied greatly, with very few snails in the cold winters. Foster (1937) and Strandine (1941) respectively recorded 15·8 g *Polygyra thyroides* (Say) protoplasm m^{-2} from Illinois flood-plains and 329 mg *Succinea ovalis* Say protoplasm m^{-2} from Illinois forests (*ca.* 4 g m^{-2} *Polygyra* and 84 mg m^{-2} *Succinea* dry weight: Mason, 1970*b*). Pomeroy (1969) determined an average biomass of 15–20 g live wt m^{-2} of *Helicella virgata* in south Australia (6·3–8·4 g dry wt m^{-2}, Mason 1970*b*). Yom-Tov (1972) recorded a biomass (dry wt of soft parts) of 0·35 g *Trochoidea seetzeni* m^{-2} on the north facing side and 0·11 g m^{-2} on the south-facing slopes of a wadi. Mason (1970*b*) calculated approximate biomasses of 3·0–7·5 g dry wt m^{-2} for *Cepaea nemoralis* (L.) and *Arianta arbustorum* (L.) in grasslands, using population data from Goodhart (1962) and Grime and Blythe (1969).

Populations of molluscs in woodlands tend to be stable, with little seasonal variations in densities (Mason, 1970*b*). In more open habitats (grasslands, dunelands) they tend to show marked seasonal variations (Bigot, 1965; Baker, 1968; Pomeroy, 1969). The population densities of *Carychium tridentatum* in woodland and *Helicella caperata* in duneland are shown in Figs 2 and 3 to illustrate this point.

From the limited data available it appears that the greatest population densities and species diversity of molluscs occur in woodland habitats, presumably because of greater habitat diversity, but the greatest biomass (of a few large species) occurs in more open habitats such as grasslands.

Most species of land molluscs live for about two years, maturing in 12–15 months. Larger species tend to mature later and live longer than smaller species (Chatfield, 1968; Hunter, 1968).

TABLE I. The population density and biomass of snail species living in beech litter, Oxford (mean of 13 sampling occasions). (Reproduced with permission from Mason, 1970*b*)

Species	No. m^{-2}	Biomass dry wt (mg m^{-2})	Biomass ash-free dry wt (mg m^{-2})
Carychium tridentatum	199·80	65·84	13·31
Cochlicopa lubrica	0·90	3·11	1·15
Columella edentula	3·37	1·88	0·94
Pupilla muscorum	0·67	0·22	0·07
Vallonia pulchella	0·67	0·90	0·25
Acanthinula aculeata	71·49	50·86	22·48
Ena obscura	3·82	52·96	23·45
Marpessa laminata	1·80	9·32	2·70
Clausilia bidentata	0·45	5·50	1·42
Vitrea contracta	39·10	22·06	9·36
Retinella radiatula	0·45	0·86	0·36
Retinella pura	24·71	21·62	10·15
Retinella nitidula	7·86	27·26	12·79
Oxychilus cellarius/alliarius	46·27	160·22	69·55
Euconulus fulvus	1·57	2·02	0·87
Punctum pygmaeum	67·16	19·39	4·17
Discus rotundatus	13·50	89·47	23·32
Vitrina pellucida	1·57	2·14	1·44
Hygromia striolata	8·54	163·66	80·04
Hygromia hispida	0·67	1·39	0·25
Total	488·65	698·91	278·07

Yom Tov (1972) has recently shown that the fecundity of the desert snail *Trochoidea seetzeni* was inversely proportional to the population density, and that the mechanism operated differently on the two opposite slopes of a wadi. Fecundity was higher on the south-facing slope of the wadi, where population was lower and predation by *Eliomys melanus* Thomas (dormouse) was higher (Yom Tov, 1970).

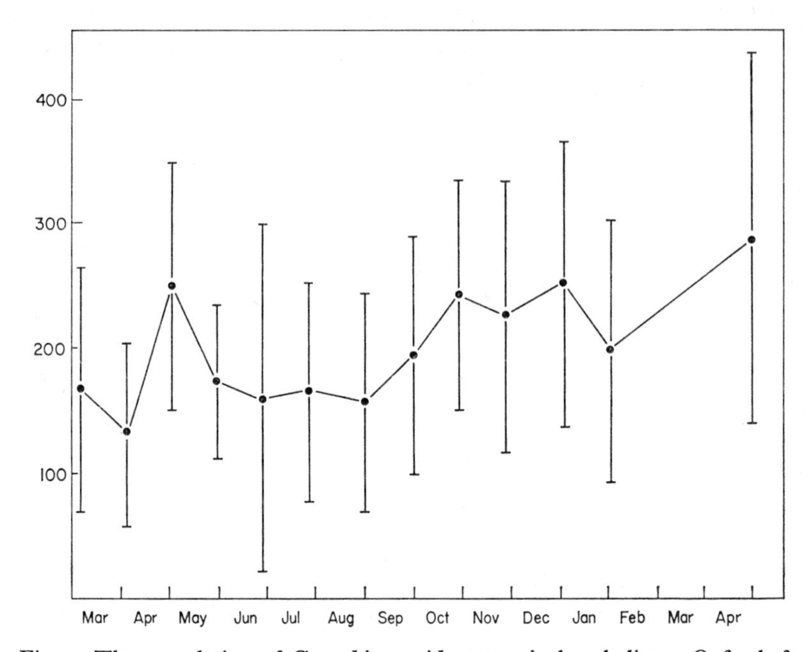

Fig. 2. The population of *Carychium tridentatum* in beech litter, Oxford, from March 1968 to April 1969 (no. per $m^2 \pm 95\%$ confidence limits). (Reproduced, with permission from Mason, 1970*b*.)

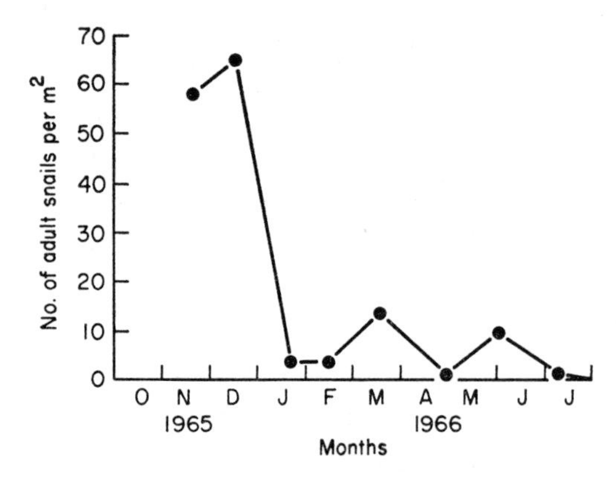

Fig. 3. The population of adult *Helicella caperata* on sand-dunes, Devon, 1965–1966. (Reproduced, with permission, from Baker, 1968)

D. General Conclusions

Population densities and biomasses of molluscs can often be very high, sometimes accounting for the majority of the animal biomass of a community (e.g. Okland, 1963; Nixon *et al.*, 1971). They are hence likely to be a functionally very important component of communities. In many species of mollusc the shell is a large proportion of the body weight and must be taken into account when comparing, for instance, feeding rates with other invertebrate groups.

It is important to note that populations in close proximity may often be very different in size and structure (McIntyre, 1970; Sutherland, 1970; Paine, 1971), and hence their impacts on nearby or contiguous habitats may be very different indeed.

III. Nutrition

Molluscs may be browsers, predators, scavengers, suspension feeders or deposit feeders (Newell, 1970), and a single species may fit into more than one of these categories. The functional anatomy of the feeding mechanisms have been dealt with by Fretter and Graham (1962), Owen (1966), Purchon (1968) and Newell (1970). However, there has been comparatively little work on the actual diets taken. The majority of nutritional studies involve marine molluscs.

A. Marine and Estuarine

1. Browsing and Grazing

The majority of chitons and gastropods in the intertidal zone are browsers or grazers. The precise methods of feeding are dependent on the type of radula and detailed information is given by Fretter and Graham (1962) and Newell (1970). Fretter and Graham (1962) give much general information on diets, which are largely of a herbivorous or microphagous nature.

Tegula funebralis was shown to browse on the diatom film, on attached benthic algae and on such detached macroscopic algae that were available (Best, 1964; Paine, 1971). The overall assimilation efficiency of *Tegula* was calculated as 70% (Paine, 1971). The assimilation efficiency of *Fissurella barbadensis* was calculated at 33·6% (Hughes, 1971), this species ingesting rock particles, sand grains, diatoms, Foraminifera, Radiolaria, small molluscs and crustaceans as well as the algae which comprise its chief food (Ward, 1966). *Littorina irrorata*, a detritus feeder, had an assimilation efficiency calculated at 45% (Odum and Smalley, 1959) and *L. planaxis*, feeding on epiphytic algae, one of 40% (North, 1954).

Southward (1964) examined the feeding of *Patella vulgata*. Starved limpets in an aquarium fed on blue-green filaments, colonial greens and diatoms, with a biomass of 12·3 mg wet weight cm^{-2} aquarium wall. In the first few hours the feeding rate was 86 mg day^{-1} g body weight $limpet^{-1}$, but during the first week it fell to 6 mg day^{-1} and in the second week to to 1·8 mg day^{-1}. Moore (1938) calculated that limpets during their first year of life required an area of 75 cm^2 cm^{-3} limpet volume for maintenance when feeding on a felt of *Enteromorpha*.

Leighton (1966) studied the feeding preferences of six algivorous molluscs. The laminarian *Macrocystis pyrifera* was generally the most preferred food, though *Norrisia norissii* (Sowerby), *Haliotis fulgens* Philippi and *Aplysia californica* Cooper, all living commonly in shallow water, preferred *Egregia laevigata*, a shallow-water kelp. *Cytoseira osmundacea* was rarely ingested, though *Haliotis corrugata* Gray would eat it in moderate quantities. The variation in preferences of closely related species (for example three species of *Haliotus*) was probably due to chemical perception.

The top-shell *Turbo cornutus* (Solander) increased in weight by 134% over a year when fed only the alga *Eisenia bicyclis*, by 73% on *Gelidium amansii* and only by 30% on *Sargassum ringgoldianum* (Ino, 1958).

Using field observations and laboratory experiments, Bakker (1959) showed that the vertical distribution of *Littorina littoralis* (L.) was correlated with the distribution of its preferred foods. Most *Littorina littoralis* were found on *Fucus spiralis* and *F. vesiculosus*. *Pelvetia canaliculata*, *Ascophyllum nodosum* and *Fucus serratus* were less attractive as food and held very few animals. There are a number of similar examples (Graham, 1955).

Aplysia spp. (sea hares) eat the types of algae that are present in their habitats, although they preferred green seaweeds (e.g. *Ulva*, *Enteromorpha*) when given a choice under laboratory conditions (Winkler and Dawson, 1963; Carefoot, 1967*a*, 1970). *Aplysia juliana* Quoy and Gaimard weighing 200 g ate 11–12 g wet weight day^{-1} of the brown alga *Undaria pinnatifida* and has been blamed for the decreased yield of this crop in Japanese waters (Saito and Nakamura, 1961). Carefoot (1967*a*, *b*, 1970) has examined in detail the nutrition of *Aplysia punctata* Cuvier and the tropical species *A. juliana* and *A. dactylomela* Rang. *Aplysia punctata* was fed on eight species of algae in the laboratory (Carefoot, 1967*a*). The maximum growth rate was obtained on *Plocamium coccineum* and this was the food most eaten in the field. *Enteromorpha intestinalis* was eaten in greater quantities than *Plocamium* but faeces production was greater and overall growth was less. *Heterosiphonia plumosa* was almost as abundant as *Plocamium* in the field, but it supported only half the growth rate of *Aplysia*. The highest absorption of food was given on *Ulva lactuca* (75% absorption), *Heterosiphonia* (71%) and *Cryptopleura ramosa* (71%), though these algae resulted in only moderate growth. With *Aplysia juliana* (Carefoot, 1970) the food preferred in both the field and laboratory was *Ulva fasciata* and this gave the highest efficiency of energy absorption (84%) and the highest growth efficiency (33%). *A. dactylomela* on the other hand grew most rapidly on *Enteromorpha*, which was scarce in the natural habitat, and this food gave the highest assimilation efficiency (68%), but not the highest growth efficiency (67%), this being obtained on *Cladophora* (84%), the food most eaten in the field. The percentage assimilation of the three species of *Aplysia* feeding on all offered diets varied from 15% (*A. juliana* on *Cladophora*) to 84% (*A. juliana* on *Ulva*), the overall mean being 57·5%.

Seaweeds differed in their growth-supporting value independently of the amount consumed and there was no noticeable difference in their chemical composition to account for this variation. Higher growth efficiencies occurred on those seaweeds which appeared structurally to be most extensively broken down.

Carefoot's studies are a detailed and important attempt to understand the factors involved in molluscan nutrition and this is an approach which is in urgent need of further research.

It is appropriate here to mention highly specialized gastropods of the order Saccoglossa which feed by sucking out the cell contents of filamentous algae. They are usually highly specialized in their food plants. For instance, *Limapontia depressa* (Alder and Hancock) fed only on *Vaucheria* and *L. capitata* (Müller) and *Acteonia cocksi* (Alder and Hancock) on *Cladophora* (Gascoigne, 1956).

The position of an animal on the sea shore may affect the amount of time available to it for feeding. For instance, Newell *et al.* (1971) observed that the winkle *Littorina littorea* fed mainly when moistened by the tide. Those winkles on the upper shore compensated for the reduced time available for feeding by having an increased rate of radula activity when immersed.

2. *Suspension Feeding*

Suspension feeding occurs in a large number of prosobranchs and their larvae. Details of the methods of ciliary feeding are given in Fretter and Graham (1962), Owen (1966), Jørgensen (1966), Purchon (1968), and Newell (1970).

The feeding of larvae has recently been investigated in some detail (Loosanoff and Davis, 1963; Walne, 1963; Davis and Calabrese, 1964; Fretter and Montgomery, 1968; Calabrese and Davis, 1970; Mapstone, 1970; Pilkington and Fretter, 1970). Fretter and Montgomery (1968) examined the feeding behaviour of nineteen species of monotocardian veligers on nineteen species of nanno- and microplankton. In the sea flagellates, and to a lesser extent diatoms, were the chief food. The cell walls could not be digested by the veligers and some cells with thick walls passed through the gut unharmed (e.g. *Chlamydomonas, Peridinium trochoideum*). In a mixed diet these cells whose walls were difficult to digest were not broken down when cells more easily digested were present. Pilkington and Fretter (1970) found that *Cricosphaera* ap *carterae* and *Exuviella baltica* were the best foods for veligers of *Crepidula fornicata* (L.), whereas *C.* ap *carterae* and *Dunaliella primolecta* were best for *Nassarius reticulatus* (L.), while *E. baltica* was consistently poorer. Growth was moderately good on *Monochrysis lutheri* and *Pyrimimonas grossii*, but metamorphosis did not occur.

Temperature also appears important, for example the larvae of *Crassostrea virginica* (Gmelin) could ingest *Chlorella* cells at 15°C but not digest and assimilate them, though they could utilize naked flagellates. At 30°C they grew well on *Chlorella* alone (Davis and Calabrese, 1964).

Walne (1963) found that the most rapid growth rate of *Ostrea edulis* L. larvae occurred with densities of 50–100 cells μl^{-1} of *Isochrysis galbana* and it seems likely that oyster larvae in the field are living on food densities which are markedly below optimal, an observation also made by Pilkington and Fretter (1970). It is thought that particulate organic matter may also be utilized to a large extent, though there is no evidence of this.

There is rather less work on the nutritional requirements of metamorphosed bivalves. Walne (1970) fed the juveniles of six species on nineteen genera of algae. Of five foods which were found to be good foods for *Ostrea edulis*, two (*Monochrysis lutherii* and *Dicrateria inornata*) were not so good for *Mercenaria mercenaria.* Those algae which proved good for larval growth also appeared good for growth of the juveniles. It is not known why the foods differ but it is not due to gross differences in chemical composition or to variations in amino acid content (note also the work of Carefoot, 1967*a*, mentioned earlier). Micronutrient content seems a likely possibility. The concentration giving best growth approximated 5000 μm^3 algal volume μl^{-1} volume of water (Walne, 1970). Allen (1962), feeding ^{32}P-labelled *Phaeodactylum* to four species of bivalves found that more than 75% of the ^{32}P filtered out of suspension was assimilated, the remainder being excreted in pseudofaeces and faeces.

The filtration rate, and hence the food intake, of bivalves is affected by a number of factors. Over a 24 h cycle, two phases of high food consumption alternated with two phases of low food consumption, during which digestion took place (Winter, 1969). Bivalves are sensitive to food concentration and to temperature (Walne, 1972) and Winter (1969, 1970) showed that with increasing temperature there was increased filtration until a threshold was reached, when an increase in particle density was not followed by an increase in the amount of particles ingested. Species living in waters with high seston contents are adapted to higher food concentrations and are unable to compensate for low concentrations by higher filtration activities. Filtration rates also increased with increased water current (Walne, 1972) and with body size (Winter, 1970).

As with gastropods, intertidal bivalves may show adaptations to the reduced periods of submersion which they experience. *Lasaea rubra* (Montagu) collected from a level corresponding to mean high water spring tides responded more quickly to wetting and had initially higher filtration rates than animals collected from a level corresponding to mean low water of neap tides (Morton *et al.*, 1957).

Shipworms, Teredinidae, are a highly specialized group of bivalves with reduced filtering ability (Perkins, Chapter 22). They bore into wood, storing and utilizing the products, and doing an immense amount of damage. It has been estimated that marine borers are responsible for 50 million dollars' worth of damage in the U.S.A., and the annual loss on the coast of British Columbia is 0·5–1 million dollars, all done by *Bankia setacea* (Tryon) (Canagaratnam, 1962). The shipworm stomach has a strong cellulolytic activity (Greenfield and Lane, 1953). On passage through the gut of *Teredo navalis* L., wood lost 80% of its cellulose and 15–56% of its hemicellulose (Dore and Miller, 1923). All shipworms obtain some food by filter feeding but the ability to do this probably decreases with a decrease in the length of the gill lamellae (Turner, 1966). Thus *Neoteredo reynei* probably feeds largely on wood, whereas *Teredora* feeds largely on plankton. A greatly elongated stomach in some species is also probably related to a diet of wood.

3. Deposit Feeding

A number of the prosobranchs (e.g. *Aporrhais*, *Turritella*, *Hydrobia ulvae*, *Macoma*) are deposit feeders. Some of them, for example *Dentalium*, transport particles to the mouth along ciliary tracts. Others feed by sucking up the deposits with a siphon (*Macoma*, *Tellina*) or utilize the material resuspended from deposits by wave action (*Donax vittatus* (da Costa)). The feeding methods are discussed in more detail by Newell (1970).

Many deposit-feeding bivalves also function as suspension feeders (Brafield and Newell, 1961) and filter feeders often obtain much of their food from bottom deposits (Marshall, 1970). Those *Tellina tenuis* that live below low water mark for most of the year probably filter water from the sand–water interface and the bottom few centimetres of the water column whereas those individuals which live higher up the beach also make use of the shallow surf water and interstitial water (Travallion, 1971). *Tellina* had a filtration rate at 15°C of 1·3–2·4 ml h^{-1} mg animal (Travallion, 1971) and *Scrobicularia plana* had a pumping rate of 0·75 ml h^{-1} mg dry wt at 16·5°C (Hughes, 1969), rates which are very much below those of filter feeding bivalves. *Scrobicularia plana* had a filtration efficiency of almost 100% with particles of the size range 4–40 μm. Copious pseudofaeces were produced which contained much organic matter available for re-assimilation (Hughes, 1969).

A number of workers have recently investigated the role of bacteria in the nutrition of deposit feeders (Newell, 1965; Adams and Angelovic, 1970; Fenchel, 1970). Newell (1965) observed that *Hydrobia ulvae* and *Macoma balthica* had dense populations in fine substrates and rather low populations in coarse substrates. Fine deposits, with greater surface area, have large populations of bacteria and also a higher content of organic nitrogen. Detailed laboratory experiments suggested that the faecal material of *Hydrobia* was subjected to a cycle in which bacteria developed on the surface, to be digested by the snail, the voided faeces then being recolonized and the cycle repeated until no more faecal material remained. *Macoma* probably behaved similarly. According to Fenchel (1970) the amphipod *Parhyalella whelpleyi* Shoem. feeds entirely on bacteria, the ingested detritus not being assimilated.

Some workers (e.g. Zobell and Feltham, 1938, 1942) have kept bivalves for long periods of time on a diet consisting solely of bacteria, though bacteria appeared unsuitable as larval food (Walne, 1958). Adams and Angeloviv (1970) fed sterilized and unsterilized detritus of *Zoster marina* to the gastropod *Bittium varium*. The snails assimilated 49% of the unsterilized and 46% of the sterilized detritus, suggesting that bacteria were not being utilized for food.

It seems unwise at the present state of knowledge to make generalizations about the role of bacteria as food for deposit-feeding molluscs.

4. Dissolved Nitrogen as a Food Source

Recent work has suggested that there is a net uptake of dissolved amino acids by aquatic invertebrates from the environment (Stephens, 1963, 1964, 1967, 1968) and that this may be an important nutritional source. However Johannes *et al.* (1969) found that the net loss of amino acids to the environment exceeded the net gain. The value of dissolved organic material as food is at present open to question.

B. Freshwater

There is much general information on the diets of freshwater molluscs (e.g. Frömming, 1956). The role of vascular plants in the nutrition of freshwater gastropods is still a matter of some controversy. The range of plant foods eaten by freshwater invertebrates was reviewed in detail by Gaevaskaya (1966). *Limnaea stagnalis* was said to feed extensively on higher plants; 40 species are known to be utilized by this animal, including living and dead material. Tsikhon-Lukanina (1958) found that consumption rates were maximum on *Ceratophyllum demersum* and *Lemna minor*, followed by *Stratiotes aloides* and *Hydrocharis morsus-ranae*. It appeared, however, that dead vascular material and algae were most satisfactory as food. Gaevskaya (1966) considered that higher plants were also of great importance as food to *Melanopsis dufouri* (Fér.), *Limnaea columella*, *L.* (*Galba*) *palustris* (Müll.) and *L.* (*Radix*) *pereger* (Kobelt) though moribund vascular material and algae may be preferred. Vascular plants were of lesser importance to nine other species investigated.

The majority of freshwater gastropods are general browsers on algae and detritus, e.g. *Limnaea truncatula* (Müll.) (Kendall, 1953), *Stagnicola reflexa* Say (Bovjerg, 1965), *Goniobasis clavaeformis* (Malone and Nelson, 1969), *Limnaea pereger* (Calow, 1970), *L. ovata* (Drap.) (Pieczynska, 1970), *Ancylus fluviatile* (Calow and Fletcher, 1972) and several other species (Monakov, 1972).

Field observations suggested that *Limnaea pereger* fed largely on epiphytic algae (Calow, 1970) and this was supported in the laboratory, where the snail assimilated filamentous greens better than diatoms. *Elodea canadensis* was not taken.

Heywood and Edwards (1962) calculated that *Potamopyrgus jenkinsi* had an assimilation efficiency of 4% when feeding on mud (particle size $<105\ \mu m$). This was a very low value, but no account was taken of a possible selectivity of food within the mud by the snail. The assimilation rate was about 7 μg organic carbon per snail per day. Calow and Fletcher (1972), using a new technique, where the changes in ratio of an inert radiotracer (^{51}Cr) to an actively assimilated tracer (^{14}C) were used to calculate assimilation efficiency, found that *Ancylus fluviatile* feeding on the diatom *Navicula* had an efficiency of 88% and an ingestion rate of 154·6 μg dry wt day^{-1}.

Monakov (1972) reviewed the Soviet literature on feeding in aquatic invertebrates. Feeding rates of snails decreased with an increase in food concentration. The assimilation efficiency of *Valvata pulchella* Studer. did not exceed 50% and was often lower than this (Tsikhon-Lukanina, 1965; Tsikhon-Lukanina and Sorokin, 1965).

Malone and Nelson (1969) observed that although *Goniobasis clavaeformis* was actively crawling over rocks throughout the day its ingestion rate in the early morning was 5 times that in the afternoon.

Freshwater bivalves appear to be non-selective filter feeders (Gale and

Lowe, 1971); for instance the stomach contents of *Lampsilis siliquoidea* (Barnes) were listed as *Volvox*, *Pleodorina*, *Microcystis*, diatoms, fragments of filamentous algae, *Euglena*, rotifers, cladocerans, various Protozoa, organic detritus and sand-grains (Churchill and Lewis, 1924). The filtration rate of *Sphaerium corneum* (L.) decreased with an increase in the concentration of food (Mitropolsky, 1966), although the mussel *Lampsilis siliquoidea* moved towards areas of high plankton density (Bovjerg, 1957). The assimilation efficiency of *Dreissena polymorpha* (Pallas) was of the order of 43–57% when feeding on bacteria (Sorokin, 1966) and that of *Hyriopsis schlegellii* 45% (Higashi, 1966).

TABLE II. The effect of population density on the biomass and fecundity of *Lymnaea elodes* (after Eisenberg, 1970)

Density (no. per pen)	Mean size (mm)	No. eggs per 10 min search	Mean no. eggs per egg mass
1058	15·24	463	20·1
1347	14·88	568	27·0
1761	14·34	277	18·5
1911	11·54	61	15·2
2068	14·28	261	18·0
2279	13·44	254	18·1
2504	12·45	139	17·4
1162[a]	20·72	14,266	46·3
1431[a]	19·18	12,184	41·2
4019[a]	16·11	10,360	39·5

[a] Spinach added.

Filter feeding also occurs in prosobranchs (Tsikhon-Lukanina, 1961*a*, *b*), the filtration rate again declining with an increase in concentration of the food. The importance of filter feeding to freshwater prosobranchs under field conditions is unknown.

Bacteria may play an important part in the nutrition of some species. Rodina (1948) was able to rear *Bithynia tentaculata* on a diet exclusively of bacteria. Calow and Fletcher (1972) showed *Planorbis contortus* (Linn.) to be exclusively a bacterial feeder, with a high assimilation efficiency of the order of 81–97%

The nutritional quality of the food of freshwater molluscs has received little attention. In the laboratory *Limnaea stagnalis* fed on lettuce laid capsules containing more eggs than snails fed on Standen's food (Steen, 1967). Eisenberg (1966, 1970) has studied the regulation of field populations of *Limnaea elodes*. The manipulation of densities of fully grown snails in pens in a pond showed regulation of density, not of this population but of the next generation of snails (Eisenberg, 1966). In a later study (Eisenberg, 1970) manipulation of the population was carried out well before the attainment of maturity and the effect of density on individual size, population biomass and fecundity was studied. Spinach was added regularly to three pens. The results are shown in Table II. In the unfed pens there was a negative relationship between the density of snails and the mean size of the snails, the total number of eggs produced and the mean number of eggs per capsule. The addition of spinach produced larger snails which weighed more, produced larger egg masses and produced an enormously greater quantity of eggs.

The data suggested that there was a normal maintainance level of nutrition at which growth and reproduction were limited. The addition of spinach added some factor which enabled the snail to realize its potential for growth and reproduction. The biochemical aspects of this interesting experiment were not examined.

C. Terrestrial

Frömming (1954, 1962) has produced much general information on the diets of terrestrial molluscs. They are, broadly speaking, generalized feeders, taking anything that comes their way (Barnes and Weil, 1945; Newell, 1971). However, closer examination reveals subtle differences in their feeding behaviour.

In the Negev Desert of Israel, *Trochoidea seetzeni* ate mainly spermatophytes (including leaves and bark), and lichens, whereas *Sphincterochila boissieri* ate chiefly soil, lichens and soil algae (Yom-Tov and Galum, 1971).

Frömming (1954) gave details of the vascular plants taken in the laboratory by three species of slugs (*Arion fasciatus* (Nilsson), *A. hortensis* Fér. and *A. intermedius*) and showed that in some cases different species of slug ate different parts of the plant, for instance *A. hortensis* took only petals, *A. fasciatus* only young leaves and *A. intermedius* (Normand) stems and leaves of *Campanula rapunculoides*. Some species of slugs specialize in feeding on fungi (Frömming, 1954) and these species tend to dominate in coniferous forests where ground vegetation is sparse (Frömming, 1958).

Pallant (1969) examined the gut contents and faeces of the slug *Agriolimax reticulatus* (Müll.) from a *Quercus* wood in England and showed that fresh green herbaceous leaves predominated in the diet, especially those of *Ranunculus repens* and *Urtica dioica* (see also Lindquist, 1941; Getz, 1959). *R. repens* and *U. dioica* were eaten most readily in food preference experiments. The assimilation efficiency of *A. reticulatus* feeding on *R. repens* was 78·4%, and an average slug (live weight 255 mg) ate 60·9 mg day^{-1} fresh weight of *Ranunculus* leaves (Pallant, 1970). The daily food intake of three species of slugs (*Arion ater* (L.), *Agriolimax reticulatus* and *Milax budapestensis* (Huzay)) feeding on grain was of the order of 11–14·7 mg dry wt (Hunter, 1968*c*). These species fed mainly in the early part of the night, and feeding was maximal at 20°C.

The diet of *Cepaea nemoralis* has been investigated in some detail (Grime *et al.*, 1968, 1970; Grime and Blythe, 1969; Wolda *et al.*, 1971). Grime *et al.* (1968) in a laboratory study of the palatability of 52 species of herbaceous plants, conducted single choice experiments in which snails were given a disc of test material and a similar disc of a reference material (*Hieracium pilosella*), while other experiments were conducted using aqueous extracts of plant material on filter paper, with filter paper moistened with water as the reference. A palatability index was calculated as mm^2 test material consumed per mm^2 reference material consumed.

Of the 52 species tested, 60% appeared highly unpalatable. Half of the remainder were less palatable than *Hieracium pilosella* and half were highly palatable. Grasses were less palatable than dicotyledons and plants from disturbed habitats tended to be more palatable than plants characteristic of undisturbed habitats. Species with soft leaves were more palatable than species with hard leaves.

In experiments with intact leaves (Grime *et al.*, 1970), *Urtica dioica* and cabbage proved more palatable than filter paper. As snails in the field often have to climb stems to obtain food Grime *et al.* (1970) ran multiple choice experiments using stems of nine species against a polythene control to determine the ease of ascent. The stems of *Heracleum sphondylium* and *U. dioica* proved attractive to the snails and this attractiveness occurred before actual physical contact, suggesting that plant odours may be important attractants.

Field studies by Grime and Blythe (1969) showed that *Cepaea nemoralis* was restricted to the south-facing slope of a Derbyshire pass, whereas *Arianta arbustorum* occurred on the floor and north-facing slope of the pass in areas dominated by nettles and *Mercurialis perennis*. In multiple choice experiments involving food potentially available in the field (Grime *et al.*, 1970), *C. nemoralis* ate living and senescent *Urtica* and *Mercurialis* and senescent *Arrhenatherum elatius* (see Table III). In the field the faeces contained mainly the latter, whereas faeces of *A. arbustorum* contained a majority of green *Urtica* and *Mercurialis*. The differences in distribution of the two species was thought to be largely a response to the different microclimates on the two slopes.

The work of Grime and his associates shows that the majority of living vascular plants are relatively unpalatable to *Cepaea*, but the palatability of many species increases with senescence. A variety of anatomical and chemical factors caused this low palatability.

Lindquist (1941) fed a number of litter types to several species of snails and measured their consumption rates. *Arianta arbustorum* showed a special preference for dead leaves of *Fraxinus excelsior* and also readily ate

TABLE III. Consumption of the main vegetation components from the Winnats Pass by *Cepaea nemoralis* in a multiple-choice experiment. (Reproduced with permission from Grime *et al.*, 1970)

Source	Species	Hours		
		0·5	7·5	16·5
South facing slope	Living foliage			
	Festuca ovina	0·00	0·00	0·00
	Festuca rubra	0·00	0·00	0·00
	Arrhenatherum elatius	0·00	0·00	0·00
	Dactylis glomerata	0·00	0·00	0·00
	Poa pratensis	0·00	0·00	0·00
	Senescent foliage			
	Festuca ovina	0·00	0·00	0·00
	Festuca rubra	0·00	0·00	0·00
	Arrhenatherum elatius	6·25	12·70	13·3
	Dactylis glomerata	0·00	0·05	0·65
North facing slope	Living foliage			
	Festuca rubra	0·00	0·00	0·00
	Holcus lanatus	0·00	0·30	0·45
	Poa trivialis	0·00	0·05	0·05
	Mercurialis perennis	0·35	0·43	0·53
	Urtica dioica	8·35	17·30	17·80
	Rumex acetosa	0·40	0·55	0·75
	Polemonium caeruleum	0·00	0·00	0·00
	Senescent foliage			
	Festuca rubra	1·65	1·90	2·20
	Holcus lanatus	0·00	0·10	0·30
	Mercurialis perennis	44·40	88·90	88·90
	Urtica dioica	2·75	20·00	20·60

The tabulated values refer to the mean percentage leaf area consumed.

litter of *Alnus glutinosa* and *Ulmus glabra*. *Fagus sylvatica* litter was not eaten and *Quercus petraea* and *Corylus avellana* were only taken sparingly. *Cepaea hortensis* had a similar diet, but was partial to *Tilia* litter and the fruiting bodies of *Psalliota campestris* and *Collybia velutipes*. *Eulota fruticum* showed a special preference for *Betula* sp. but hardly touched *Fraxinus*. *Arion subfuscus* (Drap.) and *Limax maximus* L. were omnivores with a preference for fungi. Lindquist concluded that all of the 15 species he examined in detail were omnivores, but there were six categories of omnivores depending on preferences.

Mason (1970*a*) analysed the faeces of seven species of woodland snails and used a ranking method to determine the amounts of vascular plant, chlorophyll, fungus and animal material. All species fed predominantly on higher plant material. *Hygromia striolata* (Pfeiff.) and *Arianta arbustorum* took more material containing chlorophyll than the other species. *Discus rotundatus* (Müll.) had significantly greater amounts of fungal material in the faeces, and *Discus* and *Marpessa laminata* (Montagu) took a proportion of rotting wood. *Oxychilus alliarius* (Müll.) and *O. cellarius* (Müll.) took significantly more animal material. Thus, although all species fed primarily on decaying vascular plants, there were differences in the preferred secondary foods.

Mason (1970*a*) determined the assimilation efficiency and ingestion rates of *Discus rotundatus* on a variety of foods, both living and dead, and also

TABLE IV. The assimilation efficiency of snails feeding on a variety of plant leaves at 10°C. (From Mason, 1970*a*)

Species	Food	State of food	Assimilation efficiency, % ± S.E.
Discus rotundatus	*Mercurialis perennis*	Fresh	40·36 ± 10·09
	Circaea lutetiana	Fresh	46·05 ± 6·08
	Urtica dioica	Fresh	47·70 ± 8·89
	Carpinus betulus	Litter	48·67 ± 6·26
	Castanea sativa	Litter	45·81 ± 5·60
	Fagus sylvatica	Litter	55·06 ± 3·59
	Quercus robur	Litter	58·00 ± 4·95
	Acer pseudoplatanus	Litter	44·89 ± 4·98
Helix aspersa	*Lactuca sativa*	Fresh	53·50 ± 6·04
Oxychilus cellarius	*Lactuca sativa*	Fresh	70·20 ± 4·40
Hygromia striolata	*Urtica dioica*	Fresh	52·40 ± 8·78
	Acer pseudoplatanus	Litter	38·23 ± 4·71

ran a limited number of experiments with other species. The assimilation efficiencies obtained are given in Table IV. The mean assimilation of *Discus* feeding on green vascular plants was 44·8%, there being no significant difference in the assimilation efficiency of the foods offered. The efficiency of assimilation of *Discus* eating five species of leaf litter was similarly calculated as 49·1%. With carrion as food (dead earthworm) the efficiency was very high (78·6%), similar to those recorded for invertebrate carnivores. *Urtica dioica* leaves (fresh) were ingested in greater quantities by *Discus* than living leaves of *Mercuralis perennis* or *Circaea lutetiana* and the litter of *Acer pseudoplatanus*, *Castanea sativa* and *Quercus robur* was eaten in greater quantity than *Fagus* and *Carpinus betulus*, *Acer pseudoplatanus* being eaten in greatest quantity.

Assimilation efficiencies were temperature independent, but ingestion rates, and hence absolute assimilation rates, were dependent on this factor (Mason, 1970*a*). The daily ingestion of food by snails, as a percentage of their ash-free dry body weight, ranged from 0·58% (*Discus* feeding on *Circaea*) to 5·19% (*Discus* on preferred litter), figures in close agreement to 2·5 ± 1·5% recorded by Reichle (1968) for 11 forest-floor arthropods.

D. Enzymes

The assimilation efficiencies of molluscs are generally quite high (mostly in the range 50–70%). This is probably the result of the large complement of enzymes which is a general feature of molluscan digestive systems. Myers and Northcote (1959) reported abundant carbohydrases in gut extracts of *Helix pomatia* L. and Nielsen (1962) recorded cellulase activity in seven species of woodland molluscs and found pectinases and xylanases also present. It has earlier been mentioned that the Teredinidae, *Discus rotundatus* and *Marpessa laminata* feed on wood. Many genera of gastropods and bivalves contain cellulases (reviewed in detail by Owen, 1966). Crosby and Reid (1971) have shown that relatively high cellulase levels in bivalves are related mainly to the cellulose content of the food. Chitinases have also been recorded from some species and Tercafs and Jeuniaux (1961) found a higher chitinase activity in *Oxychilus cellarius*, which feeds extensively on dead animal material, than in *Helix pomatia*.

The origin of the cellulases and chitinases has been a matter of some controversy, many workers (e.g. Florkin, 1957) concluding that the enzymes were entirely of bacterial origin. However, more recent experiments using aseptic media (Parnas, 1961; Jeuniaux, 1963) have given convincing evidence that these enzymes are, at least in part, secreted by the mollusc. Thus the digestion of complex carbohydrates is probably achieved by enzymes both from the gut of the mollusc and from the gut microflora.

IV. The Impact of Snails Within Communities

The size of many molluscan populations suggests that they may have a considerable impact within certain communities. This section will describe the few studies that have been done on this aspect.

A. The Grazing of Living Vegetation

The role of limpets in controlling littoral algae has been investigated on several occasions (Jones, 1948; Lodge, 1948; Southward, 1953, 1964; Aitken, 1962). Jones (1948) removed all the large limpets and the majority of small ones from a 10 m wide strip of shore from high water mark to low water mark. There was an initial growth of diatoms and filamentous green algae, followed by a dense cover of *Porphyra*, *Enteromorpha* and *Ulva* in the following spring, and pure stands of fucoids by the following year. At this stage spat-fall resulted in a large number of juvenile limpets colonizing the rocks underneath the fucoids. The young limpets gradually cleared the algae and within 5–7 years of the original removal of limpets the experimental area was almost clear of algae. Limpets and winkles also keep areas free of diatoms (Castenholz, 1961). These algal grazers are obviously having a profound effect on the structure of the community.

Paine (1971) has roughly calculated that *Tegula funebralis* would consume nearly all the net primary production each year in his intertidal study area in Washington. The primary production figures were preliminary, and omitted diatoms and drift algae, but it would seem that a considerable amount of allochthonous production would be required to maintain the density of herbivores.

Within freshwater systems Pieczynska (1970) has examined the role of consumers on periphyton. Chambers of nylon netting were placed round living reed-stems and *Radix ovata* were added to experimental chambers. The experimental results were preliminary, but it appeared that the snails could reduce the biomass of periphyton by about 40%.

Kehde and Wilhm (1972) studied the effect of grazing by snails on the periphyton community of laboratory streams. *Physa gyrina* was added to the streams at a density of 0·012 cm^{-2}. No reduction in the size of the standing crop was found though in an earlier study (Beyers, 1963) with a much greater biomass of snails (20 g per 1000 cm^{-2}, or *ca.* 1·04 per 1000 cm^{-2}) severe overgrazing was recorded. Preliminary experiments by Kehde and Wilhm (1972) showed a negative correlation between density of snails and net periphyton production. The chlorophyll *a* concentration of the grazed periphyton, was significantly higher than that of the ungrazed, but the diversity of the pigments was lower. Enhanced nutrient recycling due to the snails may have been responsible for the increase in chlorophyll *a*.

There is less information on the effect of grazing by molluscs on higher plants. Some species (e.g. *Achatina fulica* in the tropics and several species of slugs in temperate areas) do much damage to agricultural crops. Frömming (1952) considered that molluscs have an important effect on the production of *Lemna* and other vascular plants, but no quantitative estimates were made. Gillespie (1969) calculated that gastropods assimilated approximately 10% (consumed 20%) of the available net primary production of *Potamogeton* spp. in the Madison River, if they fed exclusively on this plant. The diet was not examined and this presumption seems unlikely; indeed the three gastropod species involved may well have been feeding entirely on the periphytic coating of the macrophytes.

It is clear that much more field experimentation is required.

B. Molluscs and Litter Decomposition

Lindquist (1941) and Frömming (1958) suggested that molluscs may be important agents in decomposition and humus formation in woodlands. Mason (1970*a*, *b*, 1971) has attempted to quantify this aspect. The populations of snails living in a *Fagus* woodland were sampled monthly to obtain detailed estimates of population density and biomass. The mean density was 489 snails m^{-2} and the mean biomass 699 mg m^{-2} (dry weight) or 278 mg m^{-2} (ash-free dry weight). At the same time the litter input on to the site was measured monthly. The annual litter input was 676 g m^{-2} (3189 kcal m^{-2}), of which 338 g m^{-2} was of beech leaves, 247 g m^{-2} was of other beech material (bud-scales, fruits, twigs and bark), and the remainder was from other trees or from the field layer (estimated at 23 g m^{-2}).

Mason (1970*a*) showed that the snails fed largely on dead plant remains. From detailed laboratory estimates of consumption rates and respiration it was calculated that the snail population consumed from 0·35–0·72% of the total litter input, 50% of which was assimilated.

The annual litter consumption of other invertebrate groups range from 1·8 to 9% of the input (Drift and Witkamp, 1958; Gere, 1963; Bocock, 1963; Berthet, 1963, 1967). Micro-organisms are probably responsible for the majority of litter breakdown (Macfadyen, 1961, 1963).

Crude calculations by Mason (1970*b*) suggested that slugs may be more important than snails in the decomposition of woodland litter.

There is no information available for the decomposition of litter by molluscs in freshwater and marine environments.

C. Biodeposition, and Sediment Changes caused by Molluscs

The production of faeces and pseudofaeces (biodeposition) by molluscs has been studied mainly in estuaries, where the process may have profound effects on the general ecology of the area. For instance Verwey (1952) calculated that *Cardium edule* in the Waddenzee deposited 100,000 t dry weight of suspended matter in one year and *Mytilus edule* deposited between 25,000 and 175,000 t in the same period. This is material made available to benthic decomposers, and is sufficient to alter markedly the substrate of the estuary.

Laboratory experiments (Haven and Morales-Alamo, 1966) showed that oysters, *Crassostrea virginica*, filter seston from the water and deposit it seven times faster than it would settle by gravity. It was estimated that oysters, on 0·405 ha of estuarine bottom, may produce up to 981 kg of faeces and pseudofaeces each week. The particle size of the biodeposit was small (95% of particles were less than 3 μm) and the organic content was low (9–23%), 4–12% being organic carbon, though this was significantly higher than in the surface sediments.

Substrate alteration by deposit-feeding molluscs has been studied by Rhoads and Young (1970) in Buzzards Bay, Massachusetts. *Nucula proxima*, *Macoma tenta* and *Yoldia limatula* were the dominant protobranch bivalves and their intensive re-working of the deposits near the surface had a number of effects. The mud surface was uncompacted and granular, consisting of faecal pellets and reworked clasts of semi-consolidated mud. The surface consisted of biogenic, sand-sized particles of low bulk-density, while the water content was greater than 60% at the surface of highly reworked sediments, compared with just over 50% without molluscs.

The effect of this large-scale deposit feeding was to produce a physical instability in the substrate. There was a high turnover rate of bottom muds by a process of resuspension by weak tidal currents. The sediment–water interface showed a high turbidity. There was a downward reworking of heavier materials (quartz, feldspar, etc.) producing an X-ray opaque layer at about 3 cm, the highly fluid sediment above this being continuously reworked. This physical instability clogs the filtering structures of suspension-feeding organisms and buries newly settled larvae or discourages them from settling. In areas with high densities of deposit-feeding organisms there are few suspension-feeders.

Both suspension- and deposit-feeding molluscs in estuaries will have a marked effect on the physical and chemical characteristics of their substrate. A similar effect is likely to occur in fresh waters with large populations of molluscs. The activity of other decomposers is likely to be much influenced. The decrease in particle size of the bio-deposited sediment will increase the microbial activity on the increased surface area and hence accelerate decomposition, particularly in those waters which contain, for instance, many *Anodonta anatina*, *A. cygnaea* or *Dreissena polymorpha*.

D. Erosion Caused by Molluscs

Because of their great abundance and their method of feeding with a file-like radula, gastropods (particularly limpets and winkles) exert a considerable erosive effect on littoral rocks. North (1954) has calculated that *Littorina planaxis* in California deepens the sandstone rock pools by 1 cm every 40 years. If *L. scutulata* Gould was also considered, the rate would increase to 1 cm every 16 years. The rate is similar to, or higher than, the effect of scouring by tides and nocturnal changes of pH, given by Emery (1946). Hawkshaw (1878) considered that limpets eroded the chalk cliffs of Dover at a rate of 1·5 mm a year, equivalent to all the other erosive forces together. Grazing gastropods therefore have a profound effect on the physical appearance of the shore (Southward, 1964).

E. Nutrient Cycling

Little work has been done on the role of molluscs in nutrient cycling, though, having shells, they may be important. Kuenzler (1961*b*) studied the phosphorus budget of a population of the mussel *Modiolus demissus* in an intertidal saltmarsh in Georgia. A large amount of particulate matter was filtered from the sea-water by the mussels. The water above the marsh contained 39,000 μgP m^{-2} of which 5480 μg m^{-2} day^{-1} entered the mussel population. The production pf pseudofaeces accounted for 4700 μgP m^{-2} day^{-1} and faeces 460 μgP m^{-2} day^{-1}. Phosphate and dissolved organic phosphorus elimination accounted for a further 283 μgP m^{-2} day^{-1}. One third of the particulate phosphate in the water was removed every day, most of which was rejected as pseudofaeces and faeces to be deposited on the surface of the marsh. This then provided raw materials for deposit feeders to regenerate phosphate and would also enhance the development of the saltmarsh. The energy flow through the mussel population was 56 kcal m^{-2} $year^{-1}$, so the role of the mussels in cycling phosphorus far exceeded their role as promoters of energy flow.

The material locked within the shell of dead molluscs must also act as a reservoir of nutrients. Mason (unpublished) found a density of up to 1000 empty shells per m^2 in the litter of *Fagus* woodland, where there were 489 live snails m^{-2}. The shells of the dead snails amounted to a biomass of 615 mg dry wt m^{-2}, in various stages of leaching and decomposition. Not only would the nutrients be leached back into the soil, but the shells would also act as a direct supply of calcium for other organisms, especially for birds during the period of egg-shell formation.

V. Discussion

Although there has been a large amount of work on the general ecology of molluscs, particularly those of economic importance as food or pests, few studies relate to the role of molluscan populations in the general functioning of communities.

Many molluscs are omnivores, but they nevertheless show preferences for food, these preferences being different in species living in the same habitat (Mason, 1970*a*). These preferences are probably linked to the quality of the food (Carefoot, 1967; Eisenberg, 1970), a quality which may be physical (due to the external structure) or chemical. Many molluscs eat litter readily, but there is no information as to what degree of decomposition they prefer. Much detailed work on the diet of molluscs is still required.

The work of Kuenzler (1961) emphasizes the need for much more work on the role of molluscs in nutrient cycling. Southward (1964) and Rhoads and Young (1970) show that molluscs may greatly alter their environment and have far reaching effects on other animals. Biodeposition, by decreasing the particle size of the sediment, will accelerate decomposition by micro-organisms.

Invertebrates have a number of effects on litter decomposition and soil formation. There is the direct action of eating and assimilating decomposing vegetation. The unassimilated matter, voided as faeces, will contain a higher microbial activity than the original litter. Thus faeces of *Limnaea pereger* respired at a rate almost twice that of the food, *Typha angustifolia* litter, from which they were produced (Mason, unpublished). Decomposition will again be accelerated. Finally the litter which was not eaten will be fragmented, a process which will accelerate leaching and increase colonization by micro-organisms. This latter activity is likely to be the most important contribution by invertebrates to the decomposition process.

The interaction between micro-organisms and molluscs is complex. Fragmentation of the litter by molluscs will increase the populations of micro-organisms, but these are also likely to act as food, exclusively so in the case of *Planorbis contortus* (Calow and Fletcher, 1972). The presence of microbial growth greatly enchanced the attractiveness of leaves to the amphipod *Gammarus lacustris* (Kaushik and Hynes, 1971) and amphipods certainly utilize micro-organisms as food (Fenchel, 1970; Hargrave, 1970). Microbial activity is greatly increased, however, despite acting as a food source (Fenchel, 1970; Hargrave, 1970) and it appears the microbial generation time is so rapid as to counteract the grazing effect of invertebrates and still colonize the increased substrate surface made available by the grazing activity.

Acknowledgements

I would like to thank Dr. J. E. Smith and the library staff of the Plymouth Laboratory, where part of this review was written.

References

Adam, W. (1942). *Bull. Mus. r. His. nat. Belg.* **18,** 1–18.

Adams, S. and Angelovic, J. W. (1970). *Chesapeake Sci.* **11,** 249–254.

Agocsy, P. (1968). *Acta zool. hung.* **14,** 1–6.

Aitken, J. J. (1962). *Ir. Nat. J.* **14,** 12–15.

Allen, J. A. (1962). *J. mar. biol. Ass. U.K.* **42,** 609–623.

Anderson, A. (1971). *J. mar. biol. Ass. U.K.* **51,** 423–437.

Anderson, S. (1972). *J. appl. Ecol.* **9,** 161–178.

Atkins, W. R. G. and Lebour, M. V. (1923). *Scient. Proc. R. Dubl. Soc.* **17,** 233–240.

Baker, R. E. (1968). *Proc. malac. Soc. Lond.* **38,** 41–54.

Bakker, K. (1959). *Archs néerl. Zool.* **13,** 230–257.

Barnes, H. F. and Weil, J. W. (1944). *J. Anim. Ecol.* **13,** 140–175.

Barnes, H. F. and Weil, J. W. (1945). *J. Anim. Ecol.* **14,** 71–105.

Berrie, A. D. (1965). *Proc. malac. Soc. Lond.* **36,** 283–295.

Berry, A. J. (1966). *J. Zool. Lond.* **150,** 11–27.

Berthet, P. (1963). *In* "Soil Organisms." (J. Doeksen and J. van der Drift, eds), pp. 18–31. North-Holland, Amsterdam.

Berthet, P. (1967). *In* "Secondary Productivity of Terrestrial Ecosystems." (K. Petrusewicz, ed.), Vol. 2, pp. 709–725. Panstowe Wydownicwo Naukowe, Warsaw.

Best, B. (1964). *Veliger* **6** (suppl.), 42–45.

Beyers, R. J. (1963). *Ecol. monogr.* **33,** 281–306.

Bigot, L. (1965). *Mém. Soc. zool. Fr.* **34,** 1–100.

Bigot, L. (1967). *Vie Milieu* **18,** 1–27.

Birch, L. C. and Clark, D. P. (1953). *Q. Rev. Biol.* **28,** 13–36.

Birkett, L. (1970). *In* "Marine Food Chains." (J. G. Steel, ed.), pp. 261–264. Oliver and Boyd, Edinburgh.

Bleakney, J. J. (1966). *Nautilus* **79,** 131–134.

Blinn, W. C. (1963). *Ecology* **44,** 498–505.

Bloom, S. A., Simon, J. A. and Hunter, V. D. (1972). *Mar. Biol.* **13,** 43–56.

Bocock, K. L. (1963). *In* "Soil Organisms." (J. Doeksen and J. van der Drift, eds), pp. 85–91. North-Holland, Amsterdam.

Bornebusch, C. H. (1930). "The Fauna of Forest Soil." Nielsen and Lydicke, Copenhagen.

Bovjerg, R. V. (1957). *Proc. Iowa Acad. Sci.* **64,** 650–653.

Bovjerg, R. V. (1965). *Malacologia* **2,** 199–207.

Boycott, A. E. (1934). *J. Ecol.* **22,** 1–38.

Boycott, A. E. (1936). *J. Anim. Ecol.* **5,** 116–186.

Brafield, A. E. and Newell, G. E. (1961). *J. mar. biol. Ass. U.K.* **41,** 81–87.

Burky, A. J. (1971). *Ecol. monogr.* **41,** 235–251.

Calabrese, A. and Davis, H. C. (1970). *Helgoländ. wiss. Meeresunters* **20,** 553–564.

Calow, P. (1970). *Proc. malac. Soc. Lond.* **39,** 203–215.

Calow, P. and Fletcher, C. R. (1972). *Oecologia (Berl.)* **9,** 155–170.

CANAGARATNAM, P. (1962). *Proc. Ceylon Ass. Adv. Sci.* **18,** 91–114.
CAREFOOT, T. H. (1967*a*). *J. mar. biol. Ass. U.K.* **47,** 565–589.
CAREFOOT, T. H. (1967*b*). *Comp. Biochem. Physiol.* **21,** 627–652.
CAREFOOT, T. H. (1970). *J. exp. mar. Biol. Ecol.* **5,** 47–62.
CASTENHOLZ, R. W. (1961). *Ecology* **42,** 783–794.
CHATFIELD, J. E. (1968). *Proc. malac. Soc. Lond.* **38,** 223–245.
CHURCHILL, E. P. and LEWIS, S. I. (1924). *Bull. Bur. Fish.* **39,** 440–471.
CLAMPITT, P. T. (1970). *Malacologia* **10,** 113–151.
COMBRINCK, C. and EEDEN, J. A. (1969). *Malacologia* **9,** 39.
CRAWFORD-SIDEBOTHAM, T. J. (1972). *Oecologia (Berl.)* **9,** 141–154.
CRISP, D. T. and GLEDHILL, T. (1970). *Archs Hydrobiol.* **67,** 502–541.
CROSBY, N. D. and REID, R. G. B. (1971). *Can. J. Zool.* **49,** 617–622.
CVANCARA, A. M. (1972). *Ecology* **53,** 154–157.
DAINTON, B. H. (1954). *J. exp. Biol.* **31,** 165–187.
DAVIS, H. C. and CALABRESE, A. (1964). *Fishery Bull. Fish Wildl. Service U.S.* **63,** 643–655.
DE WIT, W. F. (1955). *Basteria* **19,** 35–73.
DORE, W. H. and MILLER, R. C. (1923). *Univ. California Publ. Zool.* **22,** 383–400.
DRIFT, J. VAN DER (1951). *Tijdschr. Ent.* **94,** 1–168.
DRIFT, J. VAN DER and WITKAMP, M. (1958). *Archs néerl. Zool.* **13,** 486–492.
DRINNAN, R. E. (1957). *J. Anim. Ecol.* **26,** 441–469.
DUNCAN, C. J. (1959). *J. Anim. Ecol.* **28,** 97–117.
DVORAK, J. (1970). *Trans. Czech. Acad. Sci., nat. hist. Ser.* **80,** 63–117.
DZIECKKOWSKI, A. (1971). *Zck. ochr. Pryz. Polsk. Akad. Nauk* **36,** 257–286.
EISENBERG, R. M. (1966). *Ecology* **47,** 889–906.
EISENBERG, R. M. (1970). *Ecology* **51,** 680–684.
EMERY, K. O. (1946). *J. Geol.* **54,** 209–228.
FENCHEL, T. (1970). *Limnol. Oceanogr.* **15,** 14–20.
FILATOVA, Z. A. (1959). *Proc. 15th Int. Congr. Zool., London*, 1958, 221–222.
FLORKIN, M. (1957). *Proc. Intern. Symp. on Enzyme Chemistry, Tokyo*, pp. 390–392.
FOIN, T. C. and STIVEN, A. E. (1970). *Oecologia (Berl.)* **5,** 74–84.
FOSTER, T. D. (1937). *Ecology* **18,** 545–546.
FRANK, P. W. (1969). *Oecologia (Berl.)* **2,** 232–250.
FRETTER, V. and GRAHAM, A. (1962). "British Prosobranch Molluscs." Ray Society, London.
FRETTER, V. and MONTGOMERY, M. C. (1968). *J. mar. biol. Ass. U.K.* **48,** 499–520.
FRÖMMING, E. (1952). *Arch. Molluskenk.* **81,** 45–48.
FRÖMMING, E. (1954). "Die Biologie der mitteleuropaischen Landgastropoden." Duncker and Humblot, Berlin.
FRÖMMING, E. (1956). "Biologie der mitteleuropaischen Susswasserschnecken." Duncker and Humblot, Berlin.
FRÖMMING, E. (1958). *Z. angew. Zool.* **45,** 341–350.
FRÖMMING, E. (1962). "Das Verhalten unserer Schnecken zu den Pflanzen ihrer Umgebung." Duncker and Humblot, Berlin.

GAEVASKAYA, N. S. (1966). "The Role of Higher Aquatic Plants in the Nutrition of the Animals of Freshwater Basins." National Lending Library of Science and Technology, Boston Spa.
GALE, W. F. (1971). *Ecology* **52**, 367–370.
GALE, W. F and LOWE, R. L. (1971). *Ecology* **52**, 507–513.
GASCOIGNE, T. (1956). *Trans. R. Soc. Edinb.* **63**, 129–151.
GERE, G. (1963). *In* "Soil Organisms." (J. Doeksen and J. van der Drift, eds), pp. 67–75. North-Holland, Amsterdam.
GETZ, L. L. (1959). *Am. Midl. Nat.* **61**, 485–498.
GILLESPIE, D. M. (1969). *Limnol. Oceanogr.* **14**, 101–114.
GIMMINGHAM, C. T. and NEWTON, H. C. F. (1937). *J. Minist. Agric. Fish.* **44**, 242–246.
GLYNN, P. W. (1965). *Beaufortia* **12**, 1–198.
GOODHART, C. B. (1962). *J. Anim. Ecol.* **31**, 207–237.
GRAHAM, A. (1955). *Proc. malac. Soc. Lond.* **31**, 144–159.
GREENFIELD, L. J. and LANE, C. E. (1953). *J. biol. Chem.* **204**, 664–672.
GRIME, J. P. and BLYTHE, G. M. (1969). *J. Ecol.* **57**, 45–66.
GRIME, J. P., BLYTHE, G. M. and THORNTON, J. D. (1970). *In* "Animal Populations in Relation to their Food Resources." (A. Watson, ed.), pp. 73–99. Blackwell, Oxford.
GRIME, J. P., MACPHERSON-STEWART, S. F. and DEARMAN, R. S. (1968). *J. Ecol.* **56**, 405–420.
HADFIELD, M. G. (1970). *Veliger* **12**, 301–309.
HADFIELD, M. G., KAY, E. A., GILLETTE, M. U. and LLOYD, M. C. (1972). *Mar. Biol.* **12**, 81–98.
HARGRAVE, B. T. (1970). *Limnol. Oceanogr.* **15**, 21–30.
HARMAN, W. N. (1972). *Ecology* **53**, 271–277.
HAVEN, D. S. and MORALES-ALAMO, R. (1966). *Limnol. Oceanogr.* **11**, 487–498.
HAWKSHAW, C. (1878). *J. Linn. Soc. (Zool.)* **14**, 406–411.
HEPPLESTON, P. B. (1972). *J. appl. Ecol.* **9**, 235–248.
HEYWOOD, J. and EDWARDS, R. W. (1962). *J. Anim. Ecol.* **31**, 239–250.
HIGASHI, S. (1966). *Venus* **25**, 43–46.
HUGHES, R. N. (1969). *J. mar. biol. Ass. U.K.* **49**, 805–823.
HUGHES, R. N. (1970*a*). *J. Anim. Ecol.* **39**, 333–356.
HUGHES, R. N. (1970*b*). *J. Anim. Ecol.* **39**, 357–381.
HUGHES, R. N. (1971*a*). *J. exp. mar. Biol. Ecol.* **6**, 167–178.
HUGHES, R. N. (1971*b*). *Mar. Biol.* **9**, 290–299.
HUNTER, P. J. (1968*a*). *Malacologia* **6**, 369–377.
HUNTER, P. J. (1968*b*). *Malacologia* **6**, 379–389.
HUNTER, P. J. (1968*c*). *Malacologia* **6**, 391–399.
HUNTER, W. R. (1955). *Glasg. Nat.* **17**, 173–183.
HUNTER, W. R. (1957). *Glasg. Univ. Publ. Stud. Loch Lomond* **1**, 56–95.
HUNTER, W. R. (1961). *Proc. zool. Soc. Lond.* **137**, 135–171.
HUNTER, W. R. (1964). *In* "Physiology of Mollusca." (K. M. Wilbur and C. M. Yonge, eds), Vol. 1, pp. 83–126. Academic Press, New York.
HUNTER, W. R. and HUNTER, M. R. (1962). *Glasg. Nat.* **18**, 198–205.

HYNES, H. B. N. (1960). "The Biology of Polluted Waters." (Liverpool University Press.
INO, T. (1958). *Bull. Tokai reg. Fish Lab.* **22,** 33–36.
JEUNIAUX, C. (1963). "Chitine et chitinolyse, un chapitre de la biologie moleculaire." Masson, Paris.
JOBIN, W. R. (1970). *Am. J. trop. Med. Hygiene* **19,** 1038–1048.
JOHANNES, R. E., COWARD, S. J. and WEBB, K. L. (1969). *Comp. Biochem. Physiol.* **29,** 283–288.
JONES, N. S. (1948). *Proc. Lpool biol. Soc.* **56,** 60–77.
JØRGENSEN, C. B. (1966). "Biology of Suspension Feeding." Pergamon, Oxford.
KAUSHIK, N. K. and HYNES, H. B. N. (1971). *Archs. Hydrobiol.* **68,** 465–515.
KEHDE, P. M. and WILHM, J. L. (1972). *Am. Midl. Nat.* **87,** 8–24.
KENDALL, S. B. (1953). *J. Helminth.* **27,** 17–28.
KORINKOVA, J. (1971). *Hydrobiologia* **12,** 377–382.
KUCERA, C. L. (1954). *Ecology* **35,** 71–75.
KUENZLER, E. J. (1961*a*). *Limnol. Oceanogr.* **6,** 191–204.
KUENZLER, E. J. (1961*b*). *Limnol. Oceanogr.* **6,** 400–415.
KUIPER, J. G. J. and WOLFF, W. J. (1970). *Basteria* **34,** 1–42.
LADLE, M. and BARON, F. (1969). *J. Anim. Ecol.* **38,** 407–413.
LEARNER, M. A., WILLIAMS, R., HARCUP, M. and HUGHES, B. D. (1971). *Freshwat. Biol.* **1,** 339–367.
LEIGHTON, D. L. (1966). *Pacif. Sci.* **20,** 104–113.
LEWIS, J. R. (1964). "The Ecology of Rocky Shores." English Universities Press, London.
LINDQUIST, B. (1941). *K. Fysiogr. Sällsk. Lund. Förh.* **11,** 144–156.
LODGE, S. M. (1948). *Proc. Lpool. biol. Soc.* **56,** 173–183.
LOOSANOFF, V. L. and DAVIS, H. C. (1963). *Adv. mar. Biol.* **1,** 1–136.
MACAN, T. T. (1950). *J. Anim. Ecol.* **19,** 124–146.
MACFADYEN, A. (1961). *Ann. appl. Biol.* **49,** 216–219.
MACFADYEN, A. (1963). *In* "Soil Organisms." (J. Doeksen and J. van der Drift, eds), pp. 3–17. North-Holland, Amsterdam.
MAITLAND, P. S. (1965). *Proc. malac. Soc. Lond.* **36,** 339–347.
MALONE, C. R. and NELSON, D. J. (1969). *Ecology* **50,** 728–730.
MANN, K. H. (1964). *Veh. int. Verein. theor. angew. Limnol.* **15,** 485–495.
MAPSTONE, G. M. (1970). *Helgoländer wiss. Meeresunters* **20,** 565–575.
MARSHALL, N. (1970). *In* "Marine Food Chains." (J. H. Steele, ed.), pp. 52–66. Oliver and Boyd, Edinburgh.
MASON, C. F. (1970*a*). *Oecologia (Berl.)* **4,** 358–373.
MASON, C. F. (1970*b*). *Oecologia (Berl.)* **5,** 215–239.
MASON, C. F. (1971). *Oecologia (Berl.)* **7,** 80–94.
MCC.CALAN, E. (1940). *Trop. Agric., Trin.* **18,** 211–223.
MCINTYRE, A. D. (1970). *J. mar. biol. Ass. U.K.* **50,** 561–575.
MEAD, A. R. (1961). "The Giant African Snail: a Problem in Economic Malacology." University of Chicago Press, Chicago.
MITROPOLSKY, V. I. (1966). *Trans. Inst. Biol. Inland Waters Acad. Sci. U.S.S.R.* **12,** 125–128.

MONAKOV, A. V. (1972). *J. Fish. Res. Bd. Canada* **29,** 363–383.
MOORE, H. B. (1938). *Proc. malac. Soc. Lond.* **23,** 117–118.
MORGAN, N. C. (1970). *Hydrobiologia* **35,** 545–553.
MORTON, J. E., BONEY, A. D. and CORNER, E. D. S. (1957). *J. mar. biol. Ass. U.K.* **36,** 383–405.
MÖRZER BRUIJNS, M. F., REGTEREN ALTENA, C. U. VAN and BUTOT, L. J. M. (1959). *Basteria* **23,** 135–162.
MYERS, F. L. and NORTHCOTE, D. H. (1959). *Biochem. J.* **71,** 749–756.
NEGUS, C. L. (1966). *J. Anim. Ecol.* **35,** 513–532.
NEWELL, P. F. (1967). *In* "Soil Biology." (A. Burgess and F. Raw, eds), pp. 413–433. Academic Press, London and New York.
NEWELL, P. F. (1971). *In* "Methods of Study in Quantitative Soil Ecology: Population, Production and Energy Flow." (J. Phillipson, ed.), pp. 128–149. Blackwell Scientific Publications, Oxford.
NEWELL, R. C. (1965). *Proc. zool. Soc. Lond.* **144,** 25–45.
NEWELL, R. C. (1970). "Biology of Intertidal Animals." Logos Press, London.
NEWELL, R. C., PYE, V. I. and AHSANULLAH, M. (1971). *Mar. Biol.* **9,** 138–145.
NIELSEN, C. O. (1962). *Oikos* **13,** 200–215.
NIXON, S. W., OVIATT, C. A., ROGERS, C. and TAYLOR, K. (1971). *Oecologia (Berl.)* **8,** 21–30.
NORTH, W. J. (1954). *Biol. Bull.* **106,** 185–197.
ODUM, E. P. and SMALLEY, A. E. (1959). *Proc. natn. Acad. Sci. U.S.A.* **45,** 617–622.
ØKLAND, F. (1929). *Beret. om. det.* 18 *Skand. Naturforski i Kobenhavn.*
ØKLAND, J. (1963). *Nytt. Mag. Zool.* **11,** 19–43.
OWEN, D. F. (1965). *Proc. zool. Soc. Lond.* **144,** 361–382.
OWEN, G. (1966*a*). *In* "Physiology of Mollusca." (K. M. Wilbur and C. M. Yonge, eds), Vol. 2, pp. 1–51. Academic Press, New York.
OWEN, G. (1966*b*). *In* "Physiology of Mollusca." (K. M. Wilbur and C. M. Yonge, eds), Vol. 2, pp. 53–96. Academic Press, New York.
PAINE, R. T. (1971). *Limnol. Oceanogr.* **16,** 86–98.
PALLANT, D. (1969). *J. Anim. Ecol.* **38,** 391–397.
PALLANT, D. (1970). *Proc. malac. Soc. Lond.* **39,** 83–87.
PARNAS, I. (1961). *J. Cellular Comp. Physiol.* **58,** 195–201.
PIECZYNSKA, E. (1970). *Wiad. Ekol.* **16,** 133–143.
PILKINGTON, M. C. and FRETTER, V. (1970). *Helgoländer wiss. Meeresunters* **20,** 576–593.
PIP, E. and PAULISHYN, W. F. (1971). *Can. J. Zool.* **49,** 367–372.
POMEROY, D. E. (1968). *Aust. J. Zool.* **16,** 857–869.
POMEROY, D. E. (1969), *Aust. J. Zool.* **17,** 495–514.
PRATER, A. J. (1972). *J. appl. Ecol.* **9,** 179–194.
PURCHON, R. D. (1968). "The Biology of the Mollusca." Pergamon, Oxford.
RAO, M. B. and GANAPATA, B. N. (1971). *Mar. Biol.* **9,** 109–114.
REICHLE, D. E. (1968). *Ecology* **49,** 538–542.
RHOADS, D. C. and YOUNG, D. K. (1970). *J. mar. Res.* **28,** 150–178.
RODINA, A. G. (1948). *Mikrobiologiya* **17** (quoted in Gaevskaya, 1966).

SAITO, Y. and NAKAMURA, N. (1961). *Bull. Jap. Soc. scient. Fish.* **27,** 395–400.
SEGERSTRÅLE, S. G. (1965). *Bot. Gothoburg* **3,** 195–204.
SOROKIN, Y. I. (1966). *Trans. Inst. Biol. Inland Waters Acad. Sci. U.S.S.R.* **12,** 75–119.
SOSKA, G. (1968). *Ekol. Polsk. ser A.* **16,** 729–753.
SOUTH, A. (1964). *Ann. appl. Biol.* **53,** 251–258.
SOUTHWARD, A. J. (1953). *Proc. Lpool biol. Soc.* **59,** 1–50.
SOUTHWARD, A. J. (1964). *In* "Grazing in Terrestrial and Marine Environments." (D. J. Crisp, ed.), pp. 265–273. Blackwell, Oxford.
SOUTHWARD, A. J. (1965). "Life on the Seashore." Heinemann, London.
SPOONER, G. M. and MOORE, H. B. (1940). *J. mar. biol. Ass. U.K.* **24,** 283–330.
STANCZYKOWSKA, A., MAGNIN, E. and DUMOUCHEL, A. (1971). *Can. J. Zool.* **49,** 1431–1441.
STEEN, W. J. VAN DER (1967). *Archs néerl. Zool.* **17,** 403–468.
STEPHENS, G. C. (1963). *Comp. Biochem. Physiol.* **10,** 191–202.
STEPHENS, G. C. (1964). *Biol. Bull.* **126,** 150–162.
STEPHENS, G. C. (1967). *In* "Estuaries." (G. H. Lauff, ed.), pp. 367–373. American Association for the Advancement of Science, Washington.
STEPHENS, G. C. (1968). *Am. Zoologist* **8,** 95–106.
STRANDINE, E. J. (1941). *Ecology* **22,** 86–91.
SUTHERLAND, J. (1970). *Ecol. monogr.* **40,** 169–188.
TERCAFS, R. R. and JEUNIAUX, C. (1961). *Arch. Int. Physiol. Biochem.* **69,** 364–368.
THAMDRUP, H. M. (1935). *Medd. Komm. Hav. Fisk.* **10,** 1–125.
THOMAS, D. C. (1944). *Ann. appl. Biol.* **31,** 163–164.
THOMAS, D. C. (1948). *Ann. appl. Biol.* **35,** 207–227.
THORSON, G. (1966). *Neth. J. Sea Res.* **32,** 267–293.
TREVALLION, A. (1971). *J. exp. mar. Biol. Ecol.* **7,** 95–122.
TREVALLION, A., EDWARDS, R. R. C. AND STEELE, J. H. (1970). *In* "Marine Food Chains." (J. H. Steele, ed.), pp. 285–290. Oliver and Boyd, Edinburgh.
TSIKHON-LUKANINA, E. A. (1958). *Tr. Mosrybutuza* **9** (quoted in Gaevskaya, 1966).
TSIKHON-LUKANINA, E. A. (1961*a*). *Bull. Inst. Biol. Reservoir Acad. Sci. U.S.S.R.* **10,** 28–30.
TSIKHON-LUKANINA, E. A. (1961*b*). *Bull. Inst. Biol. Reservoir Acad. Sci. U.S.S.R.* **10,** 31–34.
TSIKHON-LUKANINA, E. A. (1965). *Trans. Inst. Biol. Inland Waters Acad. Sci. U.S.S.R.* **8,** 134–136.
TSIKHON-LUKANINA, E. A. and SOROKIN, Y. I. (1965). *Trans. Inst. Biol. Inland Waters Acad. Sci. U.S.S.R.* **8,** 130–133.
TURNER, R. D. (1966). "A survey and Illustrated Catalogue of the Teredinidae (Mollusca: Bivalvia)." Museum of Comparative Zoology, Cambridge, Mass.
VALOVIRTA, I. (1968). *Ann. Zool. fenn.* **5,** 245–253.
VERWEY, J. (1952). *Arch. néerl. Zool.* **10,** 172–239.
WALNE, P. R. (1958). *J. mar. biol. Ass. U.K.* **37,** 415–425.
WALNE, P. R. (1963). *J. mar. biol. Ass. U.K.* **43,** 767–784.
WALNE, P. R. (1970). *Fishery Invest. Lond.* (Ser. 2) **25** (5), 1–62.

Walne, P. R. (1972). *J. mar. biol. Ass. U.K.* **52,** 345–374.
Wara, W. M. and Wright, B. B. (1964). *Veliger* **6** (Suppl.), 30–37.
Ward, J. (1966). *Bull. mar. Sci. Gulf Caribb.* **16,** 668–684.
Wäreborn, I. (1969). *Oikos* **20,** 461–479.
Wäreborn, I. (1970). *Oikos* **21,** 285–291.
Webley, D. (1962). *Ann. appl. Biol.* **50,** 129–136.
Webley, D. (1964). *Ann. appl. Biol.* **53,** 407–414.
White, A. R. (1959). *Pl. Path.* **8,** 62–68.
Whitehead, H. (1935). *J. Animal. Ecol.* **4,** 58–78.
Williams, N. V. (1970). *Malacologia* **10,** 153–164.
Winkler, L. R. and Dawson, E. Y. (1963). *Pacif. Sci.* **17,** 102–105.
Winter, J. E. (1969). *Mar. Biol.* **4,** 87–135.
Winter, J. E. (1970). *In* "Marine Food Chains." (J. H. Steele, ed.), pp. 196–206. Oliver and Boyd, Edinburgh.
Wolda, H. (1969). *J. Anim. Ecol.* **38,** 305–327.
Wolda, H., Zweep, A. and Schuitema, K. A. (1971). *Oecologia (Berl.)* **7,** 361–381.
Yom-Tov, Y. (1970). *Ecology* **51,** 907–911.
Yom-Tov, Y. (1971). *Proc. malac. Soc. Lond.* **39,** 314–326.
Yom-Tov, Y. (1972). *J. Anim. Ecol.* **41,** 17–22.
Yom-Tov, Y. and Galum, M. (1971). *Veliger* **14,** 86–88.
ZoBell, C. E. and Feltham, C. B. (1938). *J. Mar. Res.* **1,** 312–325.
ZoBell, C. E. and Feltham, C. B. (1942). *Ecology* **23,** 69–78.

18

Aquatic Crustacea

M. Ladle

Freshwater Biological Association
Wareham
Dorset
England

I. Introduction

In preparing this review an attempt has been made to give references to works which themselves contain extensive bibliographies. It has been impossible to give a uniform coverage to all aspects of the problem of plant litter feeding by aquatic Crustacea because certain fields have attracted an inordinate amount of attention.

In the present context the terms "plant litter", "plant detritus" and "detritus" are generally considered to be interchangeable and other chapters of the present volume are regarded as defining the limits of those materials that constitute plant litter. Certain categories of material, which comprise a relatively small proportion of those referred to in earlier chapters, will play a large part in any consideration of the aquatic Crustacea. For example, decomposing lower plants, particularly algae, the faecal products of herbivorous organisms and small particles of higher plant origin in an advanced state of physical and chemical decomposition, are of great importance as food for crustaceans. These materials, together with their associated micro-organisms, form the basis of one extreme of a spectrum of litter-feeding activity which at the other extreme includes those forms which feed almost exclusively on living plant material, ingesting only small quantities of plant litter. An interesting discussion of microphagous crustaceans is included in the work of Jorgensen (1966). Gaevskaya (1969) indicates that 17 out of 27 species of "herbivorous" freshwater crustaceans considered also fed on dead plants, allochthonous material and plant detritus. The forms involved ranged from *Cladocera* to large crabs and crayfish. The review of marine coastal detritus food chains by Mann (1972) lists marine organisms which contained, on average, at least 20% of vascular plant detritus in their digestive tracts; these included cumaceans, mysids, copepods, shrimps and crabs.

It is not thought necessary in this review to argue the relative roles of plant litter and organisms associated with plant litter in the nutrition of crustaceans which (in many cases inevitably) ingest both of these components. The importance of this aspect of the problem must, however, be borne in mind and some of the more recent papers reviewed below emphasize that the relationship between plant litter and crustacean consumer is complex and often involves micro-organisms as intermediaries and their reciprocal stimulation by the activity of crustaceans, such as the example reported by Hargrave (1970*a*) in which algal production and bacterial respiration were stimulated in the presence of increasing population densities of the amphipod *Hyalella azteca* (Saussure).

II. Aquatic Crustacea which Ingest Plant Litter

In both marine and freshwater habitats, similar taxa of plant litter-feeding Crustacea occur and they can be broadly grouped into microphagous, often planktonic, forms feeding on small particles of plant detritus suspended in the water and secondly, benthic species, both macro- and microphagous, feeding on litter present in more or less organic deposits and also at the interface between sediment and water. The former group includes cladocerans and copepods and in the oceans mysids and euphausiids. There is no clear distinction between the food of many sedentary forms such as cirripedes and some amphipods and decapods which filter particles from the water; and that of planktonic forms which may ingest sediment suspended by turbulence.

The striking abundance of zooplantonic crustaceans and the susceptibility of small planktonic organisms to field and laboratory study has resulted in a corresponding emphasis on the application of scientific effort to the examination of the food and feeding mechanisms of these animals (Monakov, 1972).

A. Feeding Mechanisms

Many species of Branchiopoda, Ostracoda, Copepoda and Cirripedia together with some of the smaller members of the Malacostraca, are primarily filter feeders, using modified appendages to separate small food particles from suspension. The actual mechanisms by means of which food is collected have often been examined in great detail and the accounts of Cannon (1927, 1929, 1933*a* and *b*) (Branchiopoda, Ostracoda, Copepoda and Leptostraca), Cannon and Manton (1927, 1929) (Syncarida, Mysidacea), Crisp and Southward (1961) (Cirripedia), Fryer (1957, 1964, 1966, 1968) (Anostraca, Copepoda, Cladocera, Thermosbaenacea) Kozhova (1956) (Copepoda) and Yakovleva (1968) (Ostracoda) are a reasonable introduction to the literature of the subject. In the other major groups of aquatic crustaceans, feeding mechanisms are so varied and so poorly documented that no generalizations are warranted.

B. Forms which Ingest Plant Litter

References to the ingestion of plant detritus by aquatic crustaceans are numerous and cover most groups. The main body of evidence for plant litter feeding has usually been obtained from microscopic examination of gut contents. There is every degree of specialization in diet from species feeding almost exclusively on living macrophyte material, as is the case in freshwater crayfishes of the family Astacidae which are important consumers of macrophyte tissues (Dean, 1969; Gaevskaya, 1969), to species which feed largely on plant remains contained in the faecal pellets of other organisms, as is the case in the marine crabs, prawns and ghost shrimps cited by Frankenberg and Smith (1967).

Many anostracans ingest small suspended particles and, although some are basically algal feeders like *Streptocephalus dichotomus* Baird (Bernice, 1971) and others such as *Branchinecta gigas* Lynch are carnivores (Fryer, 1966), a number appear to consume quantities of detritus. This is probably the case with *Cheirocephalus diaphanus* Prevost which Lowndes (1933) describes as containing detritus including large items of plant origin such as moss leaves. The same is also true of *Branchinecta paludosa* (O. F. Muller), the gut contents of which were observed to contain, in addition to other things, inorganic particles and moss fragments (Fryer, 1966).

The cladocerans *Moina rectirostris* (Leydig) and *Ceriodaphnia quadrangula* (O. F. Muller) consume both algal and macrophytic detritus (Esipova, 1971) and the food of freshwater ostracods range from the tissues of higher plants and algae to benthic detritus and, in the case of *Notodromas monacha* (O. F. Muller) consists of suspended algae and surface film Neuston (bacteria, Protozoa, etc., trapped in the surface film), Luferova, 1970; Luferova and Sorokin, 1970). The marine copepod *Calanus helgolandicus* (Claus) never ingests "natural ocean detritus", but dead diatoms and the faecal material of *C. helgolandicus* itself are eaten in large quantities (Paffenhofer and Strickland, 1970).

Among the larger benthic crustaceans many species of amphipods are known to be plant litter feeders. The work of Martin (1966) on the marine species *Marinogammarus obtusatus* Dahl and *M. pirloti* Sexton and Spooner indicates that both feed predominantly on decaying algae. *Parahyalella whelplyi* Shoem is said to feed on the products of decomposition of *Thalassia testudinum* (turtle grass) (Fenchel, 1970). In both lotic and lentic fresh waters amphipods are an important group of litter feeders. Species of *Gammarus* consume large quantities of plant detritus (Minckley and Cole, 1963) and others such as *Hyalella azteca* and *Crangonyx richmondensis* (Hubricht and Harrison) feed on organic matter present in sediment deposits (Mathias, 1971). Isopods are also of importance in both marine and fresh waters. *Asellus aquaticus* L. and the closely related *A. meridianus* Rac. both feed to a large extent on allochthonous leaf litter which falls into streams and ditches (Williams, 1962; Levanidov, 1949) and *A. brevicaudus* Forbes and *A. intermedius* Forbes were observed to feed almost entirely on plant detritus in a woodland spring brook (Minshall, 1967). The marine genus *Idotea* includes many species most of which are omnivorous and at least on some occasions species such as *Idotea neglecta* G. O. Sars may feed largely on decaying marine algae (Naylor, 1955). An extreme case of plant litter feeding is found in the wood boring isopod *Limnoria lignorum* (Rathke) which is xylophagous, with particles of wood forming the major or perhaps the sole dietary source of the species (George, 1966).

The papers of Mauchline (1970*a*, *b*, 1971*a*, *b*, *c*, *d*) include references to marine mysids which, although they are predominantly carnivorous, may feed on plant litter. *Schistomysis ornata* (G. O. Sars) from Loch Etive, for example, was said at times to contain considerable quantities of fragments of terrigenous vegetable material and species of *Paramysis* and *Neomysis* also feed on plant remains to some extent.

Penaeid shrimps are closely associated with certain types of marine sediments (Williams, 1958) and include in their diet detritus and plant fragments obtained from the sediments. The freshwater crayfishes

Cambarus tenebrosus Hay and *Orconectes rusticus* (Girard) feed largely on leaf detritus (Minshall, 1967) and some tropical shore crabs are capable of separating organic matter on which they feed from muds and sands by means of specialized mouth parts (Griffin, 1971).

C. Factors Influencing Ingestion

The nature of food eaten by aquatic Crustacea often changes with the life stage involved (Klekowski and Shushkina, 1966) and also with changes in the availability of different food materials in the habitat, as occurs with *Gammarus pulex* L. which subsists almost entirely on organic detritus but consumes algae in the autumn when these are available (Scorgy, personal communication; Fig. 1). Kaushik and Hynes (1971) in their paper on the

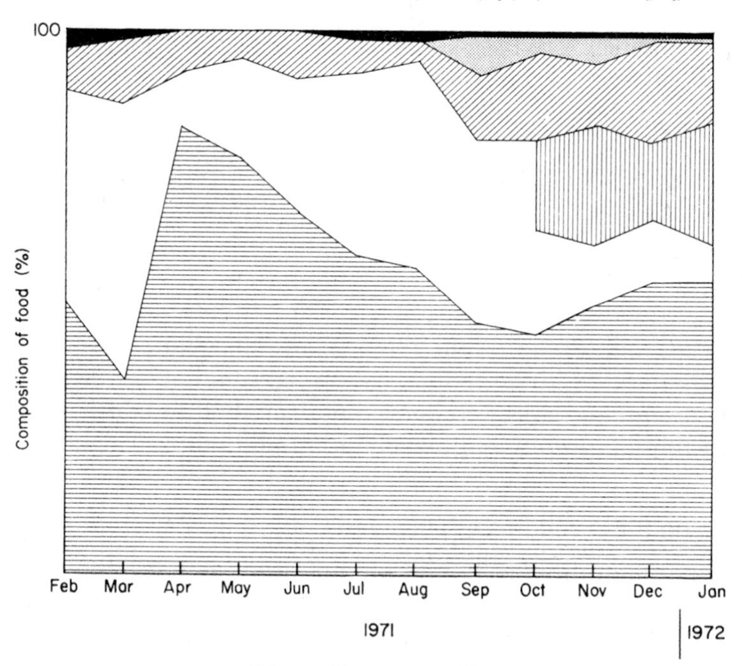

Fig. 1. *Gammarus pulex*

Composition of ingested food (mean of all size groups and substrata). After Scorgy in Progress Report of "Ecological study of the River Lambourn," October 1970–March 1972. (Unpublished report of Dept. of Zoology, University of Reading.)

Black = animal tissue.
Dots = *Cladophora*.
Oblique lines = autochthonous leaf tissue.
Fine vertical lines = allochthonous leaf tissue.
White = unidentified plant tissue.
Horizontal lines = fine detritus.

fate of dead leaves that fall into streams, considered not only the microbiological and chemical decomposition of several species of leaves under a variety of conditions but also included studies on leaf preference by invertebrates, particularly amphipods (*Hyalella azteca* and *Gammarus lacustris limnaeus* Bousfield) and an isopod (*Asellus communis* sensu Racovitza). Table I, taken from their paper, shows the effects of feeding

TABLE I. Feeding experiment with *Hyalella azteca* and *Asellus communis*. Mean values for the initial weight, loss in weight and percentage loss in weight of five species of tree leaves. (After Kaushik and Hynes, 1971)

Leaf type	*Hyalella*			*Asellus*		
	Initial wt (mg)	Wt loss (mg)	Loss as % initial wt	Initial wt (mg)	Wt loss (mg)	Loss as % initial wt
Ulmus	136·3	66·0	48·7	144·2	61·3	42·4
Acer	87·8	30·5	35·0	88·0	14·7	28·0
Alnus	187·7	18·2	9·7	198·0	30·2	15·2
Quercus	159·8	16·6	10·4	146·2	17·9	12·2
Fagus	93·1	7·1	7·6	94·3	7·4	7·7

by *Hyalella* and *Asellus* on the leaves of five different tree species. The animals showed distinct preferences for certain leaf species and elm (*Ulmus americana*) leaves were eaten more quickly than others, but leaves with microbial growth were generally preferred to those without.

D. Feeding Rates and Food Retention Times

Because the quantity and availability of plant litter as a potential food source changes in both time and space, it is not surprising that the feeding rates of many aquatic crustaceans are affected by environmental influences. A good deal is known about the effect of temperature and food concentration on the feeding rates of those small, planktonic crustaceans which can be studied experimentally. For example, *Daphnia longispina* O. F. Muller which, although it feeds largely on bacteria, includes particles of detritus in its food (Manuilova, 1958) alters its feeding rate in accordance with the concentration of food particles (Monakov and Sorokin, 1961). Ivanova (1966, 1967, 1970) used loam suspensions to study the effect of varying density and calorific value of suspended food on the filtering rates of Cladocera.

The ostracod *Cypridopsis vidua* (O. F. Muller) showed its highest feeding intensity when presented with detritus of the pond weed *Potamogeton* together with associated bacteria and Protozoa (Luferova and Sorokin, 1971).

Actual retention times of food in the gut vary greatly from species to species. *Eurycercus lamellatus* O. F. Muller, a large species of chydorid cladoceran, feeds chiefly on small (20–50 μm) particles of detritus and *Chydorus sphaericus* (O. F. Muller) feeds only on detritus (Smirnov, 1962). Basing his calculations on the rate of production of faeces by these animals, Smirnov estimated gut retention times for *Eurycercus* of as little as seven minutes.

Information on the rates of feeding of non-planktonic forms is sparse and only a few attempts have been made to overcome the technical problems involved in making such measurements. Hargrave (1972) predicted egestion by *Hyalella azteca* showing that faecal pellet production increased with increases in temperature from 5° to 20°C and that texture was an important factor affecting the rate of food intake. Rates of ingestion in feeding experiments with epiphytes, sediment and *Ulmus* leaves indicated that the rate directly reflected the ease with which food material could be "bitten". *Hyalella* would not ingest *Ulmus* leaves unless they had been softened by soaking in water for 3–4 days. Ivleva (1970) has shown a positive correlation between the rate of food consumption by the prawn *Leander adspersus* and the temperature of the water. Evidence presented by Sushchenya and Khemeleva (1967) indicates that there is a relationship between food ratio and body weight over a wide range of crustaceans from Phyllopoda to Decapoda.

III. Digestion and Assimilation

With regard to the nutrition of aquatic crustaceans, the aspects considered which relate to plant litter feeding include digestion and assimilation studies involving both natural and artificial diets. The physiology of digestion has received much attention in the larger species. A general introduction to digestion in Crustacea was given by van Weel (1970) who mentioned the crayfish genera *Astacus* and *Procambarus* as being among the few crustaceans in which digestion has been extensively studied.

A. Enzyme Production

Amylolytic activity is widespread. The amylase of *Artemia salina* (L.) has been studied (Urbani *et al.*, 1953) and these animals have also been demonstrated to produce a dipeptidase. *Gammarus roeseli* Gervais, a freshwater amphipod, has a gut exopeptidase which was the subject of work by

Ponyi *et al.* (1969) who compared the enzyme complement of this species with that of the litter feeding isopod *Asellus aquaticus*. The marine amphipods *Marinogammarus obtusatus* and *M. pirloti* which feed on decaying algae, produce in the gut fluids a non-specific esterase, a true lipase, α and β glucosidase and α and β galactosidase (Martin, 1966) but no cellulase could be demonstrated, thus confirming the observations of Agrawal (1964). Digestion has also been examined in the burrowing amphipod *Corophium volutator* Pallas which feeds almost entirely by selecting particles of litter, and the attached micro-organisms, from the muddy sediments in which it lives. *Corophium* caecal enzyme extracts exhibit powerful amylase, glycogen phosphorylase, β fructo-furanosidase (sucrase) and α-glucopyranosidase (maltase) activities; once again no cellulase activity was detectable (Agrawal, 1963). In contrast, the ability to digest cellulose has been demonstrated in the marine intertidal amphipod *Orchestia gammarella* (Pallas) which feeds chiefly on algal litter (Wildish and Poole, 1970).

Hylleberg Kristensen (1972) compared the carbohydrases produced by a number of marine invertebrates, including in his comparison several species of crustaceans which feed on detritus. His conclusion was that the enzyme spectra of the animals gave only limited information about their food sources. The omnivorous isopod *Idotea*, for example, degraded 10 different carbohydrates and *Corophium volutator* had a similar spectrum; this was in sharp contrast to the small number of enzymes produced by the barnacle *Balanus crenatus* (Brugiere) which feeds largely on zooplankton. However, a low number of carbohydrases was also found in the specialized sand grain-feeding amphipods of the genus *Bathyporeia*.

The wood burrowing isopod *Limnoria lignorum* produces its own cellulase and is able to digest both cellulose and hemicellulose (Ray and Julian, 1952) and a range of species of *Brachyura*, *Astacura* and *Palinura* have also been shown to produce cellulase enzymes (Kooiman, 1964; Yokoe and Yasumasu, 1964; Yasumasu and Yokoe, 1965).

With the exception of the above work, and some information concerning enzyme production in large carnivorous species, there is little material relating to crustacean digestion (Prosser and Brown, 1961). Even less is known about the other dietary requirements of these animals, although *Limnoria* is said to obtain amino acids from fungi living in the wood upon which it feeds (Lane, 1959).

B. Assimilation

Assimilation estimates employing techniques involving the use of both natural and artificial diets have been applied to the study of aquatic litter feeders. A general account of the assimilation of food by planktonic crustaceans is given by Pechen-Finenko (1971), and some important references are listed in Table II.

Smirnov (1962, 1965) published details of the nutrition of Chydoridae, some of which ingest large quantities of plant detritus. The earlier paper, which deals with the large species *Eurycercus lamellatus* quotes assimilation efficiencies $\left(\frac{\text{assimilation (inc. respiration)}}{\text{ingestion}}\right)$ for immature animals of that species (based on respirometry and direct experiment) in the order of 8% while mature specimens attained efficiencies of about 22%. In contrast *Polyphemus*, a large (normally predatory) cladoceran, when feeding on detritus of *Elodea* (Canadian pond weed), assimilated daily only about 0·1% of the animal's body weight (Butorina and Sorokin, 1966). In their summary of results on the nutrition of aquatic animals Monakov and Sorokin (1972) mentioned a 42% efficiency of assimilation of plant detritus by the ostracod *Dolerocypris fasciata* (O. F. Muller). In general the efficiency of detritus assimilation is lower and the level of its utilization by Ostracoda depends largely on the age and origins of the detritus. Luferova and Sorokin (1970, 1971) indicated a range of assimilation efficiencies from 27·4% to 84·6% dependent on the type of detritus presented to *D. fasciata* and *Cypridopsis vidua*.

With regard to the Copepoda, Butler *et al.* (1969, 1970) examined the feeding efficiency of the marine copepod *Calanus* as part of a lengthy study dealing with the nutrition and metabolic activities of zooplankton. The results indicated a decline in the efficiency of food utilization in winter when detritus (presumably largely of plant origin) succeeded living phytoplankton as the predominant food material.

The amphipod *Hyalella azteca* (from Marion lake) when feeding on sediment at the water–sediment interface was calculated to assimilate about 15% of the total organic matter present (Hargrave, 1970*b*, 1971) while assimilation by the litter-feeding isopod *Asellus aquaticus*, when estimated by the ash ratio method of Conover (1966), amounted to 55% in single cultures and 41% in group cultures, as opposed to 26·35% when calculated from direct measurements of production and respiration (Klekowski *et al.*, 1972). Prus (1971) published data detailing the assimilation efficiency of *A. aquaticus* giving a value of 30% as an over all annual mean. The food provided for these experiments consisted of discs cut from wet, slightly decomposed leaves of *Alnus glutinosa*. Prus also included a critical discussion of the work of Levanidov (1949) on the same subject and compared his results with those obtained for terrestrial isopods. He concluded that the assimilation efficiency value of 56% found by Levanidov was artificially high, because no precautions were taken to prevent reingestion of faeces, these being collected at the end of several months. Prus

TABLE II. Assimilation efficiencies $\left\{\frac{\text{assimilation (inc. respiration)}}{\text{ingestion}}\right\} \times 100$ of aquatic Crustacea feeding on plant litter

Species	Diet	% Assimilation	Authority
CLADOCERA			
Eurycercus lamellatus	Detritus	8–22	Smirnov, 1962
OSTRACODA			
Dolerocypris fasciata	Detritus of blue green algae	72·5	Luferova and Sorokin, 1970, 1971
Dolerocypris fasciata	Detritus of *Chlorella*	27·4	Luferova and Sorokin, 1970, 1971
Cypridopris vidua	Detritus of *Potamogeton* + bacteria and Protozoa	84·6	Luferova and Sorokin, 1970, 1971
Cypridopris vidua	Detritus of *Potamogeton* (sterile)	61·5	Luferova and Sorokin, 1970, 1971
AMPHIPODA			
Hyalella azteca	Total organic matter of surface sediment	14·9	Hargrave, 1971
ISOPODA			
Asellus aquaticus	Decaying *Alnus* leaves	26–44	Prus, 1971
DECAPODA			
Palaemonetes pugio	^{14}C-labelled *Zostera* detritus (sterile)	72·7	Adams and Angelovic, 1970
Palaemonetes pugio	^{14}C-labelled *Zostera* detritus (unsterilized)	90·9	Adams and Angelovic, 1970
Palaemonetes pugio	*P. pugio* faecal pellets (derived from *Nitschia*)	78	Johannes and Satomi, 1966
Palaemonetes pugio	Pellets reingested	40	Johannes and Satomi, 1966

states that the efficiency values obtained by the gravimetric method appear to be the most reliable and are close to those obtained for the terrestrial isopods *Oniscus asellus* L. and *Porcellio scaber* Latr. In all cases the assimilation efficiency of large winter males was only half that of ovigerous females.

Adams and Angelovic (1970) examined the assimilation of dead and decomposing *Zostera marina* (eel grass) by the grass shrimp (*Palaemonetes pugio*). Dead *Zostera* is present in large quantities in the eel grass beds of North Carolina estuaries and the shrimp was shown to assimilate labelled detritus (^{14}C) from its food as well as from associated micro-organisms and soluble organic material. Grass shrimps respire almost four times more $^{14}C\ CO_2\ mg^{-1}$ when they are fed detritus than when fed on *Zostera*, probably because they feed more readily on the former material but also because of more efficient assimilation. Johannes and Satomi (1966) also considered the assimilation efficiency of *P. pugio* when feeding on its own faecal pellets (derived from initial feeding on the diatom *Nitschia closterium*) and arrived at carbon assimilation values of between 40% and 79%.

Forster and Gabbott (1971) used artificial diets to investigate the assimilation efficiency of the prawns *Palaemon serratus* Pennant and *Pandalus platyceros* Brandt using chromic oxide as an inert reference material in the food (McGinnis and Kasting, 1964). Assimilation of plant carbohydrates was shown to be very efficient, particularly in *P. serratus* which completely assimilates wheat starch and dextrin and assimilates 90·6% of potato starch.

IV. Litter Feeding and Growth and Metabolism

The relationship between plant litter feeding and the growth, metabolism and production of crustacean consumers, has been expressed in a variety of ways. Organic matter, carbon and energy content have received much attention in recent years with particular emphasis often being placed on the use of radioactive isotopes for evaluation of these parameters.

A. Influence of Food Quality on Growth and Reproduction

Otsuki *et al.* (1969) studied the value of residues from the decomposition of cells of the green alga *Scenedesmus* as food for the cladoceran *Daphnia carinata* King, concluding that the decomposed algal cells were only capable of supporting reduced rates of growth and reproduction of the species when compared with fresh cells of the same alga. Work by Rodina (1963) showed that detritus was "excellent food upon which some Cladocera were able to go through successive generations". The Cladocera in this case were fed upon newly collected lake detritus. Microbiologically sterile suspensions of artificial detritus made by pounding up plant tissues also had a positive nutrient value for Cladocera, although no eggs were laid, presumably because of some dietary deficiency. Smirnov (1962) showed that the chydorid *Eurycercus lamellatus* grew, reaching 2·5 mm in length, attained maturity and produced broods of up to 10 young (considerably less than the maximum potential of the species, when feeding on natural sieved detritus. In a subsequent experiment the same species died when fed only on a diet of living *Chlorella*. In a comparison of the food value of various forms of plant detritus, *Moina rectirostris* and *Ceriodaphnia quadrangula* attained their highest daily growth rates when fed on detritus freshly produced from phytoplankton. The lowest rates of growth were in animals fed on detritus collected from the pond bottom. When detritus which had been treated with a weak iodine solution to kill off the microflora, was used as food, fertility and rates of growth were reduced. The numbers of young per litter decreased from 40 to 10 in *M. rectirostris* and from 35 to 7 in *C. quadrangula* (Esipova, 1969).

Comita *et al.* (1966) studied the seasonal changes in weight, calorific value and organic content of the copepod *Calanus finmarchicus* (Gunnerus) and the associated *C. helgolandicus* which has been shown to ingest readily dead phytoplankton organisms (Paffenhofer and Strickland, 1970). The weight and calorific value of the two species, increased in the winter months (stage V copepodites) when detritus was probably the main source of ingested food. *Calanus* has not, however, been fed successfully on detritus in the laboratory.

One of the few fairly complete energy budgets for an aquatic crustacean is that produced for *Hyalella azteca* by Hargrave (1971). The budget was prepared on the basis of adult (700 μg) animals feeding at 15°C on lake sediment. The results are expressed in cal amphipod^{-1} h^{-1} and indicate that of the 0·0525 cal ingested approximately 82% were egested and the remainder contributed to production, respiration and excretion in proportions of about 2·8%, 8·9% and 6·7% respectively.

B. Biochemical Aspects

Apart from the above studies, passage of materials from plant litter to aquatic crustaceans has been studied biochemically and the transfer of carotenoids of plant origin into the metabolic organization of herbivorous and detritus feeding crustaceans has been examined in brine shrimps (*Artemia salina*) (Davies *et al.*, 1965) isopods (*Idotea metallica* Bosc) (Herring, 1969) and crabs (Gilchrist and Lee, 1967). More general biochemical studies have involved analysis of the chemical composition and seasonal changes of brine shrimps, mysids and euphausiids. Jezyk and

Fenicnac (1966) concluded that the basic pattern of fatty acids in *Artemia* closely resembled that of dietary or algal fatty acids but phospholipids differed considerably from those of algae. In *Neomysis integer* (Leach) there is a loss of total lipid in animals kept in artificial environments. Marine species of mysid have a higher content of chain polyursaturated acids than *N. integer* (Morris, 1971). Raymont *et al.* (1971) give details of the biochemical composition of euphausiids.

V. Discussion

In conclusion, it seems that the bulk of plant-litter-feeding aquatic Crustacea are to be found in a relatively few taxonomic categories. A few Branchiopoda and Cladocera appear to feed to a large extent on plant remains but the bulk of the activity both in marine and freshwater habitats is in the groups Amphipoda and Isopoda and possibly also the Ostracoda although the latter have received relatively scant attention. Within the larger Malacostraca a number of species of widely different forms are known to feed partially or wholly on detritus.

Of those species examined in detail only a few have been shown to be able to digest the relatively refractory components of plant detritus (cellulose, hemicellulose, lignin, etc.) and the remainder must either digest and assimilate the small proportion of readily available products of decomposition or in many cases utilize the fungi and bacteria which are almost always associated with detrital food material (Gak, 1972; Monakov and Sorokin, 1971; Romanenko, 1970). In view of the high proportion of refractory material normally to be found in plant litter, it is perhaps surprising that most of the results so far published, which relate to crustacean assimilation, indicate efficiencies of over 20% (Table II). This may be explained by the relatively long retention time of food in the guts of most species studied, which presumably facilitates efficient digestion of all available materials. In the one or two cases where retention times are short the assimilation efficiency appears to be lower. It would be of considerable interest to have details of food retention times in some of those species which are known to produce a relatively slow-acting cellulase enzyme.

Acknowledgements

I wish to thank Dr. J. B. Buchanan, Dr. G. Fryer, Dr. C. Pinder, Mr. I. S. Farr and Mr. J. A. B. Bass for advice and help in preparing the manuscript and Mr. A. Scorgie for access to some of his unpublished work. My thanks also go to Mrs. Muriel Henley for typing the manuscript.

References

ADAMS, S. M. and ANGELOVIC, J. W. (1970). *Chesapeake Sci.* **11**, 249–254.
AGRAWAL, V. P. (1963). *J. mar. biol. Ass. U.K.* **43**, 125–128.
AGRAWAL, V. P. (1964). *Proc. zool. Soc. Lond.* **143**, 133–141.
BERNICE, R. (1971). *Hydrobiologia* **38**, 507–520.
BUTLER, E. I., CORNER, E. D. S. and MARSHALL, S. M. (1969). *J. mar. biol. Ass. U.K.* **49**, 977–1001.
BUTLER, E. I., CORNER, E. D. S. and MARSHALL, S. M. (1970). *J. mar. biol. Ass. U.K.* **50**, 525–560.
BUTORINA, L. G. and SOROKIN, Y. I. (1966). *Trudy Inst. Biol. Vnutr. Vod.* **12**, 170–174.
CANNON, H. G. (1927). *Trans. R. Soc. Edinb.* **55**, 355–370.
CANNON, H. G. (1929). *Br. J. exp. Biol.* **6**, 131–144.
CANNON, H. G. (1933*a*). *Phil. Trans. R. Soc.* B **222**, 267–352.
CANNON, H. G. (1933*b*). *Trans. R. Soc. Edinb.* **57**, 739–764.
CANNON, H. G. and MANTON, S. M. (1927). *Trans. R. Soc. Edinb.* **55**, 219–253.
CANNON, H. G. and MANTON, S. M. (1929). *Trans. R. Soc. Edinb.* **56**, 175–189.
COMITA, G. W., MARSHALL, S. M. and ORR, A. P. (1966). *J. mar. biol. Ass. U.K.* **46**, 1–17.
CONOVER, R. J. (1966). *Limnol Oceanogr.* **11**, 338–345.
CRISP, D. J. and SOUTHWARD, A. J. (1961). *Phil. Trans. R. Soc.* B **243**, 271–307.
DAVIES, B. H., HSU, W. J. and CHICHESTER, C. O. (1965). *Biochem. J.* **94**, 26.
DEAN, JACK L. (1969). *U.S. Bur. Sport. Fish. Wildl. Tech. Pap.* **24**, 3–15.
ESIPOVA, M. A. (1969). *Sb. po. prudovomu rybovodstvu, Moskva*, 79–89.
ESIPOVA, M. A. (1971). *Trudy VNII prud. rybn. kh-va.* **18**, 69–75.
FENCHEL, T. (1970). *Limnol Oceanogr.* **15**, 14–20.
FORSTER, J. R. M. and GABBOTT, P. A. (1971). *J. mar. biol. Ass. U.K.* **51**, 943–961.
FRANKENBERG, D. and SMITH, K. L. (1967). *Limnol Oceanogr.* **12**, 443–450.
FRYER, G. (1957). *Proc. zool. Soc. Lond.* **129**, 1–25.
FRYER, G. (1964). *Trans. R. Soc. Edinb.* **66**, 49–90.
FRYER, G. (1966). *Proc. Linn. Soc. Lond.* **177**, 19–34.
FRYER, G. (1968). *Phil. Trans. R. Soc.* B. No. 795, **254**, 221–385.
GAEVSKAYA, N. S. (1969). "The Role of Higher Aquatic Plants in the Nutrition of the Animals of Freshwater Basins." National Lending Library for Science and Technology. (Translation from Russian.)
GAK, D. Z. (1972). *Dokl. Acad. Sci. U.S.S.R.* **203**, 696–698.
GEORGE, R. Y. (1966). *Science, N.Y.* **153**, 1262–1264.
GILCHRIST, B. M. and LEE, W. L. (1967). *J. Zool. Lond.* **151**, 171–180.
GRIFFIN, D. J. G. (1971). *J. Anim. Ecol.* **40**, 597–621.
HARGRAVE, B. T. (1970*a*). *Limnol. Oceanogr.* **15**, 21–30.
HARGRAVE, B. T. (1970*b*). *J. Anim. Ecol.* **39**, 427–437.
HARGRAVE, B. T. (1971). *Limnol. Oceanogr.* **16**, 99–103.
HARGRAVE, B. T. (1972). *Oikos* **23**, 116–124.
HERRING, P. J. (1969). *J. mar. biol. Ass. U.K.* **49**, 767–779.
HYLLEBERG KRISTENSEN, J. (1972). *Mar. Biol.* **14**, 130–142.

IVANOVA, M. B. (1966). *Dokl. Akad. Nauk. SSSR* **170,** 724–725.
IVANOVA, M. B. (1967). *Inf. Byull. Biol. vnutr. Vod* **1,** 37–40.
IVANOVA, M. B. (1970). *Polskie Archwm. Hydrobiol.* **17,** 161–168.
IVLEVA, I. V. (1970). *In* "Marine Food Chains." (J. H. Steele, ed.), pp. 96–111. Oliver and Boyd, Edinburgh.
JEZYK, P. F. and FENICNAC, A. J. (1966). *Lipids* **1,** 427–429.
JOHANNES, R. E. and SATOMI, M. (1966). *Limnol. Oceanogr.* **11,** 191–197.
JORGENSEN, C. B. (1966). "Biology of Suspension Feeding." International series of monographs in pure and applied biology, zoology division, Vo. 27. Pergamon, Oxford.
KAUSHIK, N. K. and HYNES, H. B. N. (1971). *Archs Hydrobiol.* **68,** 465–515.
KLEKOWSKI, R. and SHUSHKINA, E. (1966). *Zh. obshch. Biol.* **29,** 199–208.
KLEKOWSKI, Z., FISCHER, E., FISCHER, Z., IVANOVA, M. B., PRUS, T., SHUSHKINA, E. A., STACHURSKA, T., STEPIEN, Z. and ZYROMSKA-RUDZKA, H. (1972). *In* "Productivity Problems of Freshwaters." (Z. Kajak and A. Hillbricht-Ilkowska, eds), pp. 749–763. Warszawa-Krakow.
KOOIMAN, P. (1964). *J. Cell. Comp. Physiol.* **63,** 197–201.
KOZHOVA, O. M. (1956). *Izv. biologo-geogr. naucho-issled. Inst., Irkutsk.* **16,** 92–120.
LANE, C. E. (1959). *Ann. N.Y. Acad. Sci.* **77,** 246–249.
LEVANIDOV, V. YA. (1949). *Trudy vses. gidrobiol. Obshch.* **1,** 100–117.
LOWNDES, A. G. (1933). *Proc. zool. Soc. Lond.* **103,** 1093–1118.
LUFEROVA, L. A. (1970). *Inf. Bull. Inst. Inland Wat. Acad. Sci. U.S.S.R.* **7,** 36–40.
LUFEROVA, L. A. and SOROKIN, Y. I. (1970). *See* Monakov (1972).
LUFEROVA, L. A. and SOROKIN, Y. I. (1971). *Trans. inst. biol. inland wat. acad. sci. U.S.S.R.* **21,** 196–203.
MANN, K. H. (1972), *Mem. 1st ital. Idrobiol.* **29** (Suppl.). 353–383.
MANUILOVA, E. F. (1958). *Rep. Acad. Sci. U.S.S.R.* **4,** 246–251.
MARTIN, A. L. (1966). *J. Zool.* **148,** 515–525.
MATHIAS, J. A. (1971). *J. Fish. Res. Bd Can.* **28,** 711–726.
MAUCHLINE, J. (1970*a*). *J. mar. biol. Ass. U.K.* **50,** 169–176.
MAUCHLINE, J. (1970*b*). *J. mar. biol. Ass. U.K.* **50,** 381–396.
MAUCHLINE, J. (1971*a*). *J. mar. biol. Ass. U.K.* **51,** 641–652.
MAUCHLINE, J. (1971*b*). *J. mar. biol. Ass. U.K.* **51,** 653–658.
MAUCHLINE, J. (1971*c*). *J. mar. biol. Ass. U.K.* **51,** 339–346.
MAUCHLINE, J. (1971*d*). *J. mar. biol. Ass. U.K.* **51,** 347–355.
MCGINNIS, A. J. and KASTING, R. (1964). *J. agric. Fd Chem.* **12,** 259–262.
MINCKLEY, W. L. and COLE, G. A. (1963). *Occ. Pap. Adams Cen. Ecol. Stud.* **10,** 1–35.
MINSHALL, C. W. (1967). *Ecology* **48,** 1:139–149.
MONAKOV, A. V. (1972). *J. Fish. Res. Bd Can.* **29,** 363–383.
MONAKOV, A. V. and SOROKIN, Y. I. (1961). *Trudy Inst. Biol. Vodokhran.* **4,** 251–261.
MONAKOV, A. V. and SOROKIN, Y. I. (1971). *Trudy Inst. Biol. Vnutr. Vod.* **22,** 37–42.

MONAKOV, A. V. and SOROKIN, Y. I. (1972). *In* "Productivity Problems of Freshwaters." (Z. Kajak and A. Hillbricht-Ilkowska, eds), pp. 765–774. Warszawa-Krakow.

MORRIS, R. J. (1971). *J. mar. biol. Ass. U.K.* **51,** 21–32.

NAYLOR, E. (1955). *J. mar. biol. Ass. U.K.* **34,** 347–355.

OTSUKI, A., HANYA, T. and YAMAGISHI, H. (1969). *Nature, Lond.* **222,** 1182.

PAFFENHOFER, G. A. and STRICKLAND, J. D. H. (1970). *Mar. Biol.* **5,** 97–99.

PECHEN-FINENKO, T. A. (1971). *Ekologiya* **3,** 64–72.

PONYI, J., BIRO, K. and P.-ZANKAI, N. (1969). *Annls Inst. biol. Tihany* **36,** 223–228.

PROSSER, C. L. and BROWN, F. A. (1961). "Comparative Animal Physiology." W. B. Saunders, Philadelphia.

PRUS, T. (1971). *Freshwat. Biol.* **1,** 287–305.

RAY, D. L. and JULIAN, F. R. (1952). *Nature, Lond.* **169,** 32–33.

RAYMONT, J. E. C., SRINIVASAGAM, R. T. and RAYMONT, J. K. B. (1971). *J. mar. biol. Ass. U.K.* **51,** 581–588.

RODINA, A. G. (1963). *Limnol Oceanogr.* **8,** 388–393.

ROMANENKO, V. I. (1970). *Mikrobiologiya* **39,** 711–715.

SMIRNOV, N. N. (1962). *Hydrobiologia* **20,** 280–294.

SMIRNOV, N. N. (1965). *Zool. J.* **44,** 1409–1411.

SUSHCHENYA, L. M. and KHEMELEVA, N. N. (1967). *Dokl. Akad. Nauk SSSR* **176,** 559–562.

URBANI, E., RUSSO, S. and ROGNONE, L. (1953). *Atti. Accad. a. nazl. Lincei, Rend. Classe. sci. fis. mat. e. nat.* **14,** 697–701.

VAN WEEL, P. B. (1970). *In* "Chemical Zoology." (M. Florkin and B. T. Scheer, eds), Vol. 5, pp. 97–115. Academic Press, New York.

WIESER, W. (1959). *Limnol Oceanogr.* **14,** 274–285.

WILDISH, D. J. and POOLE, N. J. (1970). *Comp. Biochem. Physiol.* **33,** 713–716.

WILLIAMS, A. B. (1958). *Limnol Oceanogr.* **3,** 283–290.

WILLIAMS, W. D. (1962). *Hydrobiologia* **20,** 1–30.

YAKOVLEVA, N. A. (1968). *Gidrobiol. Zh. Kiev.* **4,** 49–51.

YASUMASU, I. and YOKOE, Y. (1965). *Scient. Pap. Coll. Gen. Educ. Tokyo* **15,** 95–98.

YOKOE, Y. and YASUMASU, I. (1964). *Comp. Biochem. Physiol.* **13,** 323–338.

Part III

The Environment

19

Decomposition of Litter on the Soil Surface

S. T. Williams and T. R. G. Gray

Department of Botany
University of Liverpool
England

I. Gross Composition and Production of Litter

Before considering the influences of environmental factors on litter breakdown, it will be useful to make some comments on the formation and composition of litter. Litter production and composition are highly variable and have been comprehensively reviewed by Bray and Gorham (1964).

Annual litter fall varies considerably from one forest type to another (see Jensen, Chapter 3). Rates, expressed as t ha^{-1} $year^{-1}$, vary from 5·5–15·3 in equatorial rainforests, to 2·9–8·1 in warm temperate forests, to 1·0–6·9 in cool temperate forests. In arctic conditions the range is 0·6–1·5. Production also varies during the year but the seasonal pattern varies throughout the world. In equatorial regions, litter fall is continuous throughout the year with most in the first 6 months. If dry seasons occur, leaf fall can increase. In the warm temperate forests of Eastern Australia, leaf fall is again continuous with maxima in spring and early summer when temperature and precipitation rise. Maximum leaf fall in warm temperate forests of North America occurs in autumn as the temperature decreases. The most striking seasonal patterns occur in cool temperate forests, where autumnal cooling can lead to almost complete leaf fall in deciduous species. With gymnosperms the pattern is more variable. Finally it should be noted that litter production may vary from year to year. The ratio of highest to lowest annual litter fall varies from 2:7 for *Nothofagus* in New Zealand to 1:8 for *Acer saccharum* Marsh. in the Northern Hemisphere. Ratios can be higher for gymnosperms, a figure of 5:1 being obtained for *Pinus nigra* Arnold over a 12 year period.

Within litter, the proportion of the different components varies from one litter type to another (Table I). The major component is naturally leaf

TABLE I. Components of some litters
(From Bray and Gorham, 1964, see also Jensen, Chapter 3)

	Percentage of total litter				
	Leaf	Fruit	Branch	Bark	Other[a]
Pinus	60	11	12	14	<1
Pinus	62	17	←21→ (Branch + Bark + Other)		
Pinus	69	2	12	11	6
Picea	73	5	13	—	10
Picea-Betula	76	6	←18→ (Branch + Bark + Other)		
Betula	71	—	12	<1	16
Quercus	75	<1	15	9	—
Eucalyptus	60	15[b]	←25→ (Branch + Bark + Other)		

[a] Flowers, bud scales, fragments, epiphytes, insects.
[b] Including buds.

litter but other components are more variable. On average, the percentage of non-leaf material in forest litter is 30% for angiosperms and 29% for gymnosperms. Mineral material is present, the ash content of gymnosperms being 2–6% and angiosperms 8–14%. Organic material may be fractionated in various ways. Ether-soluble components range from 4 to 12% total dry wt, cold water-soluble matter from 3 to 14%, hot water-soluble from 3 to 9% and alcohol soluble from 3 to 13%. Hemicelluloses constitute from 10 to 19% total dry wt, cellulose 10–22%, lignin 5–8% and crude protein 2–15%.

II. Decomposition of Litter from Different Plants

It is clear that litter of different plants does not decompose at the same rate even under similar environmental conditions. This is undoubtedly due to differences in the structure and composition of their leaves and other parts.

Leaves of coniferous trees are generally decomposed more slowly than those of deciduous trees. It was estimated that a needle of *Pinus sylvestris* L. spent about 6 months in the L layer, 2 years in the F_1 and 7 years in the F_2 before being humified (Kendrick, 1959). Other parts of conifers are also resistant to decomposition. Extremely slow decomposition of softwood wood and bark was noted by Allison and Klein (1961), with many losing less than 7% of their carbon in the first 2 months after death. Allison and Murphy (1963) showed that most *Pinus* woods were decomposed more quickly than other softwoods, but still at much slower rates than hardwoods. The relative resistance of both leaves and the harder components of pine litter was reflected in the changing composition of litter in plantations of different ages (Ovington, 1962). As stands aged the proportion of harder structures, such as cones, in the litter layer, increased, while the contribution of the herbaceous ground flora decreased.

Certain non-coniferous plants also produce litter which decomposes slowly. In a study of *Calluna vulgaris* (L.) Hull litter, very small dry weight losses were recorded over long periods of time (Cormack and Gimingham, 1964). Decomposition rates varied between sites, with most rapid breakdown occurring on bare soil. In all situations, however, the litter was shed at a far faster rate than it was decomposed and hence it accumulated. Leaves of *Juncus squarrosus* L. spend about 3 years in the litter layers above acid peat, with a dry weight loss in the first year of 20–25% (Latter and Cragg, 1967). Petioles of *Pteridium aquilinum* (L.) Kuhn. lost 7–8% of their dry weight per year (Frankland, 1966). Formation of slowly decomposing mor litter in grasslands is associated with fibrous, shallow-rooting, acid-tolerant species such as *Agrostis* spp., *Festuca ovina* L., *Nardus stricta* L. and *Molinia caerulea* (L.) Moench (Barratt, 1966).

Litter of deciduous trees usually decomposes more rapidly but considerable variation occurs between different types. Melin (1930) included *Fraxinus americana* L. and *Betula papyrifera* Marsh. among fast decomposers and *Fagus grandifolia* Ehrh. among the slowest. Bocock and Gibert (1957) observed that leaves of *Betula verrucosa* Ehrh. and *Tilia cordata* Mill. were more rapidly decomposed than those of *Quercus robur* L. and *Q. petraea* (Mattusch.) Liebl. (Table II). Comparison of decomposition rates on different sites showed that they were faster on mull sites although the relative rates of various species were generally similar on all sites.

TABLE II. Disappearance of tree leaves placed on various sites (Data from Bocock and Gilbert, 1957)

	Loss in dry wt (%) in 6 months		
	Mull	Moder	Peat
Betula verrucosa (ex limestone)	82·9	26·3	27·8
Betula verrucosa (ex peat)	91·5	29·4	24·8
Tilia cordata	55·6	22·6	17·8
Quercus petraea	26·2	23·2	21·8
Quercus robur	17·4	17·0	16·7

Dry weight losses of leaves of *Quercus petraea* and *Fraxinus excelsior* L. in nylon nets on mull and moder sites were compared by Bocock *et al.* (1960). Rates of decomposition of *Quercus* were similar on both sites but *Fraxinus* disappeared more rapidly on the mull, only a few midribs remaining after 6 months. Figures after 6 months (expressed as % dry weight remaining) were *Fraxinus* on mull 5%, on moder 50%, *Quercus* on mull 85%, on moder 78%.

Disappearance of litter of a number of trees and shrubs in coarse mesh bags was followed by Bocock (1964). These bags allowed access to soil fauna, especially earthworms, and resulted in considerable differences in rates between mull and moder sites for some species. For example, the dry weights remaining after 148 days on the mull were about 10% for *Alnus glutinosa* (L.) Gaertn. and 20% for *Betula pendula* Roth. On moder the figures were 55% and 70% respectively (see Table XI).

Heath *et al.* (1966) studied breakdown rates of leaf discs from various plants held in different sized mesh bags. Three groups were recognized: (i) the fastest, in which microbial decomposition alone caused their disappearance (e.g. *Brassica oleracea* L., *Vicia faba* L.); (ii) fast at first but some persistence in small mesh bags (e.g. *Betula alba*, *Fraxinus excelsior*, *Ulmus campestris*); (iii) much slower in large or small mesh bags (e.g. *Quercus robur*, *Fagus sylvatica*). The effect of earthworms on disappearance of these leaves was most marked.

Differences in decomposition rates of leaves from the same plant were noted by Heath and Arnold (1966). "Shade" leaves, which were softer, disappeared more rapidly than "sun" leaves. This was attributed to the preference of worms for the softer leaves. It was also noted that hard leaves had a higher polyphenol content.

III. Environmental Factors and Decomposition

Consideration of environmental factors must include those which we may regard as "external" to the decomposition process (e.g. moisture, temperature) together with "internal" influences such as the chemical composition of leaves (see Miller, Chapter 4).

A. Temperature

This is undoubtedly one of the major factors controlling litter decomposition. Rate of decomposition in a tropical rainforest was found to be 8·2 t ha^{-1} $year^{-1}$ (Wanner, 1970), while in arctic litter containing lichens, *Luzula confusa* and *Poa arctica*, a rate of less than 1·2 t ha^{-1} $year^{-1}$ was estimated (Douglas and Tedrow, 1959). To a large extent such results are explained by differences in air temperatures and moisture supply. Results of various work on tropical and temperate forests were summarized by Madge (1965) (Table III). In tropical forests there is generally little or no litter accumulation with a fast turnover of organic matter in the soil. In temperate forests, little accumulates, this being particularly marked in coniferous stands. It was not possible, however, to show a clear relationship between leaf fall, litter accumulation, temperature and rainfall.

In the upper layers of soil, steep temperature gradients exist (Macfadyen, 1968). The temperature at the surface is often clearly related to air temperature, especially if the surface is shaded, and litter acts as an insulating layer on the soil. Travleev (1960) measured temperatures throughout the year in the air, in the litter and at a depth of 10 cm in the soil (Table IV). It is clear that litter temperatures were close to those of air, generally being a little lower. These average temperatures disguise the considerable short term and spatial variations which occur.

It is not surprising, therefore, that correlations have been found between rates of litter decomposition and air temperatures measured on a local or wide scale. Seasonal changes in temperature have been related to decomposition rates. Rates of carbon dioxide evolution from *Quercus alba* and *Pinus rigida* Mill. forests were assessed by measuring carbon dioxide accumulation at night during temperature inversions (Woodwell and Dykeman, 1966). Spring and summer rates were 2–3 times greater than those in winter. Seasonal temperature effects on mull and mor litter breakdown were noted by Witkamp and van der Drift (1961). Different temperatures at various altitudes result in different decomposition rates. Thus losses in weight of *Sphagnum* litter were greater in lowland than upland bogs, the temperature differences being about 10°C throughout the year (Clymo, 1965).

TABLE III. Litter fall and accumulation in various forests
(Data from Madge, 1965)

Vegetation	Rainfall (cm year^{-1})	Temperature (°C average year^{-1})	Litter fall (kg ha^{-1} year^{-1})	Litter accumulation (kg ha^{-1})	Ratio fall/ accumulation	Authors
Tropical regions:						
Broad-leaved rainforest	825	25·5	7560	4500	1·7:1	Jenny (1948) Jenny *et al.* (1949)
Broad-leaved forest	255	21	9000	14,700	1:1·6	Jenny (1948) Jenny *et al.* (1949)
Mixed rain forest	170	24·5	12,300	4300	2·9:1	Bernard (1945) Laudelout and Meyer (1954)
Moist semi-deciduous-moist evergreen forest	163	18	9400	2220	4·2:1	Nye (1961) Greenland and Kowal (1960)
Mixed dry lowland forest	120	27	5600	1700–2450	2·2:1	Madge (1965)
Temperate regions:						
Quercus woodland	75	10	1380	39,980	1:29	Jenny *et al.* (1949)
Pinus stand	75	10	2800	167,700	1:59·6	Jenny *et al.* (1949)
Mixed woodland	85	10	3100	3600	1:1·2	Witkamp and van der Drift (1961)

TABLE IV. Average temperatures in air, litter and soil (°C)
(From Travleev, 1960)

	September–November	December–February	March–June
Quercus			
Air	7·7	−1·5	10·0
Litter surface	9·0	−1·2	13·8
Beneath litter	6·2	0·8	6·8
Soil 10 cm depth	4·9	0·2	5·3
Fraxinus			
Air	10·2	−1·4	13·2
Litter surface	10·5	−1·2	20·0
Beneath litter	10·5	−1·0	16·6
Soil 10 cm depth	8·5	−1·0	14·0

Correlations of temperature with microbial respiration and carbon dioxide evolution from litter of *Pinus echinata* Mill., *Quercus alba* L. and *Acer rubrum* L. have been obtained (Witkamp, 1966*a*, *b*). A daily cycle of carbon dioxide evolution with a pre-dawn minimum and an afternoon maximum. This is correlated with the daily temperature cycle at the litter surface (Witkamp, 1969*a*; Witkamp and Frank, 1969). When the soil temperature exceeds that at the surface, thermal convection of carbon dioxide-rich air may occur. Temperature also influences mineral cycling in litter (Witkamp, 1969*b*). The influence of temperature on decomposition processes in grass litter in the laboratory was studied by Floate (1970*b*). Over the experimental period, litter lost 40% of its carbon at 30°C and 12% at 5°C. Similarly 5·4% of its nitrogen was mineralized at 30°C and −0·3% at 5°C.

Bleak (1970) showed that litter decomposition could occur under snow. When the snow depth was more than 60 cm temperatures at the soil surface ranged from −2·5 to +1·2°C. Loss in weight of grass and broad-leaved litter occurred, figures ranging from 30% for *Agropyron trachycaulum* (Link) Malte ex H. F. Lewis to 51% for *Mertensia azironica* var. *leonardi* (Rydb.) I. M. Johnst., losses given are averages of losses through two consecutive winter periods (October–June).

Forest fires also cause major changes in the temperature of the litter and the surface soil, changes in the latter being less marked. Hoffman (1924) found that surface temperatures of 850°C resulted in organic matter at 0·75 in. depth reaching 49°C, while at 1·5 in. it was only 16°C. Increasing moisture content of soil resulted in better heat penetration. Although these changes are not permanent, some less temporary effects occur. Thus, the blackened, charred areas show greater fluctuations in temperature than unburned areas (Phillips, 1930).

B. Moisture and Aeration

Relatively little attention has been paid to the moisture characteristics of litter as compared with soil. Most measurements have been expressed as percentage moisture content or percentage moisture holding capacity, rather than moisture tension.

The water-holding capacity of litter can affect soil erosion, rate of water infiltration to roots and the likelihood of forest fires (Kittredge, 1955). It was found that the field moisture capacity varied with litter type, ranging from about 150% wt water/wt dry soil (litter) in *Picea menziesii* (Mirb.) Franco (douglas fir) to 206% for *Pinus attenuata* (knobcone pine). Grass litter had a much higher capacity of 245%. Clary and Ffolliott (1969) found that the moisture holding capacity of litter varied with depth, with greatest capacities being in the amorphous H layers and lowest in the L layers. Thus when water was added, 85% was retained by the H layer and only 3·5% by the L layer. Although little attention has been given to humidity of air in litter, this will also increase in the lower layers and influence the activity of decomposers. The importance of relative humidity on fungal colonization of *Dactylis glomerata* L. (cocksfoot) culms was demonstrated by Webster and Dix (1960).

The influence of moisture on litter breakdown has been demonstrated by a number of workers. In periods of drought, decomposition of mull and mor litter was retarded and numbers of saprophagous animals in litter

TABLE V. Approximate densities per m² of some saprophagous groups in litter from mixed hardwood coppice on sandy soil before and after a dry summer (1959). (From van der Drift, 1963.)

	Mull		Mor	
	1957	1961	1957	1961
Lumbricidae	100	50	60	40
Pulmonata	20	30	5	—
Isopoda	40	10	20	5
Diplopoda	100	40	35	10
Tipulidae	80	30	5	5
Enoicyla	900	—	—	—

were reduced (van der Drift, 1963) (Table V). Interaction of temperature and moisture in controlling mull and mor decomposition was noted by Witkamp and van der Drift (1961). In arctic conditions, moisture and temperature were important factors influencing carbon dioxide evolution in the field (Douglas and Tedrow, 1959). Weight losses of *Quercus alba*, *Fagus grandifolia* and *Morus rubra* L. litters were correlated with moisture content (Witkamp, 1963). A correlation with carbon dioxide evolution in the field from litter of *Pinus echinata*, *Quercus alba* and *Acer rubrum* was also obtained (Witkamp, 1966*b*). Microbial respiration in litter bags was influenced in decreasing order by temperature, bacterial density, moisture, and age of litter (Witkamp, 1966*a*). In a laboratory study, Floate (1970*c*) found that moisture content changes had only minor effects on mineralization of *Nardus* and *Agrostis–Festuca* litter.

Aeration of much litter is obviously directly associated with moisture content. The most limiting effects on decomposition occur when litter is waterlogged, resulting in impaired gaseous diffusion and development of anaerobic conditions. This may not be a common situation in terrestrial litter.

Little information is available on gas concentrations in litter or their influence on decomposition. Brierley (1955) analysed gas in tubes placed at the junction of the humus and litter layers in a *Fagus* forest. Results showed fluctuations in oxygen content in litter from 19·5 to 20·6% (cf. 20·9% for above-ground air). The carbon dioxide content varied from 0·1 to 0·7% (cf. 0·03% for above-ground air). No correlations with depth were found. Thus the gaseous content of litter is probably often close to that of the air above it.

C. pH and Base Content

The reaction and base content of litter are very variable and depend on such factors as the leaf types in the litter, the base status of the underlying soil and the extent to which the litter has been decomposed.

Most decomposing leaf litter is acidic, with that from conifers being more acidic than that from hardwood deciduous trees. Therefore in mixed forests, hardwood litter tends to increase the base content, improve the buffering capacity and decrease the acidity. Plice (1934) studied litter from a wide range of North American trees and distinguished between hypobasic, middle and superantacid types (Table VI). These groups were distinguished by their calcium oxide content (3·6–5·7%, 2·2–3·1%, 0·3–2·1%, respectively), their antacid buffering capacity (3·5–3·98, 3·0–3·4, 1·5–2·9 mequiv. hydrogen ions inactivated by 5 g dry wt litter) and their acidity (pH 3·3–4·0, 4·1–4·5, 4·6–5·9). Only a few litter types had a pH above 4·5 (e.g. *Ulmus americana* L., *Acer platanoides* L., *Fraxinus americana*) and these had a high base content. However, the base content of some of the more acid litters was also high (e.g. *Acer saccharum*). Lutz and Chandler (1946) recognized three groups of leaves on the basis of their

TABLE VI. Base content, buffering capacity and reaction of leaf litters (From Plice, 1934)

Hyberbasic series	Middle series	Super-antacid series
1. High bases and buffer, acidity medium. *Carya microcarpa* Gray	1. High bases and buffer, acidity weak. *Carya glabra* (Mill.) Sweet *Magnolia acuminata* L. *Thuya occidentalis* L.	1. Medium bases, high buffer, acidity weak. *Acer platanoides* *Ulmus americana*
2. Medium bases, low buffer, acidity strong. *Picea canadensis* Gray, B. & B. *Picea excelsa* Link	2. Medium bases and buffer, acidity weak. *Fraxinus americana* *Betula lutea* Michx. f. Acidity strong. *Acer saccharum*	2. Low bases, medium buffer, acidity medium. *Castanea dentata* (Marsh.) Borkh. *Quercus rubra* L. sec. Duroi *Acer rubrum* *Quercus alba* *Larix europea* DC. *Larix laricina* (Duroi) K. Koch
3. Fairly low bases, low buffer, acidity strong to extreme. *Abies balsamea* (L.) Mill. *Picea maritima* *Pinus banksiana* Lamb *Tsuga canadensis* (L.) Carr	3. Fairly low bases and buffer, acidity weak. *Fagus grandifolia* Acidity strong. *Quercus Prinus* *Pinus sylvestris* *Pinus rigida* *Pinus strobus* L. *Pinus resinosa* Ait.	

calcium content, those with >2% calcium (% dry wt) (e.g. *Populus tremuloides* Michx.), those with 1–2% (e.g. *Picea abies* (L.) Karst.) and those with >1% (e.g. *Pinus sylvestris*).

Ovington and Madgwick (1957) found a broad correlation between pH of the soil and that of tree leaves and litter. This was most noticeable with hardwoods. Acid soil had litter of similar reaction but alkaline soil had litter with a pH lower than that of the soil. The F and H layers were more acidic than the L layers in 68 out of 73 conifer plantations studied. In hardwood litters there were no consistent differences between the layers.

The extent of decomposition of leaves also affects their reaction. Quite large changes occur during the initial phases of decomposition. Sjörs (1959) showed that most litters became less acidic during the first 34 days of decomposition on soil (Table VII). The initial decrease in acidity was

TABLE VII. Changes in litter reaction (pH) during early stages of decomposition (From Sjörs, 1959)

	Days on soil				
	0	5	12	21	34
Acer platanoides	5·3	4·8	4·7	4·8	5·0
Betula verrucosa	5·3	5·5	5·8	5·7	6·3
Alnus incana (L.) Moench	5·9	6·2	6·2	6·0	—
Populus tremula L.	5·6	5·8	6·3	6·5	6·8
Sorbus intermedia (Ehrh.) Pers.	4·8	5·8	6·1	6·1	6·4
Sorbus aucuparia L.	5·4	6·7	6·4	6·7	6·7
Fagus sylvatica L.	5·8	6·3	6·3	5·8	6·1
Ulmus glabra Huds.	5·9	5·8	5·9	5·9	6·4
Corylus avellana L.	6·2	6·9	6·7	6·5	6·5
Salix caprea L.	6·4	6·7	6·7	6·5	7·1
Fraxinus excelsior	6·3	7·6	7·4	6·9	7·3
Calamagrostis epigeios (L.) Roth.	5·8	5·9	5·6	5·7	6·2

almost certainly due to leaching of acidic solutes (e.g. organic acids) in the cell sap. Nykvist (1959*a*, *b*) showed clearly that material leached from litter became less acidic with time. The extent of the rise in pH depended on the aeration of the litter, the length of the leaching period and the susceptibility of the litter to decomposition. Thus in aerobic conditions, the pH of *Fraxinus* leachates rose from 5·6 to 7·6 in 24 h, and to 7·1 in 42 days. In anaerobic conditions, the corresponding values were 5·5 and 5·4. Similar, though less clear-cut, results were obtained with *Pinus* litter. Initial rises in pH were also found by Mattson and Koutler-Anderson (1954), the greatest rises occurring in litters with a high initial excess base content. However, after a few years a decrease in pH occurs as bases are removed and humic acid forms. After 14 years, humus from all the litters, with the exception of *Ulmus*, was acidic.

Base content of litter also varies throughout the year. Wright (1956) observed a marked fall in the content of calcium, potassium, sodium and magnesium in *Pinus* litter in summer. Remezov (1961) examined the effects of leaching on *Quercus* litter. The percentages remaining after 2 months were—

	K	P	N	Mg	Ca	Si	Al	Fe
%	46	53	65	65	72	83	90	100

Losses of potassium from the F layer were most marked. Latter and Cragg (1967) studying *Juncus* litter found more than 50% loss of all inorganic ions tested within 5 months, with potassium, phosphorus, magnesium and calcium losses being most marked. A complicating factor is the carriage of substances into the litter from the canopy by precipitation. Carlisle *et al.* (1966) showed that of the total amounts present in *Quercus* litter, 72% of the potassium came from the canopy.

There is evidence that the rate of litter decomposition is correlated with its base content. It is well known that many potential decomposer organisms are inactive or less active when the pH is much below 5·0. The excess base contents of freshly fallen leaves were measured by Broadfoot and Pierre (1939) and three groups were recognized, the extent of decomposition in the first 6 months increasing with base content (Table VIII). A marked

TABLE VIII. Excess base content of some tree leaves
(From Broadfoot and Pierre, 1939)

	Species	Excess base content	% decom-position in 6 months
Group 1	*Pinus strobus*	42	17
<75 mequiv excess base	*Pinus rigida*	44	21
per 100 g	*Fagus grandifolia*	70	17
	Platanus occidentalis L.	72	21
Group 2	*Quercus alba*	80	28
75–110 mequiv excess base	*Quercus velutina* Lam.	88	29
per 100 g	*Juniperus virginiana* L.	99	40
	Acer saccharum	109	33
Group 3	*Liriodendron tulipifera* L.	121	55
>110 mequiv excess base	*Robinia pseudo-acacia*	145	47
per 100 g	*Cornus florida* L.	152	43
	Aesculus hippocastanum L.	174	30

positive correlation was obtained for the last 4 months of the decomposition period (Table IX). Voigt (1965) did not obtain a clear correlation between leaf calcium content and weight or nitrogen loss but greatest nitrogen losses occurred in species with the highest calcium content (Table X). Addition of calcium carbonate also stimulated nitrogen loss (see p. 624).

TABLE IX. Relative effects of independent variables on litter decomposition (calculated from standard partial regression coefficients and expressed as percentage). (From Broadfoot and Pierre, 1939)

	0–2 months	2–6 months	0–6 months
% excess base	4	97	19
% water soluble organic matter	51	1	44
% total nitrogen	45	2	37

D. Water-soluble Organic Matter Content

This component of the litter provides a readily available energy source for decomposers and has, therefore, most influence in the initial stages of decomposition. Usually there is a rapid loss of soluble organic matter due to microbial utilization and leaching. A 50% decrease in the first 3 months was recorded for *Casuarina* litter (Burges, 1958). Sowden and Ivarson (1962) found that sugars were removed more rapidly from deciduous than from coniferous litter during the first 165 days. Broadfoot and Pierre (1939) noted a correlation between litter decomposition and the content of soluble organic matter. This was most marked during the first 2 months (Table IX). Melin (1930) suggested that water-soluble material only influenced decomposition rates for the first few weeks. The content of hot-water-soluble materials in *Fraxinus excelsior* leaves was 32% (w/w), while in *Quercus petraea* it was only 18% (Gilbert and Bocock, 1960). After a month the content of the *Fraxinus* dropped to 1·5% but in the *Quercus* the decline was more gradual.

Although it seems that amounts of these simpler carbon compounds decrease rapidly in the early stages of decomposition, they are not completely mineralized. When ^{14}C-labelled glucose was added to soil, 11% of the carbon was still present in soil after 30 days incubation in the laboratory (Mayaudon and Simonart, 1959). Such compounds may in part be resynthesized into microbial tissues or products more resistant to decay than the original (Clark and Paul, 1970). There is also evidence that microbial

TABLE X. Decomposition of litter incubated under laboratory conditions for three months at 32°C. (From Voigt, 1965)

Species	Ca in fresh litter (%)	N in fresh litter (%)	Litter alone			Litter + $CaCO_3$		
			Wt loss (%)	N deficit (%)	Final pH	Wt loss (%)	N deficit (%)	Final pH
Alnus rugosa (Du Roi) Spreng.	1·15	2·18	34·2	24·3	5·5	33·6	41·7	7·0
Liriodendron tulipifera	2·16	1·87	48·8	36·8	7·1	47·9	53·1	8·1
Cornus florida	2·20	1·63	44·6	17·8	7·2	33·4	23·1	7·4
Tsuga canadensis	0·39	1·65	36·8	12·9	6·3	23·4	11·9	6·8
Pinus resinosa	0·36	1·50	18·2	12·7	5·7	17·8	11·4	6·6
Juniperus virginiana	1·30	1·40	31·5	13·1	7·3	32·8	14·5	8·2
Hardwoods (mean)	1·84	1·89	42·5	26·3	6·5	38·3	39·3	7·2
Conifers (mean)	0·68	1·52	28·8	12·9	5·9	24·7	12·6	7·1

utilization of compounds such as glucose and lysine is inhibited by tannins which are present in litter (Basaraba, 1966; Benoit *et al.*, 1968).

E. Nitrogen Content

During decomposition, carbon is used as an energy source by decomposers while nitrogen is assimilated into cell proteins and other compounds. Thus a high nitrogen content in the original material promotes decomposition, at least in the early stages. Fresh plant materials vary considerably in their nitrogen content. Woody materials have from about 0·2–0·5% (w/w) while herbaceous plants such as grasses have from 0·5–1·5% and legumes have 1·5–3·0% (Bartholomew, 1965). Tropical forest litter has a higher nitrogen content than that of temperate forests (Nye, 1961). In the early stages of decomposition, residues with low nitrogen contents may show an increase in organic nitrogen as available nitrogen is used by initial colonizers. On the other hand, those with more nitrogen show a decline in organic nitrogen as net mineralization occurs (Bartholomew, 1965). After the initial period of breakdown, further net changes in organic nitrogen are slow and some remains throughout the decomposition process.

The significance of initial nitrogen contents to decomposition of litter has received some study (see Forbes, Chapter 23; Gray and Biddlestone, Chapter 24). Nitrogen content of deciduous leaves is somewhat higher than that of conifers. Alway *et al.* (1933) found that there was 1·5 times as much nitrogen in litter of *Acer saccharum* and *Tilia glabra* Ventenat as in that of *Pinus* spp. Nitrogen contents of a range of litters were determined by Broadfoot and Pierre (1939), figures ranging from 0·46% (w/w) for *Pinus rigida* to 2·06% for *Robinia pseudo-acacia* L. Correlation between rate of decomposition and nitrogen content was greatest during the first 2 months (Table IX). Fenton (1958) obtained figures of 1·58% for *Betula* compared with 0·62% for *Pinus radiata* Don, while Ivarson and Sowden (1959) recorded 0·79% in conifer litter and 1·01% for deciduous litter. Average values for hardwood leaves were slightly greater than those of conifers (Voigt, 1965) (Table X). Nitrogen contents of a wide range of leaves were determined by Bocock (1964); generally litter richer in nitrogen decomposed more rapidly and rates on mull were faster than on moder. The nitrogen content of slowly decomposing *Nardus* litter was 0·89% (w/w) for *Agrostis–Festuca* litter it was 1·39% (Floate, 1970*a*).

Changes in nitrogen compounds during decomposition depend to some extent on the initial nitrogen content. Little overall change in nitrogen content occurs during decomposition of some leaves. After 2 years, 85·9% of the original nitrogen remained in *Pinus sylvestris* litter and 95·9% in

Betula (Mikola, 1955). Some of this nitrogen was undoubtedly in microbial cells. Little change occurred in *Abies*, *Pinus* and *Picea* litter (Hayes, 1965). When litter in nylon nets was placed on different sites, the nitrogen content of *Fraxinus excelsior* was constant for the first 7 months on mull but increased on moder (Gilbert and Bocock, 1960). Increases in *Quercus petraea* on moder were considerable, reaching 160% in 10 months. In a

TABLE XI. Nitrogen contents of freshly fallen leaves of various trees and their decomposition. (From Bocock, 1964)

Species	Nitrogen (% dry wt)	Percentage of initial nitrogen left after 148 days[a]	
		on Mull	on Moder
Alnus glutinosa	3·06	10	55
Acer pseudoplatanus	2·52	20	50
Fraxinus excelsior	1·55	25	45
Tilia cordata	1·45	40	80
Corylus avellana	1·39	73	76
Salix sp.	1·38	45	60
Ulmus glabra	1·32	20	15
Fagus sylvatica	1·17	65[b]	80
Betula pubescens Ehrh.	1·13	42	75
Betula pendula	1·07	20	70
Aesculus hippocastanum	0·89	75	70
Quercus robur	0·79	40[b]	60
Quercus petraea	0·77	50[b]	65
Castanea sativa Mill.	0·69	75	70

[a] Approximate values from figures.
[b] After 267 days.

later study (Bocock, 1963) it was shown that nitrogen was added to the oak litter by atmospheric precipitation, insect frass and plant material falling from the canopy. Up to 25% of the inorganic nitrogen in rain was taken up by the litter. In a study of several litters, Bocock (1964) observed that the percentage nitrogen content of most increased during decomposition, with greater increases occurring on moder than on mull sites. Such additions of nitrogen have been overlooked in many previous studies of changes in nitrogen levels in litter.

Recovery of nitrogen from decomposing litter was measured by Voigt (1965). Weight losses and nitrogen deficits were more marked in hardwoods than conifers. Loss of nitrogen was sometimes increased by adding calcium carbonate and in most cases greatest nitrogen deficits were associated with relatively high pH values (Table X).

In a laboratory study of nitrogen mineralization in soil, Fenton (1958) found that additions of litter produced a marked decrease in mineralization. The depression produced by *Pinus radiata* litter was greater than that by *Betula* sp. Release of nitrate occurred with the latter but not with *Pinus*. Ivarson and Sowden (1959) noted that the initial release of ammonia was greater from deciduous than from coniferous litter. Nitrate formation-occurred in the former but not the latter and they suggested that this was due to the higher initial nitrogen content in the deciduous litter. The amount of nitrogen mineralized during decomposition of grassland litters can also be related to their initial nitrogen contents (Shaw, 1958; Floate, 1970*a*). In *Agrostis–Festuca* litter, 13·9% of the original nitrogen (1·39% w/w) was mineralized compared with 6·8% from *Nardus* (0·89% w/w) during the same period (Floate, 1970*a*).

Non-leaf materials in litter, such as twigs and bark, have a lower nitrogen content than that of leaves. Partially decayed wood persists for many years and comprised from 14 to 30% (w/w) of the litters in North American forests (McFee and Stone, 1966). Nitrogen in wood and bark of trees was determined by Allison and Klein (1961), and Allison and Murphy (1962, 1963) (Table XII). The effect of nitrogen on decomposition was

TABLE XII. Nitrogen content of bark and woods
(From Allison and Klein, 1961; Allison and Murphy, 1962, 1963)

	% nitrogen	
	Wood	Bark
Softwoods	0·051–0·227	0·038–0·390
Pines	0·038–0·130	0·048–0·179
Hardwoods	0·057–0·104	0·102–0·413

also studied. With hardwoods, a significant increase in oxidation of wood carbon resulted from additions of nitrogen; for pines, increases were much smaller (Table XIII). No effects on decomposition of bark were obtained. Allison and Klein (1961) showed that decomposition of most softwoods was so slow that the soil nitrogen supply was adequate and additions of nitrogen gave no increases in decomposition rates.

Thus the initial nitrogen content of litter can have many influences on its subsequent decomposition. Effects are most marked in earlier stages of decomposition of litters with a high nitrogen content. In later stages there is little net change and organic nitrogen becomes more resistant to decomposition. This is possibly due to the synthesis of more recalcitrant nitrogenous organic compounds by decomposers and the protective association of proteins with polyphenols.

TABLE XIII. Decomposition of wood and bark with and without addition of nitrogen (From Allison and Murphy, 1962, 1963)

	% C oxidized after 60 days (average values)			
	Wood		Bark	
	−Nitrogen	+Nitrogen	−Nitrogen	+Nitrogen
Hardwoods	30·3	45·1	22·4	24·5
Pines	16·2	16·9	8·7	8·6

F. Polyphenol Content

Plants contain a variety of polyhydroxy phenols which comprise from 5 to 15% of their dry weight. Some may be extracted with water to give tannins and they may be thus leached from litters into the soil. There is increasing amount of evidence for the importance of these substances in controlling rates of litter decomposition. After initial loss of various components, decomposition rates decrease and organic matter becomes more resistant to attack. Minderman (1968) found that decomposition of litter in the field did not equal the sum of the decay rates of its individual components. Clark and Paul (1970) showed that this could be overcome by correcting data for individual components to allow for resynthesis of secondary metabolites which were more resistant. The accumulation of condensed or polymerized polyphenols with very slow decomposition rates is an important phenomenon. These may originate from the litter itself or be synthesized by microbes. Minderman pointed out that after 10 years or more, the rates of accumulation of litter are primarily determined by such components.

The role of polyphenols in the formation of different litter types has been studied (see Lofty, Chapter 14; Harding and Stuttard, Chapter 15; Edwards, Chapter 16). Handley (1954) showed that fresh leaves of some plants contained soluble substances which resembled tannins and caused precipitation of proteins. Later he obtained complexes of protein with these extracts and determined their stability by measuring uptake of nitrogen released from them using seedlings (Handley, 1961). Resistance was most marked in complexes prepared with leaf extracts of the mor-forming *Calluna vulgaris*. Those made with *Circaea lutetiana* L. promoted seedling growth by giving a controlled release of nitrogen. It was therefore suggested these protein-precipitating substances were of fundamental importance in determining the type of litter formed.

More extractable tannins occurred in surface litter at mor than at mull sites (Coulson *et al.*, 1960). There was also a greater quantity and diversity of polyphenols in fresh, senescent and recently fallen leaves of the same plant when grown on nutrient-poor mor than on nutrient-rich mull sites. Thus an interrelationship of soil, leaf composition and litter decomposition was indicated. Davies *et al.* (1964*b*) showed that in sand cultures, shortage of nitrogen and phosphorus increased amounts of polyphenols in leaves. Extracted polyphenols have been shown to form complexes with a variety of substances. Gelatin was tanned and its stability was greatest when tanning occurred at pH 3·0–5·0 (Davies *et al.*, 1964*a*). Similar results were

TABLE XIV. Influence of tannins[a] on decomposition of various substrates

Substrate	% inhibition		Authors
	pH 4·0	pH 7·0	
Gelatin	51	30	
Gliadin	54	21	Basaraba and Starkey (1966)
Peptone	32	38	
Glucuronic acid	6	−2	
Polygalacturonic acid	30	12	
Pectin	16	25	Benoit and Starkey (1968*a*)
Hemicellulose	62	59	Starkey (1968*a*)
Cellulose	35	51	
Young *Secale* plants	53	50	Benoit *et al.* (1968)
Mature *Secale* plants	47	48	
Chitin	33[b]	—	
Saponin	39[b]	—	Lewis and Starkey (1968)
Starch	49[b]	—	
Pectin	25[b]	—	

[a] Various tannins at 2% (w/w).
[b] 1·0% tannin in soil at pH 4·2.

obtained by Basaraba and Starkey (1966) who combined tannins with gelatin and gliadin at two different pH levels. Thus complexes formed in acidic litter are likely to be more resistant. Inhibitory effects on decomposition of a number of other substances have been shown (Table XIV). Resistance is probably conferred by inhibition of microbial enzyme action and the increased recalcitrance of the complex, those formed by high molecular weight tannins being the most stable (Benoit and Starkey, 1968*b*). Thus there is good evidence for the importance of polyphenols in determining differential rates of litter decomposition, their effects being influenced by the pH of the litter and the soil nutrient status.

In the later stages of decomposition, humic materials, which also contain polyphenols, can influence decomposition. It has been shown that the action of microbial proteolytic enzymes can be inhibited or stimulated by humic acids (Ladd and Butler, 1969).

References

Allison, F. E. and Klein, C. J. (1961). *Proc. Soil Sci. Soc. Am.* **25,** 193–196.

Allison, F. E. and Murphy, R. M. (1962). *Proc. Soil Sci. Soc. Am.* **26,** 463–466.

Allison, F. E. and Murphy, R. M. (1963). *Proc. Soil Sci. Soc. Am.* **27,** 309–312.

Alway, F. J., Kittredge, J. and Methley, W. J. (1933). *Soil Sci.* **36,** 387–398.

Barratt, B. C. (1966). *Proc. N.Z. ecol. Soc.* **13,** 24–29.

Bartholomew, W. V. (1965). *In* "Soil Nitrogen." (W. V. Bartholomew and F. E. Clark, eds), pp. 285–306. American Soc. Agronomy.

Basaraba, J. (1966). *Can. J. Microbiol.* **12,** 787–794.

Basaraba, J. and Starkey, R. L. (1966). *Soil Sci.* **101,** 17–23.

Benoit, R. E. and Starkey, R. L. (1968*a*). *Soil Sci.* **105,** 203–208.

Benoit, R. E. and Starkey, R. L. (1968*b*). *Soil Sci.* **105,** 291–296.

Benoit, R. E., Starkey, R. L. and Basaraba, J. (1968). *Soil Sci.* **105,** 153–158.

Bernard, E. (1945). *I.N.E.A.C. (Brussels)* **4,** 1–240.

Bleak, A. T. (1970). *Ecology* **51,** 915–917.

Bocock, K. L. (1963). *J. Ecol.* **51,** 555–566.

Bocock, K. L. (1964). *J. Ecol.* **52,** 273–284.

Bocock, K. L. and Gilbert, O. J. W. (1957). *Pl. Soil* **9,** 179–185.

Bocock, K. L., Gibert, O. J. W., Capstick, C. K., Twinn, D. C., Waid, J. S. and Woodman, M. J. (1960). *J. Soil Sci.* **11,** 1–9.

Bray, J. R. and Gorham, E. (1964). *Adv. ecol. Res.* **2,** 101–157.

Brierley, J. K. (1955). *J. Ecol.* **43,** 404–408.

Broadfoot, W. M. and Pierre, W. H. (1939). *Soil Sci.* **48,** 329–348.

Burges, N. A. (1958). "Microorganisms in the Soil." Hutchinson, London.

Carlisle, A. A., Brown, H. F. and White, E. J. (1966). *J. Ecol.* **54,** 87–98.

Clark, F. E. and Paul, E. A. (1970). *Adv. Agronomy* **22,** 375–435.

Clary, W. P. and Ffolliott, P. F. (1969). *J. Soil. Water Conserv.* **24,** 22–23.

Clymo, R. S. (1965). *J. Ecol.* **53,** 747–784.

Cormack, E. and Gimmingham, C. H. (1964). *J. Ecol.* **52,** 285–297.

Coulson, C. B., Davies, R. I. and Lewis, D. A. (1960). *J. Soil Sci.* **11**, 20–29.
Davies, R. I., Coulson, C. B. and Lewis, D. A. (1964*a*). *J. Soil Sci.* **15**, 299–309.
Davies, R. I., Coulson, C. B. and Lewis, D. A. (1964*b*). *J. Soil Sci.* **15**, 310–318.
Douglas, L. A. and Tedrow, J. C. F. (1959). *Soil Sci.* **88**, 305–312.
Drift, van der J. (1963). *In* "Soil Organisms." (J. Doeksen and J. van der Drift, eds), pp. 125–133. North-Holland, Amsterdam.
Fenton, R. T. (1958). *Pl. Soil* **9**, 202–214.
Floate, M. J. (1970*a*). *Soil Biol. Biochem.* **2**, 173–186.
Floate, M. J. (1970*b*). *Soil Biol. Biochem.* **2**, 187–196.
Floate, M. J. (1970*c*). *Soil Biol. Biochem.* **2**, 275–283.
Frankland, J. C. (1966). *J. Ecol.* **54**, 41–63.
Gilbert, O. J. W. and Bocock, K. L. (1960). *J. Soil Sci.* **11**, 10–19.
Greenland, D. J. and Kowal, J. M. L. (1960). *Pl. Soil* **12**, 154–174.
Handley, W. R. C. (1954). *For. Comm. Bull.* No. 23.
Handley, W. R. C. (1961). *Pl. Soil* **15**, 37–73.
Hayes, A. J. (1965). *J. Soil Sci.* **16**, 121–140.
Heath, G. W. and Arnold, M. K. (1966). *Pedobiologia* **6**, 238–243.
Heath, G. W., Arnold, M. K. and Edwards, C. A. (1966). *Pedobiologia* **6**, 1–12.
Hoffman, J. V. (1924). *J. agric. Res.* **11**, 1–26.
Ivarson, K. C. and Sowden, F. J. (1959). *Pl. Soil* **11**, 237–248.
Jenny, H. (1948). *Soil Sci.* **66**, 5–28.
Jenny, H., Gessel, S. P. and Bingham, F. T. (1949). *Soil Sci.* **68**, 417–432.
Kendrick, W. B. (1959). *Can. J. Bot.* **37**, 907–912.
Kittredge, J. (1955). *J. Forest.* **53**, 645–647.
Ladd, J. N. and Butler, J. H. A. (1969). *Aust. J. Soil Res.* **7**, 253–261.
Latter, P. M. and Cragg, J. B. (1967). *J. Ecol.* **55**, 465–482.
Laudelout, H. and Meyer, J. (1954). *Trans. 5th Int. Congr. Soil Sci.* **2**, 267–272.
Lewis, J. A. and Starkey, R. L. (1968). *Soil Sci.* **106**, 241–247.
Lutz, H. J. and Chandler, R. F. (1946). "Forest Soils." John Wiley, New York.
Macfadyen, A. (1968). *In* "The Measurement of Environmental Factors in Terrestrial Ecology." (R. M. Wadsworth, ed.), pp. 59–67. Blackwells, Oxford.
Madge, D. S. (1965). *Pedobiologia* **5**, 273–288.
Mattson, S. and Koutler-Anderson, E. (1954). *Kungl. Lantbruk. Annaler* **21**, 389–400.
Mayandon, J. and Simonart, P. (1959). *Pl. Soil* **11**, 170–175.
McFee, W. W. and Stone, E. L. (1966). *Proc. Soil Sci. Soc. Am.* **30**, 513–516.
Melin, E. (1930). *Ecology* **11**, 72–101.
Mikola, P. (1955). *Comm. Inst. Forest. Fenniae* **43**, 1–50.
Minderman, G. (1968). *J. Ecol.* **56**, 355–362.
Nye, P. H. (1961). *Pl. Soil* **13**, 333–346.
Nykvist, N. (1959a). *Oikos* **10**, 190–211.
Nykvist, N. (1959b). *Oikos* **10**, 212–224.
Ovington, J. D. (1962). *Adv. ecol. Res.* **1**, 103–197.
Ovington, J. D. and Madgwick, H. A. I. (1957). *J. Soil Sci.* **8**, 141–149.
Phillips, J. F. V. (1930). *S. Afr. J. Sci.* **27**, 352–367.
Plice, M. J. (1934). *Mem. Cornell Univ. agric. Exp. Stn* **166**.

REMEZOV, N. P. (1961). *Pochvovedenie*, 1961 (7), 703–711.
SHAW, K. (1958). Ph.D. Thesis, University of London.
SOWDEN, F. J. and IVARSON, K. C. (1962). *Pl. Soil* **16**, 389–400.
SJÖRS, H. (1959). *Oikos* **10**, 225–232.
TRAVLEEV, A. P. (1960). *Pochvovedenie* 1960 (10), 92–95.
VOIGT, G. K. (1965). *Proc. Soil Sci. Soc. Am.* **29**, 756–759.
WANNER, H. (1970). *J. Ecol.* **58**, 543–547.
WEBSTER, J. and DIX, N. J. (1960). *Trans. Br. mycol. Soc.* **43**, 85–89.
WITKAMP, M. (1963). *Ecology* **44**, 370–377.
WITKAMP, M. (1966*a*). *Ecology* **47**, 194–201.
WITKAMP, M. (1966*b*). *Ecology* **47**, 492–494.
WITKAMP, M. (1969*a*). *Ecology* **50**, 922–924.
WITKAMP, M. (1969*b*). *Soil Biol. Biochem.* **1**, 167–176.
WITKAMP, M. and DRIFT, VAN DER J. (1961). *Pl. Soil* **15**, 295–311.
WITKAMP, M. and FRANK, M. L. (1969). *Pedobiologia* **9**, 358–365.
WOODWELL, G. M. and DYKEMAN, W. R. (1966). *Science, N.Y.* **154**, 1031–1034.
WRIGHT, T. W. (1956). *J. Soil Sci.* **7**, 33–42.

20

Decomposition of Litter in Soil

C. H. Dickinson

Department of Plant Biology
The University
Newcastle upon Tyne
England

I. Introduction

A superficial examination of soil suggests that it comprises a heterogeneous collection of mineral and organic materials. A closer acquaintance with this substrate assures one, however, that a delicate balance is maintained in terms of physical structure and chemical functioning, which is conducive to the efficient development of green plants. Soil is also the repository for most of the detritus from the same green plants and many aspects of this ecosystem have apparently evolved to facilitate the degradation of these remains in a rapid and economical manner.

This chapter is concerned with the decomposition which takes place within the mineral horizons of soils. The superficial, A_0, litter layers have been considered by Williams and Gray (Chapter 19). An exception to this concerns the extreme litter accumulations termed peats, which will be the subject of a later section in this chapter.

II. Sources of Plant Litter Entering Soil

A. Natural and Semi-natural Communities

Debris from primary producers accumulates on the soil surface and also within the mineral horizons along the network of roots and rhizomes. Much of the litter accumulating on the surface is considerably decomposed before the fragmented remains and soluble leachates are incorporated into the uppermost mineral horizon. Some of the litter is, however, buried fairly soon after it reaches the soil, mainly through the activity of earthworms.

A secondary addition of organic matter comprises microbial remains, bacterial cells, fungal hyphae, and animal corpses and faeces which have, in the main, been produced during the decomposition processes in the litter layer.

All these materials will be incorporated into the A_1 horizon by a variety of processes including leaching, active or passive movement or growth of organisms which then defaecate or die, mechanical disturbance of the soil due either to natural physical or agricultural processes or the activities of soil animals or in some extreme instances to the accretion of wind- or water-borne mineral and organic particles which effectively lead to an increase in soil depth. Many of these processes will operate sporadically: for example, frost heaving will occur in winter and leaching may be restricted in some climates to fairly short periods. The activity of organisms in the mineral horizons will be restricted by periodic shortages of nutrients due to such irregularities in the supply mechanisms.

The other major source of nutrients in the mineral soil are the root systems of green plants whose productivity has been discussed by Head (1971). Some indication as to the extent of root production in grasslands is given in Table I which shows the enormous biomass produced, especially in the surface horizons. Roots produce debris and soluble organic materials whilst they are growing, and following their death the quantities available increase considerably (see Waid, Chapter 6). Living roots are protected during their passage through soil by a root cap composed of parenchyma cells. These cells are short-lived, being sloughed off from the posterior margin of the cap and replaced at the tip by the activity of the apical meristem. The root hairs are also short-lived. The posterior hairs, and, shortly after, many of the epidermal and outer cortical cells, die as the root extends into the soil. Yet more debris results when lateral roots develop from the cortex, as these force their way out mechanically and chemically (Bonnett, 1969), destroying as they emerge a substantial volume of cortical tissues outside the stele.

Estimates of the quantities of such materials produced by roots are rare but Clowes (1971) has calculated that 10,000 cap cells were lost each day by one root of *Zea mays* L. growing at 23°C in water. A four month-old

TABLE I. Root biomass in some grassland sites of the central-western United States

Reference	Site and grass species	Profile depth (cm)	Root biomass (g m^{-2})
Schuster (1964)	Colorado, foothills area:	0– 30	443
	Muhlenbergia montana	30– 60	79
	Festuca arizonica	60– 91	17
	Bouteloua gracilis	91–183	9
Wiegert and Evans (1967)	Michigan, swale area:	0– 90	1018
	Poa pratensis		
	Michigan, upland area:	0– 90	685
	Poa compressa		
	Aristida purpurascens		
Dahlman and Kucera (1965)	Missouri, humid prairie:	0– 25	1575
	Andropogon gerardi	25– 56	214
	Andropogon scoparius	56– 86	112
	Sorghastrum nutans		

Secale plant was calculated by Dittmer (1937) to have produced about 241 km of roots which, if they had grown steadily, would have produced each day 118×10^6 root hairs whose individual life span was probably measured in days rather than weeks (Linsbauer, 1930). The volume of material destroyed during lateral root development is not known but it must be considerable as the numbers of such roots formed are very great. The total quantities of root debris were measured by Rovira (1956) who reported that 50 *Pisum* plants growing in quartz sand for 10 and 21 days yielded 16·7 and 31·5 mg of cell debris respectively whereas 50 *Avena* plants only yielded 6·6 and 14·8 mg. Rovira and McDougall (1967) have emphasized that these quantities do not compare with the amounts of organic material released as root exudates, which are also remarkable in the range of chemical substances they include. This chapter is not, however, concerned with the fate of soluble exudates as they do not come within the most liberal definition of litter.

These supplies of organic debris fluctuate to some extent according to the season. This is most obvious when considering deciduous woodland where the majority of the annual production of leaf litter accumulates in a very short time at the end of the growing season. Fluctuations in supply are also true for most other above ground litter sources and seasonal changes occur in root development. Kalela (1955) has reported that in a natural community of *Pinus sylvestris* the total root length doubled between May and July to reach 1000 m m^{-2}. This increase was accompanied by a dramatic, four-fold rise in the number of root tips during the same period. Both root length and number of tips then declined equally quickly until by October they were at levels below those in May. This indicated that there were substantial fluctuations in the amounts of root debris formed at each season. Williams (1969) and Rogers and Head (1969) have assessed the seasonal and biological factors affecting the development of roots of grasses and woody perennials respectively. The effect on root growth of harvesting aerial tissues has been examined by Butler *et al.* (1959) who showed that *Trifolium repens* roots responded quickly following defoliation whereas *T. pratense* and *Lotus* were less sensitive to such changes.

B. Agricultural Situations

Much of the previous section also applies in the agricultural context but in addition there are here a number of extra considerations. Cultivation frequently involves burying massive quantities of living and dead plant tissues in the mineral soil and this must create highly unusual situations for many soil organisms. Such cultivation operations will have even more impact when green or farmyard manure is ploughed in and the organisms are provided with relatively enormous quantities of readily available nutrients. Joffe (1955) has examined several aspects of the process of green manuring and he stressed the importance of the process in terms of nutrient mobilization rather than any benefit resulting from an increase in soil organic content. Alternatively some agricultural practices may lead to the opposite result whereby starvation reduces activity considerably. Burning off cereal straw and chemically destroying potato haulm removes a considerable amount of microbial food material, and harvesting substantial quantities of plant tissue, as for example sugar beet, cereals and timber, is likely to deplete the reserves of nutrients in soil.

The agricultural environment is also less closely correlated with seasonal climatic and physiological changes and debris may be created and incorporated into soil at times which are not always necessarily conducive to maximum decomposer efficiency.

III. Organisms Involved in Litter Decomposition in Soil

The major groups of litter-decomposing organisms inhabiting mineral soils are the bacteria including the actinomycetes, the fungi, the Protozoa, the nematodes, the enchytraeid worms and the lumbricid worms. An indication of the relative importance of the various groups of animals in a European grass land soil is given in Fig. 1. This is based on numbers of individuals and not on their biomass and it includes several groups, such as the spiders, centipedes and ants which are wholly or largely carnivorous. The actual number, the biomass and the relative efficiency of each group is dependent on the basic characteristics of the soil and on factors which may modify the soil environment in various ways. It has to date proved impossible to obtain reliable estimates of the relative importance of each group of organisms but Macfadyen (1963) has estimated that animals may consume 10–20% of the total organic matter supplied to the soil. This does not, however, mean that these or any other groups are relatively unimportant as many of their activities influence other organisms in a substantial manner.

IV. Factors Affecting the Activity of Soil Organisms

A. Soil Structure

Soil structure is greatly influenced by its mineral component with extreme types being dominated by high concentrations of sand or clay particles or by a very low mineral content which confers particular characteristics. Soil structure may affect the spread of microbial propagules, the growth and movement of organisms and of course the growth of plant roots. Much of the influence of soil structure is operative through its effect on water relations, temperature and aeration, but the actual physical structure of the mineral horizons may be important in limiting the movement of soil animals (Kühnelt, 1961).

The architecture of most soils involves the development of aggregations termed crumbs which are clusters of mineral and organic particles held together by forces which are more powerful than those between the crumbs. Crumb development increases pore size, rainfall penetration and drainage and the interchange between the soil and aerial atmospheres. Much speculation has centred on the nature of the forces holding the crumbs together. These probably include several of microbial origin (Griffiths, 1965). The disturbance of soil, for example by cultivation, may to some extent break up these crumbs and this may explain the finding that after such operations as much as 17 kg ha^{-1} of carbon may be rapidly oxidized with a concomitant mineralization of about 1·9 kg ha^{-1} nitrogen (Rovira and Greacen,

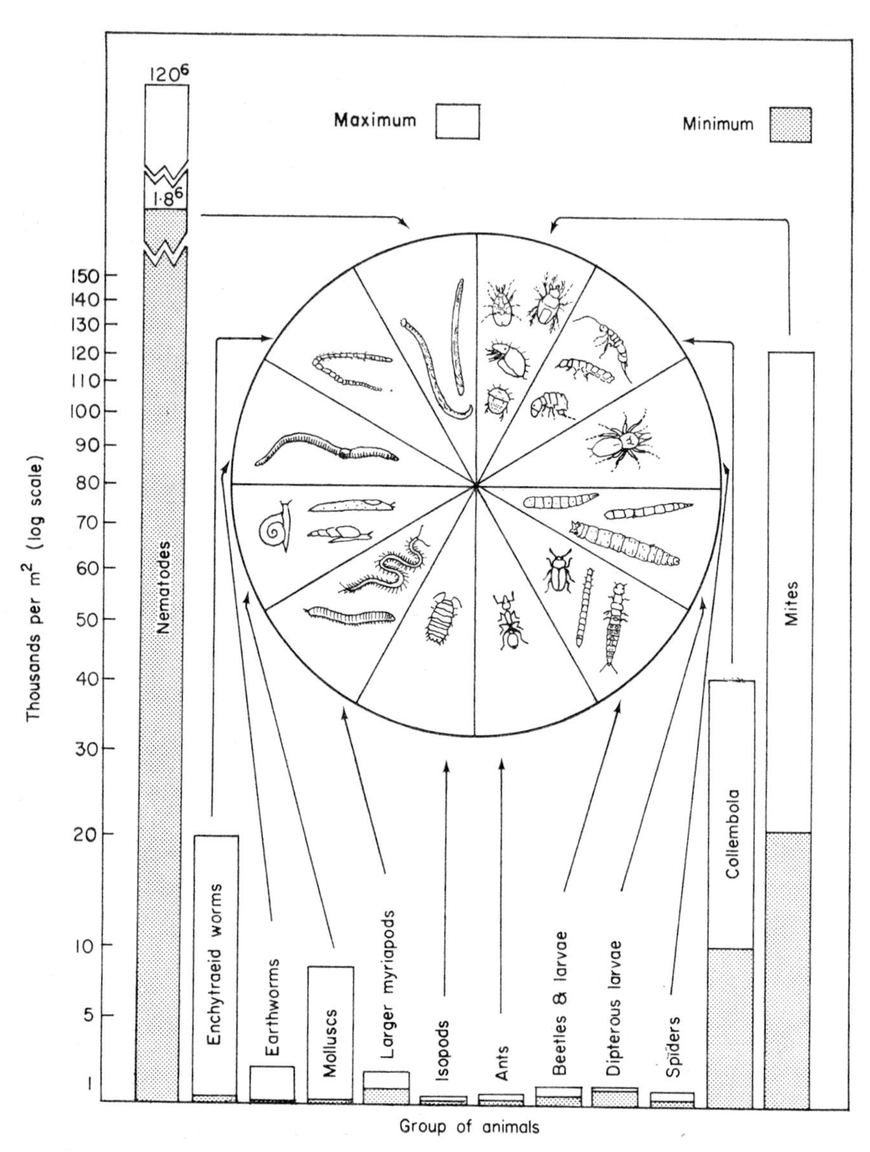

Fig. 1. Numbers of animals per m² in European grassland soil. (Data from Macfadyen, 1957)

1957). It was suggested that organic matter which had accumulated in situations within the soil where it was physically unavailable to micro-organisms was exposed by cultivation to their activities and the possible improvements in soil aeration or microbial dispersal were not considered important. Similar bursts of microbial activity have been reported when air-dry soil is re-wetted (Griffiths and Birch, 1961) and these have been ascribed to an increase in the solubility of the organic materials as a result of drying and the disruption of aggregates during this process.

B. Soil Water Relations

The most comprehensive account of soil water relations as they affect micro-organisms is by Griffin (1972). This review deals mainly with fungi and bacteria but the methods described and the concepts developed are relevant to many other soil organisms. Water is of central importance in the soil ecosystem because of the number of interactions between it and other factors which influence this environment (Fig. 2). Rain or floodwater is a major source of input to the soil not only in terms of soluble mineral

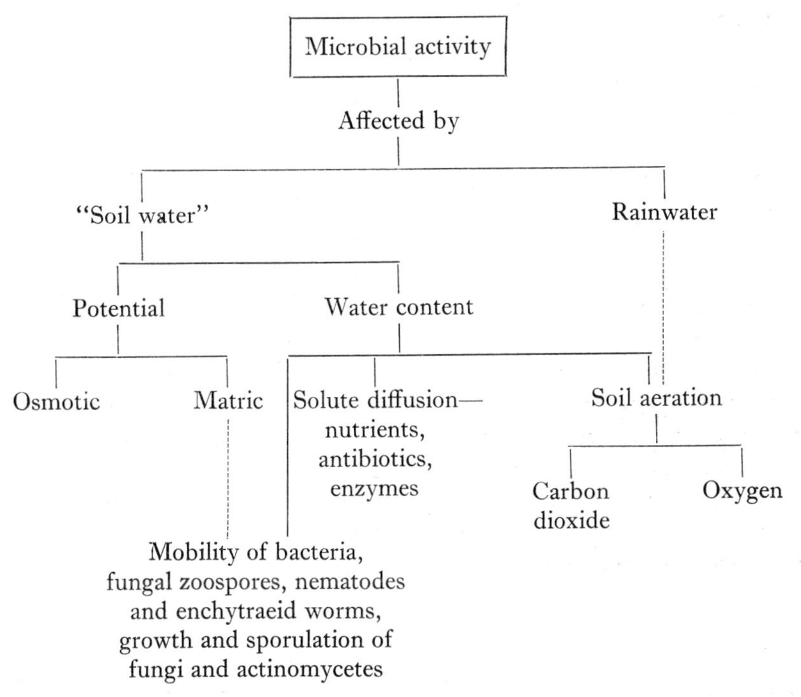

Fig. 2. Interactions between water, soil and soil organisms.
(Based on Griffin, 1972)

nutrients but also of relatively insoluble organic compounds, oxygen and microbial propagules. More dramatic effects are seen in river floodplains and salt marshes where water actually brings additions of solids which increase the effective depth of the soil profile.

Several groups of organisms, including motile bacteria, zoospore-forming fungi, Protozoa, nematodes, rotifers and tardigrades, have an essentially aquatic existence and as such they depend on liquid water for the maintenance of their activity and movement through the soil. A particularly good example of such a requirement was described by Wallace (1963) for the nematodes and Kouyeas (1964) and Lewis (1970) have provided evidence for reduced mobility of bacteria in soils where the water content fell below field capacity. Griffin (1972) considered that the influence of low water contents on bacterial motility explains their reduced activity in dry soils which, for most species, ceases altogether at −80 to −100 bar ("bar" is a measure of soil water matric potential). Some hyphal fungi were found to be capable of growing in soils which had even less water than this (Dommergues, 1962). Soils with low water potentials and high salt contents may be inhabited not only by organisms which tolerate these low potentials but also by others which have a requirement for high concentrations of specific ions (Ingram, 1957). Waterlogged soils present other problems (see Forbes, Chapter 23) and in general microbial activity is restricted and in extreme habitats peat may accumulate. The extent and type of development of several hyphal organisms has been shown to be very much dependent on the extent to which the soil pores are filled with water. Griffin (1972) has reviewed these findings as far as fungi are concerned and Williams *et al.* (1972) have shown that Streptomycetes were unable to grow through water-filled pores whereas *Micromonospora* was not restricted by such barriers. The relative humidity of the soil atmosphere may affect the behaviour of some of the animals which live in the larger spaces between soils crumbs. Andrewartha (1964) has described some animals which tolerate very dry situations and has commented on their drought-resisting mechanisms which include cryptobiosis (see also Twinn, Chapter 13), migration within soil and efficient water uptake systems. Madge (1964) and Hayes (1966) have demonstrated differences in the responses of cryptostigmatid soil mites to a graded series of relative humidities which may influence the movement of these animals between different horizons of the mineral soil and the litter layer.

C. Soil Atmosphere

Apart from water vapour there are three other biologically important components of the soil atmosphere. Oxygen is positively required by aerobic organisms for respiration and oxidative assimilation, carbon dioxide influences the pH of microhabitats and may inhibit growth of many organisms at high concentrations and volatile chemicals, as will be discussed later, have a variety of effects on different micro-organisms. Griffin (1972) has suggested that oxygen partial pressures in soil are usually in excess of 0·18 atm but in wet soils pressures as low as 0·10 atm may be common. Fungi, which are mostly obligate aerobes, are however relatively sensitive to severe reductions in oxygen concentration (Griffin, 1972) whereas many bacteria are capable of anaerobic growth. Such bacteria may become active in waterlogged soils, in which there are numerous other chemical changes, or within larger water-soaked crumbs whose centres are devoid of oxygen (Greenwood, 1968). Soils with a high organic content may also be characterized by large populations of anaerobes. Most soil animals avoid regions of low oxygen concentration.

Carbon dioxide may stimulate fungi to grow and reproduce but at oxygen concentrations of 0·21 atm it may be toxic at partial pressures above 0·10 atm. Such concentrations are, however, uncommon in most soils (Russell, 1961) and even around respiring roots the levels will scarcely reach one tenth the toxic concentration (Greenwood, 1968). The effect of carbon dioxide in solution is complicated but it undoubtedly may affect growth of micro-organisms through the changes in pH which result from its interactions in the liquid phase.

Griffin (1972) has pointed out that the combined partial pressures of carbon dioxide and oxygen in soil add up to approximately 0·21 atm. He has emphasized the importance of studying microbial responses following simultaneous inverse changes in these two gases as these may not conform to those expected on the basis of studies of isolated changes in one gaseous component.

D. Soil Physical and Chemical Characteristics

1. Temperature

Temperature affects the growth and activity of organisms directly, and in soil it is also important because of the interactions with such factors as oxygen availability and relative humidity. It has long been recognized that diurnal and seasonal fluctuations in air temperature are moderated in the deeper horizons of soil. Soil temperatures are affected by the slope and aspect of the land, the colour of the soil, the extent and nature of the vegetation cover, the persistence, depth and the degree of compaction of the litter layer and the water regime of the surface litter and the mineral soil.

Temperature is apparently more important as an ecological factor for members of the soil fauna as they can quickly respond to local changes by moving some distance through the soil. Kühnelt (1961) has described the daily vertical migrations in high alpine soils of animals seeking to avoid the high temperatures at the soil surface in the central part of the day. Low temperatures seem to be less critical as many small animals may survive even when frozen (Kühnelt, 1961; Heal *et al.*, 1967) though others, including members of the Isopoda and the Lumbricidae (see Lofty, Chapter 14) migrate downwards in the soil in winter to avoid the colder surface layers.

The majority of the soil microflora are mesophiles with optimum temperatures for growth between 25 and 37°C and effective minimum temperatures for activity between 5 and 10°C. Psychrophiles, which may be defined as organisms able to grow well at 0°C in one week (Rose, 1968), are less common though they were shown to be relatively abundant in antarctic soils (Latter and Heal, 1971). Species of *Pseudomonas*, *Flavobacterium* and *Achromobacter* are psychophilic whereas in general Gram-positive genera are not noted for this type of activity. Many fungi are known to grow at temperatures approaching 0 °C and, for example, Latter and Heal (1971) reported that nearly all the fungi they isolated from an antarctic soil were cold-tolerant whereas many of the bacteria from this soil were not. This may imply that the bacteria were surviving the cold conditions and only growing actively during short periods of warm weather. Psychrophilic organisms in soils from extreme habitats have been studied by Janota-Bassalik (1963) in peat, by Baker (1970) in antarctic soils and by Christenson and Cook (1970) and Boyd (1958) in arctic muskeg soils.

Thermophilic organisms, whose optima for growth lie between 55 and 65°C and which grow very poorly if at all below 40°C, are also common in many soils, especially those which receive additions of composted materials (see Goodfellow and Cross, Chapter 9; Cooney and Emerson, 1964; Eggins and Malik, 1969). The explanations for their presence in these ecosystems are somewhat tenuous and more knowledge is required concerning their ecological distribution and sites of active growth.

The interactions between psychrophiles and mesophiles are however of greatest significance for most temperate and sub-temperate soils whose monthly mean temperatures are normally below 15°C. Rose (1968) has stressed that growth at different temperatures may be qualitatively different, apart from any changes in growth rates. He instanced increased production of extracellular polysaccharides by certain bacteria, increases in pigment production and alterations in the end-products of fermentation as examples of changes produced by decreasing temperatures. Such changes in metabolism may account for the variations in isolation frequencies of fungi from soil incubated at different temperatures (Dickinson and Kent, 1972) and will undoubtedly alter the relative role of organisms within soil. Soil microbiologists should give more attention to the survival of organisms under cold conditions, their speed of revival from dormancy and their relative growth rates and competitive abilities within the soil at temperatures far removed from their optima.

2. pH

Soil hydrogen ion concentration is affected by the water regime, the concentration of salts and carbon dioxide in soil and the number of exchangeable cations present. It is, perhaps, the one factor in soil which more than any other varies between micro-habitats to such an extent that measurements of the bulk reaction are relatively meaningless. There are gross differences in the bacterial flora (Gray and Williams, 1971*a*), the mycoflora (Warcup, 1951) and the fauna (Kühnelt, 1961) of acid and alkaline soils but what these differences exactly mean is difficult to assess. Local accumulations of ammonia (Williams and Mayfield, 1971), carbon dioxide, clay particles, water, organic debris and micro-organisms themselves may modify the bulk pH to a significant extent. This may affect the growth of various saprophytes though others appear to be relatively tolerant of a wide range of hydrogen ion concentrations. Movement of actinomycete spores in soil has been shown to vary according to the pH of soil, probably due to differences in the extent of their absorption on soil particles (Ruddick and Williams, 1972).

3. Light

Radiation is important in terms of heating the soil but it may also have several more direct effects on the soil organisms. Many soil animals are negatively phototactic (Kühnelt, 1961) and they are therefore absent from the surface layers of soil. Some fungi may need to be exposed to light before they sporulate. The hyphae and spores of many fungi in soils which are exposed to intense sunlight are characterized by the presence of melanin in their cell walls which absorbs ultraviolet radiation and hence protects them from genetic damage (see Pugh, Chapter 10).

4. Natural Chemical Constituents

Soil contains substantial quantities of a wide variety of naturally occurring chemicals. These include simple inorganic salts, by-products released during decomposition and exudates from living organisms. These substances may be grouped as those with a beneficial effect on saprophytic activity and toxins which inhibit one or more groups of organisms. Many of these chemicals will be active in a liquid phase but there is growing evidence for the importance of a variety of volatile substances.

Inorganic chemicals, which will almost all be active in the liquid phase, will generally favour saprophytic activity either directly in that they supply an essential element or through an increase in the productivity of the primary producers or by their action in altering the soil environment. Direct benefits may be best illustrated by the action of nitrogen which may allow active saprophytic growth on cellulose-rich debris which has a high C:N ratio. Hudson (1971) and Burchill and Cook (1971) have demonstrated that urea increased the extent of fungal colonization of leaf litter and also (in the latter instance) the nature of the bacterial populations on the senescent leaves. It seems highly probable that such effects will also be operative within the soil (see Forbes, Chapter 23).

Applications of nitrochalk or superphosphate have resulted in an increase in primary productivity which in turn led to increases in the soil earthworm population (Edwards and Lofty, 1972). An increase in the population of Sphaeropsidales in fertilized sand dune soil was possibly also due to the stimulation of the herbaceous vegetation and hence the production of larger amounts of litter upon which these fungi grow (Dickinson and Kent, 1972). In this work, however, no other significant increase was recorded in the numbers of the common soil fungi over a 6 month period after fertilizer application. Domsch and Gams (1972) quote several authors who have recorded stimulation of soil fungi by addition of fertilizer and it would be valuable to determine why this effect may be demonstrated in some soils whilst being inoperative in others. Dooley and Dickinson (1970) recorded slightly larger populations of fungi in peat soils amended with high levels of sulphate of ammonia as against others where the potassium concentrations were increased. Edwards and Lofty (1972) correlated the effects of lime and ammonium sulphate in respectively increasing and decreasing earthworm populations in soils with the influence of these chemicals on the pH of the ecosystem.

Most interest in chemicals released during litter decomposition centres on those compounds that influence micro-organisms. During the progressive breakdown of organic debris many simple compounds are formed including ammonia, sulphides, methane, and carbon dioxide. Some of these compounds may affect microbial activity through their influence on soil pH or aeration but others may directly stimulate or inhibit activity. It is also known that certain decomposing materials are sources of more complex toxins including aldehydes, antibiotics, organic acids, amino acids, coumarins and glycosides (Chapman, 1965). Ethylene has also been shown to be produced in soil when the oxygen partial pressure falls due to waterlogging. Smith and Restall (1971) have detected >20 ppm of ethylene and they correlated increased production of this potentially toxic gas with changes in water potential and increased organic matter content in the soil. Beneficial compounds, such as hormones and vitamins, are released during decomposition of many types of debris. Some of these compounds are formed or released from relatively few higher plants whereas others are produced whenever decomposition occurs. Fries (1973) has given examples where compounds such as acetaldehyde and methanol from decaying litter are active in the vapour phase.

An array of biologically active chemicals produced by both living microbes and higher plants has been explored in great detail in recent years. A very long list of fungi, bacteria and actinomycetes has been shown to produce antibiotic-like compounds (Broadbent, 1968; Bilai, 1963) though relatively few of these have found commercial applications. The ecological significance of antibiotics is difficult to prove experimentally but the simple explanations seem so straightforward that more complicated suggestions must be viewed with extreme caution. It seems likely that our approach to the study of antibiotics must be refined to allow for interactions within molecular environments mediated by small quantities of transient chemicals which certainly cannot be detected in the bulk soil. Ikediugwa and Webster (1970) have demonstrated the necessity for close contact between hyphae before certain types of interaction can occur. Apart from the antibiotic producer and the sensitive organisms there are other microbes that exhibit varying degrees of tolerance to such compounds and Park (1960) has stressed that such resistance is extremely important in determining the success of organisms in soil.

Fries (1973) has listed a variety of other chemicals produced by living micro-organisms and higher plants which affect the growth and/or morphogenesis of organisms. Several of the examples given involve plant pathogens which are not strictly relevant here but it seems likely that with further study similar responses will be discovered for the less obviously economically important saprophytic organisms. Hora and Baker (1972) have linked a volatile chemical with mycostasis, whereby fungal spores in soil are prevented from germinating. They showed that actinomycetes, including *Streptomyces*, were responsible for the production of this compound.

Although most of the work on such chemicals has involved fungi and bacteria a little is known concerning the role of the soil fauna as chemical producers and the influence of the microflora on the fauna. Timonin (1961, 1962) discovered that *Musa* sp. roots attacked by the insect *Scaptoris talpa* were not infected by the fungal pathogen *Fusarium oxysporum* f. sp. *cubense*. This was due to the production of insect volatiles which inhibited the growth of *Fusarium* and other soil fungi. Giuma *et al.* (1973) have described the production of wide spectrum nematotoxins by germinating conidia of several endoparasitic fungi.

5. Sorptive Interactions

The ecological distribution of many members of the microflora within a soil is strongly influenced by interactions between the cells and the solid particles which make up the soil. Marshall (1971) has described three levels of interaction: (i) the sorption of cells onto the surface of large soil particles; (ii) the interaction of cells and soil particles of a similar size to give aggregates; and (iii) the sorption of tiny soil particles on the surface of the cells. Mechanisms to account for these interactions are as yet not fully worked out but there is no doubt that sorption alters the metabolic activity of the organisms and their extracellular enzymes. Stotzky (1968) has suggested that the failure of *Fusarium oxysporum* f. sp. *cubense* to spread in certain soils may be related to the presence of montmorillonite in these soils which through regulating the pH favours the growth of bacteria which effectively exclude the *Fusarium*.

Apart from the interactions of whole organisms with soil particles perhaps the other most important similar interaction concerns the local absorption of enzymes to form stable complexes. Skujins (1967) has provided a very good account of the origins, types and significance of the enzymes which occur in soil free from any living cytoplasm. He suggested that the origins of these proteins include living fungal, bacterial, animal and higher plant cells and also the dead cells of the same organisms. A wide range of enzymes has been detected in spite of the fact that this subject is not yet far advanced (Table II). Skujins stressed the significance of these free enzymes in soil and he recognized the existence of molecular environments on soil particles, higher plant cells and microbial cells.

TABLE II. Some enzymic activities detected in soils
(Data from Skujins, 1967)

OXIDOREDUCTASES	HYDROLASES
Catalase	Amylases
Dehydrogenases	Cellulases
Polyphenol oxidases	Lipases
	Phosphatases
	Proteinases
	Urease
TRANSFERASES	LYASES
Transaminase	Aspartic acid decarboxylase
Transglycosylases	

Such molecular environments are characterized by specific ion concentrations and a particular pH and oxidation–reduction potential, and are considered as sites of important biochemical reactions. Attempts to correlate soil enzymes with overall fertility and mineral recycling within soils have, to date, been relatively unsuccessful, because of the several parameters which affect enzyme activity. Once better methods are evolved, however, it seems likely that this approach to soil biology will become very important.

E. Biological Interactions

Within the soil ecosystem the relative efficiency of each group of micro-organism in terms of litter decomposition is determined by all the foregoing factors and last, but not least, by direct interactions between particular organisms. Many toxic or synergistic interactions may be considered in terms of the soil environment, on a bulk, micro or molecular scale, as the organisms affect one another through changes brought about within these habitats. There are, however, numerous examples whereby there is a direct effect of one organism on another. Examples of such effects include the parasitism of soil animals by fungi and bacteria, lysis of fungi by bacteria and the parasitism of fungi by other species of fungi. Such effects must be important in regulating the populations of organisms in soil.

Transport of microbial propagules within soil is an extremely important consequence of the movement of soil animals. As many bacteria and fungi are unlikely to be able to grow or swim substantial distances this transport mechanism may ensure they reach newly arrived litter supplies in a relatively short time. The deposition of propagules within faecal pellets or casts may also be important in terms of their subsequent growth.

Movement of hydrophobic *Streptomyces* spores on the surface of arthropods has been demonstrated by Ruddick and Williams (1972) and Hutchinson and Kamel (1956) have described the dispersal of soil fungi by earthworms.

Another type of interaction in the course of litter breakdown concerns the effect which organisms have on litter in terms of its attraction for succeeding groups of saprophytes. Palatibility, as has been discussed elsewhere in this book (Chapters 14, 15, 16, 19), markedly affects the choice of food by many animals and likewise it seems probable that animal activity alters the litter such as to influence the species composition and growth of the following group of micro-organisms. Some of these changes may be simple chemical transformations or additions, some may be physical alterations to the litter and some may be the result of feeding preferences for the micro-organisms themselves rather than the litter on which they have grown.

F. Effects of Pesticides and Herbicides

Most of the studies which have been carried out on the properties of pesticides, including bactericides, fungicides, nematicides, insecticides, fumigants and herbicides, have been concerned with the efficiency, the inactivation and the breakdown of these chemicals in soil. Many such compounds persist in soil and they frequently affect soil saprophytes, though fortunately not usually to the same extent as the pathogens or pests against which they are employed. In considering these compounds it seems appropriate to classify them into substances designed to operate in soil and others which are applied to above-ground parts or to seeds, but which accumulate in soil via rainwash through the crop or on debris from treated plants.

1. Fungicides

Domsch (1964) has reviewed the role of soil fungicides which are directed against pathogenic fungi residing in soil. He recognized that the boundaries between plant pathology and soil microbiology are tenuous in this field of study and he stressed the need to consider approaches whereby both pathogens and saprophytes are considered in relation to the effects of chemicals on each and the organisms on one another. Domsch listed a number of studies where the effects of fungicides on soil saprophytes had been determined but he emphasized that many of these investigations had been very general and concerned only with total numbers of organisms. In a later review (Domsch, 1970) he refined these views and listed a number of questions which are relevant to the importance of fungicides, and other introduced chemicals, in soil ecology:

(*a*) which components of the decomposer system are most essential for maintaining energy flow?

(*b*) can the succession of micro-organisms on plant remains be interrupted, leaving the tissues in a "reduced" state?

(*c*) which microbial catabolic activities are most sensitive to fungicides?

(*d*) how long do inhibitory effects persist?

Other chapters in this book are concerned with the first question and it is becoming clear that the answer will vary according to the type of litter involved and the soil environmental conditions. Domsch (1970) has commented that it is not the species composition of the decomposer groups which are significant but rather that the important functions should be maintained at all times.

The second and third questions have not been fully explored but there is some evidence for the accumulation of debris in the presence of high concentrations of fungicides where the environmental conditions would otherwise be conducive to decay and incorporation. Hirst *et al.* (1961)

have described the accumulation of an atypical surface mat of peaty debris and the poor crumb structure in orchard soils where Bordeaux mixture had been used extensively for over 30 years. More recently Pugh and Williams (1971) have analysed a similar phenomenon where Bordeaux mixture, malachite green and an organo-mercurial compound had been liberally applied to control *Fusarium nivale*. They showed that the fungicides had altered the total numbers and species balance of the fungal population in the treated soils but they did not examine any other groups of soil saprophytes. In further studies Wainwright and Pugh (1973) demonstrated that high concentrations of captan, thiram and verdasan inhibited the production of nitrate from ammonium sulphate and Alexander (1969) has reviewed a number of other instances where a similar retardation of nitrification could be ascribed to fungicides (Table III). These studies are relevant to the third question posed above as they provide data on the soil treated as an "organism" involved in saprophytic processes.

The persistence of fungicides has been reviewed in detail by Alexander (1969) and Woodcock (1971), and Cripps (1971) has described aspects of the metabolism of these compounds in soil. Alexander (1969) has discussed the problems which give rise to concern in this field. He suggested that we should concentrate on acquiring more knowledge about which organisms act on soil pesticides, how environmental conditions affect these transformations, how the structure of pesticides affects their biodegradability and what compounds are formed during their decomposition.

Soil fungicides are designed to kill selectively, or otherwise limit, populations of pathogens in soil. Alexander (1969) and Domsch (1970) have indicated that in practice they may work by altering the equilibrium between pathogens and saprophytes due to the somewhat greater tolerance exhibited by the latter group. Domsch admits, however, that our knowledge of the interrelationships between such fungicides and other groups of organisms has scarcely been begun to be explored.

2. *Insecticides*

Much of the heat in recent ecological debates has been generated by the use of persistent insecticides which seem in general to outlast fungicides and herbicides. The chief problems in this respect are DDT and its analogues and the cyclodienes including aldrin, dieldrin, heptachlor and endrin. More recent introductions, including organophosphorus compounds (e.g. parathion, malathion, diazinon) and carbamates (e.g. carbaryl), appear to pose less of a threat as they are relatively readily degraded in soil (Matsumura and Bousch, 1971). The presence of persistent insecticides, or stable toxic intermediates, in soil may frequently have little effect on the soil microbial populations unless relatively high concentrations accumulate

TABLE III. Inhibition of specific groups of micro-organisms and microbial transformations in soil by pesticides

Bacteria	Actinomycetes	Fungi	Respiration	Nitrification
Allyl alc (25 gal acre^{-1})	Allyl alc (25 gal acre^{-1})	Allyl alc (25 gal acre^{-1})	DD (3500)	Allyl alc (70)
Chlorate (250)	Captan (250)	CS_2 (500)	EDB (7600)	CDAA (12)
Chloropicrin (100)	Chlorate (250)	DD (5 ml kg^{-1})	Mylone (150)	CDEC (12)
CS_2 (300)	DD (5 ml kg^{-1})	DNBP (100)	Nabam (100)	Chlorate (90)
DD (5 ml kg^{-1})	Mylone (150)	Ferbam (50)		Chloropicrin (435)
Mylone (150)	PCP (2000)	Mylone (150)		CIPC (12)
Nabam (50)		Nabam (50)		$CuSo_4$ (50)
PCP (2000)		PCP (500)		Cyanamide (200)
Vapam (60)		TMTD (50)		DD (23 gal acre^{-1})
		Vapam (60)		DNBP (10)
				EDB (23 gal acre^{-1})
				Ferbam (187)
				Heptachlor (50)
				Methyl bromide (435)
				Monuron (25)
				Mylone (150)
				Nabam (100)
				PCP (5)
				Telone (133)
				Vapam (75)

Data from Alexander (1969). Values are ppm unless otherwise indicated.

(Table III), though it is important to note that higher organisms are generally much more susceptible to such compounds.

3. Herbicides

Many herbicides (Table III) seem to have little effect on soil respiration, nitrification, populations of micro-organisms or organic matter decomposition, especially when used at concentrations which approximate to their rates of application in the field (Fletcher, 1966; Alexander, 1969). However, some selective effects have been demonstrated and a particularly interesting study by Wilkinson and Lucas (1969) showed that herbicides altered the competitive success of various soil fungi in terms of colonization of organic debris.

This finding may be especially significant in view of the accumulation of herbicides by susceptible plants which are then killed and become available for decomposition.

Kearney *et al.* (1967), Alexander (1969) and Wright (1971) have described aspects of the metabolism of herbicides in soil.

4. Soil Fumigants

Domsch (1964) has defined these as chemicals with a generally high biotoxicity which reduce or eradicate harmful members of the soil microflora and fauna. Unlike soil fungicides there is no requirement for such chemicals to persist in soil. They rarely sterilize soil and after a short interval the soil is recolonized by surviving organisms or from deeper unaffected layers. Much of the success of such operations depends on the "beneficial" effect of the restored population in actively preventing the later introduction of the original pathogens. The "new" populations are frequently characterized by being less diverse than the original population and this state of affairs may persist for a year or even longer. The significance of such changes in populations as regards litter decomposition seems to be largely unknown.

V. The Extent of Litter Decomposition

A. Mineral Soils

1. Developing Soils

Several studies have examined the processes of microbial colonization of developing soils such as sand dunes, saltmarshes and reclaimed polders. Whilst some of these studies have had as their primary objective a study of the influence of the environmental conditions on the populations of micro-organisms their results have been useful as regards the microbial populations available for the decomposition of litter.

Studies on sand dune soils have been related to the development of a stable microflora and the influence of higher plants on the microbial colonists (Webley *et al.*, 1952; Saito, 1955; Brown, 1958; Wohlrab and Tuveson, 1965). These studies indicated that the pioneer communities in newly stabilized dunes were characterized by relatively few common species and a distinctly limited series of functional groups of organisms.

These initial communities were much influenced by the prevailing environmental conditions, amongst which moisture, and in the case of maritime dunes salinity, were most important. As the dunes stabilized and the higher plant communities became more varied and productive the microbial floras altered and dominant species were recognized which were related to an overall increase in the quantity and the availability of particular types of plant litter. The nature of the soil environment is thus modified by the plant colonizers and in turn it is probable that the microflora affects the relative success of the higher plants (Brown, 1958). Pugh *et al.* (1963) compared the cellulolytic fungi in dune soils with those below the high water mark and found that the presence of large quantities of debris in the latter situation resulted in relatively larger populations in spite of the greater salinity and transient nature of this habitat.

The microbiology of developing saltmarsh soils has been studied by Pugh (1962) who demonstrated an overall upshore increase in fungal colonization which was accompanied by selective changes in the species composition of the mycoflora. Pugh emphasized the probable importance of rhizosphere microhabitats as sites of microbial activity within which fungal species, which could tolerate the saline and frequently waterlogged and anaerobic mud environment, were able to establish themselves as saltmarsh inhabitats. A downshore increase in frequency of isolation of several other fungi was explained by a regular replenishment of spores along with the sediment from the sea water. These fungi were however unable to grow in the mud environment and their spores died after some time in the mud. Dickinson and Pugh (1965) emphasized the extreme conditions for microbial growth in such muds by their discovery that excised roots of *Halimione* were still intact and relatively sparsely colonized by fungi two years after they were deprived of fresh supplies of carbohydrate.

The reclamation of polders in the Netherlands also allows an analysis of the microbial changes which accompany the development of muds, in this instance into agricultural farmland. Pugh and van Emden (1969) have demonstrated that the progress of fungal colonization of such muds depended on the time during which the embryo soil had been dry and the type of cultivation which had been practised.

2. *Mature Soils*

Whilst there are countless studies of limited aspects of the microbiology of grassland and forest soils, relatively few attempts have been made to synthesize information so as to present a picture of the whole system of energy flow through the soil and assess the relative role of each group of micro-organisms. Clark and Paul (1970) have gone some way towards such a synthesis in their account of the microbiology of grasslands though it is unfortunate that they decided to omit reference to the microfauna of these habitats. Analyses of the characteristics of the grassland microbial populations were followed by an account of the productivity of such grasslands, the rate of litter breakdown and the fate of the humic compounds formed during the decomposition processes. Such an approach has much to commend it and perhaps soil microbiology is progressing more rapidly now that some basic principles have been established. A further analysis of the overall productivity of micro-organisms in soil has been given by Gray and Williams (1971*b*) who questioned several basic assumptions concerning the activity of fungi and bacteria in soil. They considered that most organisms in soil are normally either relatively inactive or dormant and they advanced several pieces of evidence in support of this proposal. The chief limiting factor appears to be a shortage of energy supplies, or of energy in a readily accessible form. Less readily decomposed, recalcitrant molecules are present in different amounts but these are only used by a limited number of soil microbes. Unfavourable environmental conditions, as have been discussed earlier in this paper, also account in part for the general lack of activity.

A different approach to the study of litter decomposition involves consideration of the fate of individual pieces of litter which are added to soil. Jenkinson (1965, 1966) and Grossbard (1969) have used such methods to study the decomposition of litter buried in soil and they have incorporated ^{14}C in the tissues to allow more detailed studies of their fate. Jenkinson (1965) determined that 66% of *Lolium* tissues added to soil were decomposed after six months but this only increased to 80% after four years. In a later paper he showed that the products of decomposition included a 10–12% fraction which could be detected in the biomass in soil after one year. This declined to 4% after four years (Jenkinson, 1966). He also showed that *Lolium* tops decomposed more quickly than root tissues in the first year of burial. Grossbard (1969) demonstrated that *Lolium* leaves were decomposed more quickly than those of *Dactylis* and both fungi and faecal pellets became labelled following such decomposition processes. Stotzky and Mortensen (1958) have also used labelled tissue to investigate decomposition in muck soil.

Floate and Torrance (1970) and Floate (1970*a*, *b*) have taken such an approach further in that they incubated litter and faeces with an inoculum consisting of a soil extract and they measured the production of CO_2 and the release of mineral nitrogen and phosphorus. Using this technique Floate (1970*a*) has demonstrated that at 30°C grass tissues cut monthly decomposed more quickly than those cut annually. Sheep faeces were decomposed to a lesser extent than either type but at temperatures of 10°C or below this situation was reversed and Floate (1970*b*) has emphasized the importance of the secondary decomposition of dung which may be important in maintaining nutrient cycling in cool temperate climates. The influence of temperature on grass decomposition was such that one week at 30°C was equivalent to twelve weeks at 5°C. Dooley (1970) has employed a direct examination burial technique under field conditions but the success of this technique was probably due to the very restricted decomposition processes which occurred in peat. Douglas and Tedrow (1959) have studied the decomposition of organic matter in the field in arctic soils using an evolved CO_2 technique. They found that temperature, moisture potential and soil type influenced the rate of decomposition.

B. Organic Soils

Peats may form under a variety of environmental conditions including high or low temperature, an acid or alkaline pH and nutrient rich or nutrient poor, fresh or saline, water supplies. Several aspects of the chemistry and microbiology of peats have been reviewed by Given and Dickinson (1973) who have indicated that the obvious restrictions on the decomposition of plant remains do not necessarily mean that essential mineral nutrients are not recycled. The productivity of some peat-forming communities, particularly those in transitional and rheophilous habitats, is comparable with other non-peat environments. This implies that sufficient microbiological activity is taking place to allow some selective decomposition processes to proceed.

No really convincing explanation has been advanced to account for the accumulation of peat and perhaps in view of the variety of environments and of plant communities in which it may form a unified explanation is unlikely to be found.

Little microbiological work has been carried out on the surface litter of peat, where the tendency for debris to accumulate rather than be decomposed is most evident. The surface layers of peats have been shown to have a varied microbial population though notable absentees from some or all peats include nitrogen-fixing bacteria, nitrifying bacteria, Actinomycetes, many Basidiomycotina and several groups of soil animals. Those organisms which are present may be important in that they destroy the remains of newly dead higher plant roots and microbial debris apart from any long term alterations to the peat itself. This last possibility is clearly of interest in that the organic nature of peat means that it could be expected to decompose further and even slow changes would be important over the considerable time span available for decomposition.

Reclamation of peat presents two microbiological problems. How may the populations of micro-organisms be stimulated to a level of activity comparable with a fertile loam soil and will the expanded population living in a somewhat changed environment perhaps act more vigorously on the peat itself? Studies by many workers including Nepomilvev and Kuzyakina (1965), Zverkov (1966) and Rybalkina and Kononenko (1961) have shown that cultivation and liming will bring about increases in the populations of organisms and improve the fertility of the soil. Dickinson and Dooley

(1967) have shown, however, that alterations to the environment did not dramatically alter the extent to which peat particles were colonized by fungi over a period of ten years and it seems possible that the apparent "losses" of peat during cultivation may be largely due to a progressive decline in the water table accompanied by shrinkage of the peat.

References

ALEXANDER, M. (1969). *In* "Soil Biology Reviews of Research." Natural Resources Research, UNESCO **9**, 209–240.

ANDREWARTHA, H. G. (1964). *Proc. Linn. Soc. N.S.W.* **89**, 287–294.

BAKER, K. H. (1970). *In* "Antarctic Ecology." (M. W. Holdgate, ed.), pp. 717–722. Academic Press, London and New York.

BILAI, V. I. (1963). "Antibiotic-producing Microscopic Fungi." Elsevier, Amsterdam.

BONNETT, H. T. (1969). *J. Cell. Biol.* **40**, 144–159.

BOYD, W. L. (1958). *Ecology* **39**, 332–336.

BROADBENT, D. (1968). *Pest Articles and News Summaries.* Section B, **14**, 120–141.

BROWN, J. C. (1958). *J. Ecol.* **46**, 641–664.

BURCHILL, R. T. and COOK, R. T. A. (1971). *In* "Ecology of Leaf Surface Micro-organisms." (T. F. Preece and C. H. Dickinson, eds), pp. 471–483. Academic Press, London.

BUTLER, G. W., GREENWOOD, R. M. and SOPER, K. (1959). *N.Z. Jl agric. Res.* **2**, 415–426.

CHAPMAN, H. D. (1965). *In* "Ecology of Soil-borne Plant Pathogens." (K. F. Baker and W. C. Snyder, eds), pp. 120–139. John Murray, London.

CHRISTENSON, P. J. and COOK, F. D. (1970). *Can. J. Soil Sci.* **50**, 171–178.

CLARK, F. E. and PAUL, E. A. (1970). *Adv. Agron.* **22**, 375–435.

CLOWES, F. A. L. (1971). *Ann. Bot.* **35**, 249–261.

COONEY, D. G. and EMERSON, R. (1964). "Thermophilic Fungi." Freeman, San Francisco.

CRIPPS, R. E. (1971). *In* "Microbial Aspects of Pollution." (G. Sykes and F. A. Skinner, eds), pp. 255–266. Academic Press, London and New York.

DAHLMAN, R. C. and KUCERA, C. L. (1965). *Ecology* **46**, 84–89.

DICKINSON, C. H. and DOOLEY, M. J. (1967). *Pl. Soil* **27**, 172–186.

DICKINSON, C. H. and PUGH, G. J. F. (1965). *Trans. Br. mycol. Soc.* **48**, 595–602.

DICKINSON, C. H. and KENT, J. W. (1972). *Trans. Br. mycol. Soc.* **58**, 269–280.

DITTMER, H. J. (1937). *Am. J. Bot.* **24**, 417–420.

DOMMERGUES, Y. (1962). *Anns. agron., Paris* **13**, 265–324, 391–468.

DOMSCH, K. H. (1964). *A. Rev. Phytopath.* **2**, 293–320.

DOMSCH, K. H. (1970). *In* "Pesticides in the Soil." pp. 42–46. Symposium at East Lansing State University, Michigan.

DOMSCH, K. H. and GAMS, W. (1972). "Fungi in Agricultural Soils." Longmans, London.

DOOLEY, M. (1970). *Pl. Soil* **33,** 145–160.
DOOLEY, M. and DICKINSON, C. H. (1970). *Pl. Soil* **32,** 454–467.
DOUGLAS, L. A. and TEDROW, J. C. F. (1959). *Soil Sci.* **88,** 305–312.
EDWARDS, C. A. and LOFTY, J. R. (1972). "Biology of Earthworms." Chapman and Hall, London.
EGGINS, H. O. W. and MALIK, K. A. (1969). *Antonie van Leeuwenhoek* **35,** 178–184.
FLETCHER, W. W. (1966). *Proc. 8th Br. Weed Control Conf.* **3,** 896–900.
FLOATE, M. J. S. and TORRANCE, C. J. W. (1970). *J. Sci. Fd. Agric.* **21,** 116–120.
FLOATE, M. J. S. (1970*a*). *Soil Biol Biochem.* **2,** 173–185.
FLOATE, M. J. S. (1970*b*). *Soil Biol. Biochem.* **2,** 187–196.
FRIES, N. (1973). *Trans. Br. mycol. Soc.* **60,** 1–21.
GIVEN, P. H. and DICKINSON, C. H. (1973). *In* "Soil Biochemistry." (A. D. McLaren and E. A. Paul, eds), Vol. 3, in press. Marcel Dekker, New York.
GIUMA, A. Y., HACKETT, A. M. and COOKE, R. C. (1973). *Trans. Br. mycol. Soc.* **60,** 49–56.
GRAY, T. R. G. and WILLIAMS, S. T. (1971*a*). "Soil Micro-organisms." Oliver and Boyd, Edinburgh.
GRAY, T. R. G. and WILLIAMS, S. T. (1971*b*). *In* "Microbes and Biological Productivity." (D. E. Hughes and A. H. Rose, eds), pp. 255–286. 21*st Sym. Soc. gen. Microbiol.*, Cambridge University Press.
GREENWOOD, D. J. (1968). *In* "The Ecology of Soil Bacteria." (T. R. G. Gray and D. Parkinson, eds), pp. 138–157. Liverpool University Press.
GRIFFIN, D. M. (1972). "Ecology of Soil Fungi." Chapman and Hall, London.
GRIFFITHS, E. (1965). *Biol. Rev.* **40,** 129–142.
GRIFFITHS, E. and BIRCH, H. F. (1961). *Nature, Lond.* **189,** 424.
GROSSBARD, E. (1969). *J. Soil Sci.* **20,** 38–51.
HAYES, A. J. (1966). *Pedobiologia* **6,** 281–287.
HEAD, G. C. (1971). *In* "Methods of Study in Quantitative Soil Ecology." (J. Phillipson, ed.), pp. 14–23. Blackwell, Oxford.
HEAL, O. W., BAILEY, A. D. and LATTER, P. M. (1967). *Phil. Trans. R. Soc.* B. **252,** 191–197.
HIRST, J. M., LE RICHE, H. H. and BASCOMBE, C. L. (1961). *Pl. Path.* **10,** 105–108.
HORA, T. S. and BAKER, R. (1972). *Trans. Br. mycol. Soc.* **59,** 491–500.
HUDSON, H. J. (1971). *In* "Ecology of Leaf Surface Micro-organisms." (T. F. Preece and C. H. Dickinson, eds), pp. 447–455. Academic Press, London and New York.
HUTCHINSON, S. A. and KAMEL, M. (1956). *J. Soil Sci.* **7,** 213–218.
IKEDIUGWU, F. E. O. and WEBSTER, J. (1970). *Trans. Br. mycol. Soc.* **54,** 205–210.
INGRAM, M. (1957). *In* "Microbial Ecology." (R. E. O. Williams and C. C. Spicer, eds), pp. 90–133. University Press, Cambridge.
JANOTA-BASSALIK, L. (1963). *Acta Microbiol. Pol.* **12,** 25–40.
JENKINSON, D. S. (1965). *J. Soil Sci.* **16,** 104–115.
JENKINSON, D. S. (1966). *J. Soil Sci.* **17,** 280–302.
JOFFE, J. S. (1955). *Adv. Agron.* **7,** 141–187.

KALELA, E. K. (1955). *Acta for. fenn.* **65**, 1–41.
KEARNEY, P. C., KAUFMAN, D. D. and ALEXANDER, M. (1967). *In* "Soil Biochemistry." (A. D. McLaren and G. H. Peterson, eds), pp. 318–324. Marcel Dekker, New York.
KOUYEAS, V. (1964). *Pl. Soil* **20**, 351–363.
KÜHNELT, W. (1961). "Soil Biology." Faber and Faber, London.
LATTER, P. M. and HEAL, O. W. (1971). *Soil Biol. Biochem.* **3**, 365–379.
LEWIS, B. G. (1970). *Ann. appl. Biol.* **66**, 83–88.
LINSBAUER, K. (1930). "Die Epidermis." Handbuch der Pflanzenanatomie, 4 (27).
MACFADYEN, A. (1957). "Animal Ecology: Aims and Methods." Pitman, London.
MACFADYEN, A. (1963). *In* "Soil Organisms." (J. Doeksen and J. van der Drift, eds), pp. 3–16. North-Holland, Amsterdam.
MADGE, D. S. (1964). *Acarologia* **6**, 199–223.
MARSHALL, K. (1971). *In* "Soil Biochemistry." (A. D. McLaren and J. Skujins, eds), Vol. 2, pp. 409–445. Marcel Dekker, New York.
MATSUMURA, F. and BOUSCH, G. M. (1971). *In* "Soil Biochemistry." (A. D. McLaren and J. Skujins, eds), Vol. 2, pp. 320–336. Marcel Dekker, New York.
NEPOMILVEV, V. F. and KUZYAKINA, T. I. (1965). *Izv. timiryazev. sel. '-khoz. Akad.* **1**, 71–81.
PARK, D. (1960). *In* "The Ecology of Soil Fungi." (D. Parkinson and J. S. Waid, eds), pp. 148–159. Liverpool University Press.
PUGH, G. J. F. (1962). *Trans. Br. mycol. Soc.* **45**, 560–566.
PUGH, G. J. F. and VAN EMDEN, J. H. (1969). *Neth. J. Pl. Path.* **75**, 287–295.
PUGH, G. J. F. and WILLIAMS, J. I. (1971). *Trans. Br. mycol. Soc.* **57**, 164–166.
PUGH, G. J. F., BLAKEMAN, J. P., MORGAN-JONES, G. and EGGINS, H. O. W. (1963). *Trans. Br. mycol. Soc.* **46**, 565–571.
ROGERS, W. S. and HEAD, G. C. (1969). *In* "Root Growth." (W. J. Whittington, ed.), pp. 280–292. Butterworths, London.
ROSE, A. H. (1968). *J. appl. Bact.* **31**, 1–11.
ROVIRA, A. D. (1956). *Pl. Soil* **7**, 178–194.
ROVIRA, A. D. and GREACEN, E. L. (1957). *Aust. J. agric. Res.* **8**, 659–673.
ROVIRA, A. D. and MCDOUGALL, B. M. (1967). *In* "Soil Biochemistry." (A. D. McLaren and G. H. Peterson, eds), pp. 417–463. Marcel Dekker, New York.
RUDDICK, S. M. and WILLIAMS, S. T. (1972). *Soil Biol. Biochem.* **4**, 93–103.
RUSSELL, E. W. (1961). "Soil Conditions and Plant Growth." Longmans, London.
RYBALKINA, A. V. and KONONENKO, E. V. (1961). *Pochvovedenie* **8**, 13–25.
SAITO, T. (1955). *Scient. Rep., Tohoku Univ. Ser.* 4, *Biology* **21**, 145–151.
SCHUSTER, J. L. (1964). *Ecology* **45**, 63–70.
SKUJINS, J. J. (1967). *In* "Soil Biochemistry." (A. D. McLaren and G. H. Peterson, eds), pp. 371–414. Marcel Dekker, New York.
SMITH, K. A. and RESTALL, S. W. F. (1971). *J. Soil Sci.* **22**, 430–443.
STOTZKY, G. (1968). *Adv. appl. Microbiol.* **10**, 17–54.
STOTZKY, G. and MORTENSEN, J. L. (1958). *Soil Sci. Soc. Am. Proc.* **22**, 521–524.

TIMONIN, M. I. (1961). *Pl. Soil* **14**, 323–334.
TIMONIN, M. I. (1962). *Can. J. Microbiol.* **8**, 594–595.
WAINWRIGHT, M. and PUGH, G. J. F. (1973). *Soil Biol. Biochem.* **5**, 577–584.
WALLACE, H. R. (1963). "The Biology of Plant Parasitic Nematodes." Edward Arnold, London.
WARCUP, J. H. (1951). *Trans. Br. mycol. Soc.* **34**, 376–399.
WEBLEY, D. M., EASTWOOD, D. J. and GIMMINGHAM, C. H. (1952). *J. Ecol.* **40**, 168–178.
WIEGERT, R. G. and EVANS, F. C. (1967). *In* "Secondary Productivity of Terrestrial Ecosystems." (K. Petrusewicz, ed.), pp. 3–15. Krakow, Warsaw.
WILKINSON, V. and LUCAS, R. L. (1969). *New Phytol.* **68**, 709–719.
WILLIAMS, S. T. and MAYFIELD, C. I. (1971). *Soil Biol. Biochem.* **3**, 197–208.
WILLIAMS, S. T., SHAMEEMULLAH, M., WATSON, E. T. and MAYFIELD, C. I. (1972). *Soil Biol. Biochem.* **4**, 215–225.
WILLIAMS, T. E. (1969). *In* "Root Growth." (W. J. Whittington, ed.), pp. 270–278. Butterworths, London.
WOHLRAB, G. and TUVESON, R. W. (1965). *Am. J. Bot.* **10**, 1050–1058.
WOODCOCK, D. (1971). *In* "Soil Biochemistry." (A. D. McLaren and J. Skujins, eds), Vol. 2, pp. 337–360. Marcel Dekker, New York.
WRIGHT, S. J. L. (1971). *In* "Microbial Aspects of Pollution." (G. Sykes and F. A. Skinner, eds), pp. 232–254. Academic Press, London and New York.
ZVERKOV, YU. V. (1966). *Pochvovedenie* **8**, 39–45.

21

Decomposition of Litter in Fresh Water

L. G. Willoughby

Freshwater Biological Association
Ambleside
Westmorland
England

I. The Freshwater Environment

A. Oxygen Regime and Oxygen Demand

In a consideration of the physical and chemical characteristics of freshwater environments as they relate to the decomposition of plant litter, the nature of the water itself comes under scrutiny. Seen from the point of view of the fungi the capacity of the water to store and release oxygen is of great importance, since they are predominantly aerobic. The oxygen storage capacity of fresh water at 100% saturation is approximately 10 mg $O_2\ l^{-1}$, varying with temperature and atmospheric pressure. It has been suggested (see Griffin, 1963) that truly aquatic fungi show more physiological adaptation to a limited oxygen supply (as compared with that available in sub-aerial conditions or in soil) than to any other feature of their environment. Swift-flowing streams and rivers may have a super-saturation of oxygen (e.g. 115% saturation was recorded in the River Kent, Westmorland in May 1969) but more sluggish rivers often exhibit a deficit, especially on a diurnal basis if excessive weed growth depletes the oxygen store by respiration during the hours of darkness. As a background to a study of leaf decomposition in the River Lune, Lancashire, Newton (1971) recorded occasional low oxygen values, e.g. 5·6 mg $O_2\ l^{-1}$ with a water temperature of 15°C in August 1969 and 6·2 mg $O_2\ l^{-1}$ with a water temperature of 6·6°C in November 1969. These gave calculated percentage saturation figures of only 57 and 52 respectively and showed that there were times when oxygenation was not complete in this river, despite the fact that it was apparently unpolluted above the sampling point.

The long-term extent of the oxygen demand on the aquatic environment made by leaves decaying in aerobic conditions has been studied by Chase and Ferullo (1957). *Acer*, *Quercus* and *Pinus* leaves were submerged in 19 l carboys full of water which was initially oxygenated. Successive dissolved oxygen samplings showed that the cumulative oxygen demand of *Acer* leaves rose steeply to 600 mg g^{-1} of material in 150 days, and then rose further, but less rapidly, to 740 mg in 386 days. On the other hand the oxygen demands of *Quercus* and *Pinus* were smaller but followed the same pattern, both rising to 350 mg at 150 days and then to 500 mg in 386 days. It was concluded that for aerobic decomposition in water *Acer* leaves require oxygen which amounts to 75% of their weight during the first year while *Quercus* and *Pinus* leaves require only the equivalent of 50% of their weight. Although the plotted curves were asymptotic in nature they did suggest that under the experimental conditions employed further oxygen demand would have been exerted in the second year.

Although water itself may be the sole background to plant decomposition in fresh water, any decaying lotic fragments usually soon become water-logged and sink. Regarding the physical and chemical background in the sunken condition in the natural environment the physiological versatility of bacteria as agents of decay is to be borne in mind. In particular the capacity of the aquatic bacteria to tolerate and exploit anaerobic and aerobic situations is important, since both occur in freshwater sediments.

B. The Physico-chemical Status of the Environment

In considering the physical and gaseous conditions which a waterlogged leaf or twig will encounter at the mud–water interface it is noted that the topmost few millimetres of mud have a high diffusion rate due to the stirring action of the overlying turbulent water. Below that, however, the diffusion rate is virtually molecular, within the actual sediment. A measure of the oxidizing–reducing status in the surface layers of the latter is given by the distribution of electrode potentials (Redox potentials) generally measured at pH 7 (E 7). The +0·20 to +0·30 V potential is particularly interesting and important since it marks the Fe^{++}–Fe^{+++} equilibrium. In Esthwaite Water, during the winter months, a −0·07 V potential at 4 cm depth in the mud signifies highly reducing conditions and the +0·20 isovolt lies 1 cm below the mud surface. At this time the ferric hydroxide in the surface mud (E 7 = +0·5) shows as a reddish tinge to the naked eye while below the reduced ferrous iron is grey in colour. Microbiological conditions for decomposition are fully aerobic at the mud surface, and the water shows oxygen saturation. With the onset of the summer, however, thermal stratification occurs in the lake, and the store of oxygen in the deeper water is isolated and eventually consumed. The +0·20 isovolt moves upward in the sediment and finally emerges into the water. Iron and manganese become reduced and soluble in the surface mud layers and diffuse out into the water, while phosphorus is no longer adsorbed by these cations and also comes into solution. Thus anaerobiosis of the surface mud carries a chain of parallel chemical consequences, all of which probably affect the pattern of microbial degradation. The situation is reversed when lower temperatures and high winds finally break the summer thermal stratification. Oxygen returns to the surface mud as the lake water mixes completely. In Windermere and other very large temperate lakes, although there is a summer thermal stratification the store of oxygen in the deeper water is sufficient to prevent final depletion. Therefore although the +0·20 isovolt moves up to near the mud surface it never emerges and oxidizing conditions are retained at the mud surface (Mortimer, 1971). Mackereth (1965) has pointed out that the nature of the sediments of Lake District lakes is greatly influenced by the nature of the drainage basin which the inflow waters run off. The post-glacial sediments contain 70–80% of mineral matter and the remaining 20–30% consists of organic matter. Since the rate of deposition of sediments in large lakes is very low, for example 2 mm per year in the centre of Windermere, most of the organic matter reaches a state of stability towards oxidation and aerobic microbial degradation by the time of its eventual burial. At this time it represents an accumulation of finely divided debris, largely derived from the mud surface and containing a high proportion of allochthonous plant material, but with additions from the lake biomass. The latter is considered as a fairly minor proportion of the whole; there is evidence that most of the algal matter produced in the lakes is readily oxidized at the mud surface. When the rate of deposition is high, in situations like the enclosed bays and river mouths of the larger lakes such as Windermere, and the whole area of smaller lakes such as Esthwaite Water, lacustrine deposits may include more or less intact plant litter.

Thus Franks and Pennington (1961) have described leaf fragments, catkins scales and fruits of *Betula nana* L. and *B. pubescens* Ehrh. from Esthwaite Water deposits at levels corresponding to the immediate post-glacial and up to more recent times. A high deposition rate and a correspondingly greater importance of the anaerobic decomposition cycle is characteristic of estuaries and also many streams and rivers.

An unusual kind of aquatic biotype occurs in ombrophilous bogs, the waters of which are represented by puddles, pools or even lakes. Bog waters are characterized by an extreme poverty of electrolytes, particularly calcium, a strongly acid reaction (pH 3·5–4·5 in high moor situations) and a high content of coloured humic-type materials. Because the water is unbuffered its pH is very sensitive to dissociating H^+ ions and it is not entirely clear how its acidity originates. Under these conditions even dissolved CO_2 may have a large effect. Scandinavian lakes of the bog type have a low productivity and have been termed dystrophic, as the furthest extreme from the eutrophic, productive type. Owing to the acid reaction of the water, and probably also to the lack of background ions, the decomposition of plant materials including cellulose in bogs is slow.

C. Leaching

Once immersed in water, plant litter begins to lose materials by leaching even without the activity of micro-organisms. Nykvist (1963) made a study of the organic substances leached out of tissues of seven tree species. Anaerobic, rather than aerobic, experimental systems were generally preferred, since in the latter the leachates were decomposed so rapidly by micro-organisms that they could escape detection. *Picea* and *Pinus* needles released their water-soluble organic substances at a steady rate over 20–25 days, whereas *Alnus* and *Quercus* leaves lost all theirs in the first 7 days. Amino acids from the leachates comprised α-alanine, arginine, aspartic acid, glutamic acid, glycine, leucine, lysine, serine, threonine, valine and one that was unidentified. Glutamic acid was present in the greatest amount. Of the sugars detected, fructose and glucose were found in the greatest amounts in all seven species, with xylose detected from five species and sucrose from *Betula verrucosa* Ehrh. only. Leaching of the most abundant non-volatile aliphatic acids was also investigated and these were shown to include citric and malic acids. *Fraxinus* leaves released 1·7% of their weight as malic acid; fumaric, glycollic, lactic and malonic acids were also found. According to Nykvist the rapid loss of citric and malic acids in particular, especially under aerobic conditions, explains why there is a pH rise in decaying terrestrial leaf litter. Whether or not such an effect occurs in wholly aquatic systems seems unknown at present.

Leaching of sugars from fresh woody materials, in this instance small twigs, was investigated by Willoughby and Archer (1973). Following exposure in Smooth Beck, a small Lake District stony stream, for varying periods the twigs were returned to the laboratory and agitated in sterile water in flasks for 7 days. At the end of this period total sugars in the flask water were estimated, giving a measure of the capacity of the twigs to continue to produce leachable sugars in the stream. It was shown that intact *Quercus* twigs gave an initially high sugar release, diminishing with time but still demonstrable after 12 weeks. Intact *Fraxinus* twigs release small quantities initially but a larger amount at three weeks. Decorticated twigs showed a continuous slow release of sugars over the twelve week period. The results suggested that in comparison with leaves woody materials have larger sugar reserves which take longer to leach away in the freshwater environment.

II. Gross Changes and Chemical Transformations in Allochthonous Leaves in Water

Few long-term observations have been made on the gross changes which allochthonous leaves exhibit when immersed in water in natural conditions. Newton (1971) noted that autumnal collections of leaves of five different species of trees changed their outward appearance at different rates when exposed in the River Lune, Lancashire. In a 4–5 month winter exposure period *Quercus* sp. was ostensibly unchanged while *Alnus glutinosa* (L.) Gaertn. and *Salix viminalis* L. showed considerable erosion and *Ulmus glabra* Huds. was extensively skeletonized. *Fagus sylvatica* L. seemed similar in its visible rate of decay to *Alnus*. To some extent differences in decay rates were ascribed to the mechanical structure of the leaf and it was noted that elements in the vascular strands adhered firmly in *Fagus* and *Quercus* whereas consecutive examinations of the petiole showed that they parted quickly in *Alnus*, *Salix* and *Ulmus*. Limpness and an associated slimy feel of the tissues were features of decay which were marked in the leaves of *Ulmus*, but the *Fagus* and *Quercus* leaves became brittle with the passage of time. Although breakage occurred in the mesophyll, the tissues of *Fagus* and *Quercus* frequently did not part in discrete layers, but cracked irregularly and still clung together as a whole. Eventually fragments, in which the relationships of the tissues were undisturbed, broke away from the *Quercus* leaves. Dry weight determinations for samples of decaying *Alnus* and *Salix* leaves indicated that there was an initial rapid weight loss in the first few months followed by a more gradual decline later. Some analyses of these weight losses were presented and it was shown that soluble carbohydrates were lost extremely early, 80% disappearing in the first two weeks. The changing values of cellulose and lignin composition

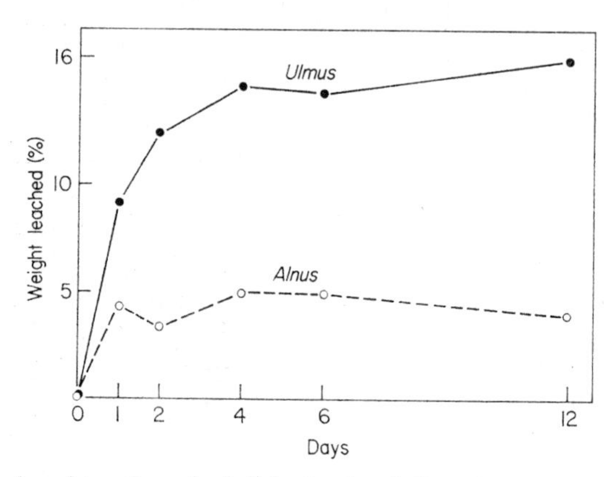

Fig. 1. Leaching from leaf disks in circulating stream water, in the laboratory at 10°C. (After Kaushik and Hynes, 1971.)

gave no clear-cut picture but examination of the data shows clearly that the percentage dry weight protein values increased with time. *Alnus* leaves collected and exposed in November had initially 16% protein, rising to 20% at the end of that month, to 22% in December and to 24% in January. The increase was not maintained later and the protein value was 22% in July.

In another study of the changes which allochthonous leaves of trees undergo when they are shed into river water, Kaushik and Hynes (1971) determined initial nitrogen and protein contents ranging from 0·7% and 4·0% for *Acer saccharum* Marsh to 2·1% and 10·8% for *Alnus rugosa* (Du Roi) Spreng, with *Fagus grandifolia* Ehrh, *Quercus alba* L. and *Ulmus americana* L. giving intermediate values. It was shown that the leaching of soluble substances, presumably sugars and some protein, took place rapidly in river water and a three or four day immersion period was sufficient to remove most of these (Fig. 1). In order to investigate long-term changes in the material, leaf disks were leached and weighed dry before being immersed for varying periods. It was shown that over an 8 week period the *Ulmus* leaves lost 12% of their initial leached weight with the rate of degradation being faster initially. The addition of 20 mg l^{-1} of N as $(NH_4)_2SO_4$ and 5 mg l^{-1} of P as KH_2PO_4 to the river water in an experimental laboratory system accelerated the rate of decomposition, particularly if both nutrients were used simultaneously. Running parallel to the pattern of weight loss was an observed increase in percentage protein content, once a slight fall had been made up. Protein content then increased rapidly but eventually tended to stabilize. Again it was noticed that there was an enhanced effect using the chemically enriched waters. In another similar experiment an attempt was made to determine whether the percentage protein increase could be ascribed to an actual uptake of exogenous nitrogen. Accordingly the amount of nitrogen found in *Ulmus* leaves at the end of each sampling time was calculated as the percentage of the initial weight. On this basis the controls neither gained nor lost. However, leaves kept in the enriched water showed significant increases, these being more pronounced when both nutrients were added. These increases varied with temperature, being more pronounced at 20°C than at 10°C, and also with the specific nature of the leaf. Thus after 9 weeks residence at 10°C in water enriched with 5 mg l^{-1} N *Acer* showed a 38·2% increase from their initial nitrogen content while *Fagus* disks showed only a 5·7% increase.

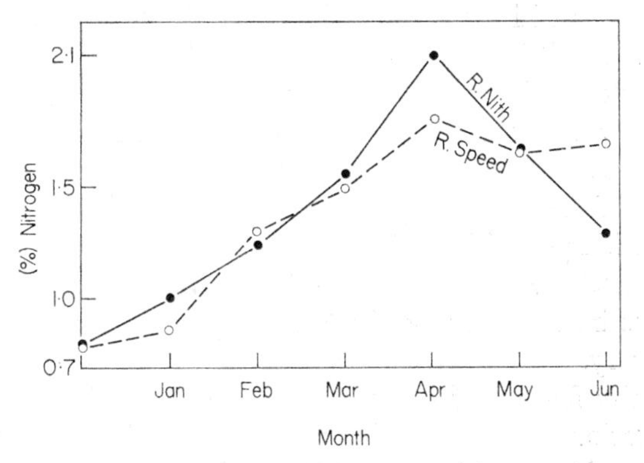

Fig. 2. Nitrogen content of *Acer* leaf disks exposed in two rivers, expressed as percentage of the ash-free weight. (After Kaushik and Hynes, 1971.)

In field experiments Kaushik and Hynes exposed leaf disks both in a clean river and in a polluted one. Discs guarded from access by the stream fauna often, but not always, showed greater weight losses than corresponding ones more freely exposed. In the River Speed, the clean river, percentage ash-free weight losses after six months unguarded exposure of *Ulmus*, *Acer*, *Quercus*, *Alnus*, and *Fagus* were 68·9, 56·8, 23·1, 21·8 and 13·0 respectively. On the other hand in the polluted River Nith, corresponding percentage ash-free weight losses after the same exposure period, and for the same species, were only 54·5, 41·9, 18·5, 10·2 and 3·3 respectively. Following the protein and nitrogen contents of the leaves during exposure for the same six month period, from January until June, *Acer*, *Quercus* and *Ulmus* showed percentage ash-free weight increases, but these were much less marked for *Alnus* and *Fagus*. From an initial nitrogen content of 1·13% the highest value recorded for *Ulmus* was 2·43% in May, in the River Nith (Fig. 2). The maximum nitrogen increase was recorded for *Acer*, with an original content of 0·78% rising to 2·1% in April, again in the River Nith. The general pattern of the results, however, suggested that nitrogen contents of leaf disks were not appreciably different in the two streams, and also that they came to have asymptotic values for each species, after about half-way through the experiment at the end of March.

III. Microbial Successions on Litter

The potential of the natural freshwater environment to provide a microbial inoculum to initiate decomposition of allochthonous plant materials is probably always high (see also Jones, Chapter 11). In part, the viable fungal spora in water can be related to truly aquatic, aero-aquatic or terrestrial fungi which suggests (Willoughby and Collins, 1966) that only a small fraction of it is likely to be involved in aquatic colonizations. The remainder is generally regarded as a wash-in spore component with no potential to develop in fresh water. However, with greater knowledge of microhabitats these views may change if, as seems possible (see Park, 1972), some of these latter fungi can be shown to have an active role in the environment. More positive information on the inoculum potential of the fungi in freshwater has been gained from several studies in which plant materials have been left exposed for varying periods and then examined in the laboratory.

A noteworthy feature of Newton's (1971) study of leaf decay in water was the utilization of natural rather than pre-sterilized material. Thus the effect of immersion on the resident phylloplane and senescent leaf micro-flora was examined, while simultaneously fresh inoculation from the aquatic environment came under inspection. A preliminary experiment with *Salix* leaves, which were plated onto nutrient agar following their immersion, indicated that aquatic species such as *Pythium* sp. and *Saprolegnia delica* Coker would colonize the immersed leaves irrespective of any competition from the established micro-flora, while at the same time the latter would decline rapidly. Long-term experiments confirmed and expanded these suggestions. Considering elements of the established micro-flora, *Aureobasidium pullulans* (de Bary) Arnaud disappeared within a month from *Alnus*, *Quercus*, *Salix* and *Ulmus* leaves but persisted for two months at its original level on *Fagus* leaves. *Epicoccum nigrum* Link was established on the leaves of all five species, being especially prominent on *Salix*, but there was a rapid decline in the two months following immersion. The Sphaeropsidales were interesting since the aquatic status of this group seems clear. *Fagus* leaves showed a particularly large Sphaeropsidales component which declined rapidly following immersion while the data suggested that aquatic colonization of *Salix* leaves may have occurred. *Fusarium* species are known to have an active role in the aquatic environment (Willoughby and Archer, 1973), in addition to their well known terrestrial role and *Phialophora* is another genus which may be considered in the same way. Before immersion *Fusarium* was found only on *Salix* leaves but subsequently it was recovered in every sampling of both these and also the *Alnus*, *Fagus*, *Quercus* and *Ulmus* leaves. *Phialophora* was present on the *Alnus*, *Quercus*, *Salix* and *Ulmus* leaves before their immersion and subsequently was also recorded in every sampling of both these and also the *Fagus* leaves. Therefore from Newton's study also there seems good evidence that *Fusarium* is an aquatic colonist, while the position of *Phialophora* is complicated by its frequent occurrence as an entity of the established microflora. Its persistence for up to nine months on exposed leaves is highly suggestive of activity rather than mere passive survival.

Observations at Smooth Beck (Willoughby and Archer, 1973), suggested that the capacity of the lotic freshwater environment to provide a fungal inoculum to initiate decomposition in allochthonous litter is very high. Spores of four aquatic Hyphomycetes, namely *Heliscus lugdunensis* Sacc.& Therry, *Lemonniera aquatica* De Wild, *Margaritispora aquatica* Ingold, and *Tricladium splendens* Ingold were found in the stream all the year round. All of these species are implicated in leaf decomposition and all, except *M. aquatica*, were shown to grow on decaying wood. *H. lugdunensis* and *L. aquatica* colonized wood very quickly while *T. splendens* occurred later, after 3–4 months' exposure. The significance of the long-term availability of sugars in decaying wood, in regard to decomposition, appears to be that early colonists such as *Fusarium* spp. and *Heliscus lugdunensis*, which can rapidly produce large pustular growths on the fresh substratum and hence show the characteristics of sugar fungi, may also persist for many months in a lesser role as the colonization proceeds further. Newton also found that aquatic Hyphomycetes were early colonists in the River Lune. In her study *Alnus* leaves yielded *Anguillospora crassa* Ingold, *Articulospora tetracladia* Ingold, *Centrospora acerina* Newhall, *Clavariopsis aquatica* De Wild, *Dimorphospora foliicola* Tubaki, *Lemonniera aquatica*, *Margaritispora aquatica*, and *Tricladium splendens* after an exposure period of only one week. *Fagus* leaves gave *C. aquatica*, *Tetracladium marchalianum* De Wild and *Varicosporium elodeae* Kegel after three weeks; *Quercus* leaves gave *T.*

splendens and *V. elodeae* after four weeks; *Salix* leaves gave *C. aquatica* and *L. aquatica* after one week, and *Ulmus* leaves gave *A. tetracladia*, *C. aquatica*, *D. foliicola* and *T. marchalianum* after three weeks. From these records *C. aquatica* appeared to be particularly consistent in its early appearance, except for the absence of a record from the first *Quercus* sampling. It is also of interest to note that Newton recorded *Cylindrocarpon–Heliscus* from every initial sampling following immersion of the leaves and these fungi, like the aquatic Hyphomycetes were never obtained from leaves sampled off the tree. It seems very possible that some, if not all, of these records might actually refer to *Heliscus lugdunensis*. Such a rearrangement of the data would therefore indicate *H. lugdunensis* to be an outstanding early colonist.

Considering colonization by aquatic Phycomycetes Newton found that *Alnus* leaves collected and exposed in August attracted *Phytophthora* and *Pythium* in one week, *Achlya* in two weeks, *Saprolegnia* after four weeks and *Dictyuchus monosporus* Leitgeb after eight months. Another batch of *Alnus* leaves collected and exposed in November also attracted *Phytophthora* and *Pythium* in a week but the first Saprolegniaceae sp. *Achlya colorata* Pringsheim did not appear until nine weeks had elapsed. *Fagus*, *Quercus* *Salix* and *Ulmus* leaves were also colonized by *Phytophthora* and *Pythium* at the first sampling following exposure and at each successive one covering in all a period of up to 10 months. As was the case with *Alnus* Saprolegniaceae were obtained less regularly on leaves of these other trees. *Salix* attracted *Achlya*, *Leptolegnia caudata* de Bary and *Saprolegnia delica* in a week but their frequency declined at later samplings. *Quercus* and *Fagus* yielded none of these species but the former did give *Plectospira*.

Newton's study showed that leaf material liberated into the aquatic environment will quickly attract a number of fungal colonists. There seems to be no suggestion that with the increasing degradation of the material, and probable concomitant changes in its biochemical nature, other and different fungi will arive and become dominant. However, an examination of her quantitative data showed that *Dimorphospora foliicola*, although it was recorded as an early colonist, could become progressively more important with the passage of time. Quantitative data were derived from counts made on the number of colonies found to grow from unit areas of the exposed material. When these counts of *Dimorphospora foliicola* were high they were so both from washed and from surface-sterilized leaves, indicating a deep-seated microhabitat for the fungus. High counts were made from *Alnus* at five and nine months, from *Fagus* at seven and nine months and from *Quercus* at eight and ten months.

The long term study of decaying wood in fresh water at Smooth Beck by Willoughby and Archer (1973) suggests that in addition to certain aquatic Hyphomycetes which were early colonists others appeared on the material later, as mentioned above. A noteworthy example here was *Dimorphosphora foliicola* which was recorded from the fifth month onwards, giving an interesting correlation with Newton's observations on leaves. It is not entirely clear at present whether aquatic Ascomycetes are late colonists or whether they merely require a prolonged period to develop fully in fresh water. At Smooth Beck *Ceratostomella* sp. was observed on five separate occasions when twigs were collected from the site after a period of exposure of 10–12 months. Twigs exposed for three, four and seven months also yielded the fungus, but only after a period of at least four further months of incubation in the laboratory. The decomposition of wood by aquatic Ascomycetes in fresh water seems sometimes to proceed in such a way that fruiting of the fungus is delayed or postponed indefinitely. Thus at Smooth Beck a species of *Massarina* imperfect state, *Anguillospora longissima* (Sacc. & Syd.) Ingold, was common, and colonized wood from two months' exposure onwards but only ever fruited to form the Ascomycete state when subjected to sub-aerial conditions. Thus in truly aquatic conditions the fungus functions solely as a decomposer, turning to the production of resistant perithecia only in response to changes in the environment.

Although it would be unwise to ignore the bacteria in any study of the microbial decomposition of plant materials in fresh water, the evidence available at present suggests that the fungi and actinomycetes have particularly active roles and may well be the more important micro-organisms under aerobic conditions. Thus Kaushik and Hynes have largely ascribed the weight losses which they observed in their leaf discs, and the parallel increases in protein and nitrogen contents, to the activities of fungi. In the River Speed fungal concentrations determined by blending and plating rose from 900–14,500 colonies per disc in January after one month of exposure, to 1400–90,000 in February and followed by a decrease to 200–9100 in June. Fungi recovered included species of *Alternaria*, *Candida*, *Cladosporium*, *Coniothyrium*, *Epicoccum*, *Fusarium*, *Penicillium*, *Phoma*, *Trichoderma* and representatives of the Mucorales. In addition species of aquatic Hyphomycetes such as *Alatospora*, *Tetracladium* and *Tricladium* were also observed on the leaf discs. Using antifungal (actidione and nystatin) and antibacterial (penicillin and streptomycin) agents they showed that addition of both arrested weight loss and percentage protein increase over a four week period, in an experimental system where soluble nitrogen and phosphorus were added to the water. When antibacterial agents were used alone weight losses and protein increases were similar to the controls but when only antifungal agents were employed these changes were greatly reduced.

IV. Cellulose Degradation

Cellulose degradation is clearly a very important aspect of the decomposition of plant material in water and certainly involves bacteria and probably actinomycetes in addition to fungi. There is evidence (Willoughby, 1961 and unpublished) that cellulose-decomposing Chytridiales exhibit a distinct speciation pattern across the profile at the margin of a lake. Thus *Nowakowskiella* species were active in the surface deposits of the lake but not in the adjacent field soils, while on the other hand active *Rhizophlyctis rosea* (de Bary and Woronin) Fischer was found only in the latter situation. So far the reasons for such distribution patterns are unknown but they may well reflect differing physiological capabilities of these fungi in distinct ecosystems. Using pure cellulose (cellophane and filter paper) as *in situ* baits for aquatic fungi there is some evidence that the colonization and degradation of this material is a function of the chemical nature of the water. Thus at Blelham Tarn a comparison of colonization in two afferent streams showed that it was much more rapid in Tock How Beck than in Wray Beck. The former carries chemical enrichment from farmyard washings while the latter has a more natural drainage arena. In July 1972 NO_3 nitrogen and PO_4 phosphorus analyses of the Tock How water were 2810 and 172 $\mu g\ l^{-1}$ respectively, as compared with only 754 and 31·5 $\mu g\ l^{-1}$ for the same ions in Wray Beck. At Tock How Beck, in addition to members of the Chytridiales, *Pythium* spp. are important early colonists of the cellulose.

Bacterial decomposition of allochthonous plant cellulose in fresh water can proceed to completion by the normal aerobic pathway. In the case of a rapid and heavy deposition of leaves and wood, however, such as may occur at the margin of a lake in a bay, the anaerobic conditions typical of deep deposition may occur when these materials are still largely undecomposed. Under these circumstances bacterial fermentation takes place and methane may be liberated to such an extent that bubbles appear. In addition to being anaerobic, methane bacteria are only active in alkaline situations, and hence do not occur in acid bogs.

Actinomycete colonization of cellulose has not so far been demonstrated by direct exposure of the pure material but the Actinoplanaceae have been shown to colonize wood in water (Willoughby, 1971). The frequency of recovery of Actinoplanaceae (see Goodfellow and Cross, Chapter 9) from agar platings of water suggests they may have a particular role in plant decomposition in the flowing waters of streams and rivers rather than in the more static waters of lakes and reservoirs. Chemically enriched streams such as Tock How Beck appear to carry most Actinoplanaceae spores, suggesting once again that the nature of the water affects their occurrence on the appropriate substratum (Willoughby, 1969). The particular niche in the freshwater environment which, more than any other, appears to favour leaf-decaying Actinomycetales is the marginal lacustrine one. On the shore of Windermere, in the English Lake District, especially in the sheltered bays, waterlogged leaves collect in thick pads which may be several centimetres deep. Fluctuation in the water level of the lake, a reflection of the total rainfall recently received into the drainage basin, occurs fairly frequently and these fluctuations lead to alternate wetting and partial drying of the marginal leaves. Aquatic Actinomycetales, known as such from their release of motile flagellated spores, were observed in the laboratory following damp incubation of these leaves and also following plating out of leaf washings onto a solidified chitin–actidione growth medium. Representatives of various motile spore producing genera of the Actinoplanaceae were recovered, comprising *Actinoplanes*, *Amorphosporangium*, *Ampullariella* and *Spirillospora* (Willoughby, 1969*a*). Couch (1949) had noted the predilection of such forms for aero-aquatic conditions, in laboratory baitings where grass leaves or pollen grains were used, when the mycelium was submerged in the substrate but the sporangia projected above the surface meniscus of the water. Since spore release from sporangia is dependent on the presence of free water we perhaps see in alternately wetted and dried marginal leaf material an almost ideal physical background for Actinoplanaceae activity. However it must not be forgotten that other observations have shown that the Actinoplanaceae will apparently also grow and fruit normally when completely submerged (Willoughby, 1971). Confirmatory evidence that the Actinoplanaceae were especially active in marginal leaf material came from actinophage studies (Willoughby, *et al.*, 1972). During these studies evidence accumulated that the detection of actinophage activity in aquatic environments was dependent on local activity of the host, when the host was present in an active growing condition so the likelihood of the detection of its phage increased. *Actinoplanes* phage, an apparently highly specific virion, was found to be present in the marginal leaves in every sampling which was made. Further strong confirmatory evidence that the peculiar physical conditions occurring in marginal leaf litter were conducive to aquatic actinomycete activity came from the isolation of two specialized morphological types, which were apparently previously undescribed (Willoughby, 1969*a*). "Spore dome" isolates bore motile spores as the terminal cells of phialides arranged to give large sporing structures ranging from the sporodochium type to the synnema type. "Spore head" isolates bore motile spores in clusters on short lateral hyphae with no suggestion of a larger fruiting unit. Thus we see the apparent evolution of particular taxa adapted to this niche in the aquatic environment.

V. Biogeochemistry of Sediments

A. Alkanes and Fatty Acids

The total effect of the freshwater environment on the decay of plant litter may be measured by reference to the state of the residual matter which eventually becomes deposited in lacustrine sediments. As the sediment becomes compressed at lower levels its biological activity diminishes and eventually disappears. At this point recent unconsolidated fine-grained muds show an organic C content which is similar to that of ancient shales. Although such material may have no recognizable microscopic structure, chemical analysis reveals groupings which give evidence of a plant origin. The alkanes are particularly informative. Higher plants yield major amounts of *n*-C_{27}, C_{29}, C_{31} and C_{33} alkanes and minor amounts of *n*-C_{26},

Fig. 3. Isomeric alkanes of $C_{31}H_{64}$. (a) *n*-hentriacontane, (b) 2-methyltriacontane (*iso*), (c) 3-methyltriacontane (*anteiso*). (After Eglinton, 1969.)

C_{28}, C_{30}, C_{32} and C_{34}. In most cases the maximum is at C_{31}. Although millions of structural isomers of C_{31} alkane are possible, extensive studies have failed to reveal more than three biologically produced ones, namely the *n* and the 2 and 3 methyl isomers (Fig. 3). For most plants the amounts of the normal alkane, *n*-C_{31}, greatly exceed those of the *iso*-C_{31} and these in their turn are more abundant than those of the *anteiso*-C_{31}. Sediments formed from material of a predominantly terrestrial origin show the characteristic higher plant alkane pattern but in sediments of entirely marine organisms or lower terrestrial plants and freshwater algae *n*-alkanes with odd numbers of C atoms between C_{23} and C_{35} no longer predominate and C_{17} (*n*-heptadecane) is then often found to be the main component of this group of compounds. Linear saturated fatty acids may also occur in the sediments and where these have an even number of carbon atoms between C_{22} and C_{32} this is regarded as denoting a contribution from higher plants. On the other hand fatty acids with an even number of C atoms below C_{20} are usually most abundant in marine or lower terrestrial organisms. Thus tetradecanoic (C_{14}:O), palmitic (C_{16}:O) and stearic (C_{18}:O) acids are the main components of the acid fraction in some recent marine sediments.

Mud Lake in Florida is surrounded by semi-tropical vegetation and in cores representing an age of a few thousand years the dominant alkane distribution is in the C_{30} region with maxima at n-C_{27} and n-C_{29} (Han *et al.*, 1968). Eocene period oil shale from Messel (near Darmstadt) is a limnic sediment deposited in small continental basins which has never been buried more than 200–300 m deep (Albrecht and Ourisson, 1971). It consists of a finely layered shale containing water, with about 30% of carbon and 35% of a clay fraction consisting almost entirely of montmorillonite. Alkanes and fatty acids have been extracted from the organic material, and in addition other very sensitive biogenic substances are present, all of which have apparently survived unchanged for more than 50 million years. The n-alkanes show a strong preponderance of compounds with odd numbers of C atoms and a maximum at n-C_{27} which indicates that the shale has a higher plant origin. Again, the linear saturated fatty acids in the Messel deposits show a strong predominance of compounds with even numbers of C atoms between C_{22} and C_{32} again suggesting a higher plant derivation. Another Eocene shale, from the Green River, Colorado, is thought to have been laid down in shallow lakes. Once again the n-alkanes show the higher plant pattern with dominance of n-C_{29} and n-C_{31}. In this deposit n-C_{17} is also very prominent, suggesting hydrocarbon production from partially degraded algae. The branched and cyclic alkanes from Green River include a peak near C_{20}, identified as phytane, which is thought to be a significant biological marker representing a chlorophyll derivative, also from algae (Johns *et al.*, 1966). Phytane occurs in quaternary post-glacial sediments at Blelham Tarn in the English Lake District (P. A. Cranwell, personal communication), and is also reported as a degradation product of *Botryococcus braunii* Kützing deposited within the last hundred years in lagoons in South Australia (Douglas *et al.*, 1969). Thus the two Eocene sediments considered have shown strong similarity in the chemical nature of some of the higher plant and even algal residues with those of more recent sediments. It lies outside the scope of this contribution to follow in detail the further change to crude oils and coal, etc. in ancient sediments, for which a marked thermal alteration is probably largely responsible. The change is accompanied by the production of a greater variety of isomeric alkanes and their generally lower C numbers, which show a weakening in the odd over even dominance, discussed above.

B. Photosynthetic Pigments

The fate of the pigments which are traditionally associated with photosynthesis in plants has been studied in freshwater lakes by Fogg and Belcher (1961). They have pointed out that although decomposition under aerobic conditions ultimately results in the destruction of chlorophylls and carotenoids, this process rarely appears to go to completion before these materials become inert in the lower anaerobic layers. In the English Lake District lakes the pigments have accordingly been deposited and preserved in the sediments at a fairly constant rate over most of the last 10,000 years of post-glacial history. During the last 150 years of deposition in Esthwaite Water however, the proportion of pigments to total organic matter has risen markedly, and this is taken to reflect the reduced oxygen status of the surface mud which now occurs every summer as the result of sewage pollution.

In their study of the persistence of higher plant pigments in lake deposits Sanger and Gorham (1970) made observations which began when the leaves were still on the parent trees. At this stage they identified seven pigments in *Populus tremeloides* Michx. (aspen), *Corylus americana* Walt. (hazel) and *Quercus ellipsoidalis* E. J. Hill (oak) comprising β-carotene, chlorophyll *a* and *b*, lutein (xanthophyll), neoxanthin, pheophytin *a* and violaxanthin. After autumnal changes *Populus* and *Corylus* had lost their entire pigment complement by the time of leaf fall but *Quercus* retained β-carotene, lutein and pheophytin *a*. This matched the observation that soil humus from oak forest contained only these three pigments. Again, in lake sediments, β-carotene and lutein were the only pigments to be found which were shown to be derived from leaves while a much larger range of different pigments was shown to be derived from decaying algae and other aquatic organisms. It was even suggested that a high ratio of lutein to the total hypophasic fraction or a high ratio of β-carotene to the total epiphasic fraction would both indicate a large allochthonous input into the lake as compared with the autochthonous contribution. Oligotrophic (unproductive) lakes might well show such ratios since the expectation is that the sedimentary matter in these would be dominated by allochthonous detritus. Although this would be a useful pointer to the past history of a lake and its surrounding vegetation, Sanger and Gorham have pointed out several possible complicating factors. For example green leaves blown into the lake by summer winds would possibly yield a larger and more varied pigment contribution to the sediments. Neoxanthin, derived from such an origin, was detected after a year of residence of leaves in the lake.

C. Dynamic Transformation in the Sediments

Recent evidence presented by Rhead *et al.* (1971) indicates that some chemical compounds of plant origin in the anaerobic sediments may undergo dynamic transformation to a greater and more rapid extent than has been suspected hitherto. These transformations are presumably mediated by micro-organisms. Rhead and his collaborators made their chemical study in the River Severn estuary. They had observed that the upper, biologically active, layers of recent aquatic sediments contain considerable quantities of saturated and unsaturated fatty acids but although these compounds are major constituents of organisms their concentration in the sediments is of the same order of magnitude as that of the sterols, which are usually only minor constituents of organisms. Therefore they argued that the fatty acids are probably subject to rapid chemical changes in the environment, and that these might be amenable to investigation. Accordingly they incubated oleic acid in the sediment at a depth of 10 cm below the surface, using the 9, 10-^{3}H, 1-^{14}C- oleic acid isotope, and recovered the labelled products of short-term diagenesis. The results showed that the labelled oleic acid disappeared rapidly, only 5% remaining in fatty acids after 9 days. Evidence for a limited change in the molecule, involving simple hydrogenation of unsaturated carbon–carbon bonds followed by loss of C_2 units without total breakdown, was sought for in loss of ^{14}C from the molecule but retention of all the ^{3}H. Lauric, myristic and to a lesser extent palmitic acids were identified as compounds of shorter chain length than the parent compound which has been formed in such a way. However, the combined long chain saturated fatty acid fraction recovered never contained more than 3% of the original label. It seemed probable that most of the oleic acid was metabolized by degradation of the carbon skeleton through β-oxidation to acetyl-Co A, followed either by conversion to CO_2 and H_2O or incorporation into the non-lipid cellular components. However, there was some evidence that total breakdown of the molecule could be followed by re-synthesis into fatty acids, and it was concluded that these compounds, as found in compacted sediments of both recent and ancient age, have not necessarily been derived directly from fatty acids originally contributed to surface layers. They are probably derived from a number of precursors through many metablic pathways.

VI. Plant Materials in Energy Budgets and Productivity Cycles

A. Lakes

In a number of studies an attempt has been made to assess the significance of the contribution of decomposing allochthonous plant materials to energy budgets, food webs and finally to productivity in the freshwater ecosystem. At Lake Marion, Canada, the suggestion was made by Efford (1969) that the decomposition of forest detritus which comes in with the inlet water proceeds in such a way that it contributes considerably to the energy budget. Total energy sources in the system are estimated at 280 g C m^{-2} year^{-1}, of which 86% is from the detritus, 10% is fixed by benthic algae, 3% by phytoplankton and 1% by aquatic plants. The surface mud, which contains the forest detritus undergoing colonization by micro-organisms, is continually ingested by benthic herbivores such as *Hyallela azteca* Saussure. These reduce the organic content from 45% to 40% and produce faeces which in their turn are quickly colonized by micro-organisms. Fish (*Salmo gairdneri* Richardson and *Oncorhynchus nerka* (Walbaum) and salamanders (*Taricha granulosa* Skilton and *Ambystoma gracile* Baird) feed on *H. azteca* and other forms and are the final products in the system. Limitation in the productivity of the lake occurs due to low temperatures during the period from October to April and a high flush through rate at the same time, the relatively high level (30%) of unusable lignin in the mud, and the short time the organic debris is at the surface layer where the most rapid breakdown occurs.

B. Streams and Rivers

In streams and rivers the evidence is even stronger that the availability of decomposing allochthonous leaves, and their palatability to the invertebrate fauna, may be very influential in respect of the final productivity assessed in terms of the resident fish population. In this connection Kennedy and Fitzmaurice (1971) compared the chemical and biological characteristics of the Irish rivers Avonmore and Inny. The former river had a pH of 6·5, a bicarbonate alkalinity of 0·2 mequiv l^{-1} and a conductivity of 44 megohm^{-1} cm^{-1}. The latter river had a higher estimated pH of 8·5 a higher bicarbonate alkalinity of 2·85 mequiv l^{-1} and a higher conductivity of 280 megohm^{-1} cm^{-1}. The River Avonmore is a moorland river running through rough sheep-grazing land with *Ulex* some stands of *Quercus* and conifer plantations. The River Inny runs through pasture land with sections bearing *Acer*, *Alnus*, *Crataegus* and *Salix* trees. The flora of the River Inny includes *Chara*, *Potamogeton* and *Ranunculus* while the River Avonmore has aquatic moss only. The two invertebrate faunas also show a marked disparity in diversity and numbers, *Gammarus* for example being abundant in the River Inny but absent from the River Avonmore. This disparity in the invertebrate trout food is reflected in the differing size and quality of the fish in these two rivers those from the River Inny being large and pink-fleshed and those from the River Avonmore being small and white-fleshed.

C. Invertebrate Feeding on Plant Leaf Materials

In a series of ingenious experiments Kaushik and Hynes made direct observations on the feeding of *Gammarus* on leaves, and showed feeding preference that was associated with both the leaf species and its condition.

The animal showed a preference for *Ulmus* disks over *Acer*, *Alnus*, *Fagus* and *Quercus*; for *Acer* over *Alnus*, *Fagus* and *Quercus*; for *Alnus* and *Quercus* over *Fagus*; and for *Quercus* over *Fagus*. In sets of *Ulmus* disks pre-conditioned for eight days in (a) water without nutrients at 10°C and (b) water with added N and P at 20°C, the total weight loss as *Gammarus* food was markedly greater in the latter. This showed that the presence of microbial growth greatly enhanced the attractiveness of the leaves.

Seen from the point of view of invertebrate animal feeding, which can eventually lead to complete degradation of the leaf material in streams, the following picture emerged. A decaying leaf declines in weight and total calorific value but the protein content rises steeply to an asymptotic level which remains constant for many weeks. During this time, and largely owing to the activities of fungi, no nitrogen is lost and when extra soluble nitrogen is available in the environment it is taken up. Therefore, mouthful by mouthful, an animal obtains no more calories from an old than a fresh leaf, but the protein content of a mouthful of decayed leaf is higher and it constitutes a more nutritious food material. Leaves from different species of trees decay at different rates and those decaying most quickly tend to be eaten first. Thus in the autumn, winter and spring, the main growing period for many stream animals, a steady supply of protein-rich food is available for them.

Another study of the relationship between decaying leaves and the invertebrate fauna was made by Mathews and Kowalczewski (1969) on the River Thames near Reading. They deposited coarse- and fine-meshed bags containing leaves of *Acer pseudoplatanus* L., *Quercus robur* L. and *Salix* spp. The coarse mesh was 3·0 mm across and allowed free access to invertebrate animals, excluding only molluscs such as *Anodonta* and *Unio*. The fine mesh was 0·27 mm across and was designed to exclude all but the smallest animals. The bags were placed on the river bottom at a depth of 1·5 m and recovered successively between February and October. On recovery all animals were removed and examined and the leaf material analysed. Weight losses in *Acer* and *Salix* leaves were more rapid than for *Quercus*. After eight months 30% of the original weight of the *Quercus* leaves remained and only 5% of the other types. On the basis of the figures obtained it was argued that the leaves of all the species investigated would have disappeared within a year. The calorific value per gram of the deposited leaves increased initially, reaching a maximum in the third or fourth month, and then declined. The percentage nitrogen content also increased with time, the increase being particularly rapid in *Acer* where material in the coarse meshed bags showed 1·43% N initially and 2·32% N after four months. An increase in the absolute amount of nitrogen was also demonstrated in some instances and in addition to its uptake from solutes in the environment mediated by the microflora, the possibility of nitrogen fixation by the latter was also considered. There was a significant difference between the total numbers of animals recovered from the leaves in the coarse- and the fine-meshed bags. Abundant animals from the coarse meshed bags included *Pisidium* sp. and leeches, together with *Asellus* and *Gammarus*. In the fine-meshed bags only the *Asellus* and *Gammarus* were found, and then very rarely, and the Tubificidae constituted the only group of any numerical importance. Although the coarse-meshed bags had the greater invertebrate fauna the leaf material did not disappear faster in these than in the fine-meshed ones. This was surprising as a weight loss from conversion into faeces (expected to be washed out subsequently) had been anticipated, since workers on soils have reported that as much as 60% of litter may be returned to the terrestrial environment in this form. To explain the disparity Mathews and Kowalczewski pointed out that soil workers appear to have generally studied the dry superficial layers rather than the wetter lower ones; the former might be expected to favour the activity of animals rather than the microflora. They also pointed out the predominant role of earthworms in the terrestrial environment and their absence in the aquatic one (see Lofty, Chapter 14). They concluded that animals seemed to have played little or no part in the disappearance of leaves in the particular aquatic situation they investigated. Thus in these studies of Mathews and Kowalczewski we see interesting correlations with those of Kaushik and Hynes, particularly in the nitrogen uptake by the leaves, and the attraction of the invertebrate fauna to this material. It seems appropriate to consider whether their final conclusion was justified in view of the occurrence of at least some specimens of the important crustaceans *Asellus* and *Gammarus* within their fine-meshed bags, and the notorious fallibility of netting in excluding animals must be borne in mind in this connexion.

D. Invertebrate Feeding on Wood

The possible contribution of decaying wood, rather than leaves, to invertebrate food webs and hence to a final productivity in terms of fish has rarely been studied. However, with the damming and flooding of large sections of the Volta River in Ghana to make a lacustrine reservoir, the opportunity to make such a study was taken by Petr (1970). He showed that submerged trees decayed slowly but in so doing they contributed to the secondary production of invertebrate animals in this water body. Nymphs of *Povilla adusta* Agnew (Ephemeroptera) were especially common and during the day these hid in burrows which were excavated inside the wood. By night the nymphs crawled out onto the wood surfaces or the rich algal periphyton which often developed on it, and even swam in the surrounding water. Examination of the stomach contents of *Povilla* indicated a mainly algae diet but in addition pieces of wood were often present in the alimentary canal. It is not clear at present whether an actual enzymatic degradation occurs or whether the role of *Povilla* and other similar animals lies merely in accelerating the physical breakdown of the substratum. Other animals of the tree "pseudoperiphyton" included a species of *Nilodorum*, a chironomid using detritus for its external tube construction. This species only occurred on the trees in quiet bays and strong wave action is thought to limit its activity. Petr pointed out the significance of the tree fauna in contributing to fish food and indicated that the consequences of two long-term changes in the environment might be reflected in faunistic changes. Firstly, a gradual decrease of nutrients in the water following the initial flooding was expected, leading to a lowered productivity of the whole ecosystem. Secondly, the gradual disappearance of the soft parts of the trees such as the bark and soft wood might be expected to eventually lead to a reduction of the burrowing forms. In this connexion the study showed that bark supported a much richer fauna than did the wood proper.

E. Detritus

The connexion between the occurrence of plant material in an advanced state of degradation as detritus, and the occurrence of benthic invertebrates has been studied by Egglishaw (1964). He made numerous simultaneous collections of the animals and the plant detritus with which they were associated on the Shelligan Burn, a stream in Perthshire, Scotland. Analysis of the results showed that the amount of plant detritus obtained from any one standard collection varied from 1 to 12 ml while the total number of invertebrate animals showed a positive correlation with this in varying from 76 to 295. The closest associations between detritus and occurrence were found for the Plecoptera (stoneflies) *Amphinemura sulcicollis* (Stephens) and *Leuctra inermis* Kempny and for Chironomidae larvae. These two stoneflies are known to feed on plant detritus, and Chironomidae larvae generally do so by preference although a small proportion of these are carnivorous. In an experimental approach the detritus was again collected, its resident animals killed by immersion in alcohol, and it was replaced in different amounts in the stream on trays. After 22 days there were significantly greater numbers of *A. sulcicollis*, *L. inermis* and chironomid larvae on the trays with plant detritus than on those without it. The study was extended (Egglishaw, 1968) to compare the relationship between plant detritus and the benthic invertebrate stock in a number of different streams, which were graded in terms of their Ca and HCO_3

levels. With low Ca and HCO_3 levels a stream could be relatively unproductive of animals even though adequate plant detritus was present. On the other hand where streams had high levels of these ions (e.g. 1·5 mequiv l^{-1} Ca^{++} and 1·43 mequiv l^{-1} HCO_3, in the Fender Burn) then they could be more productive. There was a calculated increase of 0·00688 g of bottom fauna per g of plant detritus for each increase of 100 μequiv. Ca^{++} l^{-1} in the water. The rate of decomposition of plant material in the different streams was studied in an ingenious model system by exposing rice grains for varying periods. This experiment showed that the rate of decay of the grains, and hence by implication the whole allochthonous plant input, was fastest in the streams with the high Ca levels. It was concluded that in streams, influenced by a differing chemical background, the process of plant decomposition differs correspondingly and this is reflected in variation in the amount of the benthic fauna.

References

Albrecht, P. and Ourisson, G. (1971). *Angew. Chem.* **10,** 209–225.
Chase, E. S. and Ferullo, A. F. (1957). *J. New Engl. Wat. Wks Ass.* **71,** 307–312.
Couch, J. N. (1949). *J. Elisha Mitchell scient. Soc.* **65,** 315–318.
Douglas, A. G., Eglinton, G. and Maxwell, J. R. (1969). *Geochim. cosmochim. Acta* **33,** 569–577.
Efford, I. E. (1969). *Verh. Internat. Verein. Limnol.* **17,** 104–108.
Egglishaw, H. J. (1964). *J. Anim. Ecol.* **33,** 463–476.
Egglishaw, H. J. (1968). *J. appl. Ecol.* **5,** 731–740.
Eglinton, G. (1969). *In* "Organic Geochemistry." (G. Eglinton and M. T. J. Murphy, eds), pp. 20–73. Longman, London.
Fogg, G. E. and Belcher, J. H. (1961). *New Phytol.* **60,** 129–138.
Franks, J. W. and Pennington, W. (1961). *New Phytol.* **60,** 27–42.
Griffin, D. M. (1963). *Trans. Br. mycol. Soc.* **46,** 368–372.
Han, J., McCarthy, E. D., van Hoeven, W., Calvin, M. and Bradley, W. H. (1968). *Proc. natn. Acad. Sci. U.S.A.* **59,** 29–33.
Johns, R. B., Belsky, T., McCarthy, E. D., Burlingame, A. L., Haug, P., Schnoes, H. K., Richter, W. and Calvin, M. (1966). *Geochim. cosmochim. Acta* **30,** 1191–1222.
Kaushik, N. K. and Hynes, H. B. N. (1971). *Arch. Hydrobiol.* **68,** 465–515.
Kennedy, M. and Fitzmaurice, P. (1971). *Proc. R. Ir. Acad.* B, **71,** 269–352.
Mackereth, F. J. H. (1965). *Proc. R. Soc.* B **161,** 295–309.
Mathews, C. P. and Kowalczewski, A. (1969). *J. Ecol.* **57,** 543–552.
Mortimer, C. H. (1971). *Limnol. Oceanogr.* **16,** 387–404.
Newton, J. A. (1971). Ph.D. Thesis, University of Salford.
Nykvist, N. (1963). *Stud. for. suec.* **3,** 1–31.
Park, D. (1972). *Trans. Br. mycol. Soc.* **58,** 281–290.

PETR, T. (1970). *Hydrobiologia* **36**, 373–398.
RHEAD, M. M., EGLINTON, G., DRAFFAN, G. H. and ENGLAND, P. J. (1971). *Nature, Lond.* **232**, 327–330.
SANGER, J. E. and GORHAM, E. (1970). *Limnol. Oceanogr.* **15**, 59–69.
WILLOUGHBY, L. G. (1961). *Trans. Br. mycol. Soc.* **44**, 305–332.
WILLOUGHBY, L. G. (1969). *Hydrobiologia* **34**, 465–483.
WILLOUGHBY, L. G. (1969*a*). *Nova Hedwigia* **18**, 45–113.
WILLOUGHBY, L. G. (1971). *Freshwat. Biol.* **1**, 23–27.
WILLOUGHBY, L. G. and ARCHER, J. F. (1973). *Freshwat. Biol.* **3**, 219–239.
WILLOUGHBY, L. G. and COLLINS, V. G. (1966). *Nova Hedwigia* **12**, 150–171.
WILLOUGHBY, L. G., SMITH, S. M. and BRADSHAW, R. M. (1972). *Freshwat. Biol.* **2**, 19–26.

22

The Marine Environment

E. J. Perkins

University of Strathclyde
Marine Laboratory
Gorelochhead
Dunbartonshire
Scotland

I. Introduction

Marine scientists have been concerned with the presence of organic matter in sea water since Petersen and his colleagues (Petersen and Boysen-Jensen, 1911, Boysen-Jensen, 1914 and Blegvad, 1914) suggested that detritus, derived principally from rich beds of *Zostera*, was the basic food supply in the shallow waters around Denmark. Rather unfortunately, most studies have been concerned with particles of less than 100 μm diameter, and particularly with those which are normally carried in suspension by the sea water. Except for Backlund (1945), material of a larger size remained more or less forgotten until recently.

In 1967, Darnell (1967*a*, *b*) gave the first adequate definition of detritus: "all types of biogenic material in various stages of microbial decomposition which represents potential energy sources for consumer species". Darnell further noted that while some materials may be available for consumer use immediately, other more durable components may persist for long periods and may possibly be a major source of energy in productive areas at times outside the seasons at which primary production occurs, and in areas lacking primary production, e.g. the sea bottom, at all times of the year.

For convenience, detritus may be considered to occur in two states, viz. particulate and sub-particulate of a size $\geqslant 1\ \mu m$ and $< 1\ \mu m$ respectively (Darnell, 1967*b*). Evidently, and this did not escape Backlund (1945), what the marine biologist has been accustomed to call detritus is equivalent to litter, although the definition prior to the clarification of Darnell (1967*a*, *b*) was much narrower than that applied to litter.

A review of marine biological literature leaves one in no doubt that detritus, as it has been generally understood, i.e. of a size $\leqslant 100\ \mu m$, has an importance in the economy of the sea, particularly with respect to the benthonic invertebrate species, (e.g. Wernstedt, 1942; Enequist, 1949; Tattersall and Tattersall, 1951). Nevertheless, the general absence of coherent studies of the fate and breakdown of the larger particles of plant and animal litter (or detritus), led Hedgpeth (1957), in an otherwise useful review of the ecology of sandy beaches, to state that the dependence of the true psammophiles upon particulate and suspended organic matter brought to the beach as "peloglea" or "leptopel" suggested that these substances are more important to the economy of the beach than are the irregular and seasonal strandings of such macroscopic materials as detached seaweeds and *Sargassum*. Thus he ignored the fact that macroscopic materials, so stranded, are broken down relatively readily and contribute to, or are a source of, the materials so important to the psammophilic detritus feeders, both of the shore and of the coastal sea bed. It is, therefore, essential to understand the means whereby litter is recruited in the marine environment and its fate, i.e. transport, deposition and breakdown, once it has entered the sea. In emphasizing this aspect of the problem material has been drawn from widely scattered literature in which litter is an incidental interest. This is combined with similar observations made over many years by the author and with the results of more systematic work performed since March, 1972.

II. Sources of Litter in the Sea

Although, in the sea, plant litter is derived from both marine and terrestrial sources, it is almost always accompanied by substantial amounts of dead animal material which as a source of energy and nutrients cannot be ignored. Sources are generally described qualitatively, but their influence in any one situation is variable and dependent upon a variety of factors, e.g. wind strength and direction.

A. Marine Sources

Marine, i.e. autochthonous, sources of litter must be regarded as all those sources of material which lie to the seaward of the extreme high water mark of spring tides (E.H.W.M.S.T.). Some authors (e.g. Darnell, 1967*b*), excluded marginal marsh and swamp vegetation from this grouping, but since these areas are dependent upon nutrient supplies from the offshore region for the quality and abundance of their vegetation and themselves influence the productivity of inshore waters both as feeding grounds and as sources of the external metabolites so important to the quality of these waters (Perkins *et al.*, 1963–66; Perkins, 1972) this exclusion is difficult to justify. Similar considerations apply to materials derived from beaches of sand, stone or rock.

Plant litter attributable to marine sources is derived from the following materials:

(i) attached algae which inhabit all levels of the shore and the euphotic zone. The most overt components are the fucoids and laminarians, but all make a contribution, and in sandy areas, especially in the more eutrophic estuaries, the green filamentous algae, e.g. *Enteromorpha*, may be the dominant or even sole component of the plant litter during the summer months;

(ii) flowering plants which have colonized the intertidal zone or waters of shallow seas. In temperate latitudes, this component is largely of an intertidal origin and may be dominated by the grasses *Spartina* and *Zostera*, although other plants of the saltmarsh, such as *Halimione*, *Salicornia* and *Suaeda*, evidently make an important contribution. In tropical latitudes, the development of mangroves and the extensive beds of grasses, such as *Thalassia*, which grow in the sublittoral, produce litter of a different quality;

(iii) litter (or detritus) of a much smaller initial size derived from the phytoplankton, the epiphytes growing upon the aforementioned plants and upon surfaces not colonized by these plants, and the mud flat diatoms, dinoflagellates and blue-green algae, and

(iv) faecal derivatives of all the grazing animals in the sea.

B. Terrestrial Sources

Terrestrial, i.e. allochthonous, sources of plant litter which enter the sea and the means of their recruitment are more complex, but arise in three principal ways: (i) direct natural additions at the sea's margin; (ii) fluviatile sources; and (iii) indirect additions by physical and biological processes including human activity.

1. *Direct Natural Addition at the Sea's Margin*

Except where the land and sea meet in the gentle slope of salt marshes or in an expanse of bare shingle, direct natural additions of plant litter to the sea are possible. These additions may range from the lichens, e.g. *Xanthoria*, which colonize the face of exposed cliffs, to that derived from the dense growth of trees and undergrowth on more sheltered cliffs or exaggerated erosion edges. At Vancouver Island, B.C., the latter may be sufficient to prevent passage at high shore levels: Boughey (1957) noted that, in the tropics, old coconut palms may topple into the sea.

Clearly, the most exposed situations contribute insignificant amounts of litter whereas at Vancouver Island the amounts contributed are of the order of thousands of tonnes per annum.

2. *Fluviatile Sources*

Evidently, rivers flowing into the sea contribute plant litter which has been introduced along their length. The type and quality of the natural vegetation, the size of the catchment area, the development of lakes, back-waters and slow-flowing sections, the periods at which drought and flood occur plus any seasonal changes in foliage must all influence this contribution which may range in type and quality from whole trees introduced by storm and erosion to phytoplankton and periphyton.

3. *Indirect Additions of Litter from Terrestrial Sources*

Indirect additions of plant litter to the sea arise from three sources: (i) wind-blown material, viz. leaves, pollen grains, pieces of grass and similar dried plant fragments which are readily transported by the wind; (ii) faecal material deposited by mammals, e.g. sheep, cows, rats, foxes, and birds, e.g. gulls, waders, geese and duck, which may graze or roost upon the sea's margin having first fed upon wholly terrestrial food sources (it follows that a similar transfer may take place from sea to land); and (iii) mankind makes a variety of additions to the marine environment.

(*a*) *Sewage and related materials.* Superficially the additions made by man may be largely categorized as waste disposal although this is not the sole source of plant litter which reaches the sea from this origin. The most obvious example of waste disposal is the introduction either of untreated sewage by pipeline or by the dumping, after treatment, of sewage sludge in some part of the offshore region. Formerly, untreated sewage contributed fragments of undigested plant matter, e.g. the more resistant plant vessel structures, of lettuce and corn for example; seeds, e.g. tomato and water melon; the products of technology such as toilet paper, tampons, condoms, and small articles disposed via sink, e.g. tea leaves and the remains of cigarettes. In those more "advanced" communities where the under-sink kitchen disposal unit or "garbolator" is in vogue a wider variety of kitchen waste materials is reaching the sea in increasing amounts. Additions from these sources may be very large: Glasgow Corporation dumps $1{\cdot}02 \times 10^6$ t of sewage sludge per year off the Garroch Head, Firth of Clyde (Mackay and Topping, 1970). In the period 1964–1968, waste composed of sewage sludge and dredge spoil was dumped in New York Bight at a rate of $9{\cdot}6 \times 10^6$ t per year and contributed the largest source of sediment entering the North Atlantic Ocean from the North American continent: the sewage sludge contained large amounts of human artifacts (Pearce, 1970). Like all land-based additions of litter this input is made to a relatively narrow coastal zone. When sewage is not treated before release to the rivers, then it too may have a fluviatile origin.

The world's oceans carry a large volume of shipping, of which each individual vessel has, in microcosm, the same problems of kitchen, faecal and hygienic waste disposal as a littoral town. Like the town, the shipboard solution is ejection of all unwanted material overboard. Clearly, this recruitment of litter is made to a much wider area of sea than the release of sewage from land-based installations.

In Britain, and no doubt elsewhere in the world, coastal farmers and gardeners alike view the adjacent sea as a convenient repository for "waste" faecal material, straw, tree slash, hedge clippings and weeds and therefore introduce plant litter to the sea.

(*b*) *Lumber and pulp industry*. The lumber and pulp industry, particularly that of British Columbia, is an interesting example of man's influence upon the introduction of plant litter. Except in the area around Vancouver, British Columbia has relatively few roads available for the transport of wood to the lumber and pulp mills. Instead, the sea, within a series of sheltered straits, is a great highway along which wood is transported in Davis Rafts and by barge. At their destination, however, the fiordic coastline of British Columbia is such that storage on land is difficult or impossible whereas storage in the sea is cheap and easy and is practised widely.

Two fundamental problems arise from these practices: (i) the Davis rafts may, and indeed do, break up and logs are cast ashore over wide areas and (ii) logs held in the floating stocks gradually become waterlogged and some losses due to sinking occur. Any substantial accumulations of logs from Davis rafts are collected since the wood has a considerable value.

Those not collected and those that sink represent an addition of plant litter. The contribution of the lumber and pulp industry is not of course limited to this particular proliferation of an existing problem by industrial activity. Where rivers are used to transport wood, losses of logs and bark to the sea may occur in the fluviatile efflux. In days when the Pacific North-West was less conscious of its environmental heritage, waste from saw mills (i.e. sawdust and wood rejects) could find its way into the sea. In the wood pulp industry, the production of either groundwood or chemical pulp led to significant losses of these materials to the sea.

In those areas of the world, like British Columbia, where timber is plentiful it is customary to construct most harbour and other installations from wood. Even in Britain, where wood is scarce, it is frequently used to construct jetties, groynes, slipways and moorings while in the oyster industry the limit of individual oyster layings is marked with withies. Man is not conditioned to appreciate that by raising such structures he is providing an additional supply of litter for the inhabitants of these coastal waters. Nevertheless, that he does provide it is demonstrable.

(*c*) *Synthetic and other litter*. Although the additions of sewage and that from timber industries may be very large they represent a natural, if exaggerated extension of the normal recycling of biological materials. However, since the Industrial Revolution, man has greatly modified the global environment, and unfortunately, many of his more recent additions are not so readily biodegradable as sewage, wood, cotton nets, sisal ropes, or paper bags. Synthetic materials which include nylon, terylene and courlene ropes, polythene sheets and bags, plastic cups, detergent containers and expanded polystyrene insulation contribute to the intertidal and infralittoral litter. In those areas subject to significant strandings of oil, the strandline litter may be a hellish and durable brew of unknown effect.

III. The Occurrence of Litter

While dead plants in wrack beds and the strand-line present on all but the most exposed shores (e.g. Boughey, 1957; Bousfield, 1971; Brown, 1971; Brown, D. S., 1971; Gilham, 1957; Guiler, 1951; Johnson and Snook, 1967; Lipkin and Safriel, 1971; MacNae, 1957*a*, *b*; MacNae and Kalk, 1962; Morton, 1957; Taylor, 1971) are evident examples of plant litter in the marine environment, such aggregations are not confined to high shore levels, but may occur in the infralittoral. Moreover, important though such concentrations may be, the widely dispersed, more finely divided materials are less overt, but large in amount.

Given that supply is important, litter accumulates in those situations in which turbulence is reduced, i.e. in the lee of islands, reefs, rocky outcrops and sand structures, or as a floating mass of wrack aevja in tidal pools: none accumulates upon the crests of the sheltering structures (Backlund, 1945). All headlands are characterized by individual current systems which accumulate litter. On the coast of Cumberland, intertidal accumulations are found at Selker Point and beneath St. Bees Head, at Fleswick Bay and St. Bees village. Off the north head of St. Bees infralittoral concentrations occur; on the north shore of the Solway a flood channel between Balcary Point and Heston Island holds a similar accumulation of litter which at both sites is composed predominantly of fucoid and laminarian algae, but including red and green algae, leaves, wood and hydroid perisarc also.

At Port William, Luce Bay, flood channel barblets developed in the intertidal bound shingle maintain near permanent concentrations of a predominantly algal litter on this shore.

The efflux from a tributary river to a main estuary may induce local intertidal concentrations of terrigenous litter: after a short tidal reach, the River Annan flows around Barnkirk Point into the Solway Firth; westwards from the Point a limited deposition of leaves and other terrigenous debris can occur on the midshore (Perkins, 1969).

All estuaries are characterized by an abundant litter derived from diverse sources, indeed much of the complexity of the estuarine community may arise from this abundance of organic detritus (Darnell, 1967*b*). Nevertheless, litter deposition on the open coast may not be unimportant, since mechanical damage, at least, may be expected here; the fragments produced may be removed from the shore by wave action, and thus become available to contribute to the litter supply of the offshore sea-bed and of estuaries. For example, the exposed rocky coast of northern Sutherland is interspersed with sheltered estuaries, sandy bays and inlets. Except for the Kyles of Tongue and Durness, the estuaries are insignificant. However, the foredune area of each bay and inlet may be extensive and contain large amounts of algal, terrigenous, animal and synthetic litter: each supports an abundant, active population of talitrid amphipods. Elsewhere, aggregations of litter, often containing many large individual pieces accumulate adjacent to points e.g. east of Farr Point.

A. The Occurrence of Litter in Estuaries

Estuaries must be considered in relation to the basic geomorphological type to which they belong, viz. fiordic or coastal plain.

1. Fiordic Estuaries

Fiords are characterized by a basin, or basins, of considerable depth, with a shallow bar at the entrance and a restricted development of saltmarsh at the head. Since the non-tidal circulation is effective only to the sill depth and since the restricted wind fetches are insufficient to raise sediment from the bed of the fiord by wave action the saltmarshes are starved of silt which would confer fertility; indeed, unlike the coastal plain estuaries the silt content of fiordic sediments increases with depth. Evidently, conditions in fiords are optimal for the deposition of litter on the bed of the basin, and, as the City of Oslo has found, are not suited to receive large amounts of sewage.

Fiords generally have steeply sloping sides, with either precipitous or narrow shores. In the latter case, rocky outcrops may be interspersed with variable amounts of shingle or gravel which is steep and unconsolidated at the higher levels and bound, in a more gentle gradient, at mid and lower shore levels. The autochthonous supply of algal litter may be very limited, and allochthonous litter, marine and terrigenous, predominant.

In most of the fiordic sea lochs of the Scottish west coast litter occurs in one or more strand-lines at the high water mark (H.W.M.). Such strandlines persist throughout the year at Loch Long and the Gareloch, although they shrink gradually during the summer months. The accumulation is reduced to a thin line in the more exposed situations, but may be substantial where there is a greater degree of shelter. The additions of litter are greatest in the autumn and winter months when leaves are shed by deciduous trees and algae are detached during storms; supplies of "industrial" litter are intermittent and dependent upon such events as the launching of ships.

At the bed of the Gareloch, terrigenous litter (leaves, roots, sticks and wooden materials of "industrial" origin) is predominant, but some sunken phaeophyte framents occur. Where the shoreline reduces turbulence a blanket of terrigenous litter may occur. Pearson (1970, 1971, 1972) found a widespread, but variable (~0·3 to ~7·5%), distribution of terrigenous litter, mainly twigs and leaves, on the bed of Lochs Linnhe and Eil: at its most abundant this terrigenous litter formed a layer over the basic sediment of these lochs. Although it was not observed here, fiordic situations near to pulp mills are particularly susceptible to the deposition of banks or blankets of pulp (e.g. Werner, 1968).

In the Kattegat and its associated fiords, Enequist (1949) found a widespread, but variable, distribution of plant litter; although *Zostera* was predominant, the litter included weed fragments, wood, animal remains and faecal pellets.

2. *Coastal Plain Estuaries*

Unlike fiords, coastal plain estuaries are generally rather shallow and lack a sill at the seaward end. With the exception of the stratified salt wedge estuary, the whole water column tends to be well mixed by tidal and wave action and by the non-tidal circulation. Consequently, litter masses below the low water mark (L.W.M.) are either absent or confined to larger materials e.g. sunken logs, boats and other man-made additions. Exceptions occur only in some flood channels or where the land form reduces turbulence, or in eutrophic estuaries where large mats of filamentous algae float off the marginal banks in summer, to sink after a period of floating in the surface waters. The latter may be a particular nuisance to fishermen who refer to it as "clodge".

In contrast to the fiords, saltmarshes in temperate latitudes, and mangrove swamps in the tropics are well developed in the sheltered areas of coastal plain estuaries. Similarly, the deposition of the finer silt fractions is greatest near to the H.W.M., where turbulence is least.

(a) *The temperate environment.* Before considering the individual shore, the deposition of litter in the estuary as a whole is worth consideration. Evidently, the quantity and quality of the litter present in the individual estuary is determined by its own nature and that of its hinterland. Hitherto, quantitative information has been lacking, but as Backlund (1945) showed, wrack banks may have a very variable composition. In the northern Baltic, wrack was produced predominantly by *Fucus spiralis* and *F. vesiculosus; F. serratus* only made a contribution on the west coast of Sweden. In contrast, wrack beds in the innermost parts of the Gulf of Bothnia and the Finnish Gulf were composed only of *Cladophora*, *Aegagrophila*, *Enteromorpha*, *Ranunculus* and other plants which can grow in fresh and brackish waters.

Since it is possible to identify, with relative ease and accuracy, many of the individual components of litter, even at a size of 63 μm, the relative frequency of occurrence of these components can yield information upon the transport and fate of plant litter materials in these estuaries. Broadly, the individual components can be allocated to one of five groups—algal, marine flowering plant, land plant, animal and synthetic sources. Of these, algae are best ascribed to five groups: green, fucoid, laminarian, red and miscellaneous (principally because although they may be identified readily when intact, fragments are not easy to assign specifically); the marine flowering plants to generic categories, e.g. *Limonium*, *Spartina*; the land plants to as many categories as seem advisable in the individual situation; animal sources include, e.g. crustacean exuviae, feathers, skate eggs, and synthetic sources include paper, polythene, courlene rope and cigarette filter tips. Even though this analysis introduces a bias against the contributions made by algae and marine flowering plants, the changes in contribution from each source within an estuary as a whole and within an individual shore can be assessed.

An assessment made by considering the component frequency of the strand-line litter on a number of shores along the Cumberland coast of the Solway Firth in the period 8–9.5.72 (Fig. 1*a*) showed that while the land plant components were predominant in the areas of predominantly grass marsh, algal and animal components were always present downstream of

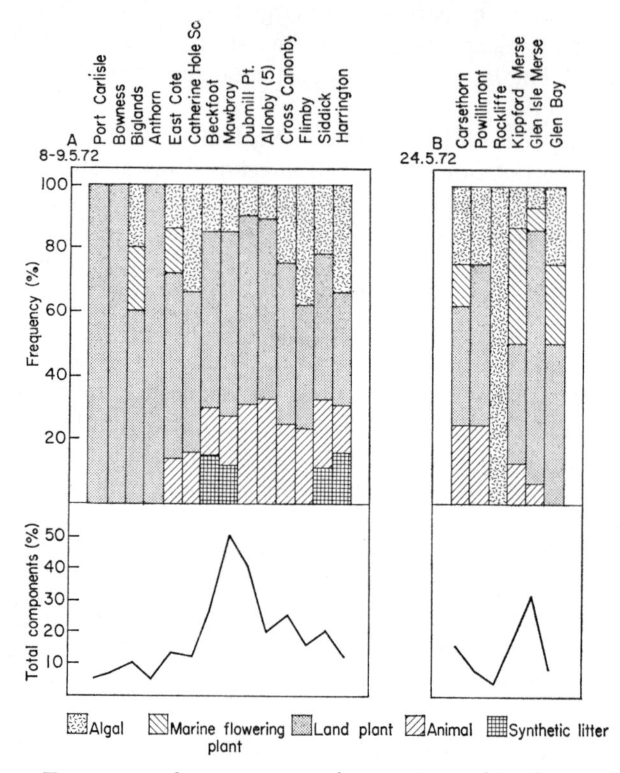

Fig. 1. Frequency of occurrence and percentage of total components of litter at stations in the Solway Firth.

East Cote, Silloth. The number of components was least in the upstream, salt marsh situations, but increased downstream and reached a maximum in the area of flood channel development at Mawbray and Dubmill Point.

In the richer salt marshes of the Urr Water, i.e. Kippford Merse and Glen Isle Merse (Fig. 1b) the litter has a more varied composition than at Port Carlisle and Bowness in the main estuary. However, the amounts deposited are insufficient to have an influence comparable with that observed in the Zwartkops estuary, S. Africa, by MacNae (1957*a*). Here, tidal wrack composed of large masses of *Zostera* and *Spartina* is deposited in a zone where the Limonietum would be expected. Plant growth is smothered by the wrack which accumulates in autumn and early winter, and disappears gradually during winter and early spring. Frequently the wrack does not disappear early enough for seeds of *Salicornia meyeriana* to germinate, and the zone remains bare throughout the spring and summer.

A complete separation of litter components occurs only infrequently on the individual shore, and an apparent greater uniformity within and between strandings is rarely confirmed by more detailed examination (Table I). Even the massive strandings of laminarian wrack which occur

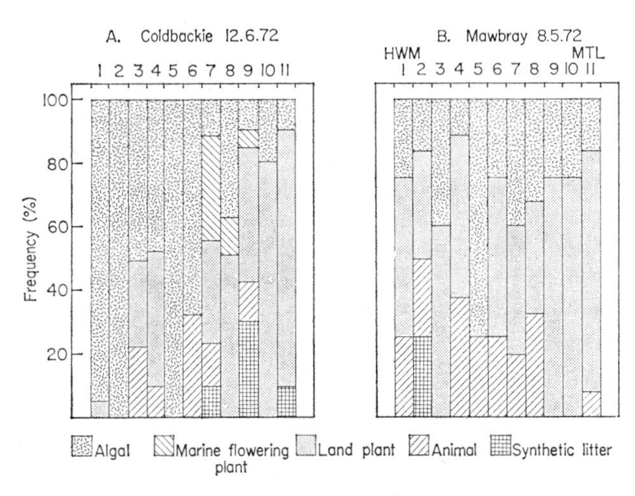

Fig. 2. Frequency of occurrence of litter components on the shores at Coldbackie, Sutherland, and Mawbray, Solway Firth.

on the coasts of Scotland are not exclusively derived from this source, although they do reflect the imbalance of the sources of supply. At Coldbackie, Sutherland, an examination of 11 strandings of litter on different parts of the shore, on 12.6.72, showed that the frequency of occurrence of the individual components was very variable (Fig. 2a). Similarly, an examination of 8 different strand-lines, and 3 situations on the sands at Mawbray, Solway Firth on 8.5.72 (Fig. 2b), revealed similar variations in the contribution made by the different components. Clearly, such results are indicative of the influence of the timing at which individual sources are likely to make a contribution to the litter supply and of the differential rates of removal by physical and biological processes. With respect to supply, deciduous trees make a major contribution only in autumn, but other sources may be equally discontinuous. During the winter of 1956–57, a massive stranding of tubes of the polychaete *Lanice conchilega* occurred at Out Head, near St. Andrews. In the Solway Firth, *Carcinus* exuviae

TABLE I. The frequency of occurrence and relative abundance of the principal categories of litter stranded at high shore levels

Site	Date	Numerical contribution of litter components (%)					Relative abundance
		Algal	Marine flowering plants	Land plants	Animal	Synthetic	
Argyll							
Loch Craignish	13.8.72	23	8	15	31	23	*Ascophyllum nodosum* ~60%, *Zostera* ~30%
Firth of Clyde							
Kennedy's Pass	22.4.72	14	0	43	43	0	Algal ~95%
Garelochhead (i)	31.8.72	12	0	63	0	25	Algal ~1%, Synthetic >90%
(ii)	31.8.72	5	0	35	0	50	Algal ~50%
Solway Firth							
Allonby (South)	16.3.72	50	0	30	20	0	Terrigenous:Marine::50:50
Beckfoot/Mawbray	16.3.72	16	0	52	16	16	Terrigenous:Marine::50:50
Carsethorn	7.3.68	12	6	41	41	0	Hydroid >75%

Cross Canonby		16.3.72	40	0	20	40	0	Algal >75%
Dubmill Point	(i)	9.5.72	9	0	64	27	0	Terrigenous > Marine
	(ii)	9.5.72	25	0	50	25	0	Terrigenous < Marine
	(iii)	9.5.72	8	0	53	39	0	Animal > 75%
Grune Point		16.3.70		0	44	56	0	Wood ~50%, Leaves ~50%
Powillimont		7.3.68	8	4	33	55	0	Hydroid >75%
Seaside		7.3.68	11	11	11	67	0	Hydroid ~60%, *Spartina* ~30%
Sutherland								
Coldbackie	(i)	12.6.72	68	0	0	32	0	Algal ~95%, Animal ~5%
	(ii)	12.6.72	9	5	45	9	32	Algal >50%
	(iii)	12.6.72	9	0	82	0	9	Algal 50%, Land Plant ~50% (Leaves 25%)
Naver Estuary	(i)	13.6.72	16	5	47	32	0	Algal > Remainder
	(ii)	13.6.72	50	0	50	0	0	Algal ~90%
Melvich Bay		13.6.72	33	0	33	17	17	Algal ~90%
Kyleside, Tongue		14.6.72	14	0	50	36	0	Algal ~50%, Leaves 25–50%, Animal Tr.
Durness	(i)	14.6.72	17	0	66	17	0	Algal ~50%, Land Plant ~50%, Animal Tr.
	(ii)	14.6.72	13	7	53	20	7	Terrigenous >50%, Marine <50%
Loch Eriboll		14.6.72	50	0	33	0	17	Algal ~90%, Land Plant ~10%

are stranded in large numbers during the month of June; at Powillimont, a massive stranding, predominantly of the hydroids *Abietinaria*, *Hydrallmania*, *Kirchenpaueria*, *Obelia*, *Thujaria* and *Tubularia* was found on 7.3.68. Such observations suggest that, taking an integrated view of a whole shore, long term differences in the frequency of individual components may be expected and indeed do occur (Fig. 3).

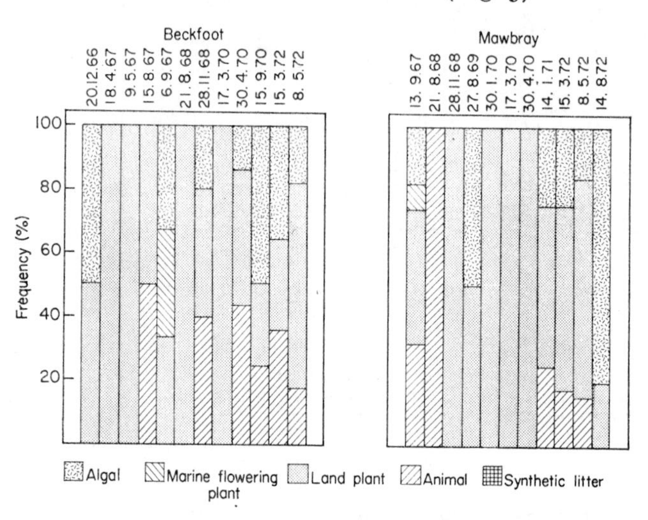

Fig. 3. Changes in the frequency of occurrence of litter components at Beckfoot and Mawbray, Solway Firth, in the period 20.12.66–14.8.72.

Litter in the strandings at H.W.M. is either of a large or a relatively large size. However, much finer material may be stranded temporarily in the ripple system of the sand flats and large amounts of such material may be deposited in flood channel barblets near the H.W.M. In salt-marshes, the strand-line at H.W.M. may be insignificant whereas a thinly scattered deposition of grass and leaf litter may occur over the whole area of grass marsh. On 13.8.72, a marked deposition of litter on the strand-line at Loch Craignish, Argyll, was accompanied by a scattered deposition of *Zostera* across the whole shore. At Siddick and Dubmill Point, Cumberland and at Durness, Sutherland fine litter may become worked in beneath large stones.

At Whitstable, Kent, large patches of finely divided vegetable matter are deposited by the equinoctial tides at the break in shore gradient where the transitions from shingle to bound shingle and bound shingle to sand flat occur. These masses were composed of wood pulp fibres and vegetable litter accompanied by some species of benthonic and littoral diatoms, thus:

Benthonic: *Melosira moniliformis, Actinoptychus senarius, Biddulphia aurita, Rhaphoneis amphiceros*—abundant; *Actinoptychus splendens, Pleurosigma aestuarii*—common, and *Auliscus caelatus, Plagiogramma gregoryanum, Synedra affinis* var. *fasciculata, Acnanthes longipes, Cocconeis scutellum, Diploneis interrupta, Navicula cyprinus, N. pygmaea, Pleurosigma angulatum, P. balticum, P. fasciola, Stauroneis salina, Surirella gemma*—occasional.

Planktonic: *Coscinodiscus excentricus, C. lineatus*—occasional. Similar strandings of fine litter have been observed in the Solway Firth and the Gareloch. A sample of such material thrown up at Powillimont, Solway Firth on 7.3.68 had a component frequency of 70% and 30% of terrigenous and animal sources respectively: a visual estimate indicated that leaf fragments formed ~75% of the total. A sample taken at Grune Point, 16.3.70, was sieved and had a median ϕ of 1·05, and a sorting coefficient of 0·53. The 1–0·5 mm fraction was the most abundant, and terrigenous sources provided the main bulk of the litter (Table II). The superficial

TABLE II. Analysis of finely divided vegetable matter stranded at Grune Point, Solway Firth, 16.3.70

		Component frequency %					Relative abundance of components (visual estimate)
Sieve (mm)	% weight	Algal	Marine flowering plant	Land plant	Animal	Synthetic	
2	0·3	0	0	67	33	0	>90% terrigenous
1	3·9	0	0	45	55	0	Wood ~ 50% leaves ~ 50%
0·5	44·5	0	0	59	41	0	Wood ~ 50%, leaves ~ 50%
0·25	40·2	0	0	43	57	0	Wood 75%
0·125	9·6	10	0	30	60	0	Wood 75%
0·0625	0·9	0	0	25	75	0	50% sand[a]

[a] General mass not easy to differentiate, but wood fibres and hydroid perisarc were positively identified.

appearance of all these materials is strongly suggestive of a predominantly algal origin: an erroneous view, as analysis showed.

Fine litter may be recovered from shore sediments. Many determinations have been made of the organic content of such soils (e.g. Clarke and Hannon, 1967; Newell, 1965), but these, for obvious reasons, do not necessarily measure or give a true indication of litter content. Of these finer materials, those of a size ⩾500 μm may be extracted readily either by sieving or flotation; indeed in the fine estuarine soils, the larger particles, *ca.* 500–1000 μm, may be composed either of organic litter alone or a mixture of organic litter and $CaCO_3$ fragments of molluscan origin.

The component frequency was studied in a series of associated samples taken from high water strandings and of fine litter ($\geqslant 500\mu$m) recovered from soil samples taken at lower levels at the same 11 stations in the Solway

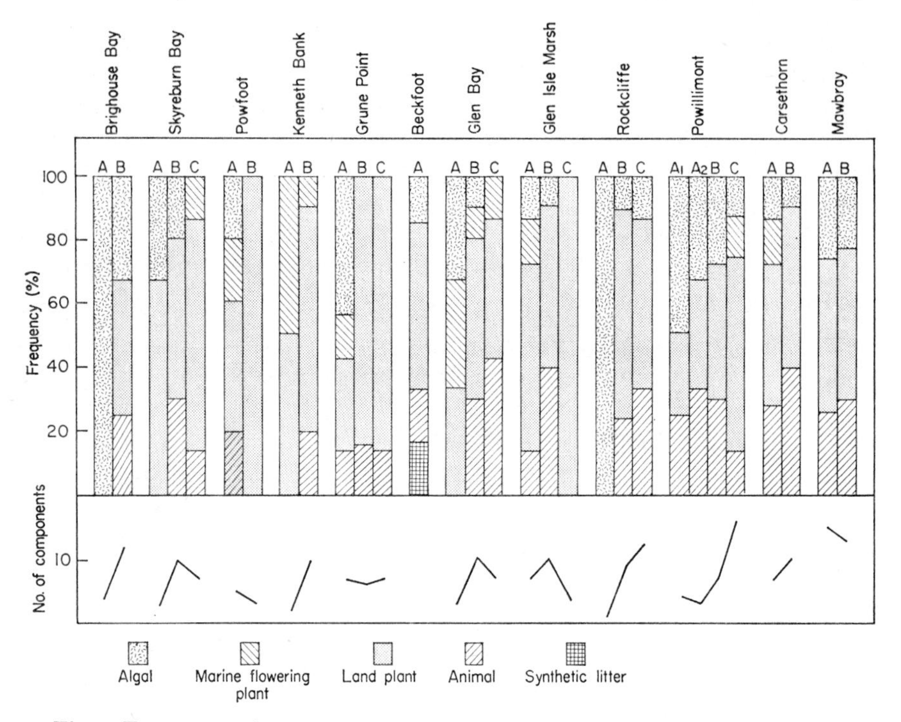

Fig. 4. Frequency of occurrence and number of components of litter at associated stations at the H.W.M. (A) and in the sediments from the mid-shore (B, C) at stations in the Solway Firth.

Firth (Fig. 4; Table III). The average number of comparable components was 6($\pm$3·9) and 8($\pm$3·7) in the coarse and fine litter respectively. Although on average, the difference between the two groups was not significant, marked differences did occur between individual pairs. At Kenneth Bank, the strand-line litter was composed of two components, *Phragmites* and grass, whereas the fine litter recovered from sediment was composed of 10 components: *Phragmites*, 7 land plants and 2 animals.

(b) *The tropical environment.* Litter strandings occur less frequently on tropical shores than on temperate shores. However, litter may be deposited among primary plant colonists on the coast of Ghana (Boughey, 1957) and at Inhaca Island, Moçambique, the infralittoral grass *Cymodocea* may be cast up in a drift line (MacNae and Kalk, 1962). Sand dunes are generally absent from tropical shores, but where they occur a strand-line and plant colonization at the foot of the dune is present as in temperate regions (Morton, 1957).

TABLE III. The fine litter content of Solway Firth intertidal sediments—14.3.72–24.4.72

Site	Shore level	Proportions of principal components source				Litter retained by 0·5 mm sieve (% dry wt)
		Algal	Marine flowering plant	Land plant	Animal	
Beckfoot	*Arenicola* "watt"	0	0	2	0	Tr
Brighouse Bay	Lower mid shore	4	0	5	3	Tr
North Lodge	*Zostera* zone	0	0	3	1	Tr
Powfoot	*Hydrobia* zone	0	0	3	0	Tr
Glen Isle Marsh	Bare sand, M.T.L.	0	0	4	0	<0·001
Grune Point	*Arenicola* "watt"	0	0	5	1	<0·001
Powillimont	*Arenicola* "watt"	2	0	4	1	<0·001
Rockcliffe	Middle shore	1	0	4	4	0·004
Rockcliffe	Rock/sand break point	2	0	6	4	0·012
Glen Bay	Bare muddy sand. M.T.L.	0	1	3	3	0·020
Skyreburn Bay	Bare muddy sand. *Salicornia* zone	2	0	5	3	0·032
Grune Point	Shingle/sand flat break point	0	0	6	1	0·038
Glen Isle Marsh	*Spartina* zone	1	0	5	4	0·040
Glen Bay	*Spartina* zone	1	1	5	3	0·043
Carsethorn	L.H.W.N.T.	1	0	5	4	0·055
Powillimont	M.H.W.M.N.T.	2	2	7	5	0·780

A very large sample (*ca.* 50 kg) of sediment taken at Carsethorn in July, 1961, was composed predominantly of litter at grade sizes >0·124 mm. Litter content of grade size >0·25 mm = 1·5%, of grade size >0·124 mm ~ 2·6% (Perkins *et al.*, 1966).

In the tropics, the deposition of litter is apparently most marked in the mangrove swamps. A strand-line at E.H.W.M.S.T. may be present as in temperate latitudes (Brown, D. S., 1971); elsewhere allochthonous litter has been deposited in the Juncetum and among *Avicennia* (Boughey, 1957; MacNae and Kalk, 1962). In Australia, rotted mangrove leaves form a sludge covered with a thick algal film. Upshore *Casuarina* may contribute a dense litter which obstructs the germination of its own seedlings (Clarke and Hannon, 1971), this litter like that of *Phragmites communis* in temperate latitudes (Haslam, 1971) exerts its influence by physical and not by chemical means. Once a pioneer stand of trees has been established, the level of the swamp rises by an accumulation of silt and organic matter, until the upper levels are immersed only by the highest tides. In Florida leaf litter is removed and peat formed mainly from *Rhizophora* and *Avicennia* roots. As in temperate latitudes, mangrove trees which fall naturally or by clearance contribute to the supply of litter (Berry, 1963).

B. The Amount of Litter Present in the Sea

Before attempting a more general assessment of the contributions of plant litter to the marine environment, some of the more pertinent points derived from studies of detritus of a size $<100\ \mu m$ are worthy of consideration. According to Parsons and Seki (1970) if a continual input (L) is made to a standing stock (X), then the change in X with time can be described by the equation,

$$X = \frac{L}{k}(1 - e^{kt})$$

where k represents the rate (time^{-1}) of decomposition of the standing stock. Given sufficient time, the system will reach an equilibrium value X_{ss}, equal to L/k. Thus, the accumulation of carbon in the surface sediment of a shallow bay will reach equilibrium in *ca.* 30 years, unlike the deep oceans which require a period of several thousand years.

Parsons (1963) stated that on a relative scale of 100 for the amount of soluble organic matter in sea water, the total organic matter present in sea water approximates to the following distribution:

> Soluble organic matter = 100, particulate organic matter = 10, phytoplankton = 2, zooplankton = 0·2 and fish = 0·002.

Thus the proportion of organic particulate matter (i.e. phytoplankton and detritus present in sea water) is small when compared with the dissolved organic matter, but is abundant in comparison with the animal matter. Sheldon and Parsons (1967) showed that there is a general inverse relationship between particle size and particle concentration in sea water. The standing stock of detrital particles in sea water may be expressed either as a dry weight per unit volume of water or as a calorific value derived directly or by calculation. According to Parsons (1963) the recorded dry weight values in the surface layers of the North Pacific ranged from 80 mg m^{-3} off Nanaimo to 6500 mg m^{-3} off La Jolla, and in the North Atlantic area from 160 mg m^{-3} in the North Sea to 750 mg m^{-3} in the Waddensea: at individual stations, values changed with depth and the time of year. Atkins *et al.* (1954) recorded values of 1·15–1·17 ppm of insoluble organic matter from the water of the English Channel. At Sapelo Island, Georgia, the standing stock of detritus at the period of mid tide ranged from 2 to 20 mg dry wt l^{-1} (i.e. 2–20 g dry wt m^{-3}); at this site adjacent to *Spartina alterniflora* marshes, the contribution of *S. alterniflora*, macrophytic algae and animal detritus was 93·9% (90·5–97·7), 4·8% (2·0–7·6) and 1·3% (0·3–6·1) respectively (Odum and de la Cruz, 1967). In the Menai Strait, Kenchington (1970) showed that detritus >190 μm was most abundant in winter and with tides of a greater range; amounts recorded ranged from <1 to 20 mg dry wt m^{-3}. The mean calorific value of hand sorted detritus was 2696 g cal g^{-1} (± 599). By use of the formula:

$$\text{Calorific value} = 4{\cdot}91 \times \%\ \text{organic matter} + 0{\cdot}327$$

Kenchington compared the material from the Menai strait with that of sedimented detritus in Southampton Water, calculated range 373–1039 g cal g^{-1} and the more general results of Parsons (1963) for the North Pacific and North Atlantic areas, calculated range 1004–3309 g cal g^{-1}.

Evidently standing stock is not a direct measure of individual contribution. A series of estimates for the possible contribution of litter to the Straits of Georgia are summarized in Table IV. Although interesting as an area subject to the impact of the pulp and lumber industry, its principal virtue for this purpose lies in the fact that Seki *et al.* (1969) have estimated the total annual contribution of organic matter introduced in the run-off, and a series of papers (Trussell, 1959, 1967) and official reports (British Columbia Hydro and Power Authority, 1966 and the U.S. Department of the Interior, 1967) permit an attempt to devise a budget of litter contribution. These estimates are order of magnitude estimates only, but they do nevertheless enable comparisons to be made, for example as between the contributions made by wood and run-off, and indicate the deficiencies of knowledge; for example, what is the contribution of macrophytic algae or what is the real direct contribution from leaves and fallen trees?

A population of approximately 120,000 contributes *ca.* 2800 t of untreated faecal waste and *ca.* 40 t of toilet paper and hygienic waste annually to the Solway Firth. Groundwood fibre is unlikely to exceed 1000 t $year^{-1}$ but the contribution of algal and terrigenous litter is difficult to assess although some estimates can be derived from the stranding of March 1970. Assuming that the strand-line deposit was that contributed during the preceding period of spring tides, and taking the average of the strandings between Silloth and Allonby, but excluding that in the flood system at Mawbray, then an estimated stranding of *ca.* 900 t occurred along the whole Solway coast from St. Bees Head, Cumberland to the

Burrow Head, Wigtonshire. Some 10 t was deposited in the Mawbray/Dubmill system, and by visual estimates strandings at East Cote/Grune Point and Powillimont/Carsethorn were of the same order; other large strandings probably occurred in the area, but an estimated total of 1,000 t was probably deposited at that particular time. However, as Backlund (1945) noted in the Baltic, this stranding was composed of materials of

TABLE IV. Estimated contribution of particulate organic matter to the waters of the Georgia Strait

Source	Form	Estimated contribution (t year^{-1})	Authority
Rivers	Particulate organic	200,000	Seki *et al.* (1969)
Rivers	Total organic	$1-2 \times 10^6$	Seki *et al.* (1969)
Wood	Activity of *Bankia setacea* upon stored wood	15,000–300,000	Perkins[a]
Wood	Harbour structures, trees on the seas margin	$\geqslant$(15,000–300,000)	Perkins[a]
Wood	Wood pulp industry—as fibre	31,000	Perkins[a]
Wood	Toilet paper, sanitary towels, disposable napkins, etc.	4000	Perkins[a]
Sewage	Faecal waste	>23,000	Perkins[a]

[a] Estimates by the author derived from information by Trussell (1959, 1967), I.D.D., British Columbia Hydro and Power Authority Survey (1966), U.S. Department of the Interior Report (1967) and by personal enquiry.

different ages. Moreover, the deposit of *ca.* 5000 t km^{-1} of fine material at Carsethorn in 1963 must modify one's view of this estimate, and both estimates indicate the scale and importance of a hitherto much neglected problem.

IV. Transport and Deposition of Litter

All materials introduced into sea water are transported in a manner which depends upon their solubility and specific gravity. Thus materials which are soluble or of a neutral buoyancy will tend to diffuse through the water column and be transported away from their source; insolubles with a specific gravity less than that of sea water will float and be influenced more strongly by the wind.

A. Transport of Soluble and Low Specific Gravity Materials

The circulation of the open ocean is essentially wind driven, thus both water and associated floating and suspended solids may be transported over long distances. Inshore, and particularly in relation to estuaries, the run-off from the land may generate reaction currents at the sea bed where dissolved and suspended solids are transported towards the land.

Terrigenous litter has a specific gravity of $\sim 0{\cdot}5$ and floats when introduced to the aquatic system, but sinks subsequently when loss of air or chemical change has reduced its buoyancy. Floating litter may be transported great distances by oceanic currents. For example, seeds of *Entada gigas* and *Mucuna urens* (Barrett and Yonge, 1958) and logs bored by teredines reach the coast of Britain by this means. Likewise, plastic litter is a global problem (Heyerdahl, 1972).

With the exception of *Sargassum* floating algae are considered to occur only near land. However, on a local scale, algal litter, particularly that with gas bladders, e.g. *Ascophyllum*, or that able to hold bubbles of photosynthetic oxygen, may float for some time before sinking. Clearly there are variations in the supply of this, and some animal litter may be wind dependent. At Pipe Clay Lagoon, Tasmania, the supply of dead molluscs and crab exuviae is dependent upon high tides coincident with easterly winds, and the area in which algal litter is deposited depends upon the prevailing northerly winds (Guiler, 1951). Prolonged easterly winds frequently introduce a non-resident algal litter into the River Blackwater, Essex (Davis and Milligan, 1964); some may be carried far upstream and deposited upon the saltmarshes.

Just as marine allochthonous sources may contribute to the supply of litter in an estuary so an estuary, or part of it, may lose litter. At Sapelo Island, Georgia, a creek which drains a *Spartina alterniflora* marsh contributes detritus to the main estuary (Odum and de la Cruz, 1967). Fluviatile particles are negatively charged and precipitate upon contact with the positively charged ions in sea water; however, positively charged organic materials are not precipitated thus and may undergo prolonged suspension.

B. Transport of Insoluble, High Specific Gravity Materials

Insoluble substances with a specific gravity which is greater than sea water, or soluble substances readily adsorbed by more dense particles behave differently from the natatory materials. In the open ocean such materials accumulate at great depth, e.g. diatomaceous and foraminiferan oozes. In the shallower waters of the Continental Shelves, however, a marked transport of insoluble, sedimentary materials may occur.

Wave action and its associated longshore currents and drift, reaction currents and an asymmetry of the tidal wave may all contribute to such transport. The last is of particular importance, for in those situations where the maximum flood current velocity exceeds the maximum ebb current velocity by $\frac{1}{2}$–$\frac{3}{4}$ knots, even at springs, a steady net transport of sediment in the direction of the flood current will always result, even though the net flow of water is in the direction of the ebb. This system is accompanied by strongly rectilinear tidal streams, and structures in the substrata result when there is a sufficient depth of sediment as in the estuaries adjoining the North and Irish Seas. In the Solway Firth, the flood channel system, with its characteristic open end towards the sea and blind end towards the land, may be traced far offshore into the Irish Sea on the one hand, and to high shore levels on the other: this shore system is particularly evident at Mawbray and Dubmill Point. Long-term records of terrigenous and marine litter strandings, which arose out of a study of wood fibres released from a paperboard plant at Siddick, near Workington, Cumberland were kept for Beckfoot and Mawbray from 1967 onwards (Fig. 3), combined with others less complete from Bowness, Grune Point, Powillimont and Carsethorn, indicate that considerable, if transient, strandings of litter are always associated with flood channel systems, and are consistent with the results obtained for silt and sea bed drifter transport by Perkins *et al.* (1963–1966).

This was reflected in the substantial stranding of March 1970, particularly in the Mawbray area (Table V). Here an estimated 10 t of litter was deposited along the 2¼ miles of coastline occupied by the system, but in the adjacent 4 miles an estimated 2·74 t was deposited. In 1963, a massive, transient deposition of fine sediment occurred at Carsethorn with the spring equinoctial tides. Assuming a litter content of 2·6% (Table III), an estimated 5 t m^{-1} of shore, or some 5000 t in all was deposited at mid-shore levels.

C. Deposition of Litter

The behaviour and deposition of plant litter upon individual shores is consistent with the transport and deposition of sedimentary materials. Moreover, the relationship between sediment transport and the movement of sea-bed drifters can be amplified to include litter.

1. Open Coast and Outer Estuary

On exposed shores, with a marked break point to a shingle or sand dune upper shore, litter of a large size is deposited near H.W.M. and the size of the individual litter component decreases downshore. Compare the

TABLE V. Litter strandings on the shores of the Solway Firth, 16–17.3.70

Site	Situation	Weight of litter deposited (kg)				Litter content (%)	
		Total wet weight	Dry weight			Dry weight: total wet weight	Dry weight: total dry weight
			Total	Sand	Litter		
Allonby (N)	1 m H.W. strand line—17.3.70	—	2·111	1·563	0·547	—	26
Allonby (S)	1 m H.W. strand line—17.3.70	1·490	—	—	—	—	—
Beckfoot	1 m H.W. strand line—17.3.70	—	1·490	1·366	0·124	—	8
Dubmill Point	(a) 1 m H.W. strand line—16.3.70	114·070	8·069	~0	8·069	7	~100
	(b) Estimated total deposited around Point	—	$4·95 \times 10^3$	—	—	—	—
	(c) Estimated deposit/1 m H.W. Strand line south of Point	0·571	0·041	—	—	—	—
Lees Scar	1 m H.W. Strand line—16.3.70	20·059	1·390	0·348	1·042	5	75
Mawbray	(a) 1 m H.W. Strand line—17.3.70	6·206	—	—	—	—	—
	(b) 1 m H.W. Strand line—17.3.70	—	1·118	0·869	0·248	—	22
	(c) 1 sq m of patch 15 m × 24 m deposited in flood barblet	—	3·525	1·101	3·517	—	69
	(d) Estimated total in 15 m × 24 m patch	—	1269	—	—	—	—
	(e) Estimated total in 17 m × 28 m patch	—	1678	—	—	—	—

sediment, in which the median ϕ decreases with distance from the H.W.M. towards the L.W.M.

Strandings of fine, granular litter frequently occur in the upper shore breaks in gradient at the time of the equinoctial tides: the sediments are usually most poorly sorted in these breaks. Intense deposition of both coarse and fine litter may occur in association with elements of the flood channel system.

2. *Inner Estuary and Saltmarsh*

The deposition of litter tends to be greatest where the greatest additions of silt occur (Table VI). In these areas, the median ϕ of the sediment

TABLE VI. The time taken for three grades of cotton thread to break when subjected to a constant strain of 50 g in the presence of intertidal soils. (Mean temperature 17 ± 3°C)

Source	Shore level	Time to break under 50*g* (days)			Silt content %	Litter content (% dry wt > 0·5 mm size)
		White machine cotton (B.S. = 778 g)	Brown cotton thread (B.S. = 1162 g)	White linen thread (B.S. = 4587 g)		
Solway Firth:						
Skyreburn Bay	*Salicornia* zone	6	—	—	56	0·032
Balcary Hotel	M.T.L.	7	13	10	67	—
Glen Bay	(a) Sand. M.T.L.	7	13	10	47	—
	(b) Seaward of *Spartina*	8	11	10	61	—
	(c) Mid *Spartina* zone	10	—	—	72	0·043
Glen Isle Marsh	Sand. M.T.L.	10	—	—	81	<0·001
Kenneth Bank	M.H.W.M.N.T.	10	—	—	54	0·001
Mawbray	(a) M.H.W.M.	19	—	—	0·8	—
	(b) E.L.W.M.S.T.	19	—	—	0·4	—
Powfoot	*Hydrobia/Corophium* zone	20	—	—	30	Tr.
Grune Point	*Arenicola* "watt"	21	—	—	17	<0·001
Brighouse Bay	Lower mid-shore	42	—	—	0·4	Tr.
Balcary Hotel	H.W.M.	36	—	—	0·4	—
Beckfoot	*Arenicola* "watt"	82	—	—	0·7	Tr.
Clyde Estuary:						
Ardmore	(a) Coarse sand—H.W.M.	19	15	41	~0	—
	(b) *Enteromorpha* zone	12	—	—	~0	—
	(c) E.L.W.S.T.	21	12	41	~0	—

Litter at Kenneth Bank > Brighouse Bay, Powfoot ≫ Beckfoot (4 pieces only).
B.S. = normal breaking strain.

increases with distance moved from H.W.M. towards L.W.M., a function of increasing turbulence. Silt deposition is greatest, not at the H.W.M., but at some level below it: in the Solway Firth, for example, this level corresponds with that occupied by the Spartinetum (Perkins *et al.*, 1963–1966).

V. Breakdown of Plant Litter

The process of decomposition of plant litter in the sea is commenced by chemical action and a mechanical breakdown by continual water movement (Darnell, 1967*b*). On the Scottish west coast, stranded algae and other litter are gradually mixed with the shingle at H.W.M. and thence abraded whenever the shingle is moved. Overwhelmingly, however, the breakdown is performed by biological agents. The decomposition of marine litter is considered in the sections dealing with wood (see Käärik, Chapter 5) and non-woody litter including algae and terrigenous materials.

A. Decomposition of Wood

Although other forms of litter may be stranded with wood at H.W.M. or sunk in the depths of a fiordic sea loch, it is considered conveniently under this heading since it must also include the deterioration of marine structures and their contribution as a source of energy. At high shore levels, wood may be very dry and breakdown may be slow except by mechanical means. Below the mean tide level (M.T.L.), however, conditions are increasingly more favourable to those organisms that can use it as a source of food.

Cellulolytic bacteria are present generally in sea water and in sediments, especially the latter. Most are aerobes and the majority belong to the *Cytophaga* and *Vibrio* groups (Scholes and Shewan, 1964, ZoBell, 1959). These, accompanied by other physiological groups of bacteria, participate in the rapid rotting of wooden piles, board boat bottoms, cordage, fish nets, lines and ropes (ZoBell, 1959). For this reason, the Japanese fishing industry replaced 75% of its nets annually in the period *ca.* 1956. The intensity of the attack depends upon the water quality and is faster in flowing rather than stagnant water (Scholes and Shewan, 1964).

Lignicolous fungi are widely distributed in temperate and tropical waters (Johnson, 1967; Jones, Chapter 11). Floras from 13 genera, e.g. *Helicoma*, *Halosphaeria*, develop upon wood blocks immersed in the sea: cellulose, pectin and starch are a preferred source of carbon for many species, but sugars may be utilized also. Attacks by the wood-boring crustacean

Limnoria lignorum, the gribble, are frequently preceded by an invasion of marine fungi and, although *Limnoria* may tunnel into fresh, fungus-free wood, it requires a balanced diet of wood and fungi (Wilson, 1960).

The principal animal agents of wood breakdown are the cosmopolitan crustaceans *Limnoria* and its associate *Chelura*, and the molluscan shipworms, e.g. *Teredo* (see Mason, Chapter 17). In the mangrove swamps of the Pandan Reserve, Malaysia, clearance has produced many dead tree stumps and logs, which near the seaward edge of the mangrove, but not within it, are inhabited by *Limnoria*, *Chelura* and the teredine *Teredo* and cylophagid, *Martesia* (Berry, 1963).

Living in the superficial layers *Limnoria* reduces wood to a friable honeycomb which readily disintegrates into small fragments. In the Gareloch, all the submarine litter present as sticks, twigs or even hardboard is utilized thus; bark is apparently unsuitable as a food and may be left as a hollow cylinder once the wood within has been consumed. Harbour structures and piles are particularly susceptible to *Limnoria* attack at the mud-line: not a surprising result in view of its association with the fungi.

Less obvious, but no less damaging, shipworms live within the depth of the wood leaving only a small insignificant hole at the surface as an indication of their presence. Although they utilize the wood as a food source (Townsley and Richy, 1965; Liu and Townsley, 1968), 50% of it may remain undigested. One species, *Bankia setacea*, can settle and grow in non-wood substrates based on agar gel and although not essential, the addition of wood flour is desirable (Trussell *et al.*, 1968). In addition to wood, teredos have destroyed hemp rope, manila cordage, jute, cork, gutta percha and plastics (Hurley, 1961; Trussell *et al.*, 1968).

In the estuaries and creeks of Essex, *Teredo navalis* is abundant in all suitable wood, including withies marking oyster beds, groyne and jetty structures, old hulks and fishing boats. Here, like the results reported by Walden *et al.* (1967) for *Bankia setacea*, the most intense infestation occurs near to the mud-line, which suggests that a relationship with micro-organisms similar to that with *Limnoria* may hold for *Teredo*; moreover rich supplies of microbenthic food are available at this level. In this estuary, *Teredo* may be succeeded by the pholad mollusc *Barnea candida* which normally burrows into stony or stiff substrata (Perkins, 1961).

The shipworms may reduce wood very rapidly. The longevity of *Bankia setacea* exceeds two years, but is usually so numerous that the wood supply is exhausted and death ensues after 6–8 months (Trussell, 1962). At 75 fathoms *B. australis* riddled a length of Australian spotted gum hardwood (*Eucalyptus maculata*) 4 m × 13 cm × 5 cm in the nine months which elapsed after its loss from the T.E.V. Rangatira and its recovery by trawler in April 1961 (Hurley, 1961). *Limnoria* may attack wood in greater numbers than the teredines, but normally takes several years to destroy a pile completely (Walden *et al.*, 1966).

B. Decomposition of Non-wood Litter

The manner in which litter is stranded and its subsequent incorporation in the sediments markedly influence the biota which colonize and consume it. Breakdown is usually rapid, but exceptions are known: in the Baltic wrack bands of *Zostera*, fucoid and other algae may be so abundant and decay so slowly that they have a geomorphological importance; some wrack deposited on swampy shores will not rot completely (Backlund, 1945). This report and those concerning the blanketing effect of fibres from pulp mills in fiords and embayments (Pearson, 1972) imply that concentration and the degree of compaction of the litter has an important influence upon its breakdown. On the other hand, terrigenous and algal litter stranded at the highest levels of shingle or saltmarsh becomes dehydrated and is inhabited by a sparse fauna lacking the talitrid amphipods. Here breakdown is slow or suspended until an unusually high tide or erosion return it to the sea, or until the shingle is colonized by terrestrial plants.

Normally, litter is deposited temporarily or at lower levels where dehydration does not occur, or brought into a dispersed association with sediment, in each, the consumer species have a ready access. The structural materials of plant litter (cellulose, lignin and biopigments) persist longest (Darnell, 1967*b*). In the Solway Firth fragments of leaf vessels occur frequently in sediment samples.

Algae rapidly begin to break down and in the absence of animals a rich compost is produced (see Frankland, Chapter 1). In some areas, e.g. Fife and Ayrshire, where large wrack beds accumulate, the wrack is removed by the truck load for use as an agricultural fertilizer. Unlike the algae, *Zostera* is very resistant to microbial decomposition: in Denmark, it has lasted for centuries when used for thatch or fencing. In the saltmarshes at Sapelo Island, Georgia, Odum and de la Cruz (1967) measured the rate of breakdown of *Distichlis*, *Juncus*, *Spartina*, *Salicornia* and crab *Uca* exposed in bags of 2·5 mm mesh. The *Uca* decomposed completely in 180 days, and in a 300 day experiment the *Distichlis*, *Juncus*, *Spartina* and *Salicornia* underwent a loss of 53, 35, 94 and 58% respectively, of particulate and soluble materials.

1. Breakdown by Micro-organisms

All marine litter is exposed to microbial degradation. Synthetic materials apparently remain unaffected for long periods (Muraoka, 1971), but a slow decomposition can occur. Jones and Le Campion-Alsumard (1970) found that polyurethane blocks submerged for *ca.* 4 years were degraded by the fungi *Corollospora maritima*, *Haligena unicaudata*, *Lulworthia purpurea* and *Zalerion maritimum*.

Animal litter is broken down by autolysis, microbial action and by the necrophagous species, i.e. animals which live upon the decaying remains of others. Darnell (1967*b*) considered the most persistent elements of animal litter to be chitinous structures, teeth, bones, scales and eye lenses. The chitinoclastic bacteria are responsible for the generally rapid decomposition of chitin (Parsons and Seki, 1970) and in some circumstances

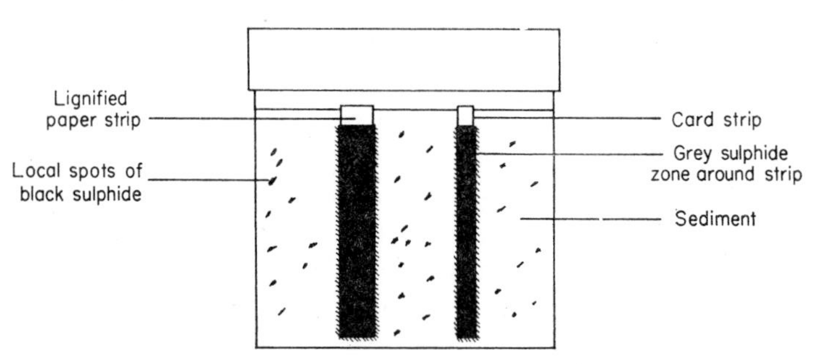

Fig. 5. A linear view of the development of blackened sulphide-containing sediment in association with paper and card strips maintained in a "sterilin" tube.

cause disease in the crabs, *Cancer* and *Carcinus* (Perkins, 1967) and in the shrimp *Crangon*.

Like wood, non-wooden litter is decomposed by cellulolytic bacteria and by the sulphate-reducing bacteria, e.g. *Desulfovibrio desulfuricans*, which produce the blackened sulphide-containing layer of marine sediments. For their development, these latter organisms require an adequate supply of moisture, organic matter and iron oxides in stable marine soils (Perkins, 1957; Baas-Becking, 1959). In these stable situations, the blackened layer is present throughout the year and the fine litter content is relatively high also (Table III). Large strandings of algal litter, e.g. *Laminaria* and *Enteromorpha*, may be accompanied by the rapid development of the blackened layer and emission of H_2S (Perkins and Abbott, 1972). Backlund (1945) noted that sulphur bacteria and purple bacteria are found characteristically with wrack aevja. It has been shown experimentally that areas of blackening occur rapidly in association with paper test strips, but that away from the immediate zone of influence of the paper the sand remains unblackened (Fig. 5).

Werner (1968) found that the lignin and cellulose of wood are converted by *Desulphovibrio desulfuricans* into a grey, fibrous sediment which in extreme cases, adjacent to pulp mills, has created blankets or beds reaching to the surface. The fermentation in this mass is accompanied by the emission of CH_4, CO_2, H_2, H_2S, and N_2. Once established such a blanket or bed may persist indefinitely. Laboratory tests at Garelochhead have shown that in the absence of sedimentary materials 50% of lignified ground-wood fibres are decomposed in *ca.* 6 months in aerobic conditions; in conditions where the sulphide-reducing bacteria developed breakdown of these fibres was negligible in 6 months. Since cotton fibres break down rapidly in the soils most liable to the development of sulphide-reducing bacteria (Table VI) it seems that some necessary factor is missing from the beds described by Werner (1968) and in the experiments performed in the absence of sedimentary materials.

Both lignicolous and non-lignicolous species of fungi contribute to plant litter decomposition. Once wrack has been stranded a growth of wrack fungi develops (Backlund, 1945); all air dried strand-line material taken from the Solway always becomes mouldy on storage. Actinomycetes, including strains which decompose laminarin and alginate, have been isolated from decomposing brown seaweeds, lagoon muds, saltmarsh soils and sands (Chesters *et al.* 1956; Wilson, 1960). Sediments with a high humus contents are more favourable to fungi than those in which the organic content is low, and those species that inhabit sand flats and salt marshes undergo a seasonal variation. However, fungi are more abundant in alkaline than in acid soils, and in sand dunes than in muds: in the latter case, this results from a lower oxygen tension (Pugh, 1961; Johnson, 1967) and since soil pH and oxygen tension are related (Perkins, 1957), oxygen tension may be important in the former case also.

During work in the Solway Firth, it has been noted frequently that some soils, notably from saltmarshes, are particularly destructive of stiff paper labels. During 1972, cottons of different breaking strain were exposed to a range of soils under a constant tension of 50 g (Fig. 6) and the time at which the cotton broke noted (Table VI). The cellulolytic activity of the individual soil was very variable and the time to breaking ranged from 6 days with soil from the *Salicornia* zone in Skyreburn Bay to 82 days with soil from the *Arenicola* "watt" at Beckfoot. These differences reflect to some degree the silt content rather than the litter content of the soil (cf. Kenneth Bank and Glen Bay, Table VI).

2. *Breakdown by Invertebrate Animals*

The reduction of litter by animals evidently occurs in several stages, commencing with the material of a large size found in the strand-line. Backlund (1945) found that the species composition of the wrack fauna was 34·3% schizophagous, i.e. species feeding upon decaying matter, 52%

carnivorous, 9% phytophagous and 4·7% parasitic. In order of abundance the schizophagous animals comprised 94·99%, the carnivores 2·88% and the phytophagous species 1·36% of the population. Only five species—viz., the dipterans *Coelopa pilipes*, *Fucomyia frigida* and the coleopterans, *Orygma luctuosa*, *Thoracochaeta zosterae* and *T. brachystoma*—were completely dependent upon wrack. The coleopterans *Ptenidium punctatum*,

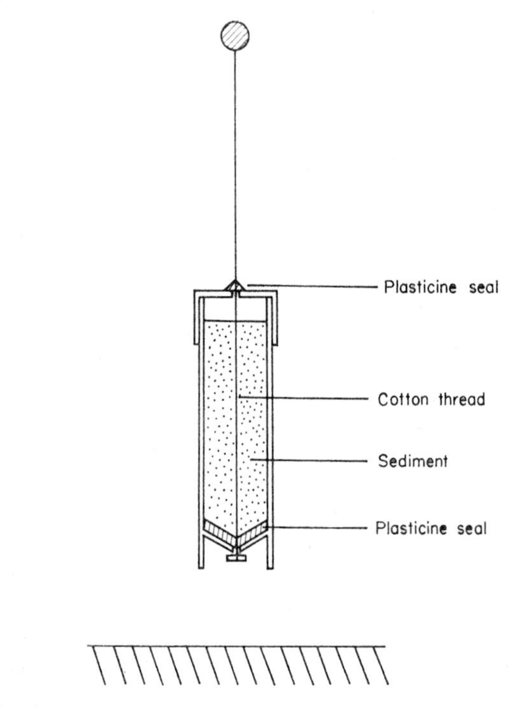

Fig. 6. Method of exposure of cotton thread to sediment as a measure of cellulolytic activity.

Acrotrichis fratercula, *Cercyon littoralis* and *C. depressus* and the dipterans *Scopeuma litorea* and *Heterochila buccata* were somewhat less dependent, but the enchytraeids *Enchytraeus albidus* and *Pachydrilus lineatus*, and the amphipod crustaceans *Orchestia gammarella* (see Ladle, Chapter 18) and *O. platensis* were not bound to the wrack as food. However, despite their abundance the trophic ecology of the schizophagous species inhabiting litter is largely unknown; yet the collembolans and enchytraeids occur in enormous numbers of evident importance. Green (1968) noted that the former are most abundant in winter when the relative humidity is high and a time lag has presumably allowed some breakdown of their food materials.

However, Crustacea are the most characteristic inhabitants of the strand-line, (Barrett and Yonge, 1958).

Dahl (1952) considered that the habitat filled by talitrid amphipods in temperate latitudes was occupied by the ghost crab *Ocypode* in the tropics. However, the talitrid amphipods are found at the Equator (Bousfield,1970, 1971) and the correct view would appear to be that expressed by MacNae and Kalk (1962), viz. that while the talitrids require cast up wrack which is scarce in tropical beaches *Ocypode* has less specialized needs and appears to replace the amphipods only because it is more conspicuous. The habit of living with buried food, noted in temperate latitudes, may not be unimportant, moreover there are indications that *Orchestia* lives deep within shingle at times of scarcity, and between periods of ingestion.

Since more is known of the trophic ecology of these amphipods their influence can be discussed to some degree. They are terrestrial rather than marine in habit and normally occupy only the highest shore levels; indeed, most species on the genus are associated with wholly terrigenous litter (Bousfield, 1970, 1971). One species, *Orchestia anomala* inhabits forest leaf litter and strand-line debris throughout the Indo-Pacific region; on the other hand the truly terrestrial talitrids, *Parochestia* spp. and *Brevitalitrus* spp., evince strongly regional endemicity (Bousfield, 1971).

In north-west Europe, the *Orchestia* spp. characteristically inhabit firmer substrata such as rock, shingle with a low sand content, and the highest levels of saltmarsh. In contrast, *Talitrus* and *Talorchestia* inhabit fine, shifting substrata such as sand and shingle with a high sand content, in which they create characteristic burrows. These differences influence the mode of breakdown of plant litter.

On the firmer beaches incorporation of large litter with the substratum is minimal. Because the litter tends to remain as a distinct layer upon the surface, the *Orchestia* spp. can only gain access to their food supply by leaving the shelter of the shingle. It was found by laboratory experiment that while *Orchestia* ingested litter at the surface of the shingle it habitually lived at a level just above that of the water surface (Fig. 7). In these conditions, the *Orchestia* rapidly consumed the litter provided and emerged from the shingle only to eat. It was noted in these experiments that while some defaecation occurred upon the litter, most faecal pellets accumulated in the bottom of the jar. Examination of a shingle bank at Coldbackie, Sutherland, where an abundance of faecal pellets were found at a depth $>0{\cdot}31$ m within the shingle, suggested that this behaviour is characteristic of *Orchestia* which inhabit shingle.

Backlund (1945) noted that *Orchestia gammarella* and *O. platensis* voraciously consumed all dead plant and animal matter which they encountered, and included paper and cloth in their diet. They consumed very large amounts of food daily, but since the faecal output was very large also, much of the intake was not utilized. Work at Garelochhead has confirmed these observations and shown that *Orchestia* may live naturally in leaf litter and that cardboard is consumed as readily as other materials. The soft tissues of leaves are consumed first, the vessels may be taken only after some delay, but as Backlund noted fungal development only became overt upon this litter once the *Orchestia* were removed. An intake of

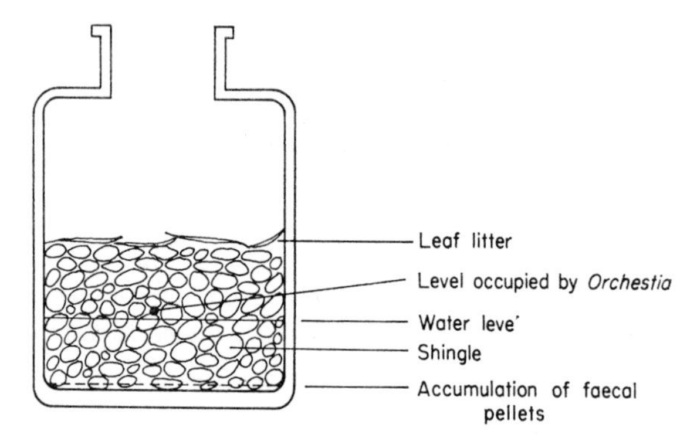

Fig. 7. Laboratory maintenance of *Orchestia* in shingle and in the presence of a plentiful supply of leaf litter.

0·81% of the dry weight of laminarian alga per *Orchestia* body weight per day or 3·42% of the wet weight of laminarian alga per *Orchestia* body weight per day was recorded.

The H.W.M. strand-line is much less evident in saltmarsh areas than on less sheltered shores, consequently the talitrid amphipods are apparently of much less importance. Although collembolans and enchytraeids are still present, and dead leaves swept into saltmarsh pools may be consumed by larvae of the caddis *Limnephilus affinis* (Sutcliffe, 1960, 1961), much of the decomposition of litter is due to the inhabitants of the sediments. The relationships between shore level and cellulolytic activity, the occurrence of fungi and humus content, and the development of the black sulphide layer and organic content, noted above, are indicative both of the importance of litter and the manner of its breakdown.

When litter is deposited upon upper shores of sand or shingle mixed with sand, it becomes invaded by and buried in blown sand; a phenomenon which is particularly apparent in sand dune areas, where a substantial zone of foredune upper shore is composed of sand mixed with large amounts of terrigenous and marine plant, animal and synthetic litter, e.g. at Coldbackie, Sutherland (Fig. 8).

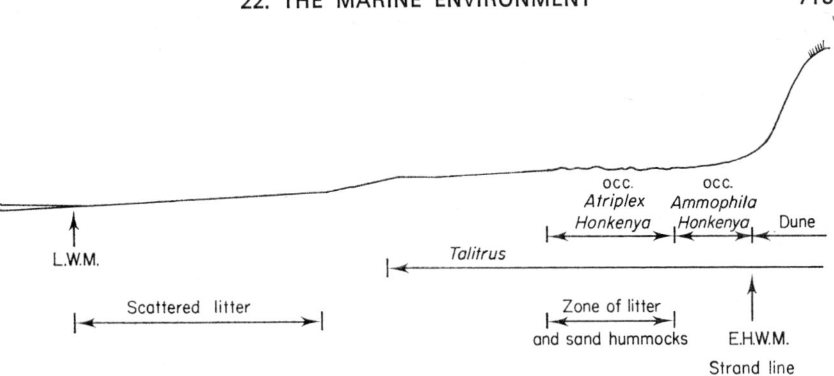

Fig. 8. Litter deposition and associated biota in the shore at Coldbackie, Sutherland, 12.6.72.

The holes made by *Talitrus* may be used as a crude measure of its abundance (crude because more than one *Talitrus* may live in a given burrow). However, by use of this measure, it can be shown that these animals tend to be aggregated around litter deposits (Table VII). Examination of the litter/sand drifts indicated that the litter, particularly algal,

TABLE VII. Distribution of *Talitrus* in bare sand and litter sand mixtures at Torrisdale Bay, Sutherland, 13.6.72

	Litter/sand "heap"		Surrounding litter free sand
	Area	Density (m^{-2})	density (m^{-2})
1.	0·5 m^2	120	0–2
2.	0·4 m^2	100	0–2
3.	2·3 m^2	$\geqslant$149	34 ± 10
Morning H.W. litter[a]		$n \times 10^2$	205 ± 86

[a] Morning H.W. litter = *Laminaria*, *Fucus vesiculosus*, *Ascophyllum nodosum*, grass, gorse, sticks, twigs, *Crangon* exuviae.

is rapidly consumed and its former presence indicated only by a brown stain comparable to the faecal compost deposited by *Orchestia* in shingle shores.

If it is assumed that the daily intake of *Talitrus* and *Orchestia* is the same for individuals of a characteristic 100 mg body weight, then on the basis of the counts made at Torrisdale Bay, Sutherland, the annual consumption of litter here by *Talitrus* alone in an area *ca.* 0·9 km^2 is ~9·6 t wet wt: an estimate which takes no account of the impact of the enchytraeids, collembolans and other insects. Nevertheless, this very approximate calculation does indicate a problem of some interest.

The influence of the talitrids in reducing the size of litter is of evident importance, since such reduction provides a greater access for the activity of bacteria and fungi, and contributes to the supply of detritus. At least

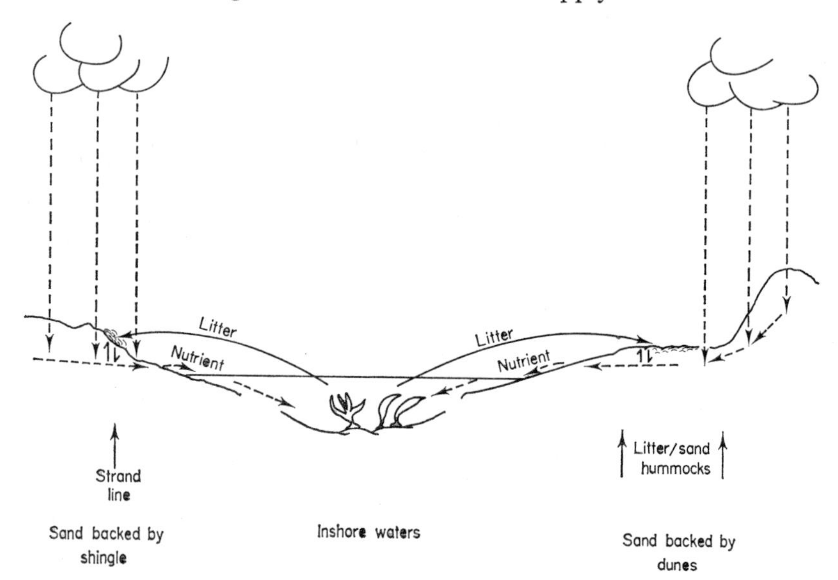

Fig. 9. The nutrient cycle in inshore waters in relation to plant litter deposition on shingle and foredune.

some of the litter ingested will be absorbed by the amphipods and excretion products will result. Shingle and dune areas are particularly influenced by rainfall and blown spray which drain from these permeable substrata in a characteristic fashion; moreover, wave energy is lost by percolation through the soil. Evidently, this circulation through the sediments provides a means whereby the nutrients and detritus derived from litter can be returned to the inshore waters. Apparently, then, the semi-terrestrial inhabitants of the high shore levels play an important part in the cycling of nutrients in the sea (Figs. 9, 10). Given this conclusion, what is the role of man who finds these areas so attractive?

The strandings of fine litter which occurred with the equinoctial tides at Whitstable, Kent, were inhabited by ciliate protozoans, *Lachrymaria olor* and *Vorticella* sp., and the harpacticoid *Alteutha interrupta*, in addition to diatoms (Perkins, 1958). *Vorticella* which occurred abundantly feeds upon bacteria, and possibly detritus and diatoms; the less abundant *L. olor* feeds upon other protozoans (Webb, 1956). The food of *Alteutha interrupta*, is unknown, but diatoms are an important source of food to harpacticoid copepods. Backlund (1945) found harpacticoid copepods living deep within wrack banks.

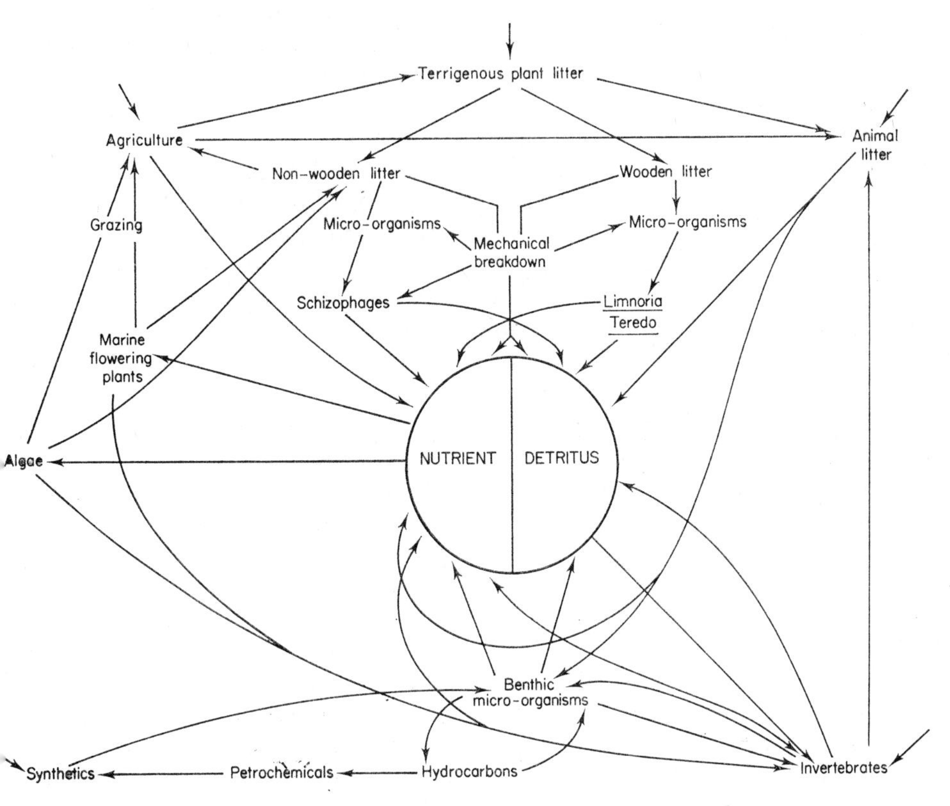

Fig. 10. A simplified trophic web in relation to plant litter deposited in the sea.

3. *Detritus as a Source of Food*

Although large amounts of detritus ($\leqslant 100\ \mu m$) are ingested by marine invertebrates, its nutritional significance remains obscure. Indeed, Darnell (1967*b*) noted that the nutritive value and role of organic detritus depend upon its natural availability, whether an organism can procure and handle it under field conditions and whether it has any nutritive value for the consumer species. Ingestion alone does not imply that it is absorbed by the animal; moreover, much inert sedimentary material and digested fragments of other foods have often been described as detritus for want of a better definition.

Zostera has had a curious influence in this connection for it led Danish workers to suggest that detritus is important, but its resistance to microbial reduction (Backlund, 1945), its indigestibility to the oyster (Yonge, 1926) and its lack of influence upon associated detritus feeders when it declined (Stauffer, 1937) left its importance in doubt. However, the work of Newell (1965), confirmed by that of Parsons and Seki (1970) showed that the bacterial biomass associated with the surface of organic and inorganic particles, rather than detrital materials themselves was an important source of food to some invertebrates. Then, Adams and Angelovic (1970) found that the gastropod mollusc, *Bittium varium,* the glass shrimp *Palaemonetes pugio* and the polychaete, *Glycera dibranchiata* were able to assimilate *Zostera* detritus; furthermore, *P. pugio* and *B. varium* obtained more nourishment from the detrital substrate than from its associated bacteria. All three species can utilize soluble carbon sources.

Khailov and Erokhin (1971) showed that at concentrations characteristic of littoral biotopes, the copepods *Tigriopus* and *Calanus* utilized external metabolites, derived from algal hydrolysate, to the extent of 13% and 1·2% of their total rations respectively. Mechanical breakdown, either by abrasion with shingle or the activities of the talitrids, and the concomitant release of these materials from plant litter may therefore be a process of some importance.

VI. Summary

In the sea plant litter is derived from terrestrial and fluviatile as well as purely marine sources. Although the importance of the finest materials, i.e. detritus, in the food webs of the sea has long been recognized, that it is derived to a significant degree from strandings of large litter has not. The composition of this litter is highly variable and few strandings are derived either from a single species or from one group of the plant kingdom. Marked changes in composition and amount occur with time, i.e. upon successive tides, and in relation to geographical location, such as along the length of an estuary or proximity relative to a flood channel system. Moreover, the nutrient status of the litter must be influenced by the amount and character of the animal component present.

On the shore, the talitrid amphipods and the ocypodid crabs play an important part in the initial breakdown of the larger particles, although these species are not exclusively responsible for this process. At low tide levels and in the infra-littoral the limnorian isopods and the teredine molluscs, well known as pests of economic importance, are influential in reducing the size of the individual litter particle. Microorganisms play an important role in this breakdown and fungi are known to be necessary to a balanced diet of the gribble, *Limnoria lignorum*. A complex ecosystem which develops in association with plant litter ensures a rapid return of nutrients, both particulate and dissolved, to the adjacent coastal waters. The importance of salt marshes to the productivity of coastal waters is now recognized, the information described above would suggest that plant litter cannot be disregarded.

References

ADAMS, S. M. and ANGELOVIC, J. W. (1970). *Chesapeake Sci.* **11**, 249–254.
ATKINS, W. R. G., JENKINS, P. G. and WARREN, F. J. (1954). *J. mar. biol. Ass. U.K.* **33**, 497–510.
BAAS-BECKING, L. G. M. (1959). *In* "Contributions to Marine Microbiology.' (T. K. Skerman, ed.). *N.Z. Oceanogr. Inst. Mem.* **3**, 48–64.
BACKLUND, H. O. (1945). *Opuscula Entomologica, Suppl.* (1945).
BARRETT, J. and YONGE, C. M. (1958). "Pocket Guide to the Sea Shore." Collins, London.
BERRY, A. J. (1963). *Bull. natn. Mus.* Singapore **32**, 90–98.
BLEGVAD, H. (1914). *Rep. Dan. biol. Stn* **22**, 41.
BOUGHEY, A. S. (1957). *J. Ecol.* **45**, 665–687.
BOUSFIELD, E. L. (1970). "The Natural History of Rennell Island, British Solomon Islands" **6**, 155–168. Danish Press, Copenhagen.
BOUSFIELD, E. L. (1971). *Steenstrupia* **1**, 255–293.
BOYSEN-JENSEN, P. (1914). *Rep. Dan. biol. Stn* **22**, 1–39.
BRITISH COLUMBIA HYDRO AND POWER AUTHORITY (1966). "The Pulp and Paper Industry of British Columbia." Vancouver.
BROWN, A. C. (1971). *Trans. R. Soc. S. Afr.* **39**, 247–279.
BROWN, D. S. (1971). *Proc. malac. Soc. Lond.* **39**, 263–279.
CHESTERS, C. G. C., APINIS, A. E. and TURNER, M. (1956). *Proc. Linn. Soc. Lond.* **166**, 87–97.
CLARKE, L. D. and HANNON, N. J. (1967). *J. Ecol.* **55**, 753–771.
CLARKE, L. D. and HANNON, N. J. (1971). *J. Ecol.* **59**, 535–554.
DAHL, E. (1952). *Oikos* **4**, 1–27.
DARNELL, R. M. (1967*a*). *In* "Estuaries." (G. H. Lauff, ed.), *Publ. Am. Ass. Advmt. Sci.* **83**, 374–375.
DARNELL, R. (1967*b*). *In* "Estuaries." (G. H. Lauff, ed.), *Publ. Am. Ass. Advmt. Sci.* **83**, 376–382.
DAVIS, D. S. and MILLIGAN, G. M. (1964). C.E.R.L. Note RD/L/N41/64.
ENEQUIST, P. (1949). *Zool. Bidrag, Uppsala* **28**, 297–492.
GILHAM, M. E. (1957). *J. Ecol.* **45**, 757–778.
GREEN, J. (1968). "The Biology of Estuarine Animals." Sidgwick and Jackson, London.

GUILER, E. R. (1951). *Pap. Proc. R. Soc. Tasmania* 1950, 29–52.
HASLAM, S. M. (1971). *J. Ecol.* **59,** 65–88.
HEDGPETH, J. W. (1957). *Geol. Soc. Am., Memoir* 67, **1,** 587–608.
HEYERDAHL, T. (1972). *The Environment This Month* **1,** 6–10.
HURLEY, D. E. (1961). *N.Z. Jl Sci.* **4,** 720–730.
JOHNSON, T. W. (1967). *In* "Estuaries." (G. H. Lauff, ed.), *Publs. Am. Ass. Advmt Sci.* **83,** 303–305.
JOHNSON, M. E. and SNOOK, H. J. (1967). "Seashore Animals of the Pacific Coast." Dover Publications, New York.
JONES, E. B. G. and LE CAMPION-ALSUMARD, T. (1970). *Int. Biodetn Bull.* **6,** 119–124.
KENCHINGTON, R. A. (1970). *J. mar. biol. Ass. U.K.* **50,** 489–498.
KHAILOV, K. M. and EROKHIN, V. E. (1971). *Okeanologiya* **11,** 117–126.
LIPKIN, Y. and SAFRIEL, U. (1971). *J. Ecol.* **59,** 1–30.
LIU, D. and TOWNSLEY, P. M. (1968). *J. Fish. Res. Bd Can.* **25,** 853–862.
MACKAY, D. W. and TOPPING, G. (1970). *Effluent and Water Treatment* **5,** 641–649.
MACNAE, W. (1957*a*). *J. Ecol.* **45,** 113–131.
MACNAE, W. (1957*b*). *J. Ecol.* **45,** 361–387.
MACNAE, W. and KALK, M. (1962). *J. Anim. Ecol.* **31,** 93–128.
MORTON, J. K. (1957). *J. Ecol.* **45,** 495–497.
MURAOKA, J. S. (1971). *Ocean Industry*, February and March 1971.
NEWELL, R. C. (1965). *Proc. zool. Soc. Lond.* **144,** 25–45.
NEWELL, R. C. (1971). "The Biology of Intertidal Animals." Logos Press, London.
ODUM, E. P. and DE LA CRUZ, A. A. (1967). *In* "Estuaries." (G. H. Lauff, ed.). *Publs. Am. Ass. Advmt Sci.* **83,** 383–388.
PARSONS, T. R. (1963). *In* "Progress in Oceanography." (M. Sears, ed.), pp. 205–239. Pergamon Press, Oxford and New York.
PARSONS, T. R. and SEKI, H. (1970). *In* "Organic Matter in Natural Waters." (D. W. Hood, ed.), Inst. Mar. Sci., U. of Alaska, Pub. No. 1, 1–27.
PEARCE, J. B. (1970). "The Effects of Solid Waste Disposal on Benthic Communities in the New York Bight." F.A.O. Technical Conference on Marine Pollution and its Effects on Living Resources and Fishing, Rome, 9–18 December 1970. FIR: MP/70/E99.
PEARSON, T. H. (1970). *J. exp. mar. Biol. Ecol.* **5,** 1–34.
PEARSON, T. H. (1971). *Vie Millieu* Suppl. No. **22,** 53–91.
PEARSON, T. H. (1972). *Proc. R. Soc. Lond.* B. **180,** 469–485.
PERKINS, E. J. (1957). *Ann. Mag. nat. Hist.* ser. 12, **10,** 25–35.
PERKINS, E. J. (1958). Ph.D. Thesis, University of London.
PERKINS, E. J. (1961). C.E.R.L. Laboratory Note No. RD/L/N.97/61.
PERKINS, E. J., WILLIAMS, B. R. H., BAILEY, M., GORMAN, J. and HINDE, A. (1963–66). "The Biology of the Solway Firth in Relation to the Movement and Accumulation of Radioactive Materials." H.M.S.O., U.K.A.E.A., P.G. Reports. 12 parts.

PERKINS, E. J. (1967). *Trans. Dumfriesshire and Galloway nat. Hist. Antiq. Soc.* 3rd ser., **44**, 47–56.
PERKINS, E. J. (1969). *Trans. Dumfriesshire and Galloway nat. Hist. Antiq. Soc.* 3rd ser., **46**, 1–26.
PERKINS, E. J. (1972). *The Environment this Month* **1**, 47–54.
PERKINS, E. J. and ABBOTT, O. J. (1972). *Mar. Poll. Bull.* **3**, 70–72.
PETERSEN, C. G. J. and BOYSEN-JENSEN, P. (1911). *Rep. Dan. biol. Stn* **20**, 3.
PUGH, G. J. F. (1961). *Nature, Lond.* **190**, 1032–1033.
PUGH, G. J. F. (1962). *Trans. Br. mycol. Soc.* **45**, 560–566.
SCHOLES, R. B. and SHEWAN, J. M. (1964). *Adv. mar. Biol.* **2**, 133–169.
SEKI, H., STEPHENS, K. V. and PARSONS, T. R. (1969). *Arch. Hydrobiol.* **66**, 37–47.
SHELDON, R. W. and PARSONS, T. R. (1967). *J. Fish. Res. Bd Can.* **24**, 909–915.
STAUFFER, R. C. (1937). *Ecology* **18**, 427–431.
SUTCLIFFE, D. W. (1960). *Proc. R. ent. Soc. Lond.* A **35**, 156–162.
SUTCLIFFE, D. W. (1961). *Trans. nat. Hist. Soc. Northumb. Durh.* **14**, 37–56.
TATTERSALL, W. M. and TATTERSALL, O. S. (1951). "The British Mysidacea." The Ray Society, London.
TAYLOR, J. D. (1971). *Phil. Trans. Roy. Soc. Lond.* B **260**, 173–213.
TOWNSLEY, P. M. and RICHY, R. A. (1965). *Can. J. Zool.* **43**, 1011–1019.
TRUSSELL, P. C. (1959). *In* "Marine Boring and Fouling Organisms." (D. L. Ray, ed.), pp. 462–468. Friday Harbour Symposia, University of Washington Press, Seattle.
TRUSSELL, P. C. (1962). *New Scientist* **13**, 432–435.
TRUSSELL, P. C. (1967). *Sea Frontiers* **13**, 234–243.
TRUSSELL, P. C., WALDEN, C. C. and WAI, E. (1968). *Material Organismen* **3**, 95–105.
U.S. DEPT. OF THE INTERIOR (1967). "Pollutional Effects of Pulp and Paper Mill Wastes in Puget Sound." Portland and Olympia, pp. 1–473.
WALDEN, C. C., ALLEN, I. V. F. and TRUSSELL, P. C. (1966). *Materials Protection* **5**, 9–11.
WALDEN, C. C., ALLEN, I. V. F. and TRUSSELL, P. C. (1967). *J. Fish. Res. Bd Can.* **24**, 261–272.
WEBB, M. G. (1956). *J. Anim. Ecol.* **25**, 148–175.
WERNER, A. E. (1968). *Pulp and Paper Mag. Canada* **69**, 127–136.
WERNSTEDT, C. (1942). *Vid. Medd. Dansk naturh. Foren.* **106**, 241–252.
WILSON, I. M. (1960). *Proc. Linn. Soc. Lond.* **171**, 53–70.
YONGE, C. M. (1926). *J. mar. biol. Ass. U.K.* **14**, 295–382.
ZOBELL, C. E. (1959). *In* "Contributions to Marine Microbiology." (T. K. Skerman, ed.). *N.Z. Oceanogr. Inst. Mem.* No. 3, 7–20.

23

Decomposition of Agricultural Crop Debris

R. S. Forbes

Department of Extra-mural Studies
The Queen's University of Belfast
Northern Ireland

I. Introduction

Crop residues, green manure crops and weeds are added to arable soils by farmers each year in season, following the pattern of nature where leaf, stem and root litter are continually added to uncultivated soils. In the artificially maintained monocultural regime of an arable field, crop plant residues and their decomposition products are important components of the soil. Crop debris is usually a principal amendment applied to the soil system and plays an important role in maintaining soil fertility, affecting both the physical structure of the soil and its nutrient status. In addition, the study of crop residues and their decomposition has been vigorously pursued with a view to the biological control of plant pathogenic organisms found on the crop debris and in the soil. The spur to research into methods of biological control is the obvious expense and serious social consequences for man and his environment of other forms of crop protection using vast quantities of chemicals.

II. Types of Debris Available for Decomposition

The majority of crop plants are either annuals, such as the cereals and peas, or potential biennials and perennials which are allowed to grow for only a single season. The runner bean, *Phaseolus coccineus* L., is an example of a South American perennial plant grown in Europe as an annual crop, while the turnip, *Brassica rapa* L., swede, *Brassica napus* var. *napobrassica* (L.) Rehb., and carrot, *Daucus carota* L., are biennials handled in the same manner. The plant breeder has bred a great number of crop plants, selecting for the useful qualities of various plant organs. Within the family Umbelliferae, for example, selection has improved the swollen root of carrot, the petioles of *Apium graveolens* L., (celery), the leaf of *Petroselinum crispum* (Mill.) Nyman (parsley), and the aromatic seeds of *Carum carvi* L. (caraway). In the great majority of crop plants only a single organ is commercially important, so that a considerable proportion of the total net production of the plant, varying with the crop, is discarded by the cultivator.

Crop debris added to soil after harvesting is not the only source of organic soil amendment provided by crop plants; as with mixed natural plant communities, extending root systems are continually sloughing off dead root hairs and other senescent root material. These, together with root exudates, maintain an active saprophytic soil population on the root surface, the rhizoplane, and in the root vicinity, the rhizosphere, which may serve to protect the plant from root-infecting soil-borne pathogens. This subject has been reviewed at length in several recent articles (Katznelson, 1965; Rovira, 1965*a*, *b*, 1969; Parkinson, 1967; Rovira and McDougall, 1967; Bowen and Rovira, 1968; Gray and Williams, 1971).

The relative amounts of organic material bound up in the roots and shoot, be it a crop plant or otherwise, are difficult to measure and vary enormously with plant variety, soil and climatic conditions, so that any attempt to compare their respective importance as substrates for decomposers in soil is problematical. In a consideration of the decomposition of crop residues, however, the underground parts of the crop plant should not be forgotten, since they will influence the composition and activity of the soil microflora and fauna, and thus affect the rate of organic decomposition.

III. Groups of Organisms Involved in Decomposition

A. Animals

While what follows is a detailed consideration of microbial decomposition and saprophytism among plant pathogens, mention should be made of the role of macro-decomposers, members of the animal kingdom. In the case of some members of the soil-borne fauna, in particular the nematodes plant and animal remains play a considerable role as a substratum for the organisms on which the animal feeds, but do not form part of the animal's diet (Nielson, 1967). Nielson (1962) suggested that the primary decomposition of plant remains is largely attributable to the soil microflora, and that, with the exception of some snails, some tipulid larvae, and possibly Protozoa, the so-called decomposing soil fauna is in reality a group of primary consumers utilizing the microflora as food.

While earthworms undoubtedly consume quite large quantities of leaf material, and similar evidence is available for some soil-dwelling arthropods, information is scarce on the question of their role in the decomposition of agricultural crop debris. This is due partly to the difficulty of quantitatively isolating these animals from the soil mass, partly to the very great problems of their identification and classification when they are separated out, and partly to the lack of techniques for studying what they are feeding on in soil. For the arthropods, it can be said that in general cultivation greatly decreases their diversity and abundance in soil. The population of arthropods in the top 20 cm of soil in plots that received inorganic fertilizers was about 18,000 per m^2, which is less than 10% of the corresponding population in permanent grassland (Russell, 1961).

B. Micro-organisms

The return to the soil of nutrients bound up in the crop plant debris stimulates decomposition by saprophytic organisms native in the soil, bringing them into active antagonism with plant pathogens on the crop and in the soil. Antagonism as a concept embraces the phenomena of competition, antibiosis and exploitation, the latter covering both hyperparasitism and predation. Competition and antibiosis seem to be more directly related to the decomposition of crop residues than does microbial exploitation, and since the latter has been comprehensively reviewed (Boosalis, 1964; Boosalis and Mankau, 1965), exploitation will not be mentioned further here. The outcome of the interaction of saprophytes and pathogens will have a profound effect on the survival of pathogenic propagules, and, through their inoculum density and inoculum potential, on disease incidence and severity in succeeding crops.

Since the recognition of the phenomenon of mycostasis in soil, attention has been given to the role of amendments in producing active microbial growth in soil. The question has become one of determining which factors overcome mycostasis, and to what degree they are effective. It has long been held that fungal plant pathogens, in evolving to the parasitic mode of life had sacrificed, to various degrees, the biochemical ability to act as saprophytes (Garrett, 1944). Simultaneously with this nutritional specialization, however, the loss of saprophytic ability has been compensated for by the development of efficient long-term resting structures. This enables many of these micro-organisms to survive in soil in the absence of a suitable host plant, awaiting the opportunity of renewed parasitism. Dormant structures, such as thick-walled chlamydospores, resting hyphae and sclerotia are resistant to decomposition by lytic bacteria and fungi. Not all soil-borne plant pathogens possess resting structures, however. Some less specialized pathogens, for example *Cephalosporium gramineum* and *Ophiobolus graminis*, rely on a saprophytic phase between pathogenic phases in their life-cycles.

The theory underlying most attempts at biological control of soil-borne plant pathogens depends on amending the soil with crop residues so that saprophytes are stimulated to active decomposition, producing a state of microbial competition and antibiosis which is detrimental to the pathogen. In this manner biological control has been effective in reducing the intensity of take-all of wheat, *Triticum aestivum* L. (Fellows, 1929); *Phymatotrichum* root rot of cotton, *Gossypium* sp. (King *et al.*, 1934); scab of potato, *Solanum tuberosum* L. (Millard, 1923); and *Rhizoctonia* disease of snap beans (Papavizas and Davey, 1960). Closely allied to the amendment of soil using crop residues is the practice of crop rotation. The effect of rotation may be due not only to "starvation" of the resting pathogen brought about by the absence of its host crop, but also to the stimulation of the antagonistic properties of the soil microflora by the presence of a non-host crop and its residues. The form and intensity of this antagonism may differ from that produced by host crops and their debris, and could conceivably prove more effective in disease control.

The process of "starving out" a pathogen resting in soil may require a break from the susceptible crop of considerable duration. In the case of farming geared to cereal production, for example, this may represent a severe financial sacrifice on the part of the farmer. Experience under English conditions shows that the amount of *Ophiobolus graminis* remaining in the soil a year after the last cereal host crop was grown is usually too little to cause serious losses in a new cereal crop, provided the fungus has not been able to survive on secondary hosts such as grass weeds or self-sown cereals (Chambers and Flentje, 1968). Quite often the *Ophiobolus* population and its vigour declines in the six months between autumn harvest and spring sowing to the extent that a spring cereal suffers much smaller disease losses than would a winter crop on the same field (Slope, 1968). In contrast, the eye spot fungus, *Cercosporella herpotrichoides* was found by Macer (1961) to survive in approximately 75% of colonized wheat straws buried in the field for three years. Again, fungi having resistant dormancy structures, such as *Fusarium culmorum*, *Helminthosporium sativum* and *Curvularia ramosa*, have survived in wheat straw buried in soil of high and low fertility for periods of more than two years (Butler, 1959). In view of the relative longevity of some pathogen propagules in soil, the aim of biological research is to discover mechanisms for increasing the rate of decline of pathogenic populations to levels which admit an economic crop return.

IV. The Environment in which Decomposition Occurs

In the first half of this century, in intensive cropping programmes it was customary to dispose of cereal stubble and other crop residues by "burning off". The use of fire was widely accepted as a useful and practical means of destroying any surface residues infected with plant pathogens (McAlpine, 1902, 1904; Carne, 1927; Pittman, 1937). As this was considered wasteful of potential soil manure, and in the long term was also detrimental to soil structure, the burning of crop debris was for a time discouraged, there being a switch to using either stock to graze and trample down post-harvest residues, or to ploughing the plant remains under. Recently, however, in the cooler wetter areas of Britain where the decomposition of organic farm waste is slow, some farmers faced with an excess of organic waste to dispose of, have reverted to the older practice of "burning off" crop residues.

A. Biological Interactions between Organisms

In order to discuss the breakdown of crop residues and their decomposition products, it is desirable to have a clear picture of the characteristics of soil, substrates and micro-organisms. Where competition is involved, where is demand most likely to exceed supply? Such is the variation possible in soil factors that generalizations become dangerous and misleading. The phenomenon of mycostasis may be considered here again, being a general concept true of most soils (see Pugh, Chapter 10). It is realized that spores introduced on or into moist unsterile soils often fail to germinate (Dobbs and Hinson, 1960). Two main ideas have been advanced to explain this widespread dormancy in soil:

(a) the nutrient deficiency hypothesis (Lockwood, 1964, 1968), which contends that the mycostatic effects observed can be wholly explained by lack of substances in soil essential for germination, and
(b) the school of thought which believes that mycostasis is due to the presence of low levels of diffusible inhibiting substances in soil (Brian, 1960; Griffin, 1962; Park, 1960, 1961, 1965), the effects of which may be annulled by application of suitable nutrients (Griffin, 1964).

Irrespective of the outcome of this debate (Griffin, 1972), nutrients essential for germination are the key to reactivation of dormant propagules, and consequently there must exist in soils generally a competitive situation due to demand for nutrients in short supply.

The term "competition" may be used broadly to denote factors favouring one species over another, as in Garrett's concept of "competitive saprophytic ability" of a pathogenic fungus (Garrett, 1970) in which "competition" is more or less synonymous with the term "antagonism" already mentioned. In a narrower and more useful sense, "competition" can be used to describe the state where there is active demand by two or more organisms for an essential prerequisite in short supply. This narrower definition will be used here, since it parallels the use of the term in wider ecological considerations (Odum, 1959; Miller, 1967).

The production by soil micro-organisms of chemicals toxic to others is well known and occurs during active microbial growth. Antibiosis, taken in the broad sense to mean the production by micro-organisms of metabolites such as alcohols, acids, specific antibiotics or respiratory carbon dioxide, which create an environment unfavourable for other members of the microflora, will again limit active decomposition of crop residues. Griffin (1962) showed that a mycostatic effect could be produced when autoclaved soils were inoculated with cultures of fungi which did not show antibiotic production in pure culture, indicating that behaviour on culture media does not necessarily indicate similar behaviour in soil. Park (1960, 1965) argued against the idea that antibiotic production confers an advantage in an antagonistic situation, but supported the view that tolerance of antibiosis is the significant feature in determining the outcome of such a struggle for survival. A good example of such a situation is given by Park (1965) who showed that *Trichothecium roseum*, which has a high antibiotic production rate, is unsuccessful in mixed cultures owing to its low tolerance to antibiosis. While it is untenable to suggest that micro-organisms, even despite their vast numbers in soil, compete for physical space, it can be said that they compete for sites free from toxic decomposition residues. Crop residues added to soil are thus both a source of available nutrients, and are a fresh physical substratum in which antibiosis is at a minimum. Certain crop residues are known to produce decomposition products which are phytotoxic. This important work has been reviewed in detail (Börner, 1960; Patrick *et al.*, 1964; Patrick and Toussoun, 1965; Lindermann, 1970), and will not be dealt with here. Garrett (1970) in mentioning this topic pointed out that in general the beneficial effects on soil of crop residue decomposition greatly outweigh the harmful. Nevertheless, work on phytotoxic decomposition products has been important in helping to unravel such problems as "sick soil" associated with replanting fruit trees.

B. Nutritional Aspects

The addition of crop residues to soil providing soluble and easily available energy materials as well as more complex substrates resistant to decomposition, results in a "microbial explosion" of activity (Burges, 1967). This initial decomposition by "pioneer" micro-organisms is of short duration, however, for very rapidly a situation builds up where competititon and antibiosis become limiting factors.

The initial burst of saprophytic activity is not restricted to obligate saprophytes. Pathogenic fungi, too, may act as pioneer colonizers of freshly dead crop debris. *Fusarium oxysporum* was recognized by Park (1959) to be a pioneer colonizer. It was found that although this fungus showed a high degree of competititive saprophytic ability when colonizing sterilized or freshly dead organic materials, previous colonization of the plant material by native soil micro-organisms for as short a time as two or three days was sufficient to prevent entry by the fungus.

The initial burst of saprophytic activity on decomposing crop residue results in the rapid utilization of soluble sugars, the immobilization of most nitrogen in the mycelium of the colonizing micro-organisms, and the build-up of antibiosis in the substrate. Crop residues have, in general, high carbon/nitrogen (C:N) ratios, and nitrogen, an essential mineral nutrient for microbial growth, tends to become limiting when carbon is present in abundant supply (Stotzky and Norman, 1961; Finstein and Alexander, 1962). The first step in either infection of a host or colonization of a dead substrate in soil by fungi, involves the germination of spores or the resumption of activity of some of the mycelium. Many organisms require exogenous nutrients for such resumption of activity (Cochrane, 1960). Exogenous carbon and nitrogen compounds are required for germination of *Fusarium roseum* conidia (Sisler and Cox, 1954), chlamydospores of *Fusarium solani* f.sp. *phaseoli* (Byther, 1965; Griffin, 1962), conidia of *Fusarium oxysporum* f.sp. *cubense* (Stover, 1959) and conidia and microsclerotia of *Verticillium dahliae* (Powelson, 1966).

Nutrient control of germination is not such a straightforward situation, however, since Byther (1965) observed the unexpected finding that conidia of *Fusarium solani* f.sp. *phaseoli* germinated with ethanol as carbon source when inorganic potassium and magnesium salts were present, and that nitrogen was unnecessary. In fact ethanol under these conditions encouraged a much higher percentage germination than given with a simple carbon source like glucose.

Recent work on volatile components of plant residues in soil has shown that they may play a part in inducing both germination (Menzies and Gilbert, 1967) and mycostasis (Balis and Kouyeas, 1968; Hora and Baker, 1970).

C. Physical Factors Affecting Decomposition

The picture is further complicated by soil physical conditions which may influence saprophytic activity. Microsclerotia of *Verticillium albo-atrum* in soil, germinated and sporulated each time after nine consecutive cycles of air-drying and rewetting, although the percentage germination decreased with succeeding germinations (Farley *et al.*, 1971). This is contrary to the findings of Kouyeas and Balis (1968), who investigated the effect on conidia of rewetting air-dried soils to a range of soil moisture tensions. Using *Arthrobotrys oligospora* as test organism, they found that as moisture tension increased from zero (i.e. water-saturated soil) the restoration rate of mycostasis increased to a maximum at 0·85 atm, and fell away again at higher tensions. They found that soil aeration was a factor involved, for water-saturated soils subjected to forced aeration proved mycostatic. Kouyeas and Balis suggested that a fungistatic factor of a volatile nature would explain both these results, since moisture and soil atmosphere "compete" for voids. Nutrient availability, however, may be involved indirectly in these two situations, where at first appearance a physical variable is in sole control of fungal activity. It has been shown by numerous experiments that drying soil increases the amount of available nitrogen (Prescott, 1920; Waksman and Starkey, 1923; Birch, 1958). This nitrogen has been identified as α-amino nitrogen (Hayashi and Harada, 1969), mineralized from peptide complex substances and structural proteins. The major sources of the nitrogen are thought to be the cell walls of micro-organisms, and plant and animal residues previously accumulated in soils.

V. The Effects of Soil Amendments on Decomposition

Despite these complications and contradictions, there is evidence based on field studies indicating that saprophytic activity in soil is governed by nutrient availability, particularly carbon and nitrogen availability as expressed in the C:N ratio. The effect of the "microbial explosion" in the initial stages of residue decomposition is both to immobilize the available nitrogen within the cells of the pioneer colonizers, and to release some of the carbon as carbon dioxide through respiration (see Williams and Gray, Chapter 19; Gray and Biddlestone, Chapter 24).

Soil amendments of high C:N ratio are used to create this situation, and have achieved effective biological control of certain diseases. In California, good control of *Phaseolus vulgareis* L. (bean) root rot caused by *Fusarium solani* f.sp. *phaseoli*, was obtained using mature straw of either barley (*Hordeum vulgare* L.) or wheat (*Triticum aestivum* L.) or maize residues (*Zea mays* L.) as a soil amendment (Snyder *et al.*, 1959); residues such as these have a C:N ratio of about 80:1. The addition of 0·3% ammonium nitrate nullified the controlling effects of the high C:N ratio amendment, thus indicating that nitrogen was the limiting factor. These workers also found that amending the soil with green barley hay, *Glycine max* (L.) Merr. (soybean), or *Medicago sativa* L. (alfalfa) residues, all of which have low C:N ratios averaging about 16:1, actually increased bean root rot in the subsequent crop. Control of root rot was apparently due to immobilization of nitrogen by the soil microflora, with which *Fusarium solani* f.sp. *phaseoli* is not able to compete. Maier (1961) found residues of barley, sorghum (*Sorghum vulgare* Pers.), and maize suppressed development of *Fusarium* root rot of pinto bean cotton crop residues had no effect on subsequent disease, and residues of low C: Nratio such as tomato (*Lycopersicum esculentum* Mill.), alfalfa and lettuce (*Lactuca sativa* L.) increased root rot severity. It is interesting to note that this disease has also been controlled using spent coffee grounds from the instant coffee manufacturing process (Adams *et al.*, 1968). It appears that an unidentified factor in the spent coffee grounds stimulated chlamydospores of *Fusarium solani* f.sp. *phaseoli* to germinate in soil where the germ tubes were quickly lysed. The influence of amending the C:N ratio on *Fusarium solani* f.sp. *phaseoli* was further demonstrated by Maurer and Baker (1964), who showed that when field soils were amended with residues of C:N ratio greater than 25:1 bean root rot was significantly suppressed. Griffin (1964) discovered that filtered aqueous extracts of soils amended with cellulose (C:N ratio higher than 25:1), and barley straw (C:N ratio 55:1), inhibited germination of conidia of *Fusarium solani* f.sp. *phaseoli* and *Gliocladium fimbriatum*. This inhibition was overcome by the addition of as little as five parts per million nitrate nitrogen to the soil during extraction. Soybean residues (C:N ratio 7:1) and alfalfa straw (C:N ratio 18:1) did not inhibit germination of the two test fungi.

Crop residues of high C:N ratio have been used to reduce soil populations of *Ophiobolus graminis*, responsible for the take-all disease of cereals

(Garrett, 1938, 1940), by starving the pathogen of nitrogen. The relationship between *Ophiobolus* and nitrogen is not straightforward, however, and it has occupied Garrett's attention for more than 30 years (Garrett, 1970). Cereal crop residues with low C:N ratios encourage the survival of *Ophiobolus* in soil. At the same time, soil amendments rich in nitrogen benefit the cereal host crop, enabling it to produce fresh roots and to grow away from the disease (Garrett, 1941, 1948).

Since cellulose, usually found with superimposed lignin, is the most abundant and durable compound in crop plant tissue, it is clear that a fungus maintaining itself by active saprophytic growth will benefit by being capable of producing cellulase. Not only this, but a facultative parasite persisting as a saprophyte should be capable of utilizing cellulose slowly, in an economical manner. Herbaceous secondary tissues are relatively easily penetrated when compared with those of woody perennials, and cellulase and soluble decomposition products may diffuse away from the sphere of activity of the cellulolytic hyphae, to be utilized by associated "sugar-fungi", themselves incapable of cellulase production, which live as commensals with cellulose-decomposing organisms (Garrett, 1963; Tribe, 1966). There is thus the possibility that a selective advantage may accrue to those fungi whose production of cellulase assures them of enough soluble products for their own energy and survival, without a surplus to be lost to commensal organisms.

Garrett (1970) sees a mechanism by which this limitation of cellulase production may be achieved in the fact that cellulase enzymes are adaptive, requiring substrate contact before induction. Hydrolysis of the cellulose immediately surrounding a particular length of fungal hypha will remove it from contact with the cellulose substrate it inhabits, so that induction of cellulase will cease. To survive, the mycelium within plant tissues must maintain active extension and branching growth, in order that continuous cellulolysis can take place. Rate of decomposition of plant tissue in soil is thus limited by the supply of mineral nutrients essential for the synthesis of microbial protoplasm. Nitrogen, the mineral element required in largest amount, may be temporarily immobilized in soil by the previous activity of pioneer decomposers, or lost to the particular soil altogether through leaching or gaseous denitrification. Decomposition of crop plants, and active survival of pathogenic fungi such as *Ophiobolus graminis*, may thus be limited by the supply of available nitrogen. Phosphate may sometimes be a limiting mineral nutrient in compost, but it is rarely in short supply in soil.

The facultative parasite *Rhizoctonia solani*, strains of which attack a wide variety of crops, is depressed in soil by addition of crop residues of high C:N ratio. Blair (1943) attributed the effect of wheat straw or dried grass amendments to nitrogen starvation through rapid utilization of available soil nitrogen by cellulose-decomposing micro-organisms. A contributing factor to the observed depression of the growth of the pathogen is the sensitivity of *Rhizoctonia solani* to carbon dioxide, which will have been respired in quantity by the activity of the cellulose decomposers. This effect is a clear example of antibiosis in action. Papavizas and Davey (1962) confirmed Blair's views on the effect of carbon dioxide. They found that *Rhizoctonia solani* was more sensitive in its pathogenic phase than in its active saprophytic phase which in turn was more sensitive than the survival phase within precolonized substrates. Further research on *Rhizoctonia* activity and survival by these workers, comparing dry and green crop residues as soil amendments, showed a similar outcome in both instances—a depression of pathogenic activity and a measure of protection for susceptible bean crops (Papavizas and Davey, 1960; Davey and Papavizas, 1960). They noted that the relationship between nitrogen and *Rhizoctonia* in soil was a complicated one. Comparison of various soil amendments in terms of their ability to inhibit saprophytic growth of the pathogen indicated that the lower the C:N ratio of the amendment, the greater was its effectiveness in this respect. The mechanism of this effect involved stimulation of antagonistic micro-organisms by the addition of nitrogen. However, of the amendments used by Papavizas and Davey, nitrogen-enriched sawdust carried more nitrogen per unit of carbon than did non-enriched oat straw (*Avena sativa* L.) and yet it was not nearly as effective. In addition, soybean residues, which had the lowest initial C:N ratio of the amendments used in their study, were rendered less effective by the addition of supplemental nitrogen.

Experimental work was carried out to clarify this situation, using soil amendments of cellulose powder and ammonium nitrate, adjusted to give C:N ratios ranging from 400:1 to 5:1 (Davey and Papavizas, 1963). Maximum inhibition of saprophytic activity occurred in those soils receiving amendments adjusted to C:N ratios in the range 100:1 to 40:1. There was less inhibition outside this range, both at higher and lower C:N ratios. This result can be explained by an interaction of two distinct and opposite effects of nitrogen on the activity of *Rhizoctonia solani*. Firstly there is the inhibitory effect reported by Blair (1943), involving decomposition of the organic material and output of respiratory carbon dioxide. The increasing proportion of nitrogen present in residues as the C:N ratio falls has the effect of increasing the rate of decomposition. At a C:N ratio of about 20:1, however, a stimulatory effect of nitrogen on *Rhizoctonia* comes into play, and the resultant depressing effect of the organic amendment upon the fungus becomes sharply reduced. Stimulation of *Rhizoctonia solani* is due to the presence of available mineralized nitrogen at C:N ratios below 15:1

(Scarsbrook, 1965), relieving the effect of nitrogen starvation suffered by the fungus in the presence of organic substrates with higher C:N ratios. There is thus a balance achieved between the effects of nitrogen immobilization and mineralization.

In the decomposition of organic residues in soil by micro-organisms, the same scheme applies with respect to nitrogen immobilization and mineralization. The majority of fungi and bacteria possess thalli having C:N ratios of about 12:1, whereas fresh crop residues generally have higher ratios. During the early stages of decomposition available nitrogen becomes immobilized in the protoplasm of colonizers, while carbon is respired and given off in gaseous form. During subsequent decomposition nitrogen continues to accumulate in the organisms carrying out the breakdown of the organic materials. The net outcome with respect to the substrate is a progressive reduction in its C:N ratio, which continues until, between ratios of 30:1 and 15:1, the rates of nitrogen immobilization and mineralization are about equal. Over longer periods, the C:N ratio of the decomposing crop residue approaches that of the micro-organisms themselves. Should the crop residue have an initial low C:N ratio of a round 8:1, as for example the green amendments used by Papavizas and Davey (1960), then nitrogen mineralization will occur from the introduction of the substrate into soil, and the ratio will gradually increase (Griffin, 1972).

The influence of soil factors, particularly nitrogen availability, on the survival of foot-rot fungi on cereal residues, has been extensively worked on by Garrett, 1940, 1944, 1966, 1967, 1971. It has been found that different species of cereal foot-rot fungi respond differently to nitrogen in the longevity of their saprophytic survival in colonized wheat straw buried in soil. *Ophiobolus graminis* and *Fusarium roseum* f. sp. *cerealis* (syn. *Fusarium culmorum*) are known to survive for longer when the nitrogen supply to the straw is not limiting, while *Cercosporella herpotrichoides* and *Curvularia ramosa* are indifferent to nitrogen supply, and the saprophytic survival of *Cochliobolus sativus* and *Phialophora radicicola* is curtailed by excess nitrogen. Garrett (1966) found that the behaviour of some of these species could be explained in terms of their ability to decompose cellulose, and the relationship of this to their linear growth rates (the cellulose adequacy index). Nevertheless the curtailing effect of a high level of nitrogen on longevity of *Cochliobolus sativus* and *Phialophora radicicola* in soil remained an unresolved problem (particularly since this nitrogen effect was not found in pure culture cellulose decomposition experiments with these two species). Garrett (1972) put forward two hypotheses to explain the differences in nitrogen responses found in the different species:

(a) that the provision of an adequate nitrogen supply for the maximum decomposition rate of straw in soil shortens survival of all the pathogen species considered, but that this effect is concealed in species with a low cellulose adequacy index (i.e. either low rate of cellulose decomposition or high rate of linear growth). This hypothesis suggests that the longevity of the primary pathogen colonizer is determined by the time taken for it to exhaust the substrate; or

(b) species with a high cellulose adequacy index, such as *C. sativus* and *P. radicicola*, may be sensitive to microbial competition when nitrogen is not limiting, and thus be displaced from the substrate before it is exhausted. The first hypothesis would suggest that the median survival periods, given the label "S50 values" (analogous to the half-life notation of radioactive isotope deay) should tend to approximate to the same value in different species when nitrogen is not limiting decomposition rate.

Garrett (1972) assembled from the published literature a total of 26 determinations of the mycelial S50 values of the six species under consideration (dormant conidia being removed from the substrate by thorough washing before viability was accessed). After taking precautions to eliminate extraneous detail by summarizing the S50 information into periods of 5 weeks, from 6–10 weeks to above 40 weeks, the general conclusion arrived at was that species whose survival period on straw in soil is shortened when nitrogen is not limiting have significantly lower S50 values than species which were indifferent or survived longer in similar circumstances. Thus the hypothesis that species whose survival is shortened by nitrogen are displaced by competing micro-organisms is the more likely of the two.

There is strong evidence for stating that nitrogen in soil acts as the principal factor limiting the rate of decomposition of crop residues, and there have been actual demonstrations of competition for nitrogen between fungi and bacteria (Marshall and Alexander, 1960, Lindsey, 1965). However, a case can be made suggesting carbon as a limiting factor in certain circumstances, though it is less well established. Powelson (1966), using a silica gel technique, found that nitrogen was not completely immobilized in alfalfa-amended or non-amended soils. On the other hand, diffusible carbon sources could not be detected by this method. He concluded that paucity of diffusible carbon substrates, rather than nitrogen, was limiting *Verticillium dahliae* germination in the soils examined.

VI. Crop Residue Decomposition and the Control of Plant Pathogens

Accepting the importance of nitrogen on decomposition, and the possible controlling effect of crop residue amendments on plant pathogens and plant disease, it should be made quite clear that this is not the whole picture. Rather, a great multiplicity of physical, chemical and microbial factors, and their complex interactions, play a part in determining the rate of crop residue decomposition and the survival of pathogen propagules. Huber and Watson (1970) after a detailed study of the effects of 27 organic amendments on specific soil-borne diseases, were unable to establish a positive correlation between C:N ratio of the crop residue and disease severity on the succeeding susceptible crop. Instead they found that disease severity was correlated with the effect of specific residues on nitrification. Working with *Fusarium solani* f.sp. *phaseoli* they found that residues such as alfalfa, soybean, pea and maize, which stimulated the biological oxidation of ammonium to nitrate nitrogen, tended to reduce the severity of bean root rot under field conditions, while addition of residues that inhibited or decreased nitrification, such as barley, generally resulted in more severe root rot. The correlation of nitrification with disease severity may indicate the importance of a specific form of nitrogen available to the crop plant or pathogen after residue incorporation. It was also shown by these workers that different crop diseases responded to different forms of nitrogen. Potato scab (*Streptomyces scabies*), wilt of potato (*Verticillium albo-atrum*) and take-all of wheat (*Ophiobolus graminis*) were decreased by ammoniacal nitrogen, but stalk rot of corn (*Gibberella saubinetii*), and bean root rot (*Fusarium solani* f.sp. *phaseoli*) were increased.

Further, it has been shown by Papavizas and Davey (1960) that the timing of planting, subsequent to soil amendment with crop residues of either high or low C:N ratio, has a significant effect on disease incidence.

These findings indicate that when considering the effect on disease organisms of adding a particular crop residue to soil the outcome is not necessarily predictable from a knowledge of the C:N ratio of the soil amendment. Disease prediction would be better based on information regarding the form of nitrogen predominantly available in the soil, the effect of the crop residue on rate of nitrification of ammonium to nitrate, and the effect of different nitrogen sources on the behaviour of the particular pathogen under consideration.

A. Flood Fallowing

Under the agricultural practice known as "flood fallowing", involving the anaerobic decomposition of crop residues under water, antibiosis may become the principal mechanism involved in pathogen control. The effects of anaerobic decomposition of soil amendments on microsclerotia of *Verticillium dahliae* were investigated in a model system by Menzies (1962). Microsclerotia were eliminated from non-amended soil under anaerobic conditions in 13 days. When the soil was amended with 1% alfalfa meal or 0·1% sucrose, microsclerotia were killed in only five days. The evidence obtained indicated that anaerobic decomposition of alfalfa or sucrose resulted in the accumulation of a fungicidal compound. Watson (1964) demonstrated that the decomposition of crop residues in soil under anaerobic conditions achieved by flooding resulted in the elimination of *Pyrenochaeta terrestris* as well as the microsclerotial *Verticillium* and apparently all species of *Fusarium*. A combination of flooding, the addition of a residue, and anaerobic fermentation was required for eradication. Soybean straw was not as effective as residues of green soybean, *Melilotus* sp. (sweet clover), maize, alfalfa, oat straw, or *Beta vulgaris* (sugar beet) subsp. *cicla* (L.) Arcangeli, in eradicating *Pyrenochaeta terrestris*. This kind of treatment was found to reduce greatly the survival of both saprophytic and parasitic fungi. Stover (1962) reporting on attempted control measures against *Fusarium oxysporum* f.sp. *cubense* causing Panama wilt disease of *Musa* sp. (banana), describes the use of flood-fallowing periods of from 3 to 18 months. Under tropical conditions the rate of decomposition of crop residues is extremely rapid. Banana pseudostems and rhizomes disappeared almost completely in three months, with the exception of highly lignified fibres, as the result of putrefactive decay under anaerobic soil conditions. The sum of flood fallowing experience for control of Panama disease indicated that the major factors favouring disease incidence lay in the characteristic of the soil, and could not be controlled by minor variations in the flooding procedures. Subtle factors of soil environment, possibly including subsurface drainage and hydrology, may have been involved in producing the variable disease incidence observed after flood fallowing.

B. Antibiotic Production

Although changes in the populations of antibiotic producing organisms have been observed during the decomposition of organic materials in soil, for example the study of the effect of green soybean residues on potato scab caused by *Streptomyces scabies* involving *Bacillus subtilis* (Weinhold, 1968; Weinhold and Bowman, 1968), and there is evidence for the occurrence of specific antibiotics in pieces of natural substrate that provide sites with a relatively high nutrient status (Wright, 1956; Norstadt and McCalla, 1969), attempts to demonstrate the presence of antibiotic substances in soil have produced only circumstantial evidence. Baker (1968) has outlined the probable reasons for this lack of direct evidence of antibiotic compounds in soil, chief of which appears to be the inactivation of specific antibiotics by adsorption onto clay colloids and humus particles. He pointed out, however, that persistence of antibiotics in the adsorbed condition may occur. A slow release may be inferred from indirect evidence, such as the reaction of the microbial flora. Lai and Bruehl (1966) have shown that saprophytic survival of *Cephalosporium gramineum* in wheat straw is aided by the production of antibiotic substances. When the pathogen eventually exhausts the food available in the straw tissues, it ceases to produce the antifungal antibiotic in effective quantities, so that other fungi become dominant on the remaining substrate, and *Cephalosporium gramineum* dies.

VII. The Saprophytic Activity of Plant Pathogens

It has been shown that a number of soil-borne plant pathogens have not totally lost their saprophytic ability to compete for decomposable substrates in soil (Garrett, 1938, 1956, 1963, 1970; Sadasivan, 1939; Walker, 1941; Butler, 1953, *a*, *b*, *c*, 1959; Park, 1958, 1959, Lucas, 1955; Macer, 1961; Blair, 1943; Papavizas and Davey, 1961; Stover, 1962; Baker *et al.*, 1967; Papavizas, 1968). Most of these studies have involved re-isolation from soils of test fungi in model systems using sterilized substrate baits. These studies have enabled the build-up of a large body of useful literature on the mechanisms of survival, and encouraged investigation of nutrient and physical conditions affecting saprophytic activity in soil. Nevertheless, the picture of the saprophytic ability of pathogenic fungi obtained is merely that of a model, and it is difficult, not to say dangerous, to extrapolate from these results to the reality of soil conditions and economic agricultural practice. Although an organism may be culturally re-isolated from soil on an introduced nutrient bait, this does not necessarily infer activity on the substrate under soil conditions. The great majority of colonies found on soil dilution plates originate from dormant spores. There is a difference between spread of a fungus from a colonized food base over soil-borne substrates, and the actual utilization of such organic matter. This is something which needs to be examined in much more detail than simple re-isolation methods allow. Really effective surface sterilization procedures would achieve a knowledge of the presence at depth in plant tissues, but criteria are needed to decide whether or not resident fungal propagules are active or dormant.

Prior occupation of a substrate is a further factor which must be considered in colonization studies. When established in wheat straw prior to burial in soil, *Trichoderma viride*, *Penicillium* sp., *Trichoderma album* and *Fusarium culmorum* prevented establishment of would-be fungal colonizers as effectively as did *Cephalosporium gramineum* (Bruehl and Lai, 1966). Because most fungal pathogens are established to some degree in living host tissues before their incorporation into the soil with crop debris, they are undoubtedly in a favourable position to survive through saprophytic activity. In the question of colonization and utilization of a substrate in soil, possession is all important, as is exemplified by *Ophiobolus graminis* (Garrett, 1970), *Fusarium roseum* "culmorum" (Cook and Bruehl, 1968; Cook, 1970), and *Cephalosporium gramineatum* (Bruehl and Lai, 1968).

Discrepancies between laboratory experiments and field studies have been highlighted by the work of Cook and Bruehl (1968) on colonization of wheat straw by *Fusarium culmorum*. They have demonstrated that whereas colonization often occurs in laboratory experiments where the sterile straw is usually presented to the test organism, colonization is rare under field conditions in the Pacific Northwest of the United States of America, where the straw is weathered and already heavily colonized by fungi of the air spora and phylloplane. Of 80 fields examined, *Fusarium culmorum* was not detected in 58, in spite of the fact that most, if not all, had been subjected to cereal culture and straw return for decades. Of the remaining fields which did show the presence of the pathogen, 13 had less than 100 *Fusarium culmorum* propagules per gram of soil. It may well be that although fungal pathogens may survive saprophytically on crop debris, their role as decomposers has been over-estimated in the past.

In conclusion, it should be apparent that while a great effort has been made to understand the complex relationship between crop, plant pathogens and saprophytic soil micro-organisms, nevertheless an enormous amount of further research is needed into achieving means of swift decomposition of crop debris, in order that accompanying pathogens may be denied at least saprophytic means of survival.

References

ADAMS, P. B., LEWIS, J. A. and PAPAVIZAS, G. C. (1968). *Phytopathology* **58**, 1603–1608.

BAKER, K. F., FLENTJE, N. T., OLSEN, C. M. and STRETTON, H. M. (1967). *Phytopathology* **57**, 591–597.

BAKER, R. (1968). *A. Rev. Phytopath.* **6**, 263–294.

BALIS, C. and KOUYEAS, V. (1968). *Annls Inst. phytopath. Benaki* **9**, 145–149.

BIRCH, H. F. (1958). *Pl. Soil* **10**, 9–31.

BLAIR, I. D. (1943). *Ann. appl. Biol.* **30**, 118–127.

BOOSALIS, M. G. (1964). *A. Rev. Phytopath.* **2**, 363–376.

BOOSALIS, M. G. and MANKAU, R. (1965). In "Ecology of Soil-borne Plant Pathogens." (K. F. Baker and W. C. Snyder, eds), pp. 374–389. University of California Press, Berkeley.

BÖRNER, H. (1960). *Bot. Rev.* **26**, 393–424.

BOWEN, G. D. and ROVIRA, S. D. (1968). *In* "Root Growth." (W. J. Whittington, ed.), pp. 170–199. Butterworths, London.

BRIAN, P. W. (1960). *In* "The Ecology of Soil Fungi." (D. Parkinson and J. S. Waid, eds), pp. 115–129. Liverpool University Press.

BRUEHL, G. W. and LAI, P. (1966). *Phytopathology* **56,** 766–768.
BRUEHL, G. W. and LAI, P. (1968). *Phytopathology* **58,** 464–466.
BURGES, A. (1967). *In* "Soil Biology." (A. Burges and F. Raw, eds), pp. 479–492. Academic Press, London and New York.
BUTLER, F. C. (1953*a*). *Ann. appl. Biol.* **40,** 284–297.
BUTLER, F. C. (1953*b*). *Ann. appl. Biol.* **40,** 298–304.
BUTLER, F. C. (1953*c*). *Ann. appl. Biol.* **40,** 305–311.
BUTLER, F. C. (1959). *Ann. appl. Biol.* **47,** 28–36.
BYTHER, R. (1965). *Phytopathology* **55,** 852–858.
CARNE, W. M. (1927). *J. Dep. Agric. W. Aust.* (2nd series) **4,** 483–488.
CHAMBERS, S. C. and FLENTJE, N. T. (1968). *Aust. J. biol. Sci.* **21,** 1153–1161.
COCHRANE, V. W. (1960). *In* "Plant Pathology—An Advanced Treatise," Vol. 2. (J. G. Horsfall and A. E. Diamond, eds), pp. 167–202. Academic Press, New York.
COOK, R. J. (1970). *Phytopathology* **60,** 1672–1676.
COOK, R. J. and BRUEHL, G. W. (1968). *Phytopathology* **58,** 306–308.
DAVEY, C. B. and PAPAVIZAS, G. C. (1960). *Phytopathology* **50,** 522–525.
DAVEY, C. B. and PAPAVIZAS, G. C. (1963). *Proc. Soil Sci. Soc. Am.* **27,** 164–167.
DOBBS, C. G. and HINSON, W. H. (1960). *In* "The Ecology of Soil Fungi." (D. Parkinson and J. S. Waid, eds), pp. 33–42. Liverpool University Press.
FARLEY, J. D., WILHELM, S. and SNYDER, W. C. (1971). *Phytopathology* **61,** 260–264.
FELLOWS, H. (1929). *Phytopathology* **19,** 103. Abstr.
FINSTEIN, M. S. and ALEXANDER, M. (1962). *Soil Sci.* **94,** 334–339.
GARRETT, S. D. (1938). *Ann. appl. Biol.* **25,** 742–766.
GARRETT, S. D. (1940). *Ann. appl. Biol.* **27,** 199–204.
GARRETT, S. D. (1941). *Ann. appl. Biol.* **28,** 14–18.
GARRETT, S. D. (1944). *Ann. appl. Biol.* **31,** 186–191.
GARRETT, S. D. (1948). *Ann. appl. Biol.* **35,** 14–17.
GARRETT, S. D. (1956). "Biology of Root Infecting Fungi." Cambridge University Press.
GARRETT, S. D. (1963). "Soil Fungi and Soil Fertility." Pergamon, Oxford.
GARRETT, S. D. (1966). *Trans. Br. mycol. Soc.* **49,** 57–68.
GARRETT, S. D. (1967). *Trans. Br. mycol. Soc.* **50,** 519–524.
GARRETT, S. D. (1970). "Pathogenic Root-infecting Fungi." Cambridge University Press.
GARRETT, S. D. (1971). *Trans. Br. mycol. Soc.* **57,** 121–128.
GARRETT, S. D. (1972). *Trans. Br. mycol. Soc.* **59,** 445–452.
GRAY, T. R. G. and WILLIAMS, S. T. (1971). "Soil Micro-organisms." Oliver and Boyd, Edinburgh.
GRIFFIN, G. J. (1962). *Phytopathology*, **52,** 90–91.
GRIFFIN, G. J. (1964). *Can. J. Microbiol.* **10,** 605–612.
GRIFFIN, D. M. (1972). "Ecology of Soil Fungi." Chapman and Hall, London.
HAYASHI, R. and HARADA, T. (1969). *Soil Sci. Pl. Nutr.* **15,** 226–234.
HORA, T. S. and BAKER, R. (1970). *Nature, Lond.* **225,** 1071–1072.
HUBER, D. M. and WATSON, R. D. (1970). *Phytopathology* **60,** 22–26.

KATZNELSON, H. (1965). *In* "Ecology of Soil-borne Plant Pathogens." (K. F. Baker and W. C. Snyder, eds), pp. 187–209. University of California Press, Berkeley.
KING, C. J., HOPE, C. and EATON, E. D. (1934). *J. agric. Res.* **49,** 1093–1107.
KOUYEAS, V. and BALIS, C. (1968). *Annls Inst. phytopath. Benaki* **8,** 123–144.
LAI, P. and BRUEHL, G. W. (1966). *Phytopathology* **56,** 213–218.
LINDERMANN, R. G. (1970). *Phytopathology* **60,** 19–22.
LINDSEY, D. (1965). *Phytopathology* **55,** 104–110.
LOCKWOOD, J. L. (1964). *A. Rev. Phytopath.* **2,** 341–362.
LOCKWOOD, J. L. (1968). *In* "The Ecology of Soil Bacteria." (D. Parkinson and T. R. G. Gray, eds), pp. 44–65. Liverpool University Press.
LUCAS, R. L. (1955). *Ann. appl. Biol.* **43,** 134–143.
MACER, R. C. F. (1961). *Ann. appl. Biol.* **49,** 165–172.
MAIER, C. R. (1961). *Pl. Dis. Reptr* **45,** 960–964.
MARSHALL, K. C. and ALEXANDER, M. (1960). *Pl. Soil* **12,** 143–153.
MAURER, C. L. and BAKER, R. R. (1964). *Phytopathology* **54,** 1425–1426.
MENZIES, J. D. (1962). *Phytopathology* **52,** 743. Abstr.
MENZIES, J. D. and GILBERT, R. G. (1967). *Proc. Soil Sci. Soc. Am.* **31,** 495–496.
MILLARD, W. A. (1923). *Ann. appl. Biol.* **10,** 70–88.
MILLER, R. S. (1967). *Adv. ecol. Res.* **4,** 1–74.
MCALPINE, D. (1902). *J. Dept. Agric. Vict.* **1,** 74–80.
MCALPINE, D. (1904). *J. Dept. Agric. Vict.* **2,** 410–426.
NIELSON, C. O. (1962). *Oikos* **13,** 200–215.
NIELSON, C. O. (1967). *In* "Soil Biology." (A. Burges and F. Raw, eds), pp. 197–211. Academic Press, London and New York.
NORSTADT, F. A. and MCCALLA, T. M. (1969). *Soil Sci.* **107,** 188–193.
ODUM, E. P. (1959). "Fundamentals of Ecology" (2nd edition). Saunders, Philadelphia and London.
PAPAVIZAS, G. C. (1968). *Phytopathology* **58,** 421–428.
PAPAVIZAS, G. C. and DAVEY, C. B. (1960). *Phytopathology* **50,** 516–522.
PAPAVIZAS, G. C. and DAVEY, C. B. (1961). *Phytopathology* **51,** 693–699.
PAPAVIZAS, G. C. and DAVEY, C. B. (1962). *Phytopathology* **52,** 759–766.
PARK, D. (1958). *Ann. Bot.* **22,** 19–35.
PARK, D. (1959). *Ann. Bot.* **23,** 35–49.
PARK, D. (1960). *In* "The Ecology of Soil Fungi." (D. Parkinson and J. S. Waid, eds), pp. 148–159. Liverpool University Press.
PARK, D. (1961). *Trans. Br. mycol. Soc.* **44,** 377–390.
PARK, D. (1965). *In* "Ecology of Soil-borne Plant Pathogens." (K. F. Baker and W. C. Snyder, eds), pp. 82–97. University of California Press, Berkeley.
PARKINSON, D. (1967). *In* "Soil Biology." (A. Burges and F. Raw, eds), pp. 449–478. Academic Press, London and New York.
PATRICK, Z. A., TOUSSOUN, T. A. and KOCH, L. W. (1964). *A. Rev. Phytopath.* **2,** 267–292.
PATRICK, Z. A. and TOUSSOUN, T. A. (1965). *In* "Ecology of Soil-borne Plant Pathogens." (K. F. Baker and W. C. Snyder, eds), pp. 440–459. University of California Press, Berkeley.

Pittman, H. A. (1937). *J. Dept. Agric. W. Aust.* (2nd series) **14,** 103–112.
Powelson, R. L. (1966). *Phytopathology* **56,** 895. Abstr.
Prescott, J. A. (1920). *J. agric. Sci. Camb.* **10,** 177–181.
Rovira, A. D. (1965*a*). *In* "Ecology of Soil-borne Plant Pathogens." (K. F. Baker and W. C. Snyder, eds), pp. 170–186. University of California Press, Berkeley.
Rovira, A. D. (1965*b*). *A. Rev. Microbiol.* **19,** 241–266.
Rovira, A. D. (1969). *Bot. Rev.* **35,** 35–57.
Rovira, A. D. and McDougall, B. M. (1967). *In* "Soil Biochemistry." (A. D. McLaren and G. F. Peterson, eds), pp. 417–463. Marcel Dekker, New York.
Russell, E. W. (1961). "Soil Conditions and Plant Growth." Longmans, London.
Sadasivan, T. S. (1939). *Ann. appl. Biol.* **26,** 497–508.
Scarsbrook, C. E. (1965). *In* "Soil Nitrogen." Agronomy Monograph No. 10. (W. V. Bartholomew and F. E. Clark, eds), pp. 481–502. Academic Press, New York.
Sisler, H. D. and Cox, C. E. (1954). *Am. J. Bot.* **41,** 338–345.
Slope, D. B. (1968). *J. natn. Inst. agric. Bot.* (*Suppl.*) **11,** 54–57.
Snyder, W. C., Scroth, M. N. and Christou, T. (1959). *Phytopathology* **49,** 755–756.
Stotzky, G. and Norman, A. G. (1961). *Arch. Mikrobiol.* **40,** 341–369.
Stover, R. H. (1959). *In* "Plant Pathology—Problems and Progress 1908–1958." (C. S. Holton, G. W. Fisher, R. W. Fulton, H. Hart and S. E. A. McCallan, eds), pp. 339–355. University of Wisconsin Press, Madison.
Stover, R. H. (1962). Phytopathological Paper No. 4, The Commonwealth Mycological Institute, Kew, Surrey.
Tribe, H. T. (1966). *Trans. Br. mycol. Soc.* **49,** 457–466.
Walker, A. G. (1941). *Ann. appl. Biol.* **28,** 333–350.
Watson, R. D. (1964). *Phytopathology* **54,** 1437. Abstr.
Waksman, S. A. and Starkey, R. L. (1923). *Soil Sci.* **16,** 137–157.
Weinhold, A. R. (1968). 1st Int. Congr. Plant Path., Abstr. p. 214.
Weinhold, A. R. and Bowman, T. (1968). *Pl. Soil* **28,** 12–24.
Wright, J. M. (1956). *Ann. appl. Biol.* **44,** 461–466.

24

Decomposition of Urban Waste

K. R. Gray and A. J. Biddlestone

Department of Chemical Engineering
University of Birmingham
England

I. The Fundamental Process

A. Introduction

Millions of tonnes of organic material, produced annually by photosynthesis in plants, are eventually degraded by microbial action and stored in the soil as humus. The degradation process normally takes place slowly, on the surface of the ground, at ambient temperature and mainly under aerobic conditions.

This natural process can be accelerated, however, by gathering the material into heaps and promoting "composting" which may be defined as the decomposition of heterogeneous organic matter by a mixed microbial population in a moist, warm, aerobic environment.

As a result of increasing human population and rising standards of living, disposal of the organic wastes generated by man is becoming a major problem. In Table I are listed the approximate annual tonnages of organic

TABLE I. Approximate production of organic waste in the U.K. (Data from Gray *et al.*, 1971*a*)

Source	Tonnes per year, fresh weight	Moisture content %, fresh weight basis
Wood-shavings and sawdust	1,070,000	—
Straws—wheat, barley, oats	3,000,000	14
Potato and pea haulms, sugar beet tops	1,600,000	77–83
Bracken, potentially available	1,000,000	—
Seaweed, potentially available	1,000,000	—
Garden wastes	1–10,000,000	—
Sewage sludge	20,000,000	95
Municipal refuse	18,000,000	20–40
Farm manures	120,000,000	85

waste produced in the U.K. (Gray *et al.*, 1971*a*). Some of the materials are extensively re-used—wood waste and straw for composite boards and animal bedding, pea haulms for silage and sugar beet tops ploughed back in. Similarly, until recently, virtually all farm manures were returned to the land. However, with recent trends towards intensive stock rearing, often under cover, many specialist farmers have insufficient land for spreading manure; this has now given rise to a serious disposal problem (Gowan, 1972).

Major difficulties are arising over the disposal of municipal refuse and sewage sludge, especially as society seeks to reduce environmental pollution (Working Parties on Refuse Collection, 1967; on Sewage Disposal, 1970; on Refuse Disposal, 1971). For refuse disposal the major method has been the various forms of tipping, together with incineration and composting (Gray, 1966; Biddlestone and Gray, 1973). The technique of tipping is now declining, however, owing to the acute shortage of tipping sites, especially near major conurbations, the shortage of labour and of top cover. Incineration of both refuse and sewage sludge is certain to be more widely practised, especially in the larger towns which can afford the high capital and running costs of these installations.

The future is likely to see a significant increase in the number of plants for composting refuse and dewatered sewage sludge. Table II illustrates the increasing organic content of domestic refuse in the U.K., due to the increase in paper packaging and the decrease in mineral dust and cinders (Staudinger, 1970). By 1975 the organic content of refuse amenable to composting will be approaching 70%. With improved processing techniques it should be possible to free municipal compost from most of its inorganic detritus. It can then be used without fear of pollution as top cover for tipping operations, for land reclamation, and either composted with, or pelletized with, additional NPK fertilizer as a soil conditioner/fertilizer for use in agriculture and horticulture.

TABLE II. Percentage composition of domestic refuse in the U.K. (Data from Staudinger, 1970)

Material	Date			
	1955	1960	1965	1975
Dust and cinders	53	51	36	18
Paper and cardboard	15	16	23	50
Vegetable and putrescible	12	12	17	13
Glass	6	6	8	6
Metal	6	6	7	6
Debris, including plastics	6	6	6	4

Accordingly composting, which satisfies the major oxygen demand of an unstable, putrescible, organic feedstock, is likely to be increasingly employed to convert a wide range of organic wastes—plant remains, animal manures, sewage sludge and the major part of municipal refuse—into a stable end product, humus, which is of value in agriculture.

B. Biochemical Aspects

The overall process flowsheet for the basic composting process is illustrated in Fig. 1. The heterogeneous organic material will normally have an in digenousmixed population of micro-organisms derived from the atmosphere, water or soil. Once the moisture content of the material is brought to a suitable level (50–60%) and it is aerated, microbial metabolism speeds up. For their growth and reproduction the micro-organisms require, in addition to oxygen and moisture, a source of carbon (the organic waste), macro-nutrients such as nitrogen, phosphorus and potassium, and certain trace elements. Energy is obtained by biological oxidation of part of the carbon from the waste. Some of this energy is used in metabolism, while the remainder is given off as heat.

Organic wastes suitable for composting vary from the highly heterogeneous material present in municipal refuse/sewage sludge to virtually

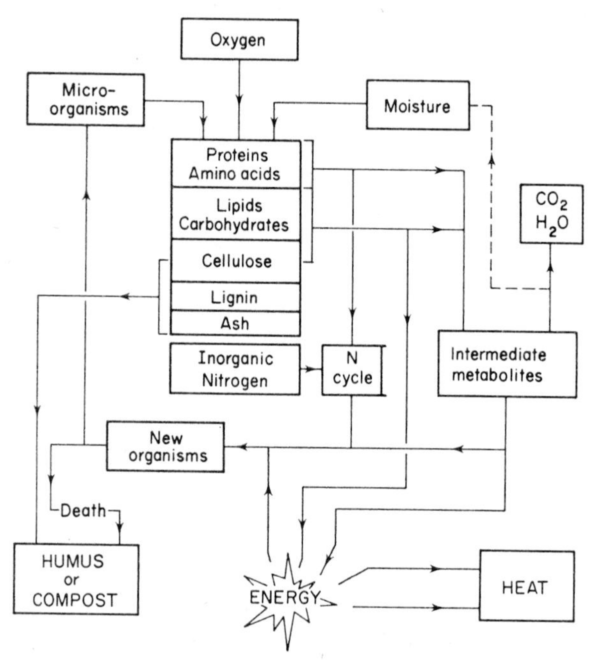

Fig. 1. The composting process.

homogeneous wastes from food processing plants. However, these wastes, whether of agricultural, industrial or municipal origin, are mixtures of sugars, proteins, fats, hemicelluloses, cellulose and lignin in a wide range of concentrations as shown in Table III.

In animals the manure composition depends on the type of animal, its feed, and, to a smaller extent, its age. In plants, the concentrations depend on the type of plant, its environment, and on its age. Fresh green material, such as young grass, contains a lot of water-soluble matter, minerals, and proteins which hold most of the nitrogen and sulphur of the plant. As the plant ages, minerals tend to return to the soil and low molecular weight compounds are converted to the polymers—hemicelluloses, cellulose and lignin.

Hemicelluloses are polysaccharides of about 50–150 sugar units which are fairly susceptible to attack by micro-organisms. Cellulose is a much larger polymer of about 1000–10,000 sugar units which is significantly more resistant. Lignin consists of a number of aromatic units linked by aliphatic side chains and is extremely resistant to enzyme attack, being the

TABLE III. Composition of organic matter (Data from Alexander, 1961; Gray, 1970)

Fraction	% in dry weight	
	Plants	Manures
Hot/cold water solubles—sugars, starches, amino acids, aliphatic acids, urea and ammonium salts	5–30	2–20
Ether/alcohol solubles—fats, oils, waxes and resins	5–15	1–3
Proteins	5–40	5–30
Hemicelluloses	10–30	15–25
Cellulose	15–60	15–30
Lignin	5–30	10–25
Minerals (ash)	1–13	5–20

last material degraded in composting. However, during composting the lignin molecule does become modified, losing some methoxy groups and aliphatic side chains and gaining carboxyl and phenolic hydroxyl groups (Alexander, 1961).

Composting is partly a breaking-down process and partly a building-up one. The key point is the cell wall of the micro-organism attacking the material. Low molecular weight materials (the water-solubles) can pass through the wall and then take part in cell metabolism, providing energy and being built up into the larger polymers. The high molecular weight components of the organic wastes cannot pass through the cell wall and are useless, as they stand, as food for the micro-organisms. In these cases the micro-organisms are able to exude extra-cellular enzymes which hydrolyse the polymers into short lengths which are the basic sugar units. Only a proportion of the micro-organisms can carry out this hydrolysis, but virtually all can assimilate the resulting fragments.

C. Microbiology

Aerobic thermophilic composting is a dynamic process which is brought about by the combined activities of a rapid succession of mixed microbial populations, each suited to an environment of relatively limited duration and each being active in the decomposition of one particular type or group of organic materials.

Some idea of the types and numbers of organisms involved in composting is given by Table IV.

TABLE IV. Organisms involved in composting (Data from Alexander, 1961)

	Genus	Numbers per g of moist compost
Microflora	Bacteria	10^8–10^9
	Actinomycetes	10^5–10^8
	Fungi	10^4–10^6
	Algae	10^4
	Viruses	?
Microfauna	Protozoa	10^4–10^5
Macroflora	Fungi, e.g. *Coprinus* spp.	
Macrofauna	Nematodes	
	Ants	
	Springtails	
	Millipedes	
	Worms	

1. Bacteria

Although the bacteria are present in vast numbers they are of small size and probably represent less than half the total microbial protoplasm. They normally need moist conditions for active growth but some types form endospores which can withstand heat and desiccation and then grow when environmental conditions are more favourable. Population changes among the bacteria during composting have not been extensively studied because of the large number of species involved and the difficulty in obtaining reproducible counts (Waksman *et al.*, 1939*a*; Webley, 1947; Yung Chang and Hudson, 1967).

2. Fungi

The thermophilic fungi in composting vegetable matter are a relatively well defined group. Fergus (1964), Pope *et al.* (1962) and Cooney and Emerson (1964) list at least eight species of thermophilic fungi indigenous to composts of hay, grass, garden wastes and stable manure which are capable of growth in the range 40–60°C. Above this temperature range they die out, reappearing, presumably by re-invasion, as the temperature falls below 60°C again. However, they prefer temperatures between 45°C and 50°C, and all of them show optimum growth at some point in this range.

The only detailed studies to date of the successional changes in fungal populations during a composting process are those published by Yung Chang and Hudson (1967), Yung Chang (1967) and Hedger (1972). These workers were able to divide the fungi that occurred into three characteristic groups. They concluded from their studies that the ability to use complex carbon sources and to thrive at high temperatures are the two important characteristics of the successful colonizers of composts.

The importance of fungi in the decomposition of cellulose in municipal wastes has been stressed by Gray (1959) who urged that the environment of composting masses should be adjusted to optimize the activity of these organisms.

3. *Actinomycetes*

The thermophilic actinomycetes in composts have not been studied as much as the fungi and their behaviour is much less predictable. However, it is known that they are capable of growth at higher temperatures than thermophilic fungi (Waksman *et al.*, 1939*b*).

Alexander (1961) postulated that the actinomycetes develop far more slowly than most bacteria and fungi and are rather ineffective competitors in the early stages of composting. They are more prominent in the later stages of the process when they become abundant or even dominant, sometimes to the extent that the surface of the composting mass takes on the white or grey colour typical of these organisms.

Some idea of the changes in numbers of the bacteria, actinomycetes and fungi during the composting of wheat straw is given in Fig. 2 which is taken from the work of Yung Chang and Hudson (1967).

Waksman and Cordon (1939) considered that no single pure culture could give as extensive and rapid decomposition as the total mixed microbial population of a compost.

D. Temperature-Time Pattern

Plant and animal remains are decomposed on the surface of the ground by psychrophilic and mesophilic organisms. When organic materials are gathered up into heaps, the insulating effect of the material leads to a conservation of heat and a rise in temperature. The maximum temperature reached, and the time taken to achieve it, depend on many process factors

—the composition of the organic wastes, the availability of nutrients, moisture content, size of heap, particle size, and the degree of aeration and agitation.

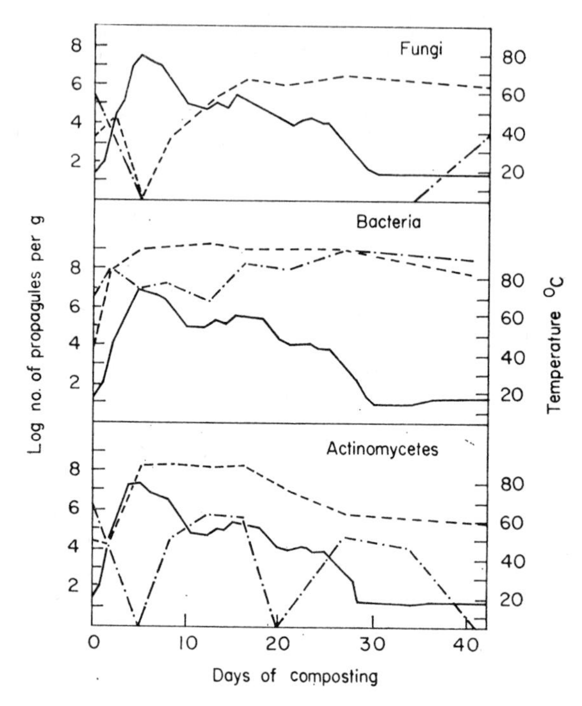

Fig. 2. Population changes in wheat straw composting (Yung Chang and Hudson, 1967.)
– – – thermophiles; —·—·— mesophiles; ——— temperature.

The typical shape of temperature–time and temperature–pH curves for the centre of a compost heap are shown in Fig. 3.

The process may be divided into four stages—mesophilic, thermophilic, cooling down and maturing.

At the start of the process the mass is at ambient temperature and is usually slightly acidic. As the indigenous mesophilic organisms multiply, the temperature rises rapidly. Among the products of this initial stage are simple organic acids; these cause the drop in pH shown in Fig. 3.

At temperatures above 40 °C the activity of the mesophiles falls off and the degradation is continued by the thermophiles; the pH becomes alkaline and ammonia may be liberated if excess readily available nitrogen is present. At 60 °C the thermophilic fungi die off and the reaction is kept going by the spore-forming bacteria and the actinomycetes. Forsyth and Webley (1948) showed that the cellulose and lignin fractions were scarcely attacked at temperatures over 60°C but waxes, proteins and hemicelluloses were readily degraded. As the rapidly degradable material becomes used up the reaction rate slackens, until eventually the rate of heat generation becomes less than the rate of heat loss from the surface of the heap and the mass starts to cool down.

In some instances, usually when composting old wastes, a number of temperature peaks occur. Yung Chang (1967), Rao and Block (1962) and Hedger (1972), all report examples of this phenomenon.

At this point in an adequately agitated heap the material is said to have reached "stability". The easily convertible materials have been decomposed, the major oxygen demand has been met, the material is no longer attractive to flies and rats and it will not give off bad odours, because the readily accessible nitrogen and sulphur have been bound up in new micro-organisms. In a mechanized composting plant this stage takes 3–5 days, after which time the material can be put into outside heaps without incurring serious risks to public health.

Once the temperature falls below 60°C the thermophilic fungi from the cooler margins of the mass can re-invade the heap centre and commence their major attack on the cellulose.

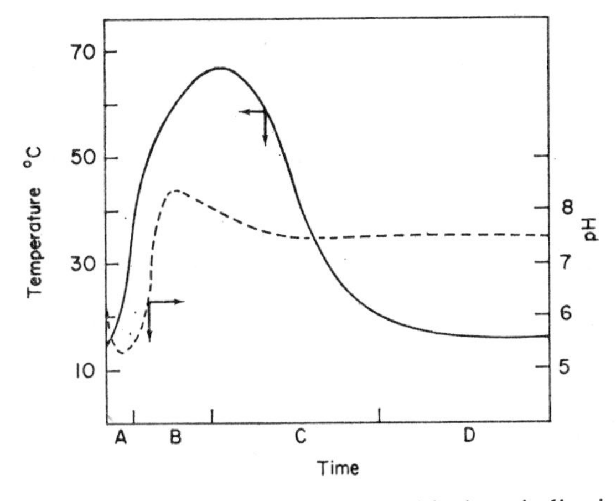

Fig. 3. Temperature and pH variation with time, indicating the stages of the composting process.

A, mesophilic; B, thermophilic; C, cooling down; D, maturing.

The hydrolysis and subsequent assimilation of the polymeric materials is a relatively slow process; hence the rate of heat generation decreases still further and the temperature falls towards ambient. At about 40°C the mesophilic organisms recommence activity, either from heat-resistant spores or by re-invasion from outside. The pH drops again slightly but usually remains on the alkaline side of neutrality.

The first three stages of the composting cycle take place reasonably quickly, being over in a matter of days or weeks. The final stage, maturing, normally requires a period of months. This takes place at ambient temperature with mesophilic organisms predominating and macrofauna appearing. Burman (1961) postulated that during the final stage of the composting process many antagonistic and predatory relationships develop among the macro and microflora and fauna; moreover, there is a pronounced formation of antibiotics which affect many of the organisms present. Heat evolution and weight loss are small. During this period complex secondary reactions of condensation and polymerization are taking place which give rise to the final end product, humus, and more particularly the stable and complex humic acids.

The composted material is said to have reached "maturity" when it will not warm up on turning, nor go anaerobic on storage, nor rob nitrogen from the soil when incorporated with it. The carbon to nitrogen (C:N) ratio of the product falls to about 10:1, the value present in the cells of micro-organisms.

There is a singular lack of information in the literature on the precise details of the biochemical changes taking place during these complex processes. This is not surprising, however, when the complexity and variability of the material, and the difficulty of obtaining homogeneous representative samples, are taken into account.

Some indication of the extent of the biochemical changes taking place in a heap is given in Fig. 4, plotted from the results of Yung Chang (1967), on composting wheat straw amended with ammonium nitrate. The straw had lost over half of its dry weight after 60 days of composting; this loss took place mainly in the first 34 days. The loss of total dry weight could be accounted for almost completely by the loss in hemicelluloses and cellulose. Cellulose degradation slowed down during the middle of the cycle, presumably because the fungi were killed off by the high temperatures. The hemicelluloses were broken down fairly steadily, however, since the hemicellulolytic actinomycetes are more tolerant of high temperatures. The ethanol-soluble fraction, which contained among other things, sugars and glucosides, only decreased from 2·30% to 1·41%, probably being continually replaced by breakdown of the polymers.

From Fig. 4 and the results of other workers, it is apparent that composting ceases after a time, even though large quantities of hemicelluloses and cellulose are still present. One explanation proposed is that the materials present in the original feed are exhausted and that the polymers determined by chemical analysis at this stage are those in the microbial protoplasm generated during composting. This view is probably only partly true. Other explanations, proposed by Stutzenberger *et al.* (1970) and Hedger (1972), are that the remaining polymers may be masked by lignin or that an accumulation of fungitoxic breakdown products, such as phenols, may occur and prevent further breakdown.

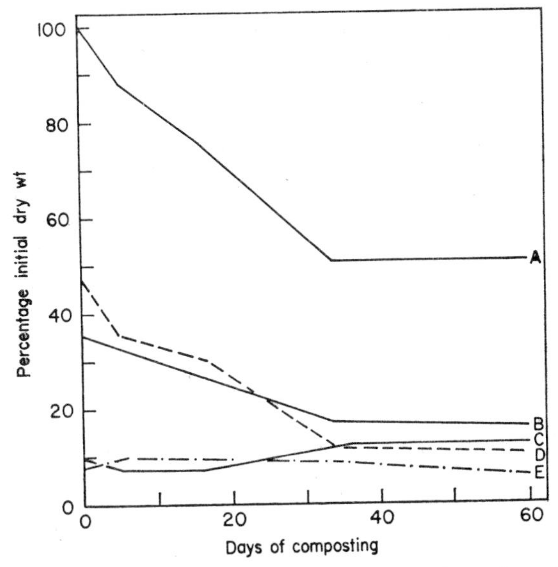

Fig. 4. The breakdown of wheat straw in composting, plotted from the data of Yung Chang (1967).

A, total; B, hemicelluloses; C, lignin; D, cellulose; E, total solubles.

It is not yet known what limits the rate of breakdown in composting when adequate air and moisture are present. This could be either the rate of enzyme attack on the polymers, or the rate at which the micro-organisms assimilate the resulting sugar fragments. The results of Stutzenberger *et al.* (1970) indicate that cellulase activity in composting is high, with a maximum in the region 65–70°C, above the temperature range for fungal activity.

The work of Yung Chang and Hudson (1967) was carried out in compost boxes. In large scale installations for treating municipal wastes, the natural process is frequently accelerated by the use of forced aeration and mechanical agitation; in one modern plant a high degree of maturity is claimed in a period of only seven days. With such accelerated processes the microbial population and the type of biochemical change may well differ from those in a slowly degrading compost heap.

II. Environmental Factors

A. Introduction

The composting of organic wastes is a dynamic and extremely complicated ecological process in which temperature, pH and food availability are normally constantly changing. In consequence, the numbers and species of organisms responsible also change markedly.

The rate of progress towards the relatively stable end product, humus, is dependent on a number of interrelated environmental factors. These include particle size, nutrients, structural strength of the material, moisture content, agitation, aeration, pH and heap size.

Nature is very obliging, so that if a mass of organic material is thrown into a heap it will eventually be turned into humus. However, it may take a very long time and may give off offensive odours *en route* due to anaerobic conditions. Accordingly, it is desirable to adopt the best operating conditions allowed by the economics of the operation in hand.

B. Preparation of Feed Material

A process flow diagram of an elaborate plant for composting municipal wastes is depicted in Fig. 5. This shows that several operations may be carried out before and after the fermentation stage. The degree of complexity of an actual plant, and the quality of the final product, depend chiefly on the level of capital investment that the local authority is prepared to make.

The diagram illustrates the stages of feed preparation which are often desirable—reception, separation, size reduction, moisture adjustment and nutrient addition.

For treating farm and garden wastes, and for preparing mushroom substrates, far simpler systems are employed.

1. Separation

It is in the long term interest of composting as a major waste disposal process that the final product, when used on the land, should be of as high a quality as possible, with a minimum of mineral matter. This is of particular importance when handling municipal wastes, the composts prepared from which can contain significant amounts of trace metals such as copper, lead, nickel and zinc. Consequently, with municipal wastes it is desirable to remove as much glass, metal, plastic and other debris as is economically possible and leave a substantially pure organic feedstock for composting. It appears to be preferable to remove the inorganic materials from the feed rather than from the final compost. Otherwise, free metals and salts are attacked by the micro-organisms and by the organic acids liberated early in the composting reaction; metal ions can then become bound into the humic acid and other organic complexes.

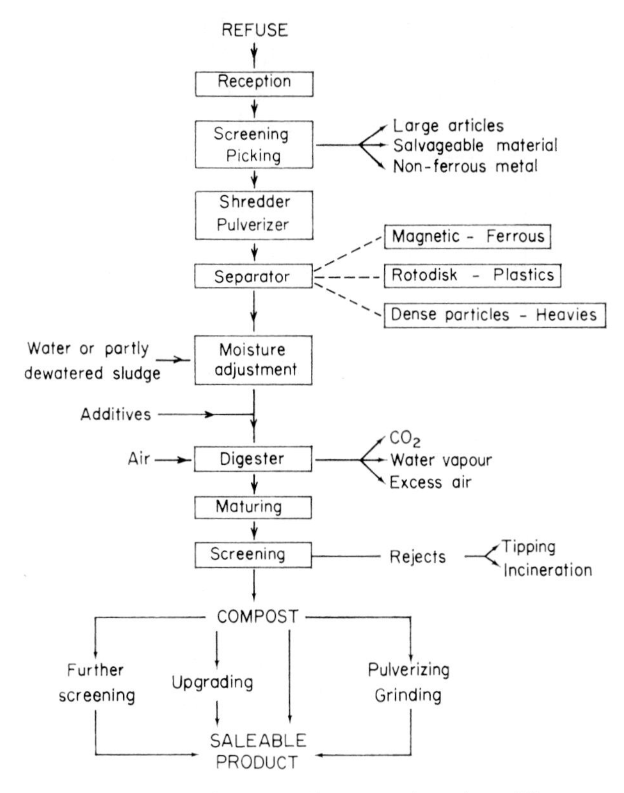

Fig. 5. Typical process flow diagram of composting plant (Gray *et al.*, 1971*b*).

2. *Size Reduction*

A fairly small initial particle size aids rapid decomposition by providing greater surface for microbial attack. It also reduces the depth for oxygen diffusion and microbial advance within the particle, aids the homogenizing of an initially heterogeneous material and improves insulation. On the other hand, too fine a particle size will involve very small inter-particle voids. These will impede the diffusion of O_2 to, and CO_2 from, the sites being attacked, especially during the thermophilic stage of composting when oxygen consumption is highest. This consideration is particularly important in heaps and windrows (elongated heaps) relying on natural aeration, rather than on a forced air supply.

practical values agree well with those deduced theoretically by Alexander (1961).

As shown in Table V taken from the data of Gotaas (1956), waste organic materials amenable to composting in fact have a wide range of nitrogen contents and C:N ratios. Some wastes, e.g. fresh grass clippings, contain more than adequate nitrogen while others, e.g. wheat straw and paper, contain very little and require supplementation for rapid decomposition.

TABLE V. Approximate nitrogen content and C:N ratios for some compostable materials (dry weight basis). (Data from Gotaas, 1956)

Material	N%	C:N
Urine	15–18	0·8
Blood	10–14	3
Mixed slaughterhouse wastes	7–10	2
Night soil	5·5–6·5	6–10
Activated sludge	5·0–6·0	6
Young grass clippings	4·0	12
Cabbage	3·6	12
Grass clippings (average–mixed)	2·4	19
Farmyard manure (average)	2·15	14
Seaweed	1·9	19
Potato haulms	1·5	25
Combined refuse	1·05	34
Oat straw	1·05	48
Wheat straw	0·3	128
Fresh sawdust	0·11	511
Newspaper	Nil	∞

Experimental data by Gray and Sherman (1970) (Fig. 7) illustrate the variation in reaction rate, as measured by CO_2 evolution, of various compostable materials. These variations reflect the availability or otherwise of readily degradable carbon substrate and of adequate nitrogen. Grass contains easily accessible nitrogen, sugars and cellulose; its C:N ratio when fresh is below 10. Young *Symphytum officinale* agg. (comfrey) is of similar composition. Wheat straw contains abundant cellulose but inadequate nitrogen; its C:N ratio is approximately 128. Sawdust contains less cellulose than straw, more lignin and very little nitrogen, giving a C:N ratio of about 500.

In practice, some composting plants handling municipal refuse use sewage sludge as a nitrogen source (Harrison, 1965; Shirrefs, 1965); one plant even employed gasworks liquor containing ammonia (Scott, 1961).

Phosphorus is the nutrient next in importance to nitrogen. Alexander (1961) stated that the N:P ratio in microbial cells is between 5:1 and 20:1, while *Azotobacter* need 1 part of phosphorus for every 5–6 parts of nitrogen that they fix from the atmosphere. Dhar (1956, 1959) was convinced of the efficacy of extra phosphatic material in composting and, according to Wylie (1960), advocated the addition of 1 part of P_2O_5 per 100 parts of

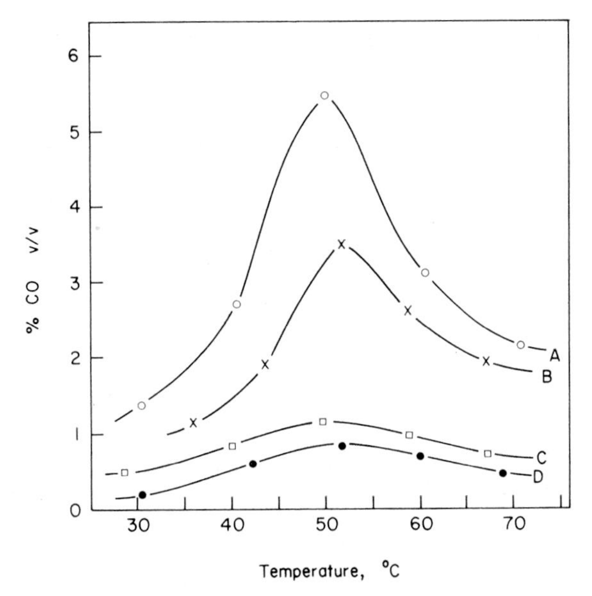

Fig. 7. Variation of decomposition rate with temperature for materials of various C:N ratios (Gray and Sherman, 1970).
A, grass; B, comfrey; C, straw; D, sawdust.

organic matter. This view is reinforced by that of Gessel (1960) who suggested the use of 15·9 kg of superphosphate per tonne of dry feed when composting wood wastes.

In the preparation of composts for mushroom-growing from horse manure, wheat straw or other crop residues and inorganic fertilizers containing N, P, K, Ca and S are frequently employed, together with an activator such as spent brewer's grains or molasses (Hayes and Randle, 1969; Smith, 1969; Ministry of Agriculture, Fisheries and Food, 1960).

In the opinion of the present authors, it would appear advantageous to optimize the concentrations of the N and P in the feed to a compost plant. This would result in a faster rate of degradation, a larger quantity of product humus due to less carbon substrate being oxidized to CO_2, and a higher nutrient value of the final product. Any extra nutrients required could be added to the feed material as sewage sludge, suitable organic wastes, basic slag or mineral or artificial fertilizers.

The other nutrients are required in substantially smaller quantities than the N and P; they would normally be present in sufficient amount in mixed waste materials of agricultural or municipal origin.

4. Other Additives: Inocula

In some circumstances the addition of other materials to the organic waste may be necessary. Soil is sometimes added as a diluent in order to reduce the peak temperature and thereby limit the loss of ammonia.

When the feed has little structural wet strength or is sticky, the addition of stiffer material may be necessary in order to hold open passages for air movement.

Exposed compostable materials normally contain a large number of different types of bacteria including actinomycetes, fungi, and other living organisms. They also usually contain, at least in admixture, all the necessary macro- and micronutrients to enable the composting reactions to take place.

Nevertheless, there has been, and undoubtedly still is, controversy over whether the inoculation of such wastes with either bacterial cultures or herbal or chemical promoters is of any assistance in hastening decomposition or in producing a better end product (Golueke *et al.*, 1954; Burman, 1961; Bruce, 1967).

The subject of the worth of additives and inocula is still unproven either way. The process of composting is so complicated, and the currently available techniques for analysing biochemical changes and compost quality are so rudimentary, that the subject may remain contentious for many years.

5. Moisture Content

Micro-organisms require a certain amount of water for their metabolism, in which it is used as a transport medium for soluble feed materials and waste products from the reaction. Organic matter can absorb a considerable amount of water, and even at 50% moisture content the particle surface often appears dry.

The upper level of moisture conducive to good composting depends greatly on the structural wet strength of the organic feed. Weak materials, e.g. paper, collapse readily on composting; in the presence of excess moisture the pores fill with water, oxygen diffusion is restricted and anaerobic conditions set in. Rigid materials, e.g. straw, retain their wet strength for a long time, until most of the wall fibres have been degraded; consequently they can be composted with initial moisture contents of as much as 85%.

In practice the optimum range is 50–70%, the lower figure applying to composting in static heaps or windrows and the upper figure being permissible in mechanically agitated systems with forced air injection.

Experimental work to evaluate this optimum range when composting ground refuse has been reported by Wiley and Pearce (1955), Snell (1957) and Spohn (1968). Wiley and Pearce concluded that optimum values lay in the range 55 to 69%. Snell, with slightly more detailed work, narrowed the range to 52–58%. Spohn reported an optimum value of about 50%.

In a subsequent paper, Wiley (1957*b*) noted that lipids—fats, oils and waxes—which are present in significant amounts in most organic wastes, are liquids at optimum composting temperatures. Accordingly, he postulated that they should be considered, together with the water content, in assessing optimum wetness for decomposition. He proposed the term "percent liquid" as being of better applicability than moisture content, where

$$\% \text{ Liquid} = \frac{100 \times (\% \text{ Moisture} + \% \text{ Lipids})}{(100 - \% \text{ Ash})}$$

During the composting process moisture is released as one of the end products of microbial metabolism. If aeration is poor in a compost heap, then this moisture can accumulate and lead to the onset of anaerobic conditions.

Where there is reasonable aeration by diffusion and the climate is temperate, little extra moisture is necessary on occasional turning of the heap. In hot climates, however, rewetting is certainly required.

In plants with forced aeration, desiccation will normally occur. Using an air flow of 0·6 m^3 per day per kg of volatile solids, approximately the minimum rate consistent with good aerobic conditions, then about 16% of the initial moisture content will be lost over 5 days if the exit gas is at 50°C. With air flows 5 times the minimum rate, which is quite reasonable practice, then considerable desiccation will occur and frequent moisture addition will be needed.

C. Agitation

The composting process can be speeded up by the judicious use of agitation. Movement of the material aids aeration, especially in heaps and windrows, introducing a fresh supply of air into the middle of large masses where diffusion alone has been insufficient to maintain high O_2 and low CO_2 levels. Agitation assists homogeneity of the composting mass, aiding an even spread of organic materials and nutrients. It will cause particles to rub together, leading to abrasion, size reduction and the exposure of unattacked material. It assists the uniformity of temperature, preventing overheating in the centre of large masses, and cooling at exposed surfaces. Some degree of agitation is vital when handling materials bearing pathogens, in order to expose all of the latter to thermal destruction.

On the other hand, too much agitation is to be avoided as it can lead to excessive loss of heat and moisture at the surfaces. Additionally, it can

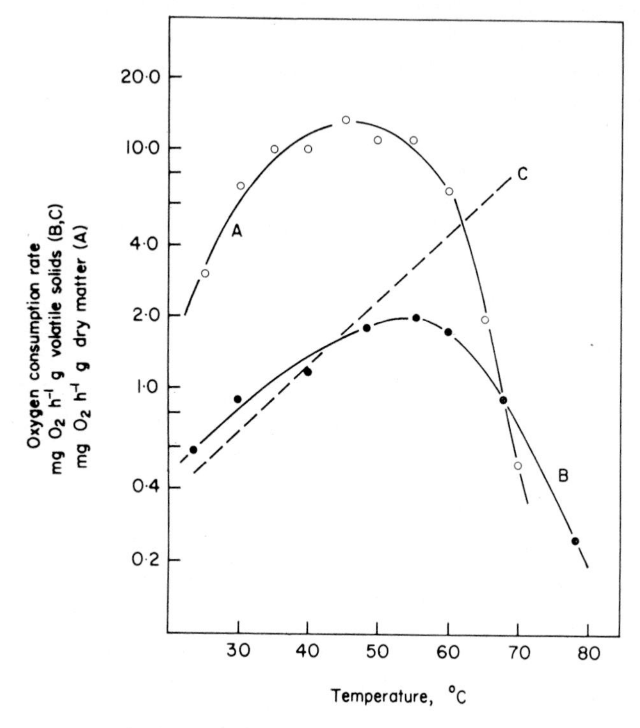

Fig. 8. The effect of temperature on oxygen consumption rate. A, from Snell (1957); B, from Gray and Sherman (1970); C, from Schulze (1960, 1962).

cause shearing of both fungal and actinomycete mycelium, thereby reducing the rate of metabolism and the degradative ability of these organisms. However, little research has been reported on the effect of agitation on the rate of composting.

Figure 8, which is considered in detail later, shows *inter alia* the effect of agitation on oxygen consumption rates in composting. Curve A depicts data from the continuously agitated reactor of Snell (1957), while curve B gives the data from an unagitated composter (Gray and Sherman, 1970). Figure 8 shows that, at the optimum temperature range of 40–55°C, the oxygen consumption, and hence the composting rate, is considerably faster in the agitated system.

Kaibuchi (1961) reported data on a small-scale mechanized pilot plant with horizontal ploughs revolving at 0·08 rpm. On a small rotary drum composter, Stoller *et al.* (1937) employed 0·13 rpm. The small drum of Schulze (1960) made 2 revolutions in 5 min at $\frac{1}{2}$ h intervals, while his continuous, batch-fed, machine of 250 l capacity was turned at 1 rpm for 5 min before and after each daily feeding (Schulze, 1962). Jeris and Regan (1968) rotated their small drum at 0·8 rpm for 4 min every 6 h; they believed that intermittent rotation was better than continuous movement.

In trials on a rotary drum "Fermascreen" without forced aeration, Sharma (1966) tried several agitation techniques, deducing the best to be periodic agitation for 5 min, either whenever the O_2 level at the centre of the mass fell below 10%, or when the centre temperature rose over 60°C (Gray, 1966; Gray and Sherman, 1970).

In full size practice, Dano Biostabilizer drums are turned continuously at 1 rpm during daytime, when they are being fed, and at 0·25 rpm during the night.

It appears that for rapid degradation, the concept of periodic agitation is the best approach. At the start of composting, immediately following pulverization, there is no need to agitate while temperatures and oxygen consumption rates are low. During the thermophilic stage and at the start of cooling, sufficient agitation should be provided to maintain high oxygen levels close to the surfaces being degraded and to keep the mass temperature below 60°C. During the remainder of the cooling stage, and in the maturing one, the need for agitation is to reduce particle size and expose fresh surfaces. Periodic disturbances of particles at intervals of several hours enables the actinomycetes and fungi to extend their mycelium for a reasonable time before disruption; such agitation is thus a compromise between the biological and the physical needs of the process.

D. Aeration

Adequate aeration is vital in composting in order to supply oxygen to the micro-organisms and to sweep away the major waste products, CO_2 and H_2O. Insufficient or maldistributed aeration leads to the onset of anaerobic conditions, with a consequent decrease in the rate of decomposition and an evolution of putrefactive odours. Too much air leads to a cooling down of the composting mass and some degree of desiccation; normally this is bad practice. However, when composting large masses of material, some cooling may be required at times in order to keep the peak temperatures within the range where fungal activity is possible, while when composting very wet feedstocks, such as animal manure slurries, the removal of water is an advantage.

When composting materials in heaps, windrows and bins using natural aeration, the diffusion of the O_2 and CO_2 results from the difference in partial pressure of each component in the atmosphere and at the site being degraded. This gas movement is opposed by the pressure drop caused by the narrow tortuous passages through the heap. It has been shown by McCauley and Shell (1956) and by Snell (1957) that this diffusion is quite inadequate to maintain good aerobic conditions during the period of peak oxygen demand, even allowing for the natural "chimney" effect with the gases being warmed in the heap, ascending and creating a draught. Accordingly, compost heaps are frequently turned to assist aeration (Kochtitzky, 1969; Gotaas, 1956; Wiley and Spillane, 1961). Improvements in natural aeration are also mentioned by Spohn (1970*a*), who advocated the building up of windrows in several stages and the use of a side-arm turning machine, and by Maier (1957), who described the mounting of heaps on perforated grids.

Forced aeration is frequently employed to speed up the rate of composting. Kershaw (1968) and Livshutz (1964) mentioned the use of perforated pipes laid under windrows, while Snell (1957) and Kaibuchi (1961) described forced aeration in bins and mechanical digesters. In an excellent paper, Spohn (1970*a*) discussed the problems arising from forced aeration of unagitated boxes—the maldistributions of air, moisture and temperature. His solution was to suck air down through the composting mass in periodic bursts. Between bursts the oxygen in the inter-particle voids fell until a pre-set level was reached, at which aeration was recommenced. The practice appeared to lead to good results.

1. Oxygen Consumption Rates

Experimental data on aeration rates and oxygen consumption have been reported by, among others, Wiley and Pearce (1955), of the U.S. Public Health Service, Schulze (1960, 1962) and Snell (1957) of Michigan State University, and Sharma (1966) and Gray and Sherman (1970) of the University of Birmingham (Fig. 8).

The results of Snell (1957) were obtained using 0·45 kg samples of ground garbage at 56% initial moisture content in constantly agitated, heated digesters. The figures plotted in Fig. 8 represent the maximum O_2 consumption reached in a run of 500 h at the appropriate temperature. In practice, only the experiment at 45 °C achieved a high degree of composting in this time, while the material in the runs at 65° and 70 °C dried out before the end. It is obvious from the curves plotted in Snell's actual paper that the oxygen consumption was not constant under isothermal conditions. For instance, during the run at 45°C the oxygen uptake started at 4·5 mg O_2 h^{-1} g dry wt, reached a peak at 13·7 after 200 h, and fell to 7·7 after 500 h. (Snell's results were reported on the basis of dry matter, not of volatile solids as are those of Wiley, Schulze and Gray. Accordingly, on a strictly comparable basis, Snell's data could be up to 50% higher than those shown in Fig. 8).

The data of Schulze were obtained from runs on an intermittently rotated horizontal drum containing about 7·3 kg of fresh ground garbage at 50–60% moisture content (Schulze, 1960), and on a larger, but similar, drum holding about 61·2 kg of material to which daily additions and subtractions were made (Schulze, 1962). From all his results, Schulze postulated that the oxygen consumption, y, in mg h^{-1} g of volatile solids, is related to the temperature, T°C, by an equation of the form

$$y = a \, . \, 10^{KT}$$

He calculated the constants to be $a = 0{\cdot}1$ and $K = 0{\cdot}028$ over the experimental temperature range 20–70°C. His equation is drawn in Fig. 8.

The results reported by Gray and Sherman (1970) over a slightly wider temperature range, came from experiments using 0·45 kg samples of equal parts of straw and grass, at 55–60% moisture content, in an unagitated, heated column. In the results shown in Fig. 8, the feed was well disintegrated in a macerator; in another set of experiments the feed was undisintegrated and lower values for oxygen uptake were obtained (see Fig. 6).

A plot by Schulze (1960) of oxygen uptake rates during an 11-day cycle showed clearly that, at a particular temperature, the uptake was significantly higher while warming up through the mesophilic and thermophilic stages than during the cooling stage.

From all these results it is apparent that the O_2 consumption at a particular instant in a composting cycle depends, not only on temperature, but on degree of agitation, feed material composition, particle size and whether the heap is warming up or cooling down. The consumption appears to increase logarithmically with temperature to a maximum within the range 45–55°C and then falls, also logarithmically, as the thermophilic fungi are killed off and the spore-forming bacteria and actinomycetes assume the dominant role.

2. *Aeration Rates*

Wiley and Pearce (1955), studied the effect of different aeration rates in constantly agitated vessels containing approximately 16·3 kg of ground garbage of 55% moisture content. The efficacy of aeration was judged from the temperature pattern during the 9-day cycle and the state of decomposition at the end. Low aeration rates, 0·24–0·40 m^3 air day^{-1} kg volatile solids, resulted in a late peak temperature and incomplete degradation; medium rates of 0·56–1·58 gave a maximum temperature after 4 days and a slow decline thereafter, yielding a satisfactory product. High rates, 2·07–4·88 yielded a peak temperature at day 4, followed by a rapid decline; a poor product resulted because cooling and desiccation had occurred. Converted to the same units as quoted earlier, these aeration rates correspond to (2·6–4·3), (6·1–19·6) and (21·8–51·8) mg O_2 supplied h^{-1} g volatile solids. Accordingly, the figures for the most satisfactory operation, at medium aeration, correspond closely with the values for peak oxygen consumption in a constantly stirred reactor quoted by Snell (1957) (Fig. 8).

E. Heat Production

Wiley (1957*a*) studied the heat production when pulverized refuse was composted and concluded that, over 8–10 day cycles, it amounted to approximately $7{\cdot}0 \times 10^6$ J kg^{-1} of initial volatile solids or $23{\cdot}2 - 27{\cdot}9 \times 10^6$ J kg^{-1} of volatile solids decomposed.

By employing very well insulated compost heaps or large masses of wastes, such a heat production can lead to high temperatures in the range 80–90°C (Rao and Block, 1962; Spohn, 1970*a*,*b*). Most experimental data, however, indicate that at such temperatures the reaction rate is low. Optimum rates are obtained at temperatures of about 55°C, at which the thermophilic fungi are exerting their maximum influence.

Provided that all the composting material is held at 55°C for several days, the destruction of pathogenic organisms should be assured. This destruction is most necessary when composting night soil, raw sewage sludge or animal manure slurries.

F. pH Control

As discussed earlier, and illustrated in Fig. 3, the pH of a composting mass changes from acid to alkaline. Several workers have investigated the effect of pH control.

Addition of acetic acid to the organic wastes appears to retard the thermophilic stage of the process (Wiley, 1956). Addition of significant amounts of calcium carbonate or calcium hydroxide speeds up the reaction and results in greater decomposition and loss of nitrogen, presumably as ammonia (Wiley, 1956; University of California, 1953).

However, the usual object of composting is to maximize both the quantity of product humus and its nitrogen content. These aims far outweigh any consideration of marginally faster rate of reaction obtained by the use of excessive quantities of alkali. Accordingly, it would appear that, for most composting feedstocks, control of pH is neither necessary nor desirable.

G. Heap Size

Small heaps have high surface/volume ratios and hence much of the material has to act as insulation. At least 1 tonne of wastes is needed to ensure that a reasonable proportion of the heap reaches a satisfactory temperature. In cold climates some 3 tonnes are necessary to reach thermophilic temperatures.

For heaps composting under natural aeration conditions the material should not be piled over 152 cm high or 244 cm wide, otherwise the diffusion of air to the centre will be seriously impeded. The heap can be elongated into a windrow of any convenient length.

H. Resumé of Ideal Conditions

There are now sufficient experimentally determined data in the literature on the microbiological, chemical and physical parameters of composting for reasonably accurate process design of composting plants. Optimum values of the important parameters are summarized in Table VI.

The outstanding problem is how to translate these parameters into a low cost, but rugged and reliable, process plant.

III. Practical Processes

The composting of organic waste materials has been practised for many centuries by farmers and gardeners in many parts of the world. Probably the outstanding example has been that of the Chinese in the river deltas who, by returning to the soil their crop residues, human wastes and alluvial mud swept down the rivers and canals, have been able to support very high population densities for many hundreds of years.

A. Historical

Composting, as practised by the Chinese, had changed very little over the centuries, being essentially a small-scale batch operation in heaps. With the adoption of the process this century by the Western world, some progress has been made in the understanding of the fundamental reaction and its application to large-scale waste treatment.

The interest in composting in the West probably stems from an extended visit to China, Japan and Korea in 1909 by Professor F. H. King (1927)

TABLE VI. Optimum values of major composting parameters
(Data from Gray *et al.*, 1971*b*)

Parameter	Value
C:N ratio of feed	30–35:1
C:P ratio of feed	75–150:1
Particle size	1·3–3·3 cm for agitated plants and forced aeration 3·3–7·6 cm for windrows, unagitated plants and natural aeration
Moisture content	50–60%
Air flow	0·6–1·8 m^3 air day^{-1} kg volatile solids during thermophilic stage, being progressively decreased during cooling down and maturing
Maximum temperature	55°C
Agitation	Short periods of vigorous agitation, alternating with periods of no agitation which vary in length from minutes in the thermophilic stage to hours during maturing
pH control	Normally none desirable
Heap size	Any length but not over 152 cm high or 244 cm wide for heaps and windrows using natural aeration. With forced aeration, maximum size depends on need to avoid overheating.

of the U.S. Department of Agriculture. His carefully recorded observations were used by Sir Albert Howard, a British botanist employed by the Indian Government. After several years of experiment, Howard found that his Indore method of heap composting gave optimum results in terms of the vegetable and animal wastes, the supply of labour and the climatic conditions available in his district (Howard and Wad, 1931).

In the years immediately following 1931 the Indore method was taken up by plantation industries, farms and gardens in many parts of the world. As a result, a number of minor changes were incorporated which increased the output per man from the process. This improved form involved pits 91 cm deep, vegetable wastes being put on in 15 cm layers, followed by manure in 5 cm layers, a thin sprinkling of earth and wood ashes, and watering. The layering was continued until a height of 152 cm was reached. Vertical aeration vents, about 10 cm in diameter, were then made with a crowbar every 91–122 cm across and down the heap of wastes. The latter were turned twice, after 2–3 weeks and after about 5 weeks, the material being ready in 90 days (Howard, 1943).

Shortly before Howard's compost work in India, Hutchinson and Richards (1921) carried out extensive investigations at Rothamsted

Experimental Station on the production of artificial farmyard manure. They successfully composted straw with chemical additives and patented their process and activator under the name of "Adco".

B. Mechanical Processes

The modern approach to composting has arisen in response to a need for controlled and hygienic disposal processes to deal with the vast quantities of solid organic waste.

The past three decades have seen the introduction of some 20–30 different processing schemes for composting municipal wastes, each with varying degrees of success (Gray *et al.*, 1973). All the processes comprise some of the stages given earlier in the typical process flow diagram (Fig. 5).

Equipment for feed preparation and product finishing are very similar in most of these processes. The major difference arises in the fermentation stage, which has been attempted in all manner of pits, heaps, cells, bins, digesters, silos and drums. This variety is a reflection on the lack of research into the subject, or incomprehension of the research results, and the difficulties in following the course of the reaction by meaningful chemical or physical parameters and in evaluating the quality of the product.

Although often practised because of its cheapness, refuse composting in windrows with natural aeration is a very inefficient process. Little control is possible over temperature distribution, aeration and moisture content, while space requirement is large and adequate gaps must be allowed between windrows for turning operations.

Many of the defunct methods, and even some of the existing methods of refuse composting appear to have been designed solely on a materials handling basis. They have made little real attempt to provide for the needs of the microbial population. During the 1950s new designs were brought out which attempted to take note of at least some of the major parameters (Davies, 1961). These plants mainly attempted to improve aeration, either by intermittent agitation or by forced air supply, and used rotating drums, vertical digesters or multideck houses (Fig. 9). In recent years some new designs are starting to show a keener appreciation of the *modus operandi* of composting. Two such examples are Spohn's pulsed aeration method and the Nusoil digester (Fig. 10), both of which claim to reach full compost maturity without the necessity for windrowing (Pittwood, 1970; Spohn, 1970*b*, 1972).

For the composting of farm and garden wastes, and the production of mushroom substrates, relatively crude processing schemes are still employed. Agitation is rare but simple forced aeration is sometimes used (Giffard, 1972; Biddlestone and Gray, 1973).

Fig. 9. Multi-deck compost house for 350 t day^{-1} of municipal refuse, Bangkok. (Reproduced by permission of Clarke Chapman–John Thompson Ltd.)

Fig. 10. Module of Nusoil Digester for municipal refuse. (Reproduced by permission of Nusoil Ltd.)

C. Compost Composition

Typical composition ranges of compost produced from municipal and farm wastes are given in Table VII (Gotaas, 1956). Municipal compost tends to be higher in ash, and lower in organic matter and the major nutrients, than compost prepared from farm or garden wastes.

TABLE VII. Chemical analyses of composts (Data from Gotaas, 1956).

	% by weight, dry basis		
Substance	Municipal Composts		Garden/farm Composts
Organic matter	25	to	50
Carbon	8	to	50
Nitrogen (as N)	0·4	to	3·5
Phosphorus (as P_2O_5)	0·3	to	3·5
Potassium (as K_2O)	0·5	to	1·8
Calcium (as CaO)	1·5	to	7·0
Ash	65	to	20

Composts vary quite widely in composition. Those prepared from municipal wastes have analyses tending to the left of the table, those from garden/farm wastes have compositions tending to the right.

In addition to these major constituents, composts, especially those prepared from municipal wastes, contain many trace elements. Although some of these are essential to plant life, there is some concern lest heavy metal trace elements should be present in excess and lead to a build-up of toxicity in the soil (Gray, 1968).

D. Compost Uses

The major use of composting is likely to be in torrid countries in which the rate of oxidation of soil humus is far higher than in temperate climes. Desert sands can be stabilized and eroded lands brought back into cultivation by the importation and spreading of humus. Tree belts can be established in compost to reduce wind erosion and to influence the local micro-climate. For many underdeveloped countries, unable to afford the Western aids to agriculture, the careful composting and return to the soil of plant and animal wastes is the only present hope for improving crop production.

In the U.K. the major application could well be in the treatment of animal manure slurries by conjoint composting with straw and other solid wastes. The application of composting to meeting the disposal of municipal wastes in the U.K. is likely to be relatively small compared with incineration (Fulbrook *et al.*, 1973). However, with the possible organization of refuse disposal on a county basis there are likely to be more plants installed, possibly a central one in each county, in order to supply top dressing for the reclamation of derelict and waste lands, for landscaping and use in parks and sportsfields. Appropriate uses for such municipal compost in agriculture in the United States and in Germany are described by Tietjen and Hart (1969). Another area for the use of compost lies in meeting the rising demand for substrates for mushroom growing.

Tests have been carried out using compost as a supplementary fuel in cement kilns. The compost has a stable calorific value while the resultant ash merely forms part of the final clinker which is finally ground to give the cement product (Skitt, 1972).

Another application of compost lies in the manufacture of composition boards for building. Careful control of the composting process can result in compost with fairly long fibres suitable for admixture with wood particles in making low density boards (Skitt, 1972).

There is undoubtedly immense scope for improvements in composting methods, in the production of good quality products and in transportation and spreading of the compost when used in agriculture.

IV. Further Information

The subject of composting is vast and complicated, with numerous facets from many academic and practical disciplines. For further information the reader is referred to the World Health Organization monograph on the subject by Gotaas (1956), the appropriate chapter in the recent Report of the Working Party on Refuse Disposal (1971), and the literature review by the Compost Studies Group at the University of Birmingham (Gray *et al.*, 1971*a*,*b*, 1973).

Acknowledgements

The authors gratefully acknowledge the financial support of the Agricultural Research Council, the Jack and Mary Pye Trust, several municipal corporations and several compost plant manufacturers, for research on various aspects of composting.

References

ALEXANDER, M. (1961). "Introduction to Soil Microbiology." John Wiley, New York.

ARDITI, A. (1967). M.Sc. Thesis, University of Birmingham.

BIDDLESTONE, A. J. and GRAY, K. R. (1973). *The Chemical Engineer* **270,** February, 76–80.

BRUCE, M. E. (1967). "Common Sense Compost Making." Faber and Faber, London.

BURMAN, N. P. (1961). *In* "Town Waste put to Use." (P. Wix, ed.), pp. 113–130. Cleaver-Hume Press, London.

BURMAN, N. P. (1969). Private communication.

COONEY, D. G. and EMERSON, R. (1964). "Thermophilic Fungi." Freeman, San Francisco.

DAVIES, A. G. (1961). "Municipal Composting." Faber and Faber, London.

DHAR, N. R. (1956). *Proc. natn Acad. Sci. India* **25** (A, 4), 211.

DHAR, N. R. (1959). *Nature, Lond.* **183,** 513.

FERGUS, C. L. (1964). *Mycologia* **56,** 267–284.

FORSYTH, W. G. C. and WEBLEY, D. M. (1948). *Proc. Soc. appl. Bact.* **3,** 34–39.

FULBROOK, F. A., BARNES, T. G., BENNETT, A. J., EGGINS, H. O. W. and SEAL, K. J. (1973). *Surveyor* 12th January, pp. 24–27.

GESSEL, S. P. (1960). *Compost Sci.* **1,** (1), 26–29.

GIFFARD, W. H. (1972). *J. Soil Asscn* **77,** (1), 27–31.

GOLUEKE, C. G., CARD, B. J. and MCGAUHEY, P. H. (1954). *Appl. Microbiol.* **2,** 45–53.

GOTAAS, H. B. (1956). "Composting." Monograph No. 31, World Health Organization, Geneva.

GOWAN, D. (1972). "Slurry and Farm Waste Disposal." Farming Press, Ipswich.

GRAY, K. R. (1966). *Br. chem. Eng.* **11,** 851–853.

GRAY, K. R. (1968). *Public Cleansing* **58,** 331–334.

GRAY, K. R. (1970). Unpublished data.

GRAY, K. R. and SHERMAN, K. (1970). *Public Cleansing* **60,** 343–354.

GRAY, K. R., SHERMAN, K. and BIDDLESTONE, A. J. (1971*a*). *Process Biochemistry* **6,** (6), 32–36.

GRAY, K. R., SHERMAN, K. and BIDDLESTONE, A. J. (1971*b*). *Process Biochemistry* **6,** (10), 22–28.

GRAY, K. R., BIDDLESTONE, A. J. and CLARK, R. (1973). *Process Biochemistry* (in press).

GRAY, W. D. (1959). *Rhodesian Engr* Sept., 30–33.

HARRISON, F. (1965). *J. Inst. Sew. Purif.* (6), 525–531.

HAYES, W. A. and RANDLE, P. E. (1969). *Rep. Glasshouse Crops Res. Inst.* 1968, 142–147.

HEDGER, J. N. (1972). Ph.D. Thesis, University of Cambridge.

HOWARD, A. and WAD, Y. D. (1931). "The Waste Products of Agriculture." Oxford University Press, London.

HOWARD, A. (1943). "An Agricultural Testament." Oxford University Press, London.
HUTCHINSON, H. B. and RICHARDS, E. H. (1921). *J. Minist. Agric. Fish.* **28**, 398–411.
JERIS, J. S. and REGAN, R. W. (1968). *Compost Sci.* **9**, (1), 20–22.
KAIBUCHI, Y. (1961). *Proc. ASCE, J. San. Eng. Div.* **87**, Paper SA6, 101.
KERSHAW, M. A. (1968). *Process Biochemistry* **3**, (5), 53–56.
KING, F. H. (1927). "Farmers of Forty Centuries." Jonathan Cape, London.
KOCHTITZKY, O. W. (1969). *Compost Sci.* **9**, (4), 5.
LIVSHUTZ, A. (1964). *World's Poultry Sci. J.* **20**, 212–215.
MAIER, P. P. (1957). *Proc. 12th Industrial Waste Conf., Purdue University*, Series 94, 590–595.
MCCAULEY, R. F. and SHELL, B. J. (1956). *Proc. 11th Industrial Waste Conf., Purdue University*, Series 91, 436–453.
MINISTRY OF AGRICULTURE, FISHERIES AND FOOD (1960). "Mushroom Growing." Bulletin No. 34, H.M.S.O., London
PITTWOOD, A. S. (1970). Private communication.
POPE, S., KNAUST, H. and KNAUST, K. (1962). *Proc. 5th Int. Conf. on Sci. Aspects of Mushroom Growing, Philadelphia.*
RAO, S. N. and BLOCK, S. S. (1962). *Developments in Ind. Microbiol.* **3**, 326–340.
SCHULZE, K. L. (1960). *Compost Sci.* **1**, (1), 36–40.
SCHULZE, K. L. (1962). *Appl. Microbiol.* **10**, 108–121.
SCOTT, J. (1961). *In* "Town Waste put to Use." (P. Wix, ed.), pp. 29–52. Cleaver-Hume Press, London.
SHARMA, A. C. (1966). M.Sc. Thesis, University of Birmingham.
SHIRREFS, W. R. (1965). *Municipal Eng.* p. 2705.
SKITT, J. (1972). "Disposal of Refuse and Other Waste." Charles Knight, London.
SMITH, J. (1969). *Process Biochemistry* **4**, (5), 43–52.
SNELL, J. R. (1957). *Proc. ASCE, J. San. Eng. Div.* **83**, Paper 1178.
SPOHN, E. (1968). *Städtehygiene* (6).
SPOHN, E. (1970*a*). *Wass. Abwass.* (5).
SPOHN, E. (1970*b*). *Compost Sci.* **11**, (3), 22–23.
SPOHN, E. (1972). *Compost Sci.* **13**, (2), 8–11.
STAUDINGER, J. J. P. (1970). "Disposal of Plastic Waste and Litter." Monograph No. 35. SCI, London.
STOLLER, B. B., SMITH, F. B. and BROWN, P. E. (1937). *J. Am. Soc. Agron.* **29**, 717–723.
STUTZENBERGER, F. J., KAUFMAN, A. J. and LOSSIN, R. D. (1970). *Can. J. Microbiol* **16**, 553–560.
TIETJEN, C. and HART, S. A. (1969). *Proc. ASCE, J. San. Eng. Div.* **95**, Paper SA2, 269–287.
UNIVERSITY OF CALIFORNIA, BERKELEY (1953). "Reclamation of Municipal Refuse by Composting." Technical Bulletin No. 9, Series 37.
WAKSMAN, S. A., CORDON, T. C. and HULPOI, N. (1939*a*). *Soil Sci.* **47**, 83–113.

WAKSMAN, S. A., UMBREIT, W. W. and CORDON T. C. (1939*b*). *Soil Sci.* **47**, 37–61.
WAKSMAN, S. A. and CORDON, T. C. (1939). *Soil Sci.* **47**, 217–225.
WEBLEY, D. M. (1947). *Proc. Soc. appl. Bact.* **2**, 83–89.
WILEY, J. S. and PEARCE, G. W. (1955). *Proc. ASCE, J. San. Eng. Div.* **81**, Paper 846.
WILEY, J. S. (1956). *Proc. 11th Industrial Waste Conf., Purdue University*, Series 91, 334.
WILEY, J. S. (1957*a*). *Proc. 12th Industrial Waste Conf., Purdue University*, Series 94, 596–603.
WILEY, J. S. (1957*b*). *Proc. ASCE, J. San. Eng. Div.* **83**, Paper 1411.
WILEY, J. S. and SPILLANE, J. T. (1961). *Proc. ASCE, J. San. Eng. Div.* **87**, Paper SA5, 33–52.
WORKING PARTY ON REFUSE COLLECTION (1967). "Refuse Storage and Collection." H.M.S.O., London.
WORKING PARTY ON SEWAGE DISPOSAL (1970). "Taken for Granted." H.M.S.O., London.
WORKING PARTY ON REFUSE DISPOSAL (1971). "Refuse Disposal." H.M.S.O., London.
WYLIE, J. C. (1960). *In* "Waste Treatment." (P. C. G. Isaac, ed.), pp. 349–366. Pergamon Press, Oxford.
YUNG CHANG and HUDSON, H. J. (1967). *Trans. Br. mycol. Soc.* **50**, 649–666.
YUNG CHANG (1967). *Trans. Br. mycol. Soc.* **50**, 667–677.

Author Index

Numbers in italics indicate pages where references are given in full; numbers in **bold** *type indicate the opening page of an author's contribution in this book.*

D

E

M

N

O

S

T

U

V

Y

Z

Subject Index

B

C

M

N

T

Systematic Index

D

G

H

I

M

S